Higher Excited States of Polyatomic Molecules

VOLUME III

Higher Excited States of Polyatomic Molecules

VOLUME III

Melvin B. Robin

AT&T Bell Laboratories
Murray Hill, New Jersey

1985

ACADEMIC PRESS, INC.

(Harcourt Brace Jovanovich, Publishers)

Orlando San Diego New York London
Toronto Montreal Sydney Tokyo

ACADEMIC PRESS, INC.
Orlando, Florida 32887

United Kingdom Edition published by
ACADEMIC PRESS INC. (LONDON) LTD.
24–28 Oval Road, London NW1 7DX

LIBRARY OF CONGRESS CATALOGING-IN-PUBLICATION DATA

(Revised for vol. 3)

Robin, Melvin B.
Higher excited states of polyatomic molecules.

Includes bibliographies and index.
1. Molecular spectroscopy. 2. Spectrum, Ultraviolet. 3. Electronic excitation. I. Title.
QC454.M6.R6 541.2'8 73-9446
ISBN 0-12-589903-3 (v. 3)

PRINTED IN THE UNITED STATES OF AMERICA

85 86 87 88 9 8 7 6 5 4 3 2 1

Contents

Preface

This is Volume III in the series of books on higher electronic excitations in polyatomic molecules. In this, attention is focused on excitations beyond 50 000 cm^{-1} (*6.2 eV*) in molecules containing four or more atoms, always with two goals in mind. First, I shall assign the various transitions on the basis of their orbital characteristics, and second, I shall demonstrate the relatedness of transitions in otherwise unrelated systems. The 50 000-cm^{-1} limit is a natural barrier between the much-studied and well-understood valence transitions of the sort $n \rightarrow \pi^*$, $\pi \rightarrow \pi^*$, $d \rightarrow d$, ligand $\rightarrow$ metal, *etc*., on the one hand, and on the other, the higher excited states involving both Rydberg excitations and the excitations to σ^* MO's about which there is so much confusion. Volumes I and II considered such higher excited states in all classes of organic, inorganic and biological molecules, as gleaned from the literature and personal correspondence through 1974. Volume III commences at that point and carries forward into 1985; being a supplement rather than a revised edition, Volume III is to be read in conjunction with Volumes I and II, rather than instead of them.

Following the philosophy behind Volumes I and II, that behind Volume III has been to produce a research monograph which is not only absolutely current, but which also goes beyond the literature in its attempt to synthesize a coherent explanation of what is found there. Success in this not only gives unity to the field, but uncovers many unsuspected relationships between spectra of different molecules and between the results of different types of spectroscopy. Given the intimate ties between spectroscopy, quantum chemistry and photochemistry, Volume III is naturally rich in material of interest to workers in each of these fields.

However, the accent has been placed much more on spectroscopy and quantum chemistry.

Progress in the area of higher excitations in polyatomic molecules has been impressive in the last decade; were it otherwise, there would be no need for the present Volume. In the optical realm, a vast amount of spectroscopic data has been accumulated, most of which is totally new. Such optical studies in the vacuum-ultraviolet region have been eclipsed in part by the latest electron-impact energy loss work in which the advantages of variable impact energy, variable angle, unlimited spectral range and very high resolution have proved to be a very powerful combination indeed. Bound excitations involving inner-shell orbitals in the vicinity of their ionization potentials are now recorded routinely using both synchrotron radiation and electron impact on a wide variety of gaseous and solid samples. The flood of inner-shell spectral data which has become available in the last decade is given considerable weight in Volume III, for the similarities to and differences from the outer-shell spectra are instructive.

Since the previous Volumes were written, the totally new technique of multiphoton ionization (MPI) spectroscopy has blossomed, yielding a tremendous amount of spectroscopic information relevant to the vacuum-ultraviolet region. Yet another class of excitations which is becoming more understandable involves the resonant capture of an incident electron by a neutral molecule in either its ground or excited states. Such temporary-negative-ion (TNI) resonances can involve both Rydberg and antibonding valence MO's, just as in the optical spectra of neutral molecules. The presence of molecular TNI resonances adds another dimension to the study of the higher excited states of polyatomic molecules.

Most importantly, very complete *ab initio* calculations on both valence and Rydberg excited states of polyatomic molecules now are commonplace. These are especially useful in assigning the components of Rydberg transitions the degeneracies of which have been broken by the aspherical symmetry of the molecular core, and in assessing the mixing of Rydberg and valence configurations. The latter is closely connected to the role of excitations to σ^* MO's in polyatomic spectra, and it is hoped that the prominence given such excitations in Volume III will encourage more theoretical work in this area.

The first two Chapters of this Volume are used to define several terms used in later Chapters, to discuss the general aspects of the newer experimental techniques relevant to higher excitations, and to put forward several qualitative ideas constructed to fit the data as I see it. These ideas are amenable to testing by *ab initio* calculations in large bases, and I encourage theorists to get involved and help refine, improve, disprove, expand, *etc.* our understanding in this area. These Chapters are then followed by discussions of the spectra of the various molecular classes as defined in the earlier Volumes. My intention for each molecular class in Volume III is to build upon what already has been established in Volumes I and II. In order to make a smooth connection with what already has been discussed in the earlier Volumes, frequent references will be made to the earlier text where relevant. References such as [**I**, 210] and [**II**, 60] in Volume III, for example, refer to Volume I, page 210 and to Volume II, page 60, respectively, while [**I**, M47] for example, refers to reference M47 in Volume I. In order to make Volume III as timely as possible, Addenda have been added listing papers appearing on the scene too late to be incorporated into the text.

M. B. Robin
February 8, 1985
Murray Hill, New Jersey

Acknowledgments

Special thanks must be given first to AT&T Bell Laboratories for its support of basic research efforts of this sort. Both the philosophical and technical support rendered my work by Bell Laboratories are exemplary, and their faith in the ultimate value of this effort is appreciated. I am extremely lucky in working with Ms. Patty McCrea, whose virtuosity at the keyboard and constant good nature in the face of repeated changes are substantial ingredients of this book. Moreover, the opportunity to discuss aspects of molecular spectroscopy with Bell Labs colleagues at the spur of the moment was an important stimulus, and I owe thanks to John Tully, Krishnan Ragavachari and Frank Stillinger in this regard. Finally, the drudgery of library work was considerably eased and the timeliness of the final product enhanced by all those who were so kind in sending me reprints and preprints of their work.

This book was written for Charles and Dixon.

CHAPTER I

A Catalog of Orbitals and Excitations in Polyatomic Molecules

I.A. Definitions

It is necessary that we begin with a general description of the various types of one-electron and two-electron excitations discussed in this Volume for this will allow us to define certain terms which others have used in contrary ways, and so will avoid confusion and misunderstanding. Referring to Fig. I.A-1, we first focus upon the various orbital classifications, and then upon the various classes of electron excitation.

The occupied MO's of a molecular ground state in which the AO components have their maximal principal quantum numbers [H ($n = 1$); C ($n = 2$); Cl ($n = 3$); *etc.*] are labelled generically by the symbol ϕ_i^o in the Figure. These are the valence MO's responsible for forming chemical bonds and/or lone-pair orbitals, and herein are called the *outer-shell orbitals*. The complementary virtual-orbital set of *antibonding valence*

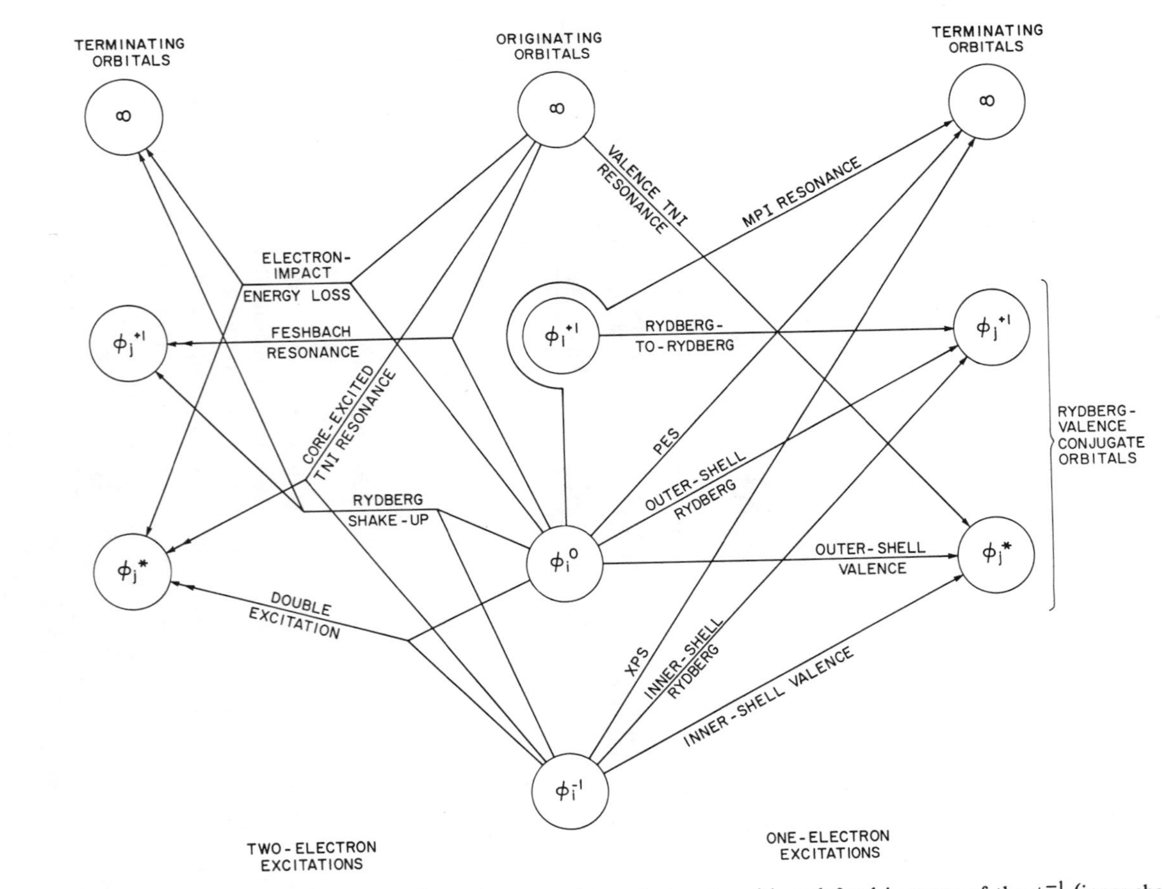

Fig. I.A-1. Schematic display of various types of one-electron and two-electron transitions defined in terms of the ϕ_i^{-1} (inner-shell), ϕ_i^{o} (outer shell), ϕ_j^* (antibonding valence), ϕ_j^{+1} (Rydberg) and ∞ (free electron) orbitals.

MO's is labelled ϕ_j^*, and involves orbitals often designated as σ^*, π^*, *etc.*, in the spectroscopic literature. If the MO in question involves AO's having principal quantum numbers smaller by 1 or more than those in the outer-shell orbital set, then they are labelled as ϕ_i^{-1} in Fig. I.A-1, and are termed *inner-shell orbitals* in the text. The ϕ_i^{-1} orbitals typically are of the sort $1s_C$, $2p_{Si}$, $4d_I$, *etc.*, and unlike the orbitals of the ϕ_i^o set, have no virtual-orbital complements. The orbital set having AO components with principal quantum numbers larger by 1 or more than those in the outer-shell set are labelled in Fig. I.A-1 as ϕ_j^{+1} and constitute the Rydberg manifold of the molecule [I, 8]. In a Rydberg state, the nuclei and all remaining occupied MO's form the *core* about which the Rydberg electron revolves. Members of the ϕ_j^{+1} Rydberg orbital set may have conjugate-orbital partners (Section I.D and [I, 24]) among the members of the set of antibonding valence orbitals ϕ_j^*. Finally, the orbitals designated as ∞ in Fig. I.A-1 represent free electrons of unspecified energy either ionized from the cationic core or incident upon the neutral molecule.

It should be pointed out that the definitions of orbital shells given above are based upon atomic concepts, whereas we wish to apply them to molecules. Because of this, the distinction between the outer-shell and inner-shell orbitals in molecules may be somewhat blurred, as may be that between the valence and Rydberg concepts for the terminating orbitals. Moreover, even if the atomic concepts themselves were linearly independent and unambiguous, still they could be mixed more or less in the molecule. In practice, there is a general sense of order and understanding gained by the classification scheme given in Fig. I.A-1, however, one must be prepared to find occasional instances of mixed character.

Using the definitions of orbital types in Fig. I.A-1, we proceed to a discussion of the various relevant excitation types. Throughout this Volume, the excited configuration of spin multiplicity s resulting from the promotion of one electron from orbital ϕ_i to ϕ_j is written as ${}^s(\phi_i, \phi_j)$; if s is unspecified, then the configuration is a spin singlet. Any excitation terminating at ϕ_j^* is termed *valence*, whereas any excitation terminating at ϕ_j^{+1} is termed *Rydberg*. Promotion of an electron from a component of the ϕ_i^o set to one of the ϕ_j^* valence MO's is called an *outer-shell valence excitation*. The analogous transition from an inner-shell ϕ_i^{-1} orbital to ϕ_j^* is termed an *inner-shell valence excitation*. As discussed in Section I.C, inner-shell and outer-shell valence excitations may occur both above and below the ionization potentials $(\phi_i^o)^{-1}$ of the originating orbitals ϕ_i^o. If the terminating orbital instead is a ϕ_j^{+1} Rydberg orbital, then the transitions from ϕ_i^o and ϕ_i^{-1} are called *outer-shell Rydberg* and *inner-shell Rydberg* excitations, respectively (Section I.B). Transitions between ϕ_j^{+1} orbitals of the Rydberg set are termed *Rydberg-to-Rydberg* transitions. Molecular excitations originating at outer-shell MO's generally are observed at frequencies between 15 000 and 400 000 cm^{-1} (*2 and 50 eV*), whereas the molecular excitations originating at the inner-shell MO's are observed at frequencies above 400 000 cm^{-1}, and as high as 80 000 000 cm^{-1} (*10 000 eV*). A complete bibliography of atomic and molecular inner-shell excitation studies to 1981 has been prepared by Hitchcock [H65].

Ionization potentials of the outer-shell ϕ_i^0 and inner-shell ϕ_i^{-1} orbitals are conventionally measured by photoelectron spectroscopy (PES) and by X-ray photoelectron spectroscopy (XPS), respectively (Fig. I.A-1). Multiphoton excitation in which an electron is first promoted resonantly from ϕ_i^0 to ϕ_j^{+1} and then from ϕ_j^{+1} to ∞ is monitored by multiphoton ionization (MPI) spectroscopy (Section II.A). The ionic state resulting from the loss of an electron from orbital ϕ_i^0 is written as $(\phi_i^0)^{-1}$.

With an electron or high-frequency photon incident upon a neutral molecule, several types of two-electron transitions are possible. An electron impact energy-loss results when an incident electron scatters inelastically from a target molecule, leaving a second electron in a level of higher energy, Fig. I.A-1. If the incident electron instead has the proper energy to bind to one of the vacant ϕ_j^* antibonding valence MO's so as to form a temporary-negative ion (TNI), the transition is called a *valence TNI resonance* (Section II.B). It is also possible that the incident electron inelastically excites an outer-shell Rydberg excitation and then is bound to the excited molecule by occupying one of the ϕ_j^{+1} Rydberg orbitals; this is called a *Feshbach resonance.* While there is evidence for two-electron valence promotions of the type $\infty + \phi_i^0 \rightarrow \phi_j^* + \phi_k^*$, hybrid excitations such as $\infty + \phi_i^0 \rightarrow \phi_j^* + \phi_j^{+1}$ are unknown. TNI resonances in which an incident electron resides in a Rydberg orbital while the core remains unexcited do not occur. Inner-shell ionization as monitored by XPS frequently is accompanied by the excitation of a second electron from ϕ_i^0 to a ϕ_j^* or ϕ_j^{+1} orbital in the positive ion. Two-electron transitions such as these are called *valence shake-up* and *Rydberg shake-up* excitations, respectively (Section II.C). Cederbaum *et al.* [C16] predict that outer-shell shake-up excitations will be plentiful in the 200 000–400 000 cm^{-1} (*25–50 eV*) regions of polyatomic molecules, however, very few outer-shell transitions have been assigned as such to date. According to a calculation on F_2 [D16], a two-electron hybrid excitation in which one electron is excited to an antibonding valence orbital and a second is excited to a nonconjugate Rydberg level is predicted to fall at 80 000–120 000 cm^{-1} (*10–15 eV*), however such a neutral-molecule hybrid excitation has yet to be observed experimentally. Though two-electron excitations to bound valence orbitals most often are assigned to broad bands observed just beyond the inner-shell ionization potentials, the evidence for such assignments usually is weak. Another type of molecular excitation commonly observed in the spectra of liquids and solids between 120 000 and 200 000 cm^{-1} (*15 and 25 eV*) involves a multielectron promotion called a plasmon or collective excitation (Section II.D); there is no certified example of a plasmon in a vapor-phase molecule [**I**, 35].

The revival of interest in molecular excitations in the vacuum-ultraviolet and x-ray regions has led to the appearance of many review articles. Those covering specific subtopics are acknowledged in the appropriate places in the text, while those of a more general nature are listed here [A31, K30, K31, M45, S16, S18, S60]. These are recommended to the reader wanting a brief overview of the field.

I.B. Inner-Shell and Outer-Shell Rydberg Excitations

The Splitting of Hydrogenic Rydberg States

The simple concept of hydrogenic Rydberg series in molecular spectra must be refined somewhat to account for the molecular core no longer being the impenetrable point charge that it is in the hydrogen atom. Because the hydrogenic orbitals of the same n but different l differentially penetrate the electron shells screening the Rydberg electron from the nuclei, these levels are split with an ordering roughly depending upon the quantity $n + l$ [A24, F20]. Thus, decreasing penetration in the series 3s, 3p, 3d places 3s below 3p, and 3p below 3d in the Rydberg ladder. To the extent that the Rydberg orbital of a given n and l is penetrating, the Rydberg electron often can sense that the positively-charged core is less than spherically symmetric, and this acts to lift the degeneracy of levels having $l \geqslant 1$; the azimuthal sublevels are characterized by the quantum numbers λ. This "core splitting" is evident for example in the unequal energies of the (π, 3pxy), and (π, 3pz) Rydberg configurations of acetylene. In molecules of high symmetry, transitions between degenerate orbitals (such as $1\pi_u \rightarrow 3pxy$ in acetylene) give rise to multiplet splitting depending upon how the λ values of the $1\pi_u$ and 3pxy shells are coupled. If the ground state of a molecule is a closed shell, then the open-shell Rydberg configurations of a given n, l, λ are split into singlet-triplet pairs by the exchange interaction, leading to states characterized by quantum numbers n, l, λ, s. If however, the excitation originates at an orbital having $l \geqslant 1$ on a heavy atom, then spin-orbit coupling will act to further mix the singlet and triplet configurations.

All other things being equal, separations between low-lying molecular Rydberg levels are largest for $\Delta n = 1$, significantly smaller for $\Delta l = 1$ and somewhat smaller yet for $\Delta\lambda = 1$ and $\Delta s = 1$ [**II**, 12]. In regard penetration splitting, core splitting, multiplet splitting and exchange splitting, the magnitudes depend largely upon the spatial overlap of the Rydberg orbital and the core orbitals, and so these splittings go to zero as n goes to infinity. Two other types of splitting are possible which may or may not shrink to zero with increasing n:- the spin-orbit splitting and the Jahn-Teller splitting. Each of the six types of possible Rydberg splitting is discussed briefly below.

It is well known that penetration decreases in the order s, p, d, f, *etc.* In the atom, this is explained in terms of the centrifugal potential which keeps electrons of large l away from the region of small r. In molecules where this makes no particular sense the same ordering holds, and so one argues that even though l formally may not be a quantum number, the angular nodal properties of the Rydberg orbitals which characterize the various levels in the atom are still present in the lower symmetry molecules and so act in the same way to keep the Rydberg electron more-or-less away from the core.

The splitting of the degeneracy of n,l,λ Rydberg configurations (where $l \geqslant 1$) by the nonspherical symmetry of the core has been discussed from an electrostatic point of view by Lindholm [**I**, L25]. In the case of the $n = 3$ Rydberg levels of ethylene, Mulliken [M82, M83] suggests instead that the splittings will reflect the ordering of their precursors in the core, whereas McMurchie and Davidson [M53] calculate a somewhat different ordering and suggest that the levels are ordered according to the angular momentum of the Rydberg orbital. Though there is still very little experimental data in polyatomic molecules against which to test these ideas, there is now a considerable cache of *ab initio* calculations of np and nd core splittings which can serve the same purpose. Limiting the discussion to the lowest np set of levels, the electrostatic approach predicts that npπ levels will lie above npσ levels in linear systems and above npσ and npσ' levels in planar systems. Indeed, the calculations do predict this ordering in acetylene [B15, D14, W27], ethylene [B15, M53], methyl radical [Y4], ammonia [P26], water [C25, P26, W2], hydrogen sulfide [M11, R29], methylene radical [R41], *trans*-1,3,5-hexatriene [N14], boron trichloride [I6], formaldehyde [H14, L20], thioformaldehyde [B86] and benzene [H25]. Not all planar systems behave so nicely, however, for 3pπ is calculated *not* to be the uppermost 3p orbital component in butadiene [B77, P26], formic acid [D17] and diimide [V12]. Moreover, substitution of one or more methyl groups for one or more of the hydrogens of the planar systems strongly affects the np sublevel ordering. Thus the npπ

sublevel is no longer uppermost in allene [D19] or dimethyl ether [W2] and is the *lowest* sublevel in propane [R18], diborane [E7], acetone [H47] and thioacetone [B71]. In the nonplanar class, only ethane [B76] is anomalous, for 3pπ is uppermost here even though the core contains out-of-plane hydrogen atoms.

It is surprising that there is any order whatsoever to the pattern of np sublevels when one considers that the electrostatic picture will be altered by the ever-present Rydberg/valence mixing (Section I.D). It is a further surprise that the np core-splitting pattern for a given molecule is calculated to be independent of whether the transition originates in the outer-shell or inner-shell orbitals; only in thioacetone [B71] is the ordering predicted to depend upon the originating MO.

At the Hartree-Fock level, the singlet-triplet splitting is twice the exchange integral between open-shell orbitals and so depends strongly on the spatial overlap of the originating valence MO and the terminating Rydberg orbital. Inasmuch as penetration promotes this spatial overlap, the singlet-triplet splitting should increase with increasing penetration and thereby with increasing Rydberg term value [J7, R18, V12]. This hypothesis can be tested using the extensive singlet-triplet Rydberg assignments in water given by Chutjian *et al.* [C25], Table I.B-I. There is a general confirmation of the suspected relationship between term value and singlet-triplet splitting in this Table, however, the $^1(1b_1, 3pa_1)$ and $^1(1b_1, 4pb_1)$ configurations are very much out of line, possibly owing to misassignments. A most interesting decrease of the exchange splitting in the (4p, 5s) Rydberg states of the alkyl bromides with increasing size of the alkyl group [F6] is discussed in Section IV.B, and the spatial-overlap explanation behind this should be applicable to all Rydberg states originating at a single atomic center.

Working from the point of view that increased term values translate into increased singlet-triplet splittings, one is led directly to the conclusions that perfluorinated molecules should have relatively large singlet-triplet Rydberg splittings as should inner-shell Rydberg excitations. In regard the latter, we feel that the expectation will not be realized however, because the singlet-triplet splitting involves the exchange integral between, say $1s_C$ and a ϕ_j^{+1} orbital whereas the exhalted term value in such cases comes from the antishielding (*vide infra*) which implies that ϕ_j^{+1} penetrates the outer-shell valence orbitals, but not $1s_C$.† In the case of

† In fact, Friedrich *et al.* [F33] argue that the exchange splitting is *smaller* for inner-shell Rydberg configurations, and that this is the reason behind the larger term values for inner-shell excitations (*vide infra*).

perfluorinated molecules, the large Rydberg term values relate to the large atomic term value on the F atom [I, 51], implying an increased overlap of ϕ_j^0 and ϕ_j^{+1} orbitals.

TABLE I.B-I

CORRELATION OF RYDBERG TERM VALUES AND SINGLET-TRIPLET SPLITS IN WATER[a]

Configuration	Term Value, cm^{-1}	Singlet-Triplet Split, cm^{-1}
$^1(1b_1, 3s)$	42 100	3200
$^1(3a_1, 3s)$	40 760	3140
$^1(1b_1, 3pb_2)$	28 400	1610
$^1(3a_1, 3pa_1)$	21 600	3140
$^1(1b_1, 3pa_1)$	21 000	240
$^1(1b_1, 3pb_1)$	19 800	2820
$^1(1b_1, 4s)$	16 900	1050
$^1(1b_1, 3db_2)$	14 300	1290
$^1(1b_1, 3da_1)$	11 700	80
$^1(1b_1, 4pb_2)$	11 400	240
$^1(1b_1, 4pb_1)$	9830	1690
$^1(1b_1, 5s)$	8780	640

[a] Reference [C25].

The multiplet splitting pattern also depends in part but not solely upon the exchange integrals. One can say only that multiplet splitting for Rydberg configurations will be smaller than for valence configurations, and will go to zero as n increases.

Because spin-orbit coupling involves the electrostatic potential gradient which is large only close to the nuclei, clearly this effect can be significant only for the open-shell core of a molecule built of one or more heavy atoms. Being a core effect, it will persist undiminished as n increases in a Rydberg series. In an instructive study, Felps *et al.* [F6] show that in the series from HBr to C_3H_7Br, the spin-orbit coupling constant maintains its value of $1280 \pm 50\ cm^{-1}$ in the (4p, 5s) state as appropriate for the core hole localized on the Br atom, whereas the exchange splitting drops rapidly in this series as the Rydberg orbital delocalizes over the alkyl group

(Section IV.B) and the 4p, 5s exchange integral shrinks. As with all excitations in closed-shell molecules, Rydberg excitations to open-shell upper states will be split into singlet-triplet pairs when exchange is dominant and into spin-orbit components when spin-orbit coupling is dominant. Since transitions to these pairs of levels have very different intensity ratios in the two extreme cases, the observation of the experimental intensity ratio can be used to assess the importance of spin-orbit coupling relative to exchange splitting [S28].

In regard Jahn-Teller splitting in Rydberg states, we point out that the requisite degenerate upper state can arise either from degeneracy in the Rydberg orbital, or in the originating orbital, or both. In the first case, though formally Jahn-Teller susceptible, the degeneracy is in an essentially nonbonding orbital and so there is very little electronic stabilization to be gained by geometric distortion of the core. Whatever small Jahn-Teller activity there is at low n will go to zero as n increases. In the opposite situation, the degeneracy is in the bonding orbitals of the core, and a large (static) Jahn-Teller splitting results which will closely resemble that in the relevant PES band. This splitting will remain constant with increasing n.

Term Values and Antishielding

In Volumes **I** and **II**, the generalizations that the (ϕ_i^o, ns) Rydberg term values decrease with increasing alkylation of a particular central chromophore, and that for a given molecule or its isomers the (ϕ_i^o, ϕ_j^{+1}) Rydberg term value is independent of the originating orbital ϕ_i^o were repeatedly used. While there is full support for the first premise in the spectroscopic data of the last 10 years, it is now clear that the second premise is not true as stated. In anticipation of this, we note that Maria *et al.* [**II**, AD126] have demonstrated a linear relation between ionization potential and Rydberg term value in rare gases, hydrogen halides and methyl halides, Fock *et al.* have shown a similar relationship in the N_2, CO, BF series [F18] and Spence [S65] has demonstrated the same thing for Feshbach resonances (Section II.B).† It is a short step from these

† There is considerable data supporting the proposal of Maria *et al.* [**II**, AD126] that the molecular term value varies linearly with the relevant ionization potential. Note however that this relationship holds only within a class of closely related molecules, that fluorination of the molecules in a particular class destroys the expected linear relationship [R35] and that the linear relationship does not hold when transitions originating at inner-shell and outer-shell MO's are considered simultaneously.

examples to imagine in a given molecule that the Rydberg term values will increase as the ionization potentials of the originating MO's increase, rather than remain constant as postulated earlier.

In fact, the (ϕ_i^o, ϕ_j^{+1}) Rydberg term value should change with the nature of the originating orbital ϕ_i^o in an interesting manner, as a moment's reflection will show. Viewing the term value as an excited-state ionization potential, it is clear that its value will depend upon the effective nuclear charge (Z_{eff}) binding the Rydberg electron to the core. In fact, the term value equals $RZ_{eff}^2/(n-\delta)^2$. If we take $n = 3$; $\delta = 1$ and a term value of 25 000 cm^{-1} as appropriate for an outer-shell transition to 3s, then the corresponding $Z_{eff} = 0.96$. This is the effective nuclear charge appropriate to the situation in which the hole is in a valence MO ϕ_i^o composed of 2s and 2p AO's. Now according to Slater's rules [E15], if the hole instead is in the 1s orbital, then the nuclear shielding constant for a 3s orbital decreases by 0.15,† so that Z_{eff} increases to 1.11. This shielding effect in turn leads to an inner-shell 3s term value (ionization potential) of 33 400 cm^{-1}, which is 30 % larger than if the hole were in an outer-shell valence orbital. According to Slater's rules, an argument such as this will apply to 3s and 3p orbitals, but not to 3d.

The effective quantum number for a Rydberg orbital, $n - \delta$, is related to the measured term value, TV (cm^{-1}), through the relationship [E8]

$$n - \delta = (109\ 734/TV)^{1/2}$$

and from this an average squared radius can be calculated for an orbital with azimuthal quantum number l as

$$\langle r^2 \rangle = (n-\delta)^4 \left\{ 5/2 - 3/2(n-\delta)^{-2}[l(l+1) - 1/3] \right\}$$

When applied to the term values of 25 000 and 33 400 cm^{-1} quoted above as typical for outer-shell and inner-shell excitations to 3s, the effect of the higher effective nuclear charge is seen to reduce $\langle r^2 \rangle$ from 23.66 a_0^2 to 14.08 a_0^2. Though compressed in cross section by almost 50 %, the terminating level of the inner-shell excitation is still very much a Rydberg orbital.

† In estimating $Z_{eff} = Z - \sigma$ for a 3s orbital, Slater's rule prescribes that each electron in a 2s or 2p orbital contributes 0.85 to the nuclear shielding constant σ, whereas each electron in 1s contributes 1.0.

Rephrasing the argument given above, the Rydberg electron is electrostatically bound to the core owing to the imperfect shielding of the nucleus by the valence electrons. The shielding of the 3s orbital from the nucleus is more efficiently performed by the inner-shell electrons (1s) than the outer-shell electrons (2s and 2p) because they are less easily penetrated by 3s. Consequently, when the Rydberg configuration is changed from an outer-shell one such as (2p, 3s) to an inner-shell one such as (1s, 3s) the shielding of the 3s orbital from the nucleus decreases and the term value rises. This increased binding of the Rydberg electron as the hole level descends, we call "*antishielding*". The antishielding effect results in an exhaltation of the inner-shell Rydberg term value which will be largest for highly penetrating ns orbitals, least for nonpenetrating nd and nf orbitals, and intermediate for np.

Turning to the experimental outer-shell and inner-shell term values, Table I.B-II, one does see the general trend of larger term values for inner-shell Rydberg (ϕ_i^{-1}, 3s) configurations, however it is not as large as expected, and there are a few notable exceptions. In particular, the (ϕ_i, 3s) term value of methane actually *decreases* on going from $\phi_i^0 = t_2\sigma$ to $\phi_i^{-1} = 1s_C$, whereas for the corresponding transitions in silane there is no change, and a slight decrease is noted in ethane. The reasons behind these exceptions are as yet obscure (however, see below). It is also seen in Table I.B-II that the (ϕ_i^{-1}, 3s) term value is generally larger when ϕ_i is on the more electronegative atom, however formaldehyde and acetaldehyde show the opposite behavior. Parallel data for (ϕ_i^0, 3p) purposely has not been listed because this configuration is split into several components by the asymmetry of the core in most molecules, and it is not clear that the same components are being compared in the outer-shell and inner-shell spectra.

The simple antishielding picture is more appropriate to an atom than a molecule, for in the latter, the hole orbitals can differ not only in their shielding but also in their spatial distribution. Thus, for inner-shell excitation, the positive change in largely localized at the excited atom, whereas it can be much more delocalized for an outer-shell transition. The localization aspect of the antishielding phenomenon can be probed only by a series of excited-state calculations. Friedrich *et al.* [F33] also noticed the term value difference between corresponding outer-shell and inner-shell Rydberg configurations, and attributed this to the difference in the excited-state exchange integrals. While the relative importance of exchange *versus* antishielding remains an open question, one should note that the exchange effect at most can equal the singlet-triplet split in an outer-shell Rydberg configuration, *i.e.*, 2000 cm^{-1}. This simple approach in which the exhaltation of the term value is attributed to an increased

TABLE I.B-II

COMPARISON OF INNER-SHELL AND OUTER-SHELL (ϕ_i, ns) RYDBERG TERM VALUES

Molecule	(ϕ_i^{o}, 3s)[a] Outer-Shell Term Value, cm^{-1}	(ϕ_i^{-1}, 3s)[b] Inner-Shell Term Value, cm^{-1}	References
CH_4	31 600	29 600	[T19, **I**, 112]
SiH_4	26 700[c]; 26 800[d]	26 600[e]	[D27, F33, R31]
C_2H_6	29 500	28 900	[H56, **I**, 125]
C_2H_4	27 400	28 900	[E1, H56, T19, **II**, 24]
C_2H_2	26 000	26 800	[E1, H56, **II**, 107]
CH_3F	33 000	37 100; 42 700[f]	[H57, H61, W20, **I**, 180]
CH_3Cl	27 580	31 860	[H59, **I**, 166]
CH_3Br	28 940	29 800	[H57, H61, **I**, 166]
CH_3I	27 310	29 400	[H61, **I**, 166]
C_6H_6	22 900	25 000	[E1, H56, H58, Section XIX.A]
H_2O	41 400	46 000	[A7, W19 **I**, 259]
CH_3OH	34 070	32 300; 38 700[g]	[W19, **I**, 259]
CH_3OCH_3	25 730	30 200	[W19, **I**, 259]
NH_3	35 760	40 300	[W19, **I**, 210]
CH_3NH_2	31 200	33 100; 36 200[h]	[W19, **I**, 210]
PH_3	31 000	34 400	[F33, **I**, 234]
H_2CO	30 477	34 800; 32 000[i]	[H64, **II**, 80]
CH_3CHO	27 508	29 900	[H64, **II**, 80]
NCCN	28 400	32 200[j]	[C43, H63]

[a] The originating MO ϕ_i^{o} is the outermost filled MO in the molecular ground state, unless otherwise specified.

[b] The originating MO ϕ_i^{-1} is $1s_C$ unless otherwise specified.

[c] ($2t_2$, 4s). [g] ($1s_O$, 3s).

[d] ($3a_1$, 4s). [h] ($1s_N$, 3s).

[e] ($2p_{Si}$, 4s). [i] ($1s_O$, 3s).

[f] ($1s_F$, 3s). [j] Both ($1s_C$, 3s) and ($1s_N$, 3s).

effective nuclear charge possibly abetted by differences in exchange energies assumes that the orbital composition *vis-a-vis* Rydberg/valence mixing remains constant. Inasmuch as there is no reason to expect that this always will be so, some irregular behavior of the exhaltation is expected (and observed; *cf.* Section III.A).

Progress in identifying the very high members of molecular Rydberg series is aided by the higher resolution of modern spectrometers, and in some cases by the use of lasers in two-color Rydberg-to-Rydberg absorption experiments. However, there is considerable rovibronic congestion as one nears the ionization limit and so care must be exercised in this region. Dehmer points out an illustrative example [D12] which should promote caution on the part of molecular spectroscopists working on the higher Rydberg states close to the ionization limit. In the series Xe, HI, CH_3I each within approximately 8000 cm^{-1} of its ionization limit, the spectra stand in obvious 1:1:1 correspondence. However, work at higher resolution on HI shows that many of the "peaks" are in fact just chance coincidences of otherwise unrelated rotational transitions! Since most vacuum-ultraviolet spectra are rotationally unresolved, one must acknowledge an element of uncertainty when placing "peaks" in long Rydberg series or comparing them in different molecules.

The Rydberg term table of [**I**, 315] has been recalculated using a somewhat more accurate value of the Rydberg constant, and certain numerical errors have been removed [B61].

Rydberg States in Ions

A Rydberg electron in a molecular positive ion experiences a formal nuclear charge equal to +2 and according to the Rydberg formula, the Rydberg term value consequently will be 4 times larger than that in the corresponding neutral molecule. On the other hand, the penetration (δ) in the positive ion is not necessarily as large as that in the neutral, and this too will influence the term value in the positive ion. Calculations on the ^2A_g state of the ethane cation [R17] are generally instructive on this point. The combined action of nuclear charge and penetration in the ethane cation result in (ϕ_i^o, 3s) and (ϕ_i^o, 3p) term values which are 2.8 times as large as those in the neutral system. Interestingly, the δ's are considerably smaller in the positive ion, implying behavior which is more hydrogenic than that in the neutral. To the extent that the proposed linear dependence of the term value on the ionization potential holds [**II**, AD126], we are led to expect the ratio of positive-ion to neutral-molecule ionization potentials also to be 2.8. Experimental data on a number of polyatomic systems [A20] leads to an ionization potential ratio of 2.9 ± 0.1. This rough multiplicative factor of 2.9 can be of value in identifying Rydberg excitations beyond a neutral molecule's lowest ionization potential even when that of the corresponding cation is unknown.

The question of Rydberg levels in anions is moot. In this case, the optical electron does not experience a $-1/r$ potential and thus the only bound levels of the system are the antibonding valence MO's. Consequently, the electronic spectra of anions such as TiO_4^{4-} and BF_4^- (Chapter XXI) will not display Rydberg transitions.

Rydberg Intensities

In assembling the data for Volumes **I** and **II**, it was noticed that for every certified Rydberg excitation, the measured oscillator strength f did not exceed 0.08 per degree of spatial degeneracy. Since that time, a tremendous amount of new data as well as highly accurate calculations on several polyatomic molecules have come forth in quantitative support of this upper limit on f. Consequently, we are confident in proposing this intensity criterion as one of the primary standards for distinguishing between strongly allowed valence excitations and strongly allowed Rydberg excitations. By this measure, the previous claims that the bands at 67 400 cm^{-1} (*8.36 eV*) in formic acid ($f = 0.3$), for example, or at 78 000 cm^{-1} (*9.67 eV*) in cyanogen ($f = 2.0$) are Rydberg excitations as suggested by their frequencies are strongly contradicted by their intensities.

The upper limit on f of 0.08 per degree of spatial degeneracy has been empirically determined from the many experimental and theoretical studies of outer-shell Rydberg spectra. Undoubtedly, there is an upper limit as well for inner-shell Rydberg spectra. However, since very few molecular inner-shell intensities have been measured, one cannot say more than that the upper limit will be far smaller than 0.08. In ethylene and acetylene [B15] the calculated upper limit to f for Rydberg transitions from $1s_C$ appears to be 0.004 per degree of degeneracy, and this figure is supported as well by calculations on the $1s_B$ spectra of BF_3 and BCl_3 [I6]. Moreover, since the f values for Rydberg series members decrease as n^{-3}, one predicts qualitatively that for a series of molecules such as CH_4, SiH_4, GeH_4 and SnH_4, the $1s_C \rightarrow 3p$ transition of methane will be most intense, followed by $1s_{Si} \rightarrow 4p$ of silane, $1s_{Ge} \rightarrow 5p$ of germane and $1s_{Sn} \rightarrow 6p$ of stannane. Such an effect may be operative in CH_4 and CCl_4, where the $1s_C \rightarrow 3p$ transition of CH_4 is much stronger than the $1s_C \rightarrow 4p$ transition of CCl_4 [H59].

The relatively low oscillator strengths of fully allowed Rydberg excitations imply relatively long radiative lifetimes ($f = 0.08$ implies $\tau = 10^{-7}$ sec). In fact, it is this long lifetime which allows Rydberg states to appear as resonances in MPI spectra. However, there is evidence that the more penetrating Rydberg MO's assume more of the antibonding valence character of their conjugate orbitals and this in turn leads to a lifetime shortened by predissociation. Thus in ammonia, all Rydberg excitations terminating at ns are predissociated whereas those terminating at nd are not, and so the latter are longer lived [B34, F27]. Several experiments demonstrate molecular Rydberg lifetimes in the μsec regime [K44], however, it must be pointed out that the shortest lifetime ever directly measured for a molecular excited state (0.070 psec) involves a Rydberg state of benzene [W18].

The point was made in the discussion above that the term values and the singlet-triplet splittings of Rydberg configurations depend upon the spatial overlap of the originating valence MO and the terminating Rydberg orbital (*vide supra*). Inasmuch as the electric-moment matrix element for an electronically allowed Rydberg excitation also is dependent upon this spatial overlap, one expects that the allowed Rydberg intensity should increase with increasing penetration, *i.e.*, term value. Though there is little data on this point, it is interesting to note that the $1b_1 \rightarrow 3sa_1$ transition of water has a very large term value of 42 100 cm^{-1} and an oscillator strength (0.060) quite close to the Rydberg upper limit (0.08). On the other hand, the (n_O, 3s) term values in alcohols decrease monotonically with increasing bulk of the alkyl group [**I**, 258] whereas the $n_O \rightarrow 3s$ oscillator strengths vary in a much more complicated way [O1].

I.C. Inner-Shell and Outer-Shell Valence Excitations

As explained and illustrated so abundantly in Volumes **I** and **II**, the key to understanding many of the excitations in vacuum-ultraviolet spectra rests in recognizing that the most prominent bands often are the first few Rydberg excitations converging upon the first few ionization potentials. With the availability of successive ionization-potential values from photoelectron spectra, it is trivial to calculate term values for such transitions (ionization potential minus excitation energy) and make secure outer-shell Rydberg assignments on the basis of well-established term-value trends [**I**, 51]. With slight adjustment, the same approach also is successful in assigning inner-shell Rydberg spectra at hundreds eV energy (Section I.B). With this new point of view, the pendulum of fashion has swung in the vacuum ultraviolet. Prior to the publication of Volumes **I** and **II**, our understanding of most excitations in the vacuum ultraviolet was based upon calculations using valence-orbital basis sets, which of course resulted in all features being assigned as valence transitions. Molecular Rydberg transitions certainly were known at the time, but were recognized only when atomically sharp and when they formed long, convergent series. The concept of the term value and its use throughout all spectral regions now has pushed the pendulum far to the other side, so that at present almost every upper state observed in the vacuum ultraviolet is assigned as Rydberg! Statistically, this is probably much closer to the truth than the opposite approach, however as Prof. Lindholm points out, the question remains, "Where *are* the valence excitations which theory predicts to lie in this region, along with the Rydberg excitations?" For the most part, this question has been answered as regards the $\pi \rightarrow \pi^*$ excitations of planar systems, though there are still many gaps to be filled in the $\pi \rightarrow \pi^*$ catalog at higher frequencies. Since intuitive arguments, intensity considerations and band-shape analysis as well as *ab initio* calculations all have served to identify $\pi \rightarrow \pi^*$ bands in the vacuum ultraviolet, the greatest need at the moment is for understanding the role of σ^* orbitals in both optical and X-ray excitations. In the past, spectroscopists have used the "σ^*" symbol to denote any virtual orbital of local or global σ symmetry, without further specifying its nodal pattern or spatial extent. In this work, we take σ^* to be an antibonding valence orbital in zeroth order; in certain cases where this zeroth-order orbital may be mixed with Rydberg orbitals of the same symmetry, the result will be identified as a valence/Rydberg hybrid.

$\mathscr{A}$-Band Definitions

Let us expand the definitions of the originating (ϕ_i^o) and terminating (ϕ_j^*) MO's somewhat. The outer-shell valence orbitals ϕ_i^o we divide further into np lone-pair orbitals, orbitals which have local σ symmetry, $\phi_i^o(\sigma)$, and those which have π character and are spread over two or more centers, $\phi_i^o(\pi)$. The virtual-orbital complements of the last two MO's are the antibonding MO's $\phi_j^*(\sigma^*)$ and $\phi_j^*(\pi^*)$, respectively. Using these zeroth-order orbitals, the one-electron excited states can be classified into three general types, as in Table I.C-I. Regardless of the originating MO, transitions terminating at ϕ_j^{+1} are Rydberg excitations. Excitations to V states involve promotions between valence orbitals which are complementary bonding–antibonding pairs, while the $\mathscr{A}$ bands arise from valence excitations between orbitals occupying different parts of space either because of their differing spatial orientations or spatial extents. The

V valence states require considerable configuration interaction for a proper description, whereas the $\mathscr{A}$ states do not.

TABLE I.C-I

CLASSIFICATION OF ONE-ELECTRON EXCITED STATES

Originating MO		Terminating MO	
	ϕ_j^{+1}	$\phi_j^*(\pi^*)$	$\phi_j^*(\sigma^*)$
ϕ_i^{-1}	Rydberg	$\mathscr{A}$	$\mathscr{A}$
$\phi_i^{o}(n\text{p})$	Rydberg	$\mathscr{A}$	$\mathscr{A}$
$\phi_i^{o}(\pi)$	Rydberg	V	$\mathscr{A}$
$\phi_i^{o}(\sigma)$	Rydberg	$\mathscr{A}$	V

It is now possible to extend our understanding of valence spectra beyond the $\pi \rightarrow \pi^*$ excitations by considering the classification and term values of the valence $\mathscr{A}$ bands. The prototypical $\mathscr{A}$ band is found in the alkyl halides $R-\ddot{X}$, and is due to the excitation of an electron from the np lone-pair orbital on $\ddot{X}$ to the $\sigma^*(R-X)$ antibonding MO involving the halogen and the radical bound to it, Fig. I.C-1A. Simply termed an $\mathscr{A}$ band in the past, we now classify this excitation as an $\mathscr{A}(n\text{p}, \sigma^*)$ band. Though $\mathscr{A}(n\text{p}, \sigma^*)$ bands are well known as such in outer-shell spectra, the $\sigma^*(R-X)$ MO is active as well in the inner-shell spectra of the same molecules, and so these are called $\mathscr{A}(\phi_i^{-1}, \sigma^*)$ bands, with the inner-shell excitation originating at the ith inner-shell orbital ϕ_i^{-1}.

What now if the halogen and its lone pairs in $R-\ddot{X}$ are replaced by $-\ddot{O}H$, $-\ddot{S}H$, $-\ddot{N}H_2$ groups, *etc.*, so as to form alcohols, thiols, amines, *etc*? There is no choice but to follow the scheme of Fig. I.C-1A and to consider $\mathscr{A}(n\text{p}, \sigma^*)$ and $\mathscr{A}(\phi_i^{-1}, \sigma^*)$ bands in these substances as well. In fact, such $(n\text{p}, \sigma^*)$ configurations are valence conjugates to the lowest $(n\text{p}, (n+1)\text{s})$ Rydberg configurations in such systems. Experience shows that whereas in the halides there is very little mixing among these conjugates so that $(n\text{p}, \sigma^*)$ has a clear and well-defined existence, in several of the $R-\ddot{Q}H_x$ systems, the $(n\text{p}, \sigma^*)$ configuration is completely miscible with the members of the conjugate Rydberg manifold and dissolves completely in them so that there is no residue which can be called "$(n\text{p}, \sigma^*)$". This appears to be the case in H_2O, for example. See Section I.D for a general discussion of Rydberg/valence mixing.

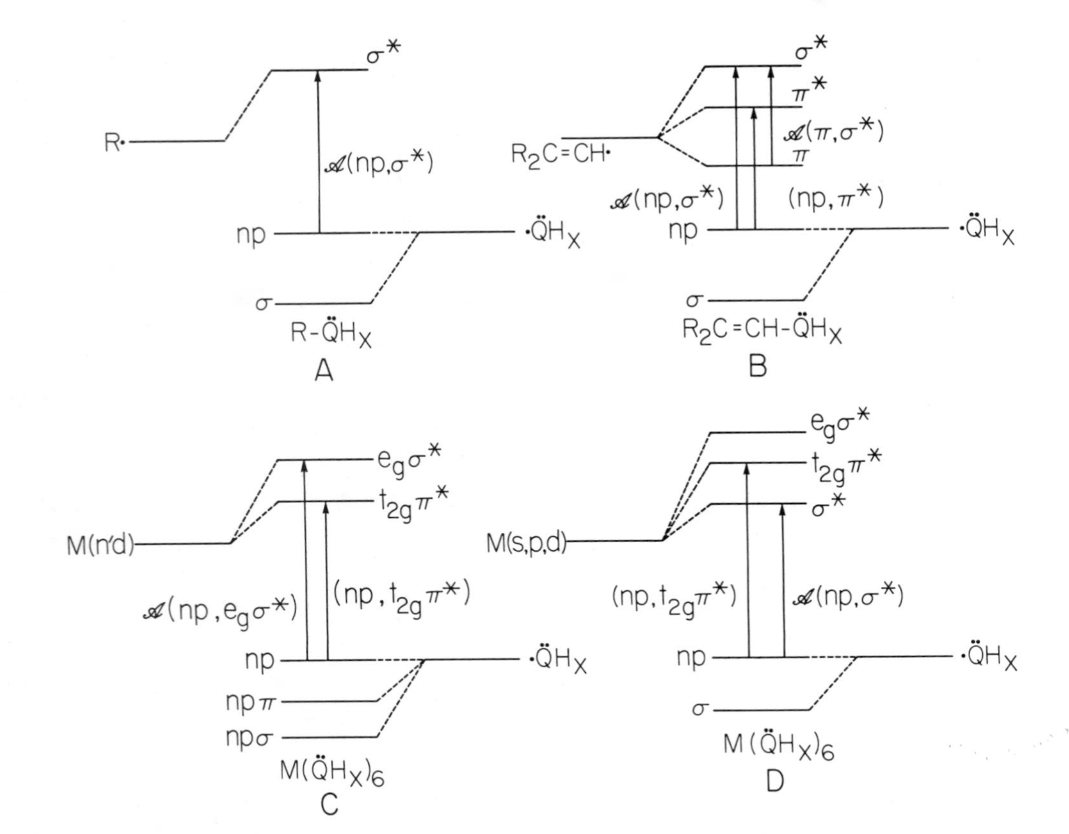

Fig. I.C-1. Examples of 𝒜-band transitions in various chromophoric groups. Orbitals labelled "*n*p" are lone-pair orbitals on Q, whereas all other MO's are either bonding or antibonding between Q and the R or M radicals, except those labelled "Rydberg".

When the $-\ddot{Q}H_x$ group is bonded directly to an unsaturated center as in $H_2C{=}CH\ddot{Q}H_x$, the $\mathscr{A}$-band possibilities are manifold, Fig. I.C-1B. As in $R{-}\ddot{Q}H_x$, there is an $\mathscr{A}(n\text{p}, \sigma^*)$ band present. However the $\mathscr{A}(n\text{p}, \sigma^*)$ band in unsaturated systems generally is not the lowest-frequency $\mathscr{A}$ band, for there generally are occupied π MO's above np in these compounds. In such a case, the lowest $\mathscr{A}$ band is $\mathscr{A}(\pi, \sigma^*)$, Fig. I.C-1B, in an obvious notation. Two other transitions are possible here:- $\pi \rightarrow \pi^*$ and $n\text{p} \rightarrow \pi^*$. The first of these is encompassed in the well-known V_n notation of $\pi \rightarrow \pi^*$ excitations, and otherwise does not fit into the $\mathscr{A}$-band scheme. Inner-shell valence excitations in unsaturated systems are written as $\mathscr{A}(\phi_i^{-1}, \sigma^*)$ and $\mathscr{A}(\phi_i^{-1}, \pi^*)$, with the latter being lower in general. In unsaturated systems, $\mathscr{A}(n\text{p}, \sigma^*)$ bands generally are observed only for halides, in which they are low-lying; in other unsaturated systems they often are covered by $\pi \rightarrow \pi^*$ and Rydberg excitations and so are not visible, though they are undoubtedly present (if not totally dissolved in the Rydberg manifold). If the $-\ddot{Q}H_x$ radical itself is part of a double bond, *i.e.*, $R{=}\ddot{Q}H_x$ (acetone, for example), then the $\mathscr{A}(n\text{p}, \pi^*)$ band becomes the well-known $n_Q \rightarrow \pi^*$ band, and $\pi \rightarrow \sigma^*$ is no longer low-lying.

In saturated systems bearing $-\ddot{Q}H_x$ groups in which the central atom is metallic [for example, Fe_2Cl_6 or $U(OCH_3)_6$], the lowest antibonding orbitals available for valence excitations may be components of the nd or nf shells, Fig. I.C-1C. When dealing with nd orbitals in octahedral symmetry, $t_{2g}\pi^*$ and $e_g\sigma^*$ levels result from σ-bond formation with the nd orbitals and $\mathscr{A}(n\text{p}, t_{2g}\pi^*)$ and $\mathscr{A}(n\text{p}, e_g\sigma^*)$ bands are observed in the valence spectrum. This situation prevails in WF_6, for example, wherein the 2p $\rightarrow$ 5d excitations appear as valence $\mathscr{A}$ bands while transitions from np to nf will be part of the Rydberg spectrum. In contrast, in UF_6, nf is part of the valence shell and so $\mathscr{A}$ bands of the sort $\mathscr{A}(n\text{p}, n\text{f}\delta)$ are observed. The $\mathscr{A}(n\text{p}, t_{2g}\pi^*)$ band and others like it in compounds carrying a central metal atom are the familiar ligand-to-metal charge transfer excitations of ligand field theory. Systems of this sort can be the spawning grounds for giant resonances (*vide infra*).

In nonmetals of high formal oxidation state (SF_6, for example), Fig. I.C-1D, the valence σ^* MO's involve the s and p orbitals of the central atom and transitions from np to both nd(t_{2g} and e_g) and to nf(a_{2u}, t_{2u} and t_{1u}) are Rydberg in nature. The Rydberg/valence character of the transitions in the highly oxidized systems are readily sorted out on the basis of the observed term values, which are about twice as large for $\mathscr{A}$ bands as for the corresponding Rydberg transitions.

Since the σ^* MO's are far more active in the inner-shell spectra of molecules, it is easiest to see how they influence molecular spectra by comparing the inner-shell spectra of several molecular hydrides with that of their isoelectronic united atom. Thus in Fig. I.C-2, the 2p molecular spectra of HCl and H_2S are compared with that of Ar, revealing a number of differences. Interpretation of these spectra requires an allowance for the $2p_{3/2} - 2p_{1/2}$ spin-orbit splitting in each case. First, we note that the most prominent single feature in the molecular spectra, a very broad transition about 50 000 cm^{-1} below the $(2p)^{-1}$ ionization limit, does not appear in the spectrum of the isoelectronic atom, and hence must involve molecular σ^* MO's, which is to say, they are $\mathscr{A}(2p, \sigma^*)$ bands. Next, the sharp $2p \rightarrow 4s$ transition of the atom is barely detectable in the molecular spectra, suggesting appreciable mixing between the $(2p, a_1\sigma^*)$ and $(2p, 4sa_1)$ conjugate configurations. Finally, whereas $2p \rightarrow 4p$ is absolutely forbidden in the atom, it is an allowed transition in the molecular case. Once beyond the transitions to 4p, the Rydberg spectra of the atom and its isoelectronic molecular hydrides are quite comparable, except for certain small core splittings in the latter. The 2p spectra of PH_3 (Fig. VI.B-1) and SiH_4 (Fig. IX.A-1) closely follow the pattern of molecular transitions described above, with the lowest excitations being $\mathscr{A}(2p, \sigma^*)$ bands.

$\mathscr{A}$-Band Term Values

In the earlier Volumes (and in this work as well), we depend strongly on the regularity of the term values for the assignment of Rydberg transitions. This approach is natural since the term value $R/(n - \delta)^2$ appears prominently in the Rydberg formula describing the energetics of Rydberg series members. Alternatively, the Rydberg term value can be interpreted as the Rydberg excited-state ionization potential. For a valence excited state (ϕ_i^o, ϕ_j^*), the term value has no meaning in the Rydberg sense since there is no convergent series of valence states, however, it does have meaning when viewed as the ionization potential of an electron in ϕ_j^* when promoted from ϕ_i^o.

Several simple predictions can be made regarding the nature of $\mathscr{A}$ bands and their term values:-

1) Whereas the term value of the $(np, (n + 1)s)$ Rydberg configuration is understandable in terms of the core penetration of an essentially nonbonding $(n + 1)$s Rydberg orbital, in the case of an (np, σ^*) configuration the term value will strongly reflect the antibonding nature of the σ^* MO.

2) In a compound $R\ddot{X}$, the R–X bond strength will depend upon the σ/σ^* split. Thus there will be an *inverse* relationship between R–X bond

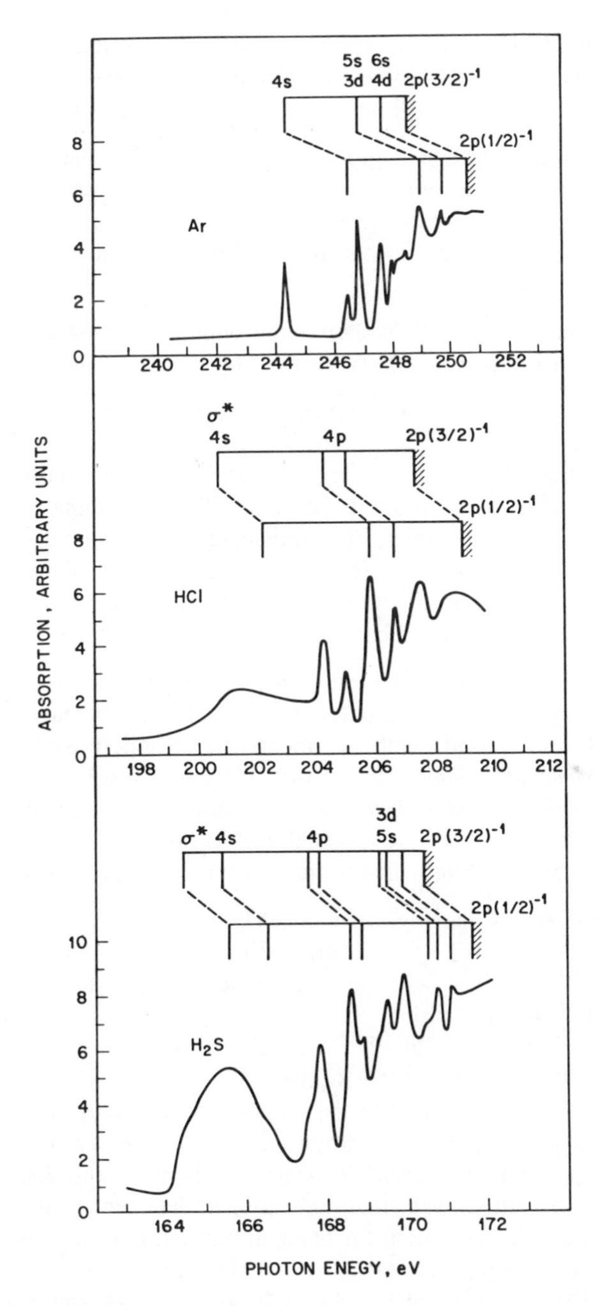

Fig. I.C-2. X-ray absorption spectra in the 2p regions of the isoelectronic systems Ar, HCl and H_2S in the gas phase [S27]. In each case, the spin-orbit coupling of the $2p^5$ core configuration is allowed for.

strengths and (np, σ^*) term values in a series of constant R and variable X, as depicted in Fig. I.C-3. A relationship of this sort had been hinted at earlier by Ishiguro *et al.* [I6] and by Robin and Kuebler [I, R18].

3) The larger the alkyl R group in RX, the weaker the R–X bond. As in *2*) above, this would tend to increase the (np, σ^*) term value. On the other hand, large R groups strongly stabilize the $(n\text{p})^{-1}$ ionic state by hyperconjugation, thereby decreasing the (np, σ^*) term value. The latter factor will be dominant.

4) Fluorination will act in a direction opposite to that of alkyl groups, and will increase the (np, σ^*) term values by stabilizing σ and σ^* MO's.

5) Ionization from σ^* will depend in part upon the effective nuclear charges of the AO's in the σ^* MO. These will increase as the hole descends so that the (1s, σ^*) term value will be substantially larger than that for (np, σ^*). This parallels the effect of antishielding upon Rydberg states (Section I.B), however, the antishielding increment will be much larger for a σ^* MO than for a Rydberg orbital since it is more "penetrating." Exchange effects too are responsible in part for the larger term values of inner-shell $\mathscr{A}$ bands.

6) Outer-shell $\mathscr{A}$(np, σ^*) excitations for overlap reasons will be among the weaker transitions in the outer-shell spectrum, whereas for the same reason, $\mathscr{A}$(1s, σ^*) bands will be among the strongest transitions in the inner-shell spectrum. Unless the $\mathscr{A}$(np, σ^*) bands are low-lying, they will be lost in the more intense Rydberg absorption of the outer-shell spectrum. The (np, σ^*) upper states are usually strongly dissociative, and under strong spin-orbit coupling will split into several overlapping continua.

7) In molecules having a halogen atom adjacent to pi-electron unsaturation, there will appear both $\mathscr{A}$(np, σ^*) and $\mathscr{A}$(π, σ^*) bands in the outer-shell spectrum, with separations nearly equal to the difference of $(n\text{p})^{-1}$ and $(\pi)^{-1}$ ionization potentials. Each occupied π MO in the molecule will be responsible for a specific $\mathscr{A}$(π, σ^*) band. $\mathscr{A}$(np, π^*) bands also will be found in these systems.

In order to test the possibility that valence term values may show trends which are revealing, excited-state and ionization data have been assembled on the $\mathscr{A}$(np, σ^*)-band transitions of halomethanes, Table I.C-II. In this, care was taken to assure that the proper halogen lone pair $(n\text{p})^{-1}$ ionization potential was taken for calculation of the term value. One sees in the Table that for excitation from the various halogen AO's into σ^*(C–X), the (ϕ_i, σ^*) term value indeed drops as the energy of ϕ_i descends, in accord with the antishielding concept. Moreover, a hole in 1s_C is about equivalent in antishielding to a hole in 2p_Cl, but is much less

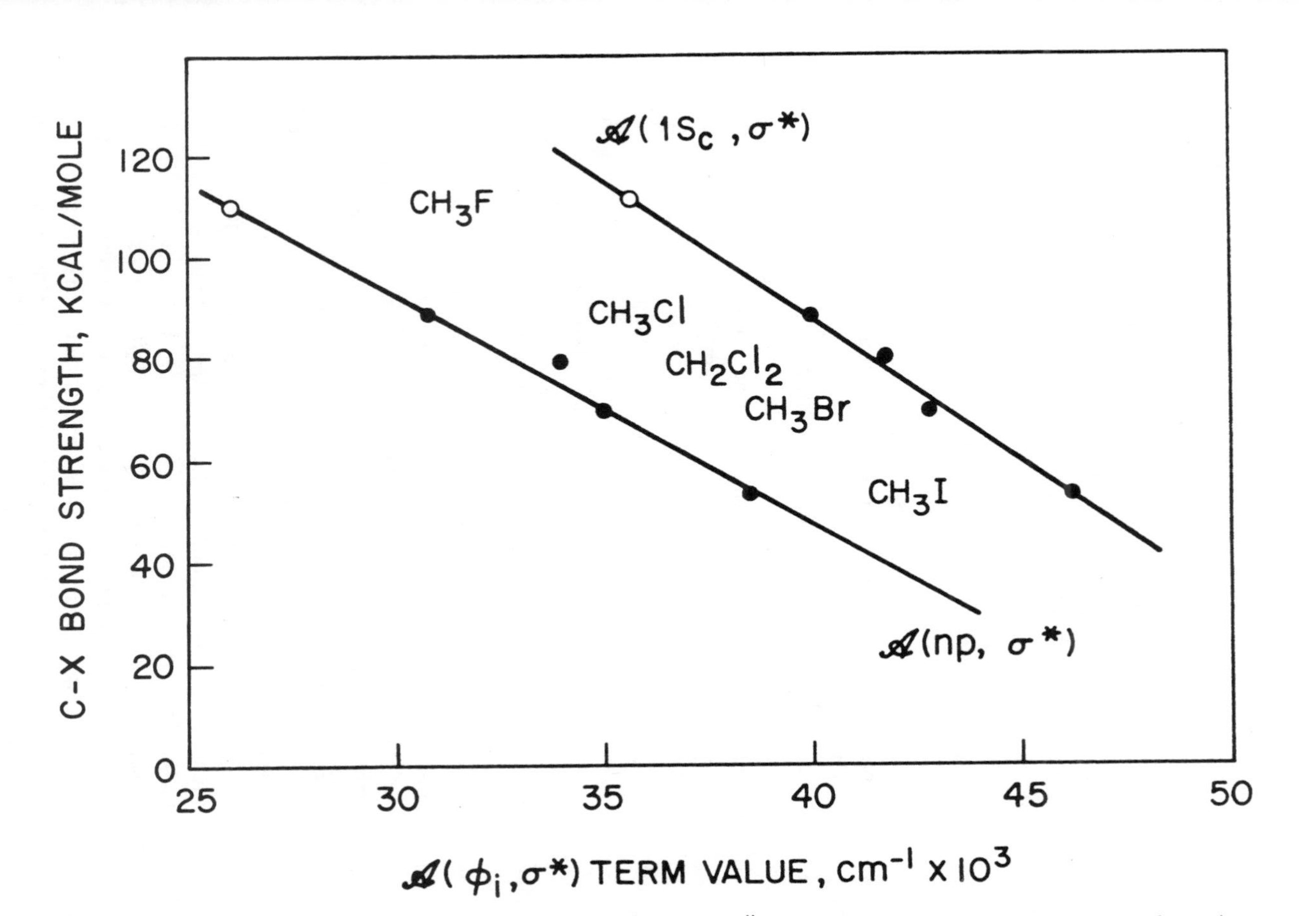

Fig. I.C-3. Diagram illustrating the inverse relationship between the R–$\ddot{X}$ bond strength and the corresponding $\mathscr{A}(n\mathrm{p}, \sigma^*)$ and $\mathscr{A}(1\mathrm{s}_\mathrm{C}, \sigma^*)$ term values in a one-electron orbital scheme.

antishielding than when ϕ_i^{-1} is $2s_{Cl}$ or $1s_{Cl}$. For a given ϕ_i, there is also an increase of the (ϕ_i, σ^*) term value as the methane molecule is successively halogenated, Tables I.C-II and I.C-III. This can be explained as due to interactions between C—X bond orbitals which split the σ^*(C—X) manifold into several components, the lowest of which is successively stabilized upon stepwise halogenation. This stabilization is most evident in the valence TNI resonances of the chloromethanes, in which a resonance measures the energy necessary to attach an electron to the molecule via the σ^* MO's [B85]. Thus between CH_3Cl and CCl_4 there is a monotonic decrease of the lowest valence TNI resonance from 27 800 cm^{-1} (*vert., 3.45 eV*) to a figure which is actually just slightly negative.

Support for several of the other expectations stated above is abundant in the data of Tables I.C-II and I.C-III. Thus looking first at the methyl halides, one notes the increasing $(n\text{p}, \sigma^*)$ term values in the series CH_3Cl, CH_3Br and CH_3I wherein the C—X bond strengths are decreasing, Fig. I.C-3. From this simple relationship we can extrapolate to a $(2\text{p}, \sigma^*)$ term value of 26 800 cm^{-1} in CH_3F. An identical trend is seen as well in the $(1s_C, \sigma^*)$ term values of the methyl halides, with the value of the $(1s_C, \sigma^*)$ term value for a particular compound being approximately 8000 cm^{-1} larger than for $(n\text{p}, \sigma^*)$. The extrapolated value of $(1s_C, \sigma^*)$ for methyl fluoride is 36 500 cm^{-1}. Because there is a monotonic increase of the lowest $(n\text{p}, \sigma^*)$ term value as one goes from CH_3X to CX_4 for X = Cl, Br and I, it is surprising that the $(4\text{p}, \sigma^*)$ term value decreases on going from SiH_3Br to $SiBr_4$, Table I.C-III.

The effects of alkylation on $(n\text{p}, \sigma^*)$ term values are seen most clearly in the iodides, Table I.C-III. In the alkyl iodides, the $\mathscr{A}(5\text{p}, \sigma^*)$ band falls consistently at 38 500 cm^{-1} (*vert., 4.77 eV*) while the ionization potential decreases from 77 030 cm^{-1} (*advert., 9.550 eV*) in methyl iodide to 71 860 cm^{-1} (*advert., 8.909 eV*) in cyclohexyl iodide. As a result of this overwhelming stabilization of the cation by the hyperconjugative alkyl groups, the $(5\text{p}, \sigma^*)$ term value steadily drops from a value of 38 500 cm^{-1} in methyl iodide to 33 360 cm^{-1} in cyclohexyl iodide. In the alkyl chlorides, the same trend is evident, though with a smaller slope. Making the R group unsaturated has no noticeable term-value effect. The depressing effect of the alkyl group on (ϕ_i, σ^*) term values is evident as well in the (ϕ_i^{-1}, σ^*) term values of HCl [$(2p_{Cl}, \sigma^*)$, 47 500 cm^{-1}], CH_3Cl [$(1s_C, \sigma^*)$, 40 000 cm^{-1}] and C_2H_5Cl [$(1s_C, \sigma^*)$, 38 700 cm^{-1}].

Comparing data in Table I.C-II, one sees that the difference between outer-shell $\mathscr{A}$-band term values and inner-shell $\mathscr{A}$-band term values in the methyl halides is typically 10 000 cm^{-1}, whereas for inner-shell and outer-shell Rydberg term values for the lowest Rydberg states, this difference is more like 0—2000 cm^{-1} and is even negative occasionally

TABLE I.C-II

$\mathscr{A}(\phi_i, \sigma^*)$-BAND TERM VALUES IN THE HALOMETHANES, CH_xX_{4-x}

	$(1s_C, \sigma^*)$	$(1s_X, \sigma^*)$	$(2s_X, \sigma^*)$	$((n-1)p_X, \sigma^*)^a$	$(np_X, \sigma^*)^a$	$((n-1)d_X, \sigma^*)^a$
CH_3F	36 500[b]	-	-	-	26 800[b]	-
CH_3Cl	40 000	46 000	44 000	42 400	32 200	-
CH_2Cl_2	41 100	-	53 500	42 200	34 000	-
$CHCl_3$	46 800	-	58 400	49 600	36 000	-
CCl_4	50 800	-	57 600	51 200	38 000	-
CH_3Br	42 700	-	-	-	35 000	44 400[c]
CH_3I	45 600	-	-	-	38 530	48 800[d]

[a] The highest filled halogen lone-pair orbital is assigned quantum number n.

[b] Extrapolated values from Fig. I.C-3.

[c] Term value for the $\mathscr{A}(3d_I, \sigma^*)$ band.

[d] Average of the two $(4d_I, \sigma^*)$ spin-orbit components having term values of 49 200 and 48 400 cm^{-1}.

(Table I.B-II). Thus the antishielding on going from the outer shell to the inner shell has a far larger effect on valence σ^* term values than on the conjugate Rydberg term values, in qualitative accord with Slater's rules. This statement holds not only for $\mathscr{A}$-band transitions in halides, but for transitions to π^* in ketones and to σ^*(X—H) in the second-row hydrides. As with the inner- and outer-shell Rydberg excitations, exchange also has been put forward to explain the difference of inner- and outer-shell π^* valence term values [E1]; for valence excitations, the exchange effect will have a larger magnitude, but will not necessarily be a larger fraction of the term-value difference (exhaltation).

It seems likely that the generalizations quoted above for the monohalides will hold as well for polyhalides, but this is not as simple to prove since

TABLE I.C-III

OUTER-SHELL $\mathscr{A}(n\text{p}, \sigma^*)$-BAND TERM VALUES IN SATURATED, HALIDE-CONTAINING MOLECULES

Molecule	Frequency, cm^{-1}	Ionization Potential, cm^{-1}	Term Value, cm^{-1}	References
CH_3Cl	58 800	90 980	32 200	[G5, H72, H73]
CH_2Cl_2	58 800	92 800	34 000	[H73]
$CHCl_3$	57 200	93 200	36 000	[H73]
CCl_4	57 100	95 100	38 000	[C12, H73, R47, R48]
C_2H_5Cl	59 000	89 020	30 000	[H73]
n-C_3H_7Cl	58 600	87 840	29 200	[I, R30]
i-C_3H_7Cl	57 700	87 270	29 600	[I, R30]
n-C_4H_9Cl	58 200	87 760	29 600	[I, R30]
CH_2FCl	62 500	94 700	32 200	[H73]
CHF_2Cl	66 200	101 600	35 400	[D43, H73]
CF_3Cl	65 300	104 900	39 600	[D43, H73, S12]
CF_2BrCl	61 000	104 000	43 000	[H73]
$CHCl_2F$	57 100	96 800	39 700	[H72, H73]
CF_2Cl_2	56 500	99 200	42 700	[H72, H76, K20, M63]
$CFCl_3$	55 860	96 000	40 140	[H72, H73, H76, K20, V21]
C_2F_5Cl	66 700	104 500	37 800	[D43, H73]
CH_3CF_2Cl	63 900	100 800	36 900	[D43, H73]
CF_3CHCl_2	57 100	96 800	39 700	[G34]
CF_3CCl_3	57 500	95 200	37 700	[T22]
$CF_2ClCFHCl$	57 500	96 800	39 300	[D43]
$CFCl_2CF_2Cl$	55 000	97 190	42 200	[D43, H73]
$CF_3CHBrCl$	59 500[a]	98 400	38 900	[T22]
$CH_3OCF_2CHCl_2$	57 500	92 800	35 300	[T22]
HCF_2OCF_2CHFCl	57 500	98 400	40 900	[T22]
$GeCl_4$	56 800	97 800	41 000	[C12]
$SnCl_4$	50 200	97 600	47 400	[C12]

TABLE I.C-III (continued)

Molecule	Frequency, cm^{-1}	Ionization Potential, cm^{-1}	Term Value, cm^{-1}	References
PCl_3	57 000	94 450	37 450	[I, 241]
CH_3Br	50 000	85 000	35 000	[C9, G5]
CH_2Br_2	45 500	85 600	40 100	[C9]
$CHBr_3$	41 200	84 500	43 300	[C9]
CBr_4	39 500	83 800	44 300	[C9, R47]
CF_2Br_2	44 000	90 200	46 200	[D44]
CF_3Br	48 850	96 800	47 950	[D44, R48]
CF_2ClBr	48 800	95 400	46 600	[D44]
$CF_3CHClBr$	49 000[b]	90 660	41 660	[T22]
SiH_3Br	52 400	89 000	36 600	[C11]
$SiBr_4$	51 300	87 500	36 200	[C28]
$GeBr_4$	40 800	87 100	46 300	[C28]
$SnBr_4$	38 000	88 700	50 700	[C28]
CH_3I	38 500	77 030	38 530	[G5, I, B45]
C_2H_5I	38 500	75 380	36 880	[I, B45]
n-C_3H_7I	38 500	74 610	36 110	[I, B45]
i-C_3H_7I	38 500	74 120	35 620	[I, B45]
t-C_4H_9I	37 000	73 240	36 240	[I, B45]
C_5H_9I	38 500	73 200	34 700	[I, B46]
$C_6H_{11}I$	38 500	71 860	33 360	[I, B46]
CI_4	25 800	76 000[c]	50 000	[R47]
CF_3I	37 700	85 580	47 880	[I, 157]
C_2F_5I	37 000	85 620	48 600	[R35]

[a] A $3p_{Cl} \rightarrow \sigma^*(C-Cl)$ excitation.

[b] A $4p_{Br} \rightarrow \sigma^*(C-Br)$ excitation.

[c] No ionization potential data available; value taken from X-α scattered wave calculation.

one must be careful to correlate each $\mathscr{A}$-band with the proper ionization potential when computing the term value.

The term-value effects of replacing hydrogen by fluorine on several valence excitations are summarized in Table I.C-IV. One sees that in the case of the alkyl halide $\mathscr{A}$ bands, fluorination shifts the outer-shell $\mathscr{A}$-band term value to higher frequencies by up to 12 000 cm^{-1}, but an average value is more like *ca.* 8000 cm^{-1}. A similar positive increment

TABLE I.C-IV

FLUORINATION EFFECTS ON VALENCE-STATE TERM VALUES

Upper-State Configuration	Hydrogenated Compound	Term Value, cm^{-1}	Fluorinated Compound	Term Value, cm^{-1}
$(2p, \sigma^*)$	SiH_4	35 000	SiF_4	45 500
$^3(\pi, \pi^*)$	$H_2C{=}CH_2$	49 950	$F_2C{=}CF_2$	47 200
$^1(\pi, \pi^*)$	$H_2C{=}CH_2$	23 050	$F_2C{=}CF_2$	13 600
$^3(\pi, \pi^*)$	$H_2C{=}CHCl$	49 200	$F_2C{=}CFCl$	47 400
$^1(\pi, \pi^*)$	$H_2C{=}CHCl$	26 700	$F_2C{=}CFCl$	20 200
$(3p, \sigma^*)$	CH_3Cl	31 000	CF_2HCl	35 400
			CF_3Cl	39 500
$(3p, \sigma^*)$	CH_2Cl_2	35 900	CF_2Cl_2	43 100
$(3p, \sigma^*)$	$CHCl_3$	36 000	$CFCl_3$	40 400
$(3p, \sigma^*)$	C_2H_5Cl	30 000	CH_3CF_2Cl	36 900
			C_2F_5Cl	37 800
$(4p, \sigma^*)$	CH_3Br	34 000	CF_3Br	46 800
$(4p, \sigma^*)$	CH_2Br_2	40 000	CF_2Br_2	46 200
$(5p, \sigma^*)$	CH_3I	39 000	CF_3I	46 200
$(1s, \pi^*)$	$CH_3CH{=}CHCH_3$	46 000	$CF_3CF{=}CFCF_3$	54 800

applies to the inner-shell $(2p, \sigma^*)$ term values in the SiH_4/SiF_4 pair, the only one for which inner-shell data is available. Interestingly, the opposite effect holds for the π^* MO term value of olefins, where fluorination weakly decreases the $^3(\pi, \pi^*)$ term value and strongly depresses it in $^1(\pi, \pi^*)$. Using the arguments originally put forward to explain the perfluoro effects in olefins [I, B59], one argues here that the inductive effect of the fluorination acts to increase the (π, π^*) term value and that the C–F antibonding character of the π^* MO acts to decrease it. The latter is dominant, apparently. In the case of Rydberg term values, the antibonding factor does not play a significant role, and so the fluorination acts in the opposite manner as regards the term value. A sampling of $\mathscr{A}$-band term values in highly fluorinated species such as SF_6, XeF_6, *etc.*, Table I.C-V, reveals $\mathscr{A}$-band term values as large as 80 000 cm^{-1} for inner-shell transitions.

TABLE I.C-V

TERM VALUES FOR $\mathscr{A}$ BANDS IN PERFLUORINATED SYSTEMS

Molecule	Transition	Term Value, cm^{-1}	References
BF_3	$1s_B \rightarrow \sigma^*$	− 18 500	[I6]
CF_4	$1s_C \rightarrow \sigma^*$	40 000	[B66]
NF_3	$n_N \rightarrow \sigma^*$	40 100	[I, 228]
	$1s_N \rightarrow \sigma^*$	52 000	[B63, S44, V26]
	$1s_F \rightarrow \sigma^*$	52 000	[B63, S44, V26]
SiF_4	$2p_{Si} \rightarrow \sigma^*$	45 500	[F34, S60]
	$1s_F \rightarrow \sigma^*$	43 500	[F34, S60]
SF_6	$\phi_i \rightarrow \sigma^{*a}$	48 000[b]	[G45, S19]
	$2p_S \rightarrow \sigma^*$	67 800	[H60]
	$2s_S \rightarrow \sigma^*$	58 900	[H60]
	$1s_F \rightarrow \sigma^*$	53 200	[H60]
MoF_6	$\phi_i^o \rightarrow \sigma^{*a}$	72 000[c]	[M27, M36]
WF_6	$\phi_i^o \rightarrow \sigma^{*a}$	65 000[c]	[M27, M36]
XeF_6	$4d \rightarrow \sigma^*$	80 000	[N23]

[a] Transitions originating at several outer-shell MO's of the molecule.

[b] Average term value for several $\mathscr{A}(\phi_i^o, \sigma^*)$ bands, Table VIII.B-I.

[c] Average term value for several $\mathscr{A}(n\text{p}, \pi^*)$ bands, Table XXI.A-I.

The $\mathscr{A}$-band term values of the halogenated olefins (Table X.C-I) and benzenes (Section XIX.C and Table XIX.C-I) again are quite regular, with the $\mathscr{A}(\pi, \sigma^*)$ and $\mathscr{A}(n\text{p}, \sigma^*)$ term values remarkably alike in a given compound (*cf.* the appropriate Sections for further discussions on these).

The valence transitions remaining largely hidden from view in the vacuum ultraviolet are the N → V ($\sigma \rightarrow \sigma^*$) and the higher $\mathscr{A}(\pi, \sigma^*)$ and $\mathscr{A}(\sigma, \pi^*)$ varieties. The general feeling here is that valence excitations in the vacuum ultraviolet are directly dissociative, leading to the broad underlying continua so often seen in vacuum-ultraviolet spectra. The results of MPI spectroscopy (Section II.A) support this view. On the other hand, our hopes for observing $\sigma \rightarrow \sigma^*$ excitations are buoyed by the confident assignment of both $\sigma \rightarrow \sigma^*$ (107 000 cm^{-1}, *13.3 eV*) and $\sigma \rightarrow \pi^*$ (123 000 cm^{-1}, *15.3 eV*) transitions in acetylene. The bands are at high frequency, broad, beyond the first ionization potential and more or less autoionized [H28]. Further, the $\sigma \rightarrow \sigma^*$ transitions in the hydrogen halides [J1, S70] have been identified in the 75 000–85 000 cm^{-1} (*9.3–10.5 eV*) region as vibronically sharp but highly nonvertical transitions. The analogous $\sigma \rightarrow \sigma^*$ bands in other halogen-containing molecules have not been identified as of yet, but should be close by.

I.D. Rydberg/Valence Conjugates; Existence, Mixing and Ordering

Coexistence of Rydberg/Valence Conjugates

The question of the mixing of Rydberg/valence conjugate orbitals and the real or imagined existence of such conjugate pairs remain unsettled in the literature. The problem centers upon the fate of subtractive LCAO-MO's of the form $\eta_A - \eta_B$ as the centers A and B coalesce in the united atom [G36, M4, M80, M81, S51]. Though the proper way to handle this problem is still in question, a practical approach is simply to dismiss such orbitals from consideration whenever the $\eta_A\eta_B$ overlap approachs unity [S51]. Generalizing, we take the pragmatic view that if R_{AB} is much larger than the equilibrium bond length R_e, then the proper MO's are $\eta_A \pm \eta_B$ where η_A and η_B are valence orbitals, and $\chi_A \pm \chi_B$ where χ_A and χ_B are Rydberg orbitals. In contrast, when R_{AB} is much shorter than R_e, then the only proper forms are $\eta_A + \eta_B$ and $\chi_A + \chi_B$ for the valence and Rydberg MO's, respectively. In the most interesting region in the vicinity of R_e, the prescription is to take $\eta_A \pm \eta_B$ as the valence MO's, but only $\chi_A + \chi_B$ for the Rydberg set.

The pragmatic prescription given above leads to two immediate consequences. First, when considering the Rydberg spectra of dimer-type molecules such as diazabicyclooctane, diones or *trans*-dibromoethylene, one should *not* try to construct molecular Rydberg orbitals by taking the subtractive combination of local Rydberg orbitals on the equivalent sites. The subtractive combinations are discouraged because they generate a superabundance of levels the presence of which is not supported by experiment, and more importantly, because the subtractive orbital combination tends to "disappear" as n and the overlap increase. On the other hand, for the allowed additive combination, one might just as well think of an in-phase linear combination of local functions as equivalent to a single-center Rydberg function. The second consequence of our approach is that whereas $\eta_A - \eta_B$ is inappropriate for the Rydberg set at R_e, the $\chi_A - \chi_B$ combination in the valence set nonetheless is still valid, along with $\chi_A + \chi_B$. In this way, we have preserved the mutual existence of $\eta_A + \eta_B$ and $\chi_A - \chi_B$ as Rydberg/valence conjugate orbitals in the vicinity of R_e.

Now this approach is not accepted by Mulliken [M80, M81], who claims that the subtractive virtual valence MO $\chi_A - \chi_B$ has no existence independent of the Rydberg orbital $\eta_A + \eta_B$, so that there cannot be discrete transitions to conjugate valence and Rydberg states. Our position is that this is not correct, though in several instances it may appear to be so since one or the other transition of a conjugate pair often is not reported experimentally. However, it may be that the valence excitation in question is at very high frequency, autoionized, overlapped or dissociated so as to make it difficult to observe experimentally. Additionally, the valence excitation may have no specific existence as such because it is so thoroughly mixed with the conjugate Rydberg manifold that no specific level contains it as a major component. This last situation, discussed further below, is quite different from Mulliken's scenario, though the end result is the same. In other cases, the valence excitation has been identified but the conjugate Rydberg excitation is a component of the core-split Rydberg multiplet which has not been resolved or has not been identified positively as to symmetry.

In support of the simultaneous reality of conjugate Rydberg and valence states, we first quote the numerous highly accurate calculations on a variety of molecules which predict their existence. Thus for example, in HF, a molecule which Mulliken mentions explicitly, Bettendorf *et al.* [B40] calculate that the $\mathscr{A}(2p\pi, \sigma^*)$ valence band will be centered at 84 040 cm^{-1} (*10.42 eV*) with the upper state being repulsive, whereas the conjugate Rydberg transition to (2pπ, 3s) is predicted to fall at 105 400 cm^{-1} (*13.07 eV*) with a bound upper-state surface. These predicted transitions of HF are closely parallel to the conjugate $\mathscr{A}(5p\pi, \sigma^*)$ valence and 5pπ $\rightarrow$ 6s Rydberg bands of methyl iodide, about which there is no question. Interestingly, X-α calculations on H_2S which do not employ the LCAO format at all nonetheless predict the separate existence of $(5a_1, \sigma^*)$

and ($5a_1$, $3p\sigma$) conjugate excited states [R29]. Finally, we mention that the inner-shell spectra of several second-row hydrides, Figs. I.C-2, VI.B-1 and IX.A-1, clearly and indisputably show that conjugate valence and Rydberg excitations are seen in the same spectrum. Note also that conjugate valence and Rydberg excitations are well-documented in atomic spectra [L29].

Conjugate and Nonconjugate Mixing

Rydberg/valence mixing in a molecule can occur in two ways depending upon whether the mixing takes place at the MO level (mixing of conjugates) or at the CI level (mixing of nonconjugates). In the former case, the two configurations which are mixed [(ϕ_i^o, 3s) and (ϕ_i^o, σ^*), for example] differ by only a one-electron promotion and consequently they can be mixed by considering a one-electron operator and the 3s/σ^* overlap distribution. This is equivalent to mixing at the MO level. In the second case, the nonconjugate Rydberg/valence configurations are of the sort (π, 3s) and (n_O, π^*), and though of the same overall symmetry, they differ by a two-electron promotion. These can be mixed only as configurations, and only by interelectron exchange and repulsion. An excellent example of nonconjugate mixing is that between (n_O, 3s) and (π_2, π_3^*) in formamide (Section XV.A).

In a molecule of low symmetry, a given σ^* MO may have several Rydberg orbitals of different l but the same symmetry, which are formally conjugate to it. If there is one conjugate Rydberg orbital which most closely resembles σ^* and has the largest overlap with it, this orbital is termed the *prime conjugate*. In methane, the prime Rydberg conjugate to $a_1\sigma^*$ is $3sa_1$, whereas in water the prime conjugate to $b_2\sigma^*$ is $3db_2$ rather than $3pb_2$ as judged by their electron distributions (Section VII.A).

Valence Miscibility with the Rydberg Sea

Having stated our point of view regarding the reality of distinct valence and Rydberg conjugate states, we next consider the possibility of their mixing. Mies [M57] considers theoretically the configurational mixing of a valence-state of the Cd atom with a Rydberg manifold which is instructive. In this case, the $5p^2$ valence-state interloper lies beyond the ionization limit of the $5s^2$ ground state, and is in interaction with the (5s, nd) Rydberg manifold. Though this is a nonconjugate mixing, it nonetheless shows, Fig. I.D-1, how widely the amplitude of the valence level is spread among the discrete (5s, nd) and continuum (5s, ϵd) levels,

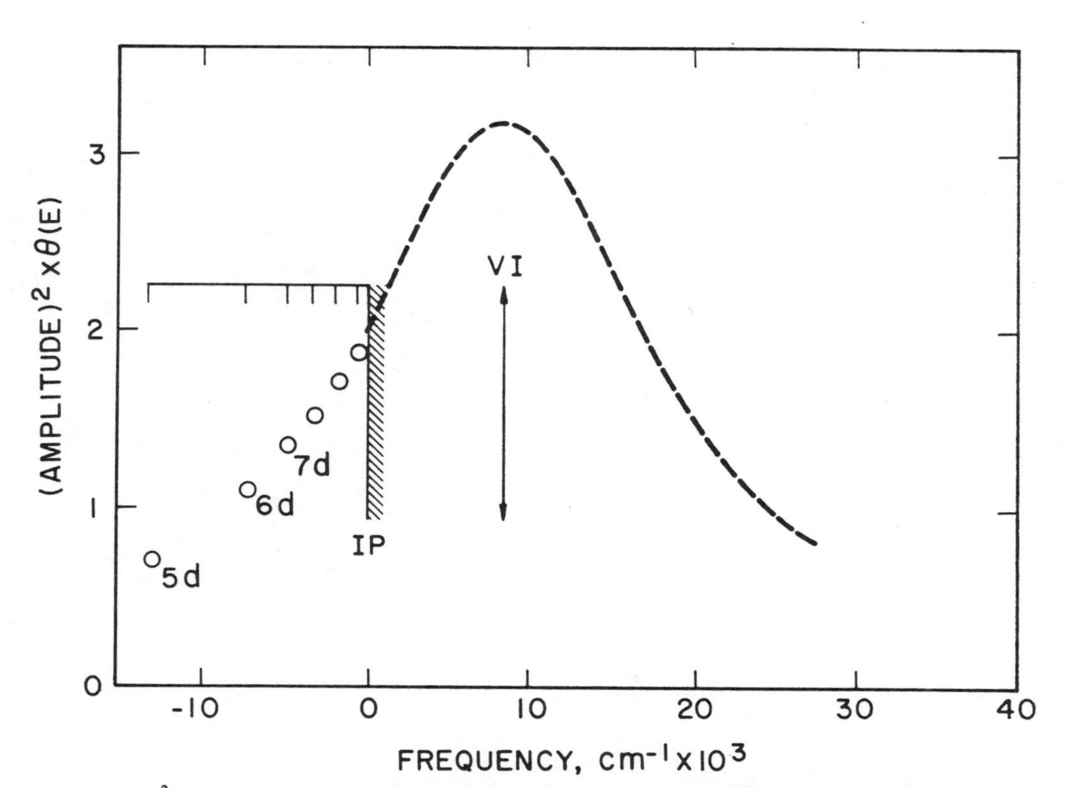

Fig. I.D-1. Mixing of the $5p^2$ valence interloper into the (5s, *n*d) Rydberg and continuum manifold of Cd(I) [M57]. The position of the valence interloper (VI) before mixing is shown by the arrow, while the amplitude of the VI across the spectrum after mixing is shown by the dashed curve and discrete points.

rather than just being mixed into its nearest neighbors. Similar effects are calculated for the molecular case. Thus for example, Buenker and Peyerimhoff [B75, P24] explain that in ethylene the (π, π^*) valence state and the prime conjugate $(\pi, 3dxz)$ Rydberg state are mixed so as to form approximately 50:50 mixtures in each. The higher component in which (π, π^*) and $(\pi, 3dxz)$ are out-of-phase, in a sense is the valence conjugate of the $(\pi, 4dxz)$ Rydberg level and again there is strong mixing between these. In this way, the (π, π^*) valence character is propagated throughout the $(\pi, ndxz)$ Rydberg manifold. In this example, the lowest level retains sufficient (π, π^*) character to be labelled as such, however, this is not necessarily always the case. As a counterexample, the $(1b_1, a_1\sigma^*)$ valence configuration of water is calculated to be so thoroughly mixed into the $(1b_1, nsa_1)$ Rydberg sea of levels that it no longer contributes enough to any one state to merit calling that state "$(1b_1, a_1\sigma^*)$". This extreme miscibility of the valence level and its conjugate Rydberg manifold is most apparent in the outer-shell spectra of the first-row and second-row hydrides, excepting the hydrogen halides. Note too that the miscibility of σ^* valence orbitals with the conjugate Rydberg sea can be selective, for in H_2S, the $a_1\sigma^*$ MO is totally miscible as in H_2O, whereas the $b_2\sigma^*$ MO is not. It is also possible that the extent of miscibility of the (ϕ_i^o, ϕ_j^*) valence configurations in the conjugate Rydberg sea is dependent upon the nature of the ϕ_i^o orbital.

It is a fascinating question as to when an interloping valence level is totally dissolved in the Rydberg sea and when it is only dampened by it. Looking at the spectra, it seems that transitions to such distinctly valence levels still appear when they are energetically at the edges of the sea and so only get their feet wet, *i.e.*, when the interlopers precede the Rydberg parade and when they follow it as shape resonances (Section II.C). In the intermediate region, they find themselves in a high density of compatible Rydberg levels and so readily can lose their valence identity. This is reasonable in light of the inherently small mixing element that exists between a large Rydberg and a small valence function, meaning that wavefunction mixing can be effective only when the levels are near-degenerate. A valence interloper will avoid dissolution by falling below the Rydberg parade, *i.e.*, by having a term value significantly larger than that of the lowest conjugate Rydberg level. This means going to heavy-halogen $\mathscr{A}(np, \sigma^*)$ bands, fluorination of the molecule or exciting from inner-shell orbitals (Section I.C).

Symmetry is another important factor in determining the miscibility of a valence level with the conjugate Rydberg sea. Note that the Rydberg/valence mixing of conjugates will depend upon the overlap of these orbitals, and if the conjugate Rydberg orbitals have very high angular momentum then they will not penetrate the core and so will have a greatly diminished tendency to mix with the particular antibonding valence MO. Generalizing, an antibonding valence MO of s symmetry will be most likely to dissolve in the conjugate *n*s Rydberg sea, whereas p-symmetry and d-symmetry antibonding MO's will be much less apt to dissolve in their less penetrating Rydberg manifolds. An antibonding MO of f symmetry offers the extreme case, for here the *n*f Rydberg orbitals are quite nonpenetrating and there will be very little tendency indeed for dissolution. However, the symmetry of the molecule also must be high if the f-type valence orbital is to retain its valence character, for only in high-symmetry systems will the *n*s, *n*p and *n*d Rydberg seas have symmetry representations different from that of *n*f. Beautiful examples of these principles at work are discussed in Chapter III for alkanes and in Section XXI.A for UF_6.

Owing to the tendency for certain valence levels to dissolve in the conjugate Rydberg manifold, the Rydberg/valence nature of a nominally valence configuration cannot be assessed correctly by calculation unless the basis set is rather complete in the conjugate Rydberg orbitals. This is in contrast to the approach often used of implementing the Rydberg basis only when concerned directly with calculating the higher Rydberg states.

Mixing Effects and Unmixing

Because the properties of Rydberg and valence states are so different, configurational mixing of these two types of states can produce profound effects. The consequences of Rydberg/valence mixing are best illustrated by calculations on the $^1(\pi, \pi^*)$ and $^1(\pi, 3dxz)$ configurations of ethylene [B75]. Calculations show that the mixing matrix element between Rydberg and valence configurations generally is largest when the two configurations differ by a one-electron promotion, *i.e.*, when they are conjugate. In the case of ethylene, mixing with the $^1(\pi, n dxz)$ manifold lowers the $^1(\pi, \pi^*)$ frequency by *ca.* 7000 cm^{-1}, and results in a final V-state orbital which has 45–50 % Rydberg character. Rydberg/valence mixing may also be vibronic, as in the NO molecule [B60], rather than purely electronic.

For reasons of spatial overlap, the $\pi \rightarrow \pi^*$ valence excitation of ethylene is far more intense than the conjugate $\pi \rightarrow 3dxz$ Rydberg excitation, and consequently the calculated f value for the $\pi \rightarrow \pi^*$ excitation decreases from 0.485 to 0.293 upon admitting $n dxz$ orbitals into the calculation. Though the "missing" oscillator strength is donated to the Rydberg excitations, it is not very noticeable in any one of them since it is distributed among all the $\pi \rightarrow n\text{d}xz$ transitions. The importance of Rydberg/valence mixing on the f values in atoms also has been noted [L29].

As expected, the values of $<x^2>$ for the out-of-plane x coordinate computed for valence and Rydberg MO's vary sharply depending upon orbital type. Thus for the π^* valence MO in the $^3(\pi, \pi^*)$ state of ethylene, $<x^2>$ is 3.88 a_0^2, while for 3dxz in $^1(\pi, 3\text{d}xz)$ it swells to 53.6 a_0^2; the π^* MO of mixed character in $^1(\pi, \pi^*)$ has the intermediate value 24.3 a_0^2. Another measure of spatial extent is offered by the Coulomb self-energy integral J_{jj} in the (ϕ_i, ϕ_j) configuration. As shown in Table I.D-I, the value of J_{jj} decreases with increasing ϕ_j orbital size, and in fact can be used as a criterion on which to base a Rydberg/valence assignment. One sees that valence excitations have computed upper orbital self-energies in the 73 000–89 000 cm^{-1} (*9–11 eV*) range whereas the Rydberg orbital self-energies occupy the 0–32 000 cm^{-1} (*0–4 eV*) range, depending upon their term values. Interestingly, the π^* self-energy of the $^1(\pi, \pi^*)$ valence-shell configuration, 32 500 cm^{-1} (*4.03 eV*), is far below that normally assigned to valence states, but is at the top of the Rydberg range, a clear indication of its heavy Rydberg admixture. In contrast, the π^* self-energy for the $^1(1\text{s}, \pi^*)$ level, 93 600 cm^{-1} (*11.6 eV*), is solidly in the valence camp, as appropriate for a purely valence configuration. With the exception of the $^1(\pi, \pi^*)$ level of ethylene, one see that the J_{jj} self-energies of (ϕ_i, ϕ_j) configurations are very much independent of the open-shell orbital ϕ_i.†

The highly anomalous self-energy value for the π^* MO in the $^1(\pi, \pi^*)$ state of ethylene can be understood in terms of simple valence-bond theory. In the valence-bond representation, the V state of ethylene is strongly ionic, involving resonance structures of the form

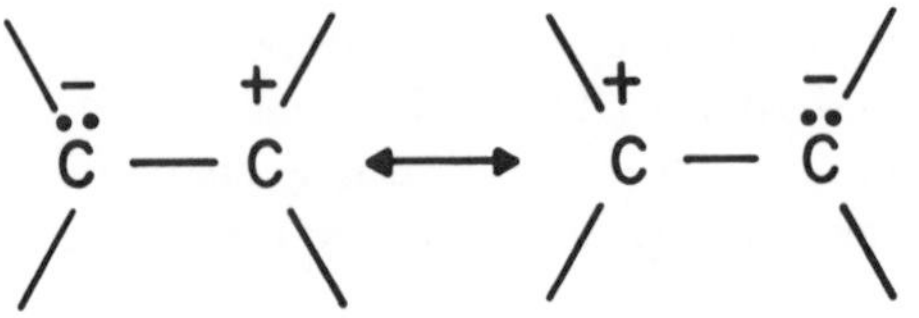

```
\-    +/       \+    -/
 ¨              
 C ─ C   ←→    C ─ C
                    ¨
/     \       /     \
```

† Yet another possible measure of Rydberg/valence character is the virial $-V/2T$. As applied to the excited states of acetylene [D15], the virial is 0.5–0.7 for valence MO's, but 0.9–1.0 for Rydberg orbitals, depending upon penetration.

In these structures, the Coulombic repulsion of the two 2pπ electrons on one center is relieved in part by a radial expansion of the π^* wave function so as to allow in-out correlation. Since allowed electronic transitions can take place only to such charge-separated states, we expect in general that intense N → V excitations ($f \geqslant 0.2$) will involve considerable upper-state Rydberg/valence mixing (conjugate and nonconjugate), with a concomitant decrease of oscillator strength. It was factors such as this which operate on V states which prompted us to distinguish them from $\mathscr{A}$ states, Table I.C.-I.

TABLE I.D-I

COULOMB SELF-ENERGY VALUES (J_{jj}) CALCULATED FOR RYDBERG AND VALENCE CONFIGURATIONS

Molecule	Configuration (ϕ_i, ϕ_j)	J_{jj}, cm^{-1} (*eV*)	References
C_2H_4	$^1(\pi, \pi)$	94 400 (*11.7*)	[B15, B75, P24]
	$^3(\pi, \pi^*)$	88 700 (*11.0*)	
	$^1(\pi, \pi^*)$	32 500 (*4.03*)	
	$^1(\pi, 3s)$	29 800 (*3.70*)	
	$^1(\pi, 3p\sigma)$	28 100 (*3.48*)	
	$^1(\pi, 3d\pi)$	20 000 (*2.48*)	
	$^1(1s, \pi^*)$	93 600 (*11.6*)	[B15]
	$^1(1s, 3s)$	29 200 (*3.62*)	
	$^1(1s, 3p\pi)$	25 200 (*3.13*)	
	$^1(1s, 3d\pi)$	20 600 (*2.56*)	
	$^1(1s, 4p\pi)$	13 400 (*1.66*)	
C_2H_2	$^1(1s, \pi^*)$	95 200 (*11.8*)	[B15]
	$^1(1s, 3s)$	28 700 (*3.56*)	
H_2S	$^1(2b_1, 3b_2\sigma^*)$	74 600 (*9.25*)	[P24, S47]
	$^1(2b_1, 4px)$	24 600 (*3.05*)	
	$^1(2b_1, 4dxz)$	19 100 (*2.37*)	
	$^1(2b_1, 4dxy)$	18 400 (*2.28*)	
SiH_4	$^1(2p, \sigma^*)$	81 500 (*10.1*)	[F33]
	$^1(2p, 4s)$	37 300 (*4.63*)	
	$^1(2p, 3d)$	19 800 (*2.45*)	

The large spatial differences between Rydberg and valence orbitals lead to large differences of singlet-triplet splittings which are subsequently modified by Rydberg/valence mixing. This effect will be observed most readily for Rydberg transitions, for in this case, the singlet-triplet split is normally 2000–3000 cm^{-1} for the lowest Rydberg excitations and shifts of only a few thousand cm^{-1} induced by valence mixing will appear as anomalies. Thus the unexpectedly large split between the $^3(1t_2, 3s)$ and $^1(1t_2, 3s)$ states of methane (7000 cm^{-1}, Section III.A) possibly is due to mixing with the conjugate valence configuration $(1t_2, a_1\sigma^*)$. There is very little evidence to guide one here, but it seems reasonable to assume that Rydberg/valence mixing will act to make the Rydberg singlet-triplet split larger and the valence singlet-triplet split smaller. As is most clearly illustrated by the case of ethylene discussed above, the Rydberg/valence mixing in the singlet and triplet manifolds in general will be quite different; Nakatsuji [N11] states that electron correlation will act to promote Rydberg/valence mixing more in the singlet manifold. This follows as well from the energy-denominator argument given above for the situation where the singlet valence interloper is below the conjugate Rydberg stack and the triplet valence interloper is even more removed. However, if the interloper is a singlet shape resonance, then its triplet will be even closer to the Rydberg manifold and so may have more Rydberg character than the singlet.

Provided the Rydberg/valence mixing is not too extreme, it is possible to experimentally deperturb these levels. Thus, as described in Section II.D, any of the well-known external perturbations which are so effective in helping to distinguish valence from Rydberg excitations also are effective in removing the effects of Rydberg perturbations on valence transitions. Interacting Rydberg and valence levels also can be deperturbed by first producing an inner-shell hole (*cf.* shake-up excitations, Section II.C) which shifts the offending Rydberg levels far out of resonance with the valence level of interest.

Experimental proof of the complete dissolution of a valence orbital in the Rydberg sea is difficult, as is determining whether a distinct σ^* level exists among the Rydberg levels, or if it exists above them as a shape resonance. Two possibilities exist for solving this problem. First, a high-quality calculation can settle the question in any particular case. Note however that the Rydberg/valence nature of a nominally valence configuration cannot be assessed correctly by calculation unless the basis set is rather complete in the conjugate Rydberg orbitals. Second, the comparison of $(\phi_i, 3s)$ and $(\phi_i)^{-1}$ Franck-Condon envelopes can be used as a qualitative guide to conjugate mixing in outer-shell spectra. It is generally true that the Franck-Condon envelopes of the lowest Rydberg excitation in a molecule and the corresponding photoelectron band are identical [I, 73]. However, exceptions do occur. In several instances, the lowest (ϕ_i, ns) level is mixed with (ϕ_i, σ^*) such that it is dissociative and rapidly deRydbergizes, whereas this is not true for the $(\phi_i)^{-1}$ ionic state. Consequently, in these cases the Franck-Condon envelopes for $\phi_i \rightarrow 3s$ excitation will be noticeably different from those for $(\phi_i)^{-1}$ ionization in the photoelectron spectra.

H_2O is an excellent example of a molecule showing extreme differences of Rydberg and PES band profiles due to Rydberg/valence conjugate mixing, for the $1b_1 \rightarrow 3sa_1$ transition is very nonvertical with a halfwidth of 6000 cm^{-1} whereas the $(1b_1)^{-1}$ band of the PES is exceedingly vertical, with an overall width of less than 400 cm^{-1}, Fig. VII.A-1. Similar but less extreme band-profile differences in CH_4, H_2S and HCOOH suggest that the lowest Rydberg levels in these molecules are strongly mixed with their valence conjugates. On the other hand, such mixing is much weaker in NH_3 as judged by the similarity of absorption and PES band envelopes, and effectively is nonexistent in the alkyl chlorides, bromides and iodides, for example.

If $\phi_i \rightarrow a_1\sigma^*$ is not observed below the Rydberg bands originating at ϕ_i, and does not mix with them as evidenced by the close similarity of the $\phi_i \rightarrow 3s$ and $(\phi_i)^{-1}$ band shapes, then it seems safe to conclude that $a_1\sigma^*$ is suspended somewhere in the $(\phi_i)^{-1}$ ionization continuum as a shape resonance. This argument does not hold for σ^* MO's which are less than totally symmetric, for in these cases the conjugate Rydberg orbitals are only weakly penetrating and do not mix very well with σ^* regardless of their relative frequencies.

In addition to configurational mixing, it is now well documented that the mixing and unmixing of Rydberg and valence conjugates in molecules can be induced by changes of internuclear distance.† Thus, for example, in each of the hydrides CH_4, C_2H_5, SiH_4, NH_3, PH_3, CH_3 and H_2O it has been computed that at the ground-state geometry, the first excited state of the same spin multiplicity is overwhelmingly Rydberg, as measured by any criterion. However, as a hydrogen leaves the molecule, the Rydberg state in question rapidly evolves into a valence state of the system, passing through all degrees of mixture on the way to total separation. In this case, the increasing bond length leads to a decreasing $\sigma - \sigma^*$ split. Con-

† This follows of course, from our discussion on wavefunction collapse precipitated by small changes of the effective potential (*vide infra*).

sequently, the σ^* level quickly drops below the band of conjugate Rydberg levels, the mixing is turned off, and the eventual dissociation products have valence configurations [M4]. It also may happen that at the ground-state equilibrium geometry, the $a_1\sigma^*$ valence MO is above the block of Rydberg levels, so that $\phi_i \rightarrow a_1\sigma^*$ is a shape resonance. In this case, as the relevant bond is lengthened, $a_1\sigma^*$ first comes into resonance with the Rydberg sea and assumes a strong Rydberg character before passing out the other side and becoming valence again at very large distances. These aspects of Rydberg/valence mixing are of extreme importance to a discussion of photochemistry in the vacuum ultraviolet, and the interested reader is referred to a number of papers in which the subject is explored in depth [E10, E11, M4, S14, S15]. The rapidly changing Rydberg/valence character of a potential surface raises a valid question of semantics for the photochemist who is concerned with its broadest reaches, however this is not as often confounding for the spectroscopist who works closely in the vicinity of the ground-state R_e.

Wavefunction Collapse and Giant Resonances

There is a fascinating development taking form in atomic spectroscopy which appears to have strong parallels in the area of molecular Rydberg/valence mixing. Consider the effective radial potential in an atom (V_{eff}) as consisting of a Coulombic term ($-1/r$) and a centrifugal term $\left\{\frac{l(l+1)}{r^2}\right\}$, as plotted in Fig. I.D-2 for the 4f orbital of lanthanum in the 6s5d4f configuration. As Connerade explains [C39, C42], in region A, the Coulombic part of the potential is dominant, however as r decreases, the centrifugal effect acts to raise the potential, as in region B. Once into region B however, the electron then is able to penetrate the core more effectively, thereby experiencing a higher effective nuclear charge and so lowering the potential again, as in region C. Finally, at very short distances, region D, the $1/r^2$ dependence of the centrifugal potential overcomes the $-1/r$ dependence of the Coulombic potential and V_{eff} arises abruptly. The net result of these two competing potential terms in the case of orbitals having $l \geqslant 2$ is a double-well effective potential. The hump in region B is called the centrifugal barrier, but in fairness it must be noted that it is due equally to the effect of penetration, and does not even appear in the hydrogen atom where penetration is not possible.

Fig. I.D-2 shows the double-well V_{eff} for 4f orbitals in the excited states of both the lanthanum and barium atoms, and as expected for neighboring atoms in the periodic table, the curves are semiquantitatively alike. What

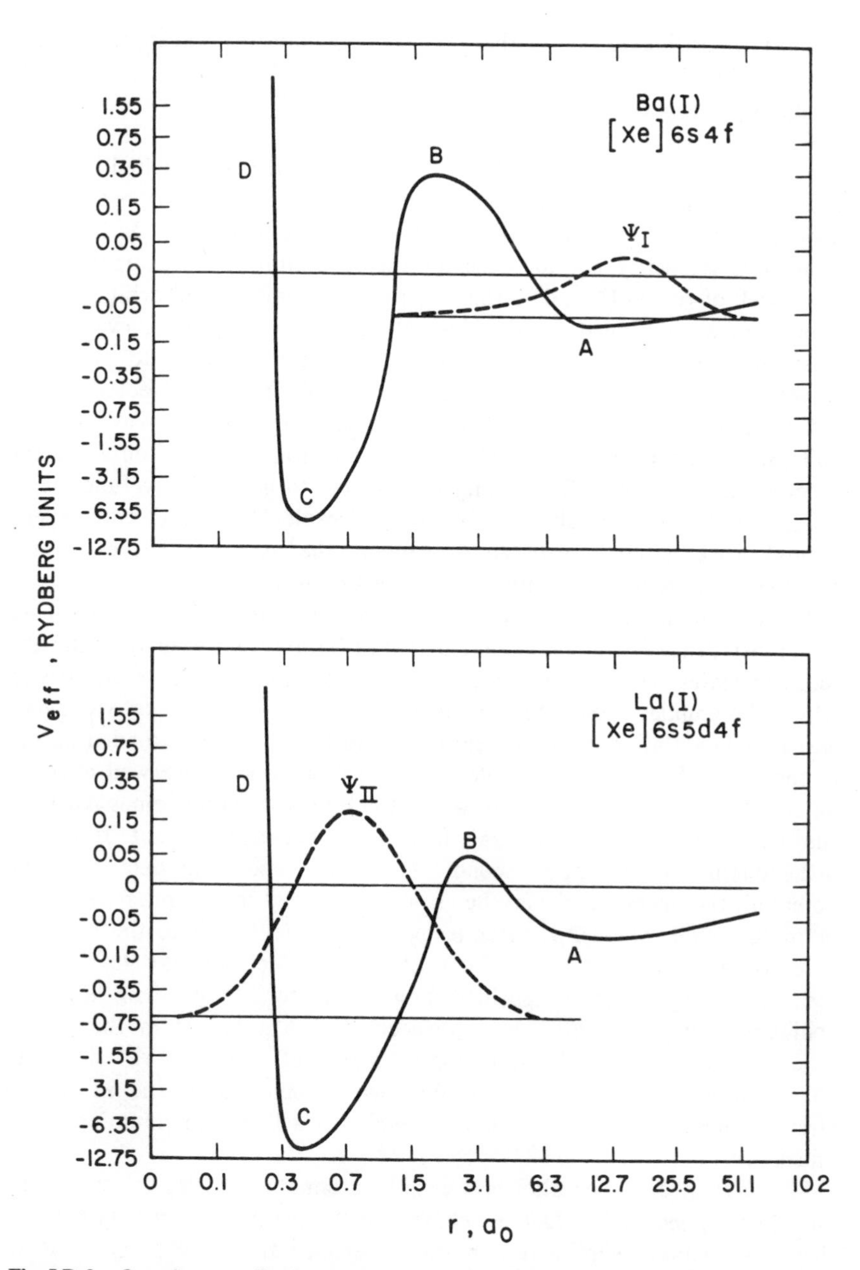

Fig. I.D-2. One-electron effective potentials for the 4f orbitals in Ba(I) and La(I) (*full curves*) and the corresponding outer-well (Ψ_I) and inner-well (Ψ_{II}) wavefunctions (dashed curves) [C39].

is most surprising to see is that the corresponding 4f wavefunctions are exceedingly different, with that in the [Xe]6s4f configuration of Ba occupying the outer well (Ψ_I) and peaked at *ca.* 15 a_o, whereas that in the [Xe]6s5d4f configuration of La occupies the inner well (Ψ_{II}) and is peaked at less than 1 a_o!! The 4f orbital in the Ba excited state is properly termed a Rydberg orbital because of its size and because its energy is dictated largely by the $-1/r$ part of the potential. On the other hand, this is clearly not the case for the 4f orbital of La trapped in the inner well of region C, and so this is termed a "nonRydberg" orbital [C39]. Moreover, as is known for double-well potentials in general, even slight dissymmetries between the two well profiles are sufficient to place the wavefunction totally in one well, and so small changes of potential are sufficient to shift the electron from one well to the other. Wavefunction collapse results from the transfer of the electron from the outer to the inner well. Only with fine tuning is one able to get the wavefunction to occupy both wells simultaneously. Considerable theoretical and experimental work is now available to support the notion of large-l wavefunction collapse under relatively small perturbation.

In our attempt to apply this type of reasoning to molecules [R36] we first must discard the centrifugal potential because we wish to apply the ideas to states where l may equal zero or to low-symmetry molecules in which the angular momentum is quenched [N17]. In its place, we invoke a spatial modulation of the potential due to the spatially extended nature of the core. Thus, in a molecule MX_n we propose a double-well effective potential along the M−X line in which the repulsive hump comes not from the combination of centrifugal force and penetration, but from the orthogonality to the M−X bond. Depending upon the details of the potential, the lowest level may be strongly localized in the inner well (Ψ_{II}) or in the outer well (Ψ_I); after proper orthogonalization to the occupied core orbitals, these are seen to be antibonding valence orbitals (σ^*, π^*, δ^*, *etc.*, depending upon unspecified angular factors) and their Rydberg conjugates (ns, npπ, ndδ, *etc.*), respectively. Arguing as in the atomic case, we expect that relatively small changes of potential will suffice to change drastically the nature of the lowest bound level, and in certain circumstances lead to a wavefunction which is a linear combination of Ψ_I and Ψ_{II}.

We have forced a parallel between the atomic and molecular potentials and then argued for analogous behavior in the two models largely because there are notable similarities in the experimental observations on Rydberg/nonRydberg spectra in atoms and on Rydberg/valence mixing and ordering in molecules. Following the description of the atomic spectra given by Connerade [C39] we note the following analogies:

i) In atoms, the two states of the double-well potential are small radius (nonRydberg) and large radius (Rydberg). In molecules, the two states of the double-well potential are small radius (antibonding valence) and large radius (Rydberg).

ii) As the effective potential for an atom is varied, the lowest level in the outer well may collapse into the inner well. As the M—H bond in a molecule is continuously lengthened, the lowest Rydberg level may be rapidly transformed into an antibonding valence level.

iii) In light atoms, the inner well does not contain a bound level, but with increasing atomic number the virtual level falls until it is resonant in the inner well and the wavefunction collapses. In comparing the spectra of molecular species, the prominence of σ^* valence levels is always greater in the heavier species. Such antibonding valence levels usually are not significant in the spectra of hydrides, but always are significant in the spectra of the corresponding halides, the more so on going from fluorides to iodides (Section I.C).

iv) Excitations from the inner shells of atoms lower the energies of nonRydberg levels and so promote wavefunction collapse. In a given molecule, excitations to σ^* are lower in the inner-shell spectra than in the outer shell (Section I.C).

v) Atomic resonances to inner-well levels (*giant resonances*) are much more intense than transitions to outer-well levels (Rydberg excitations) for reasons of spatial overlap. Inner-shell molecular transitions to valence levels are much more intense than transitions to Rydberg levels for reasons of spatial overlap.

vi) Atomic transitions to inner-well levels appear in solids whereas those to outer-well Rydberg levels generally do not. This parallels exactly the behavior of valence and Rydberg upper states in molecules (Section II.D).

vii) For certain shapes of the effective potential in atoms, the inner-well and outer-well levels are mixed and both are occupied simultaneously. This is termed "partial collapse." In the molecular case, there is considerable evidence as well for Rydberg/valence mixing in many molecules.

viii) When the outer-well wavefunction in an atom collapses to the inner well, the next higher outer-well level moves down and then plays the role of the lowest outer-well level. Energetically, this lowest level can be close to the position of the lowest level before collapse, however, by its node count it is one higher. Thus there is a semantic problem as to whether the lowest outer-well function after collapse should be assigned quantum number n or $n + 1$. Exactly the same problem has been pointed out in regard the numbering of Rydberg levels when preceded by a conjugate valence level (*cf.* [**I**, 26], for example).

ix) In atoms, the inner-well levels may be below (as a bound level) or above the relevant ionization limit (as a resonantly localized continuum level). A similar situation holds for the valence states of molecules.

In view of the many similarities between the atomic and molecular cases, it is important to point out an essential difference as well. In the atomic case, a double-well effective potential is inappropriate for s and p symmetries because of their zero or small l values. In contrast, s and p electrons in molecules may move in a double-well effective potential with distinct inner-well and outer-well wavefunctions provided the inner-well wavefunction does not dissolve completely in the sea of outer-well functions.

In atoms, the most dramatic spectral consequence of wavefunction collapse is the appearance of giant resonances, most often of the d → f variety. Examples of d → f giant resonances in polyatomic molecules are sparse but nonzero. To date, the best example is that of the 83 000-cm^{-1} (*10.3-eV*) band of cyclopropane (Fig. III.E-1), with the 84 000-cm^{-1} (*10.4-eV*) band of cyclohexane (Fig. III.E-1) and the 129 000-cm^{-1} (*16 eV*) band of neopentane (Fig. III.C-1) following close behind. Yet another possible example of a molecular giant resonance is the 105 000-cm^{-1} (*13.0-eV*) band of UF_6 (Section XXI.A). Clearly, molecular giant resonances can be found as intense valence features at high frequencies in the widest variety of systems, provided their symmetries are high.

Much of what we are saying here is either explicitly or implicitly contained in Dehmer's original discussion [**I**, D6] of the double-well model for molecular excited states. What we want to emphasize is that the inner well is associated with antibonding valence MO's and that relatively small differences of the shape of the effective potential can make the lowest level Rydberg, antibonding valence or a mixture of these.

Rydberg / Valence Ordering

Having named the orbitals in a most general way in the previous Section, it is then necessary to discuss their ordering. The question of Rydberg/valence ordering is of prime importance in understanding the higher excited states of polyatomic molecules. This question is closely related to that of the mixing of Rydberg orbitals and their conjugate valence orbitals [**I**, 24] and the possible deficiencies of a classification scheme built upon the presumption that these are distinct orbital types. It is clear from the many examples discussed in the text that there is no single answer to the question of relative Rydberg and valence conjugate ordering. Examples abound of ϕ_j^{+1} below ϕ_j^* and of ϕ_j^* below ϕ_j^{+1}. In a few molecules, both orderings appear in different spectral regions, depending upon the originating orbital. Our aim here is to display the spectrum of behavior exhibited by the widest selection of molecules, and rationalize this behavior after the fact.

Arbitrarily, we have divided the range of observed Rydberg-valence conjugate ordering of saturated systems into three regimes, Fig. I.D-3. Shown in each regime are the energies of the one-electron Rydberg (ϕ_j^{+1}) and valence (ϕ_j^*) conjugate orbitals, drawn with respect to the appropriate ionization limit (IP). As described in Section I.A, the "inner" and "outer" orbital labels refer to the shells in which the excitations originate. To be more specific, we take ϕ_j^{+1} as the lowest *n*s Rydberg orbital and ϕ_j^* as the totally symmetric, conjugate valence orbital. In accord with observation (Sections I.B and I.C), the exhaltation of the valence (ϕ_i, ϕ_j^*) term value on going from outer-shell excitation to inner-shell excitation is drawn in Fig. I.D-3 as much larger than that for the Rydberg (ϕ_i, ϕ_j^{+1}) term value.

In Regime X, the ϕ_j^{+1} level is far below the σ^* level for outer-shell excitations. Though the term value exhaltation due to antishielding for σ^* is much larger than for ϕ_j^{+1} in the inner-shell spectrum, ϕ_j^{+1} remains the lower level in the inner-shell spectrum. The outer-shell and inner-shell spectra of the alkanes are clear examples of molecules in Regime X.

In Regime Z, the picture is the reverse of that in Regime X. Here the σ^* level is far below the ϕ_j^{+1} level in the outer-shell spectrum, and the inner-shell exhaltations work to increase this separation still further. Examples of molecules with Regime Z spectra are the alkyl iodides, bromides and chlorides (Chapter IV).

In the intermediate region, Regime Y, the σ^* and ϕ_j^{+1} orbitals are close to one another and more or less mixed, depending upon the penetration of the Rydberg manifold. Most interestingly, because antishielding affects σ^* much more than ϕ_j^{+1}, there is a region (labelled Q in Fig. I.D-3) in which ϕ_j^{+1} is below σ^* in the outer-shell spectrum but σ^* is below ϕ_j^{+1} in the inner-shell spectrum. Note however, that ϕ_j^{+1} and σ^* may be difficult to distinguish in any case in Regime Y. Examples of molecules in this Regime include second- and third-row hydrides and many fluorinated systems with transitions originating at $2p_F$.

In unsaturated systems where in general the π^* orbitals are low and the *n*d or *n*f conjugate Rydberg orbitals are high-lying, the molecules will be in Regime X even though their ϕ_j^{+1}/σ^* ordering for other symmetries may be appropriate to Regimes Y or Z. For example, in ethylene the 3s

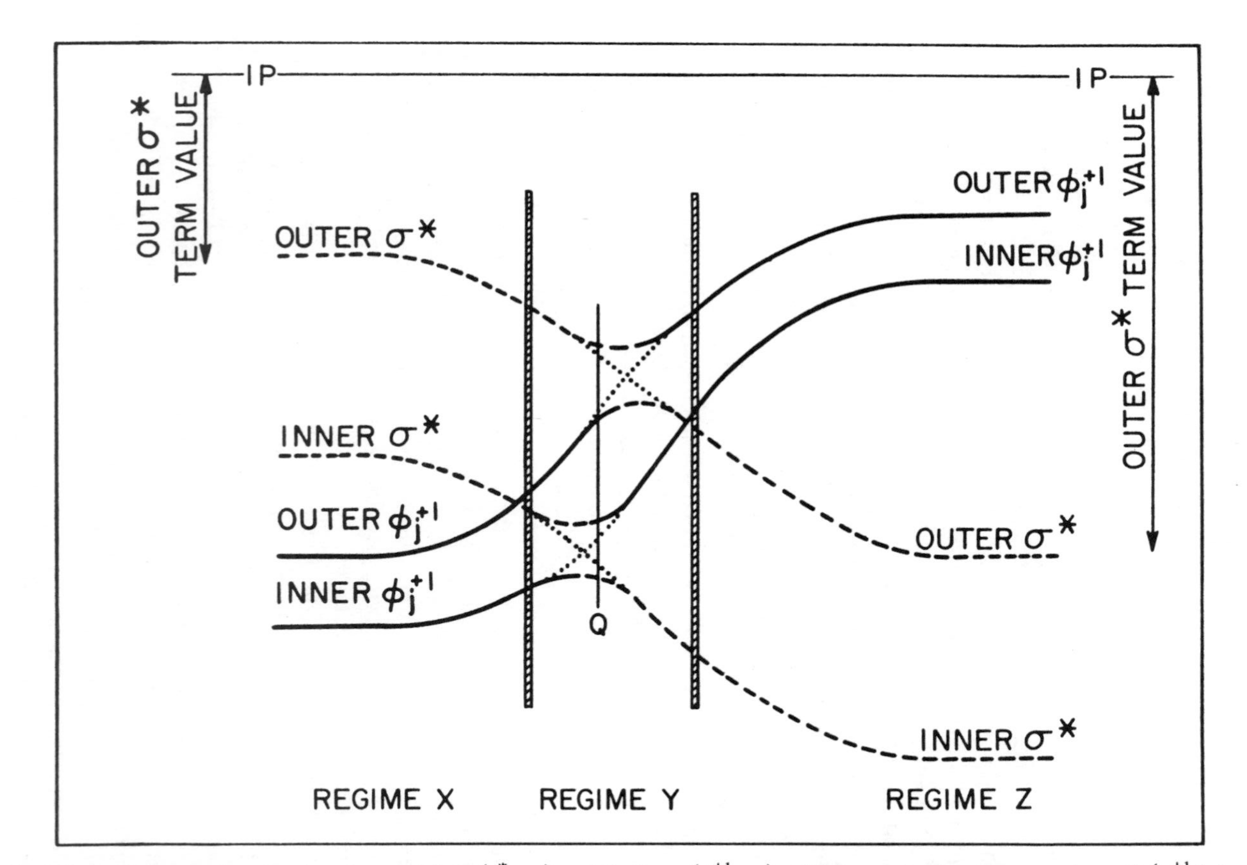

Fig. I.D-3. Ordering of conjugate valence orbitals (σ^*, - -) and Rydberg (ϕ_j^{+1}, —) orbital pairs in the extreme regimes X (ϕ_j^{+1} below σ^*) and Z (σ^* below ϕ_j^{+1}), and in the strongly overlapped regime Y. At point Q, the σ^*/ϕ_j^{+1} ordering is different in the inner- and outer-shell spectra.

Rydberg orbital is below its conjugate $a_1\sigma^*$ orbital (Regime X), whereas 3dxz is above its conjugate $1b_{3g}\pi^*$ orbital (Regime Z).

I.E. Addendum

J. P. Connerade, M. Pantelouris, M. A. Baig, M. A. P. Martin and M. Cukier, On the resonantly localized 5,ϵf continuum state in thorium, *J. Phys. B: At. Mol. Phys.* **13**, L357 (1980).

J. P. Connerade, M. W. D. Mansfield, M. Cukier and M. Pantelouris, Experimental evidence for the existence of a 5,ϵf resonantly localized continuum state in uranium, *J. Phys. B: At. Mol. Phys.* L235 (1980).

J. P. Connerade, A general formula for the profiles of "giant resonances", *J. Phys. B: At. Mol. Phys.* **17**, L165 (1984).

J. P. Connerade and M. Pantelouris, On the profiles of the "giant resonances" in Gd and GdF_3, *J. Phys. B: At. Mol. Phys.* **17**, L173 (1984).

T. C. Chang, C. S. Hsue, P. A. Ruttink and W. H. E. Schwarz, Compromise orbital calculations by an improved MC-GBT approach. Core excited states of HF, to appear in *Chem. Phys.*

CHAPTER II

New Sources of Spectral Information

The various Sections of this Chapter deal with certain experimental techniques and phenomena either not covered previously in Volumes **I** and **II**, or for which important new aspects have been developed.

II.A. Multiphoton Ionization Spectroscopy

Only those aspects of the multiphoton ionization (MPI) technique which impact directly on the higher excited states of large molecules will be discussed here; broad reviews of this rapidly expanding field are available elsewhere [J12, P13]. Briefly, the MPI experiment involves focusing visible/ultraviolet laser pulses into a gas filling an ionization cell and scanning the laser frequency while monitoring the laser-induced ionization current. Though ionization generally occurs at all laser frequencies, this current increases by factors of approximately 10^5 whenever two or more photons are exactly resonant with higher states of the target molecule and are absorbed, symmetry permitting. This multiphoton resonance is followed by the absorption of another one or two photons so as to ionize the excited electron. As examples of the spectra obtained with the MPI technique, the components of the 5p → 6s Rydberg transitions of methyl iodide-h_3 and -d_3 are shown in Figure II.A-1 and the $4a_2 \rightarrow 3sa_1$ transition of norbornane is presented in Fig. II.A-2. The first of these illustrates transitions which appear both as two-photon resonances in the MPI spectrum and as one-photon resonances in conventional absorption experiments. The second Figure illustrates a transition in norbornane which is quite intense in the MPI spectrum as a two-photon resonance, but which is otherwise absent in the one-photon absorption spectrum for symmetry reasons.

Three aspects of MPI spectroscopy are relevant to our discussion:

(*1*) Most MPI resonances in large molecules involve two or three photons. Using visible/ultraviolet laser pulses in the 15 000–40 000 cm^{-1} (*2–5 eV*) region, levels between 30 000 and 80 000 cm^{-1} (*4 and 10 eV*) can be accessed as two-photon resonances while levels between 45 000 and 120 000 cm^{-1} (*6 and 15 eV*) are attainable as three-photon resonances. Thus, using the MPI technique with an air path and quartz optical elements, resonance levels in the target molecule can be reached which are beyond the lithium fluoride cutoff!

(*2*) In molecules of high symmetry, the electric-dipole selection rules for two- and three-photon resonances often will differ from those for one-photon excitation. For example, the $g \rightarrow u$ selection rule for one-photon absorption in centrosymmetric molecules becomes $g \rightarrow g$ and $u \rightarrow u$ for two-photon absorption. In a situation such as this, the one-photon-allowed transitions are forbidden as two-photon resonances, whereas the allowed two-photon resonances are forbidden as one-photon transitions. Clearly, MPI will reveal the presence of many new transitions in molecules of high symmetry, as illustrated in Figs. II.A-2 (norbornane) and II.A-3 (benzene).

(*3*) Current flows in the MPI cell only when the resonance state absorbs one or more photons so as to ionize the molecule. This occurs when the rate of optical up-pumping from the resonance state is larger than that of other processes (fluorescence and/or dissociation) which deplete the population of the resonance state. Experimentally, it is found in almost all cases that the up-pumping rate is dominant, and ionization occurs if the resonance level is a Rydberg state, but not if it is valence. Thus virtually all molecular MPI resonances involve Rydberg levels. In the particular case of benzene shown in Fig. II.A-3, the valence transitions from $^1A_{1g}$ to $^1B_{1u}$ and $^1E_{1u}$ do not appear in the MPI spectrum of the vapor, but the Rydberg transition to (πe_{1g}, 3s) does. The general lack of valence resonances in MPI spectra is due to the fact that they involve exciting the optical electron to a high level which is strongly antibonding, and which unavoidably leads to the molecule's dissociation before it can be ionized. In the case of a Rydberg-state resonance, though the energy level is

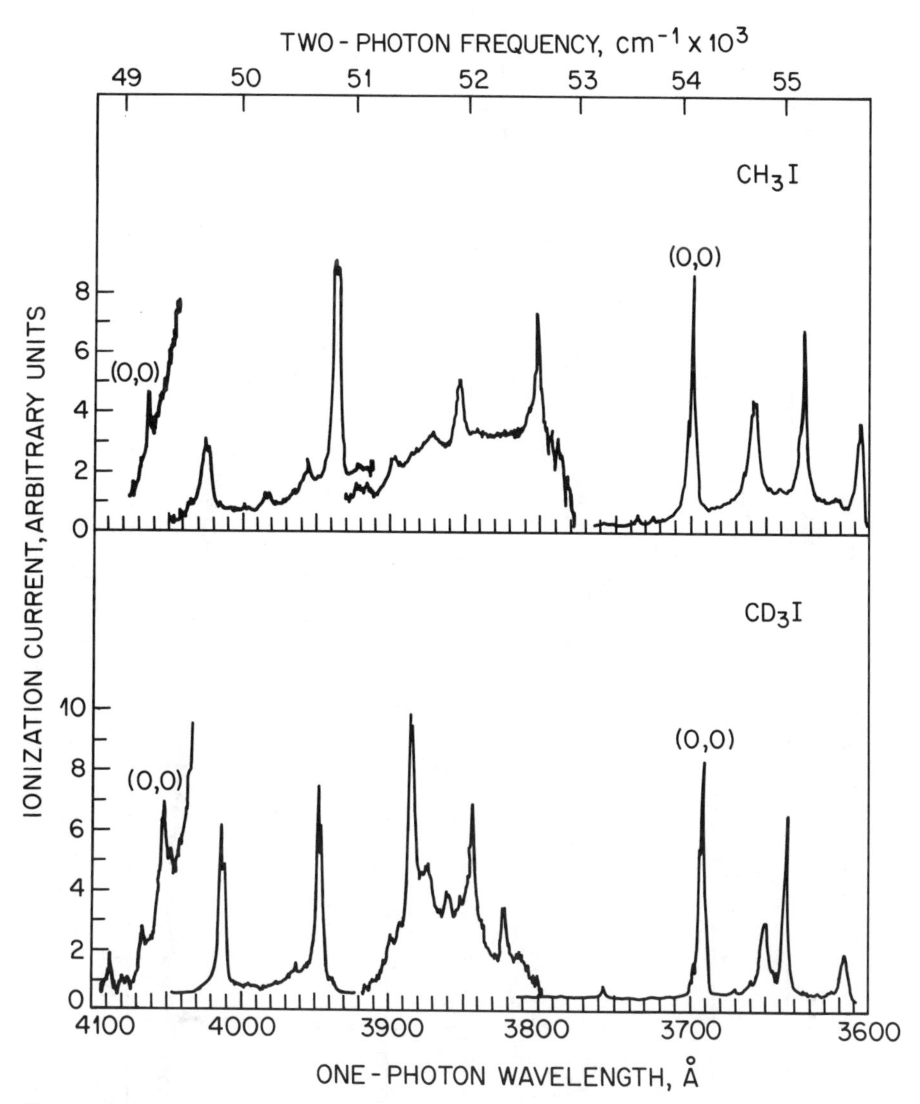

Fig. II.A-1. The multiphoton ionization spectra of methyl iodide-h_3 (*upper*) and of methyl iodide-d_3 (*lower*) at pressures of 60 Torr [R35]. The 5p → 6s resonances are at the two-photon level.

equally high or higher, the energy is invested in promoting an electron into a nonbonding orbital and so the molecule is far less likely to dissociate before it can absorb further photons and be ionized.

The Rydberg nature of a particular MPI resonance can be tested using an external perturbation, just as with a one-photon absorption band

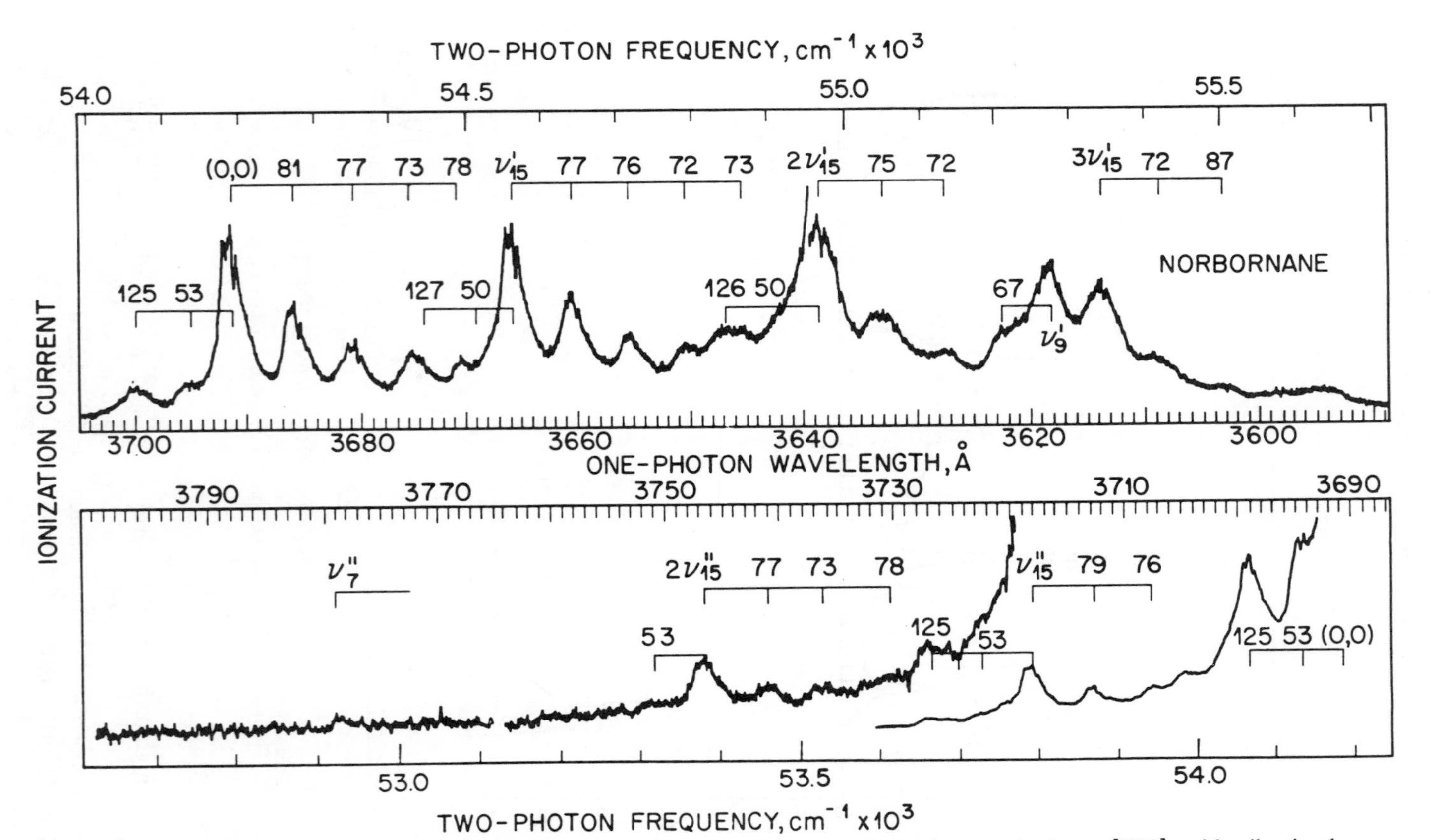

Fig. II.A-2. The multiphoton ionization spectrum of norbornane vapor, resonant at the second photon [H29], with vibrational progressions and sequences as indicated.

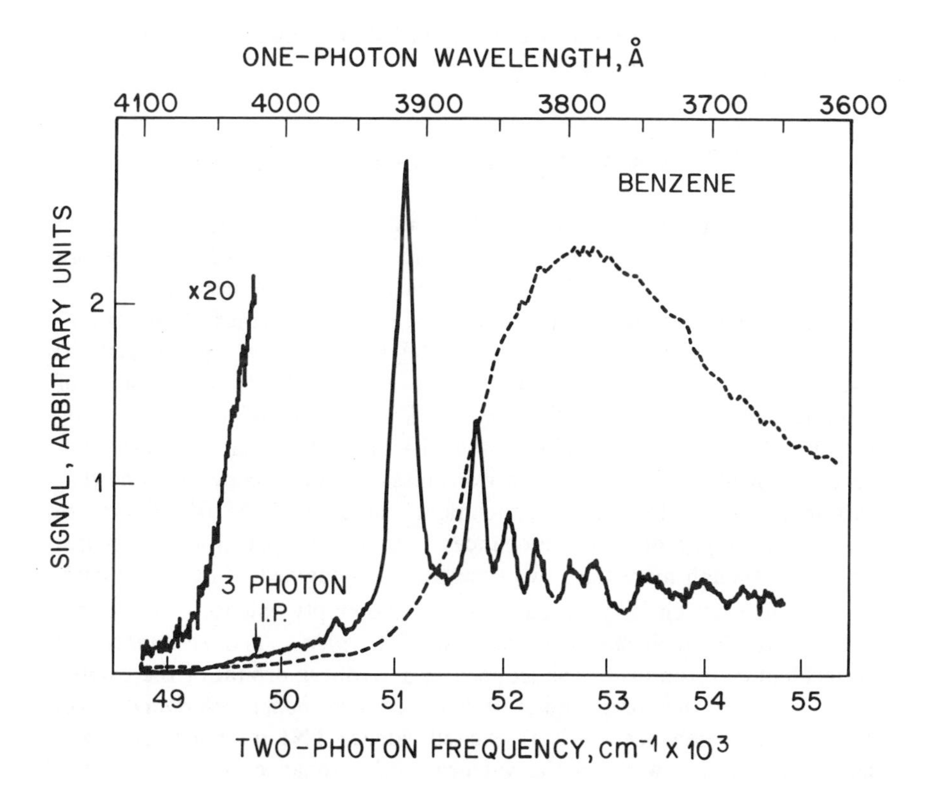

Fig. II.A-3. The multiphoton ionization spectrum of benzene vapor (full curve), resonant at the ($e_{1g}\pi$, 3s) Rydberg level on the second photon [J10]. The dashed curve is the two-photon fluorescence efficiency of the same transition for a 3% solution of benzene in n-hexane [S34].

(Section II.D). So, for example, the two-photon $n_O \rightarrow 3s$ Rydberg transition of acetaldehyde in the MPI spectrum broadens considerably upon application of 80 atm of He perturber gas [H30], and in Fig. II.A-3 one sees that the sharp structure of the two-photon $\pi e_{1g} \rightarrow 3sa_{1g}$ Rydberg band of benzene vapor becomes a featureless mass when the benzene is in solution. Information on the symmetry of the MPI resonance state in the vapor phase often can be gleaned by studying the excitation using circularly and linearly polarized light.

Using the focused output of a pulsed dye laser, it is possible to do not only MPI spectroscopy, but four-wave sum mixing as well, which also can reveal the presence of multiphoton resonances in the mixing gas [I4]. Third-harmonic generation in rare gases shifts laser-pulse frequencies from the visible region into the vacuum ultraviolet and so would allow high-resolution absorption studies in this region [M59], however very little has been done with this to date.

II.B. Temporary-Negative-Ion Resonances

Most neutral atoms and molecules are able to bind an incident electron of appropriate energy for times of the order of $10^{-15} - 10^{-10}$ sec before decaying by either dissociation or electron emission. Transitions to such quasi-stationary states of the negative ion are termed "*temporary-negative-ion*" (TNI) resonances. The TNI resonances of atomic systems have been reviewed by Schulz [S25] and Heddle [H33], while those of diatomics are reviewed by Schulz [S26] and those of polyatomics by Jordan and Burrow [J16]. The frequencies of TNI resonances most often are ascertained by direct electron-transmission spectroscopy or somewhat more indirectly by dissociative attachment studies [C23, M10]. There are two general types of TNI resonances observed in polyatomic systems, Fig. I.A-1, though more than two types can be imagined. In the first case, the incident electron is captured in one of the empty valence orbitals, ϕ_j^*, of the unexcited molecule, and so is called a *valence TNI resonance.* If the incident electron instead is trapped in a Rydberg orbital of the excited target molecule, the compound state is called a *Feshbach resonance.* Like the Feshbach resonance, the core-excited valence TNI resonance is a two-electron excitation, whereas the valence TNI resonance is a one-electron excitation.

Valence TNI Resonances

Representing the ground state of the target molecule as N, the valence TNI resonance configuration is written as $^2(\mathrm{N}, \phi_j^*)$, with the configuration N playing the role of the core. Valence TNI resonances are found at frequencies between 0 and 70 000 cm^{-1} (*0 and 9 eV*) with widths between 400 and 25 000 cm^{-1} (*0.05 and 2.5 eV*). The detectability of TNI resonances in general depends upon the upper state being relatively long lived ($\geqslant 10^{-15}$ sec) thereby leading to narrow resonances; vibronic structure often is seen in the longer-lived excitations. The π^* MO's of unsaturated systems are often active in valence TNI resonances, as are the σ^*(C–X) MO's of halogen-containing molecules [P25] and the σ^*(C–H)

MO's of saturated molecules.† The positions of several π^* valence TNI resonances in unsaturated systems are shown in Fig. II.B-1 along with the complementary π MO energies taken from photoelectron spectra. Though the Figure shows a surprisingly good match between the π orbital splittings in the positive ions and the π^* orbital splittings in the negative ions, such is not always the case. Thus in methane (Section III.A), the $2a_1\sigma - 1t_2\sigma$ split of 72 000 cm^{-1} in the positive ion is significantly larger than the 40 000-cm^{-1} split between the $3a_1\sigma^*$ and $2t_2\sigma^*$ orbitals in the negative ion. It is clear from the TNI data already collected on pi-electron systems that whereas alkyl groups strongly stabilize π-cationic states, they strongly destabilize π^*-anionic states. Note that in the fluoroethylenes, the π^* MO electron affinities are observed to increase with fluorination, however the observed transitions are very nonvertical whereas calculations show the adiabatic electron affinities actually decrease rapidly with fluorination [P1]. There is too little data on σ^* valence TNI resonances to allow generalization, except to note that fluorination can result in a very low-lying σ^*(C–F) MO [H78].

Referring to the effective potential of Fig. I.D-2, one notes that if the core is neutral, there is no long-range attraction and whatever potential well exists for the molecule-electron system in general will not contain a bound level of negative energy. In fact, only a resonance of positive energy is possible, with the wavefunction localized in the vicinity of the molecular bonds, *i.e.*, in an antibonding MO. Such a description of a valence TNI resonance state from the effective potential point of view closely resembles that of the neutral-molecule shape resonance (Section II.C). Indeed, it has been said that the virtual orbitals occupied in shape resonances are the same as those occupied in valence TNI resonances [D9, T15]. We agree with this up to a point; the comparison is a fair one for valence orbitals only, and only if one recognizes that the

† Our general conclusion on studying the outer-shell excitation spectra of saturated species such as H_2O, NH_3, CH_4, SiH_4, *etc.*, is that the valence excitations into σ^* MO's are never seen, and in fact in H_2O and NH_3, calculations reveal that the lowest σ^* MO's are totally dissolved in the sea of Rydberg levels and have no distinct existence as such. From this, one might be tempted to think that ^{2}(N, σ^*) TNI valence configurations also would not exist in these systems. However, it must be remembered that though the (ϕ_i^o, σ^*) valence configuration is dissolved in the (ϕ_i^o, ϕ_j^{+1}) Rydberg manifold when the optical electron is bound to a core of + 1 charge, the Rydberg manifold does not exist in the case of an electron bound to a neutral core, as in a TNI valence resonance. Consequently, a σ^* MO may be most important in the valence TNI resonance spectrum of a molecule while being almost irrelevant to its neutral-molecule spectroscopy!

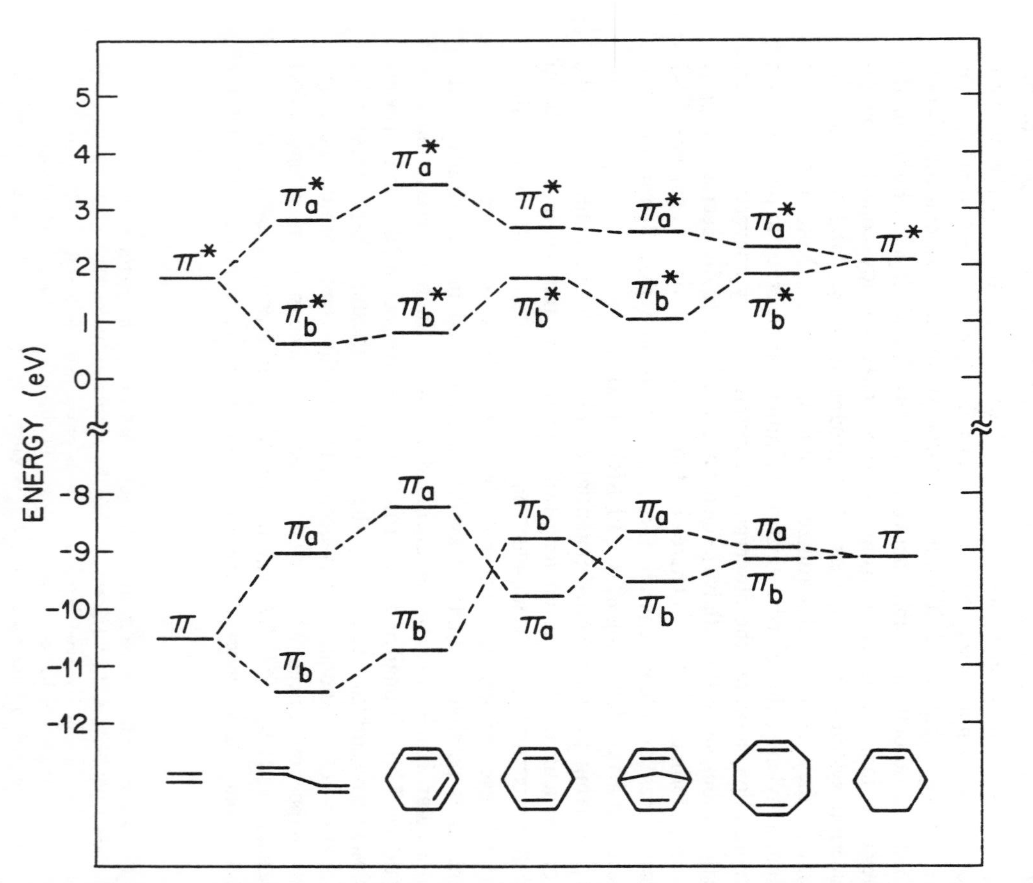

Fig. II.B-1. Pi-electron energy levels of various unsaturated systems as determined by photoelectron and temporary-negative-ion spectroscopies [J15].

energy of the antibonding MO can be rather different in the negative ion and in the neutral molecule due to screening. In general, one might expect that the negative term values of TNI resonances would be larger than those of the related shape resonances.

Valence TNI excitations also are known in which the incident electron excites a valence transition in the core while being bound in a valence MO. Thus, Bader *et al.* [B3] invoke the existence of such an electron/valence-excited exciton complex in order to account for the electron-transmission maxima observed in neat-crystal spectra of pi-electron systems. The evidence for core-excited valence TNI resonances is strongest in molecules where an inner-shell electron is excited to the π^* valence orbital in which the incident electron also is resident [K21]. A large number of such core-excited inner-shell valence TNI resonances are reported which are 3000−10 000 cm^{-1} (*0.4−1.3 eV*) below the parent $\mathscr{A}(1s, \pi^*)$ bands, and which have been assigned $^2(1s, \pi^{*2})$ configurations. Apparently the adverse $\pi^*\pi^*$ self energy can be overcome by strong correlation in these states, and possibly by a further ballooning of the π^* MO to more closely approach Rydberg dimensions.

Feshbach Resonances

The second type of TNI resonance involves capture of the incident electron in a Rydberg orbital [W11]. This is at first sight unlikely, since the target molecule is electrically neutral whereas Rydberg orbitals follow from the attraction of an electron to a positively-charged core with a $-1/r$-like potential. While this is so, binding of the incident electron is possible nonetheless if the target is first excited to a Rydberg state by the impacting electron and the incident electron is then captured in the same or a different Rydberg orbital. The rationalization here is that the two optical electrons can be bound when both are in large-radius Rydberg orbitals, for each can be attracted by the unshielded positive ion if they will correlate their motions so that they are always on opposite sides of the core. Such TNI Rydberg states in which two electrons occupy one or more Rydberg orbitals are normally referred to as *Feshbach resonances* [S26]; we will use this term and *Rydberg TNI resonances* interchangeably. The importance of electron correlation (also called dynamic screening) in stabilizing the Feshbach configuration is well documented, for example, by the calculations on H_2O (Section VII.A) by Jungen *et al.* [J22]. In fact, calculations on several systems by Weiss and Krauss [W11] show that the binding of the incident electron in $^2(\phi_i, ns^2)$ configurations is due totally to the correlation of the electrons in ns. Clearly, by this mechanism there will be no TNI resonance involving the binding of a Rydberg electron to a valence excited state of the neutral molecule.

The Feshbach configuration $^2(\phi_i, ns^2)$ in which an electron is promoted from valence orbital ϕ_i to Rydberg orbital ns while the incident electron also is bound in ns has as its parent configuration the $^1(\phi_i, ns)$ Rydberg state of the neutral molecule. Because the optical electrons in a Feshbach state are in largely nonbonding orbitals, the vibronic structure in such a Rydberg TNI excitation is often very much like that of the neutral-molecule transition to the parent state. If the parent state is dissociative, then the Feshbach resonance can be so lifetime broadened as to be unobservable. Spin-orbit and Jahn-Teller splittings which are characteristic of the parent Rydberg transition also will appear with high fidelity in the corresponding Feshbach resonance, lifetime permitting. Thus, calculations on the $^2(1b_1, 3s^2)$ Feshbach state of water [W11] show that it dissociates into OH and H^- ground-state fragments in a way which closely parallels the dissociation of the $^1(1b_1, 3s)$ parent Rydberg state. Though less well-known, inner-shell Feshbach resonances also have been observed in polyatomic molecules (*cf.* Section III.A, for example).

Heddle [H33] mentions a simple model for the Feshbach state:- place two electrons in a common Rydberg orbit about a positively charged core such that the two electrons are always diametrically opposite one another. In that case, the potential term in the Hamiltonian is $-\dfrac{2e^2}{r} + \dfrac{e^2}{2r}$ and the energies of the system are $-9R/8n^2$, whereas with the incident electron at infinity, the energies are $-R/n^2$. Thus the Feshbach state of a given n has an energy $R/8n^2$ below that of the neutral-molecule Rydberg state which in turn has a term value of R/n^2.† Taking a typical molecular term value of 35 000 cm^{-1} for the lowest parent Rydberg excitation places the daughter Feshbach excitation at about 4500 cm^{-1} lower frequency. The frequency difference between the parent and Feshbach configurations is here termed ***the Feshbach decrement***.

Looked at another way, the frequency of a Rydberg TNI resonance will equal the Rydberg frequency to the parent state, diminished by the electron affinity of the molecule in the Rydberg parent state. This latter is known for several atomic systems having an ns electron outside a positive core (H, Li, Na, K, *etc.*) to be 4000−5000 cm^{-1} (*0.5−0.6 eV*). Arguing, as we have done so often in the past [**I**, 63] that the binding energy of an

† Allowance for penetration changes n^2 to $(n - \delta)^2$ in each of the expressions given above.

ns Rydberg electron is largely independent of the electronic and molecular structure of the core, leads one to expect that the $(\phi_i, n\text{s})$ molecular Rydberg state electron affinity similarly will be 4000–5000 cm^{-1}. Thus it is concluded once again that the Feshbach resonance frequency will be *ca.* 4000–5000 cm^{-1} less than that of the Rydberg excitation to the corresponding parent state.† In agreement with this, it is observed repeatedly in atoms and molecules that their $^2(\phi_i, n\text{s}^2)$ Feshbach resonances are approximately 4000 cm^{-1} lower in frequency than the transitions to the parent Rydberg states [S25, S26]. However, there is some evidence as well that the Feshbach decrement increases with increasing frequency of the parent-state excitation.

Read [R8, R9] proposes that the two-electron term value of the $^2(\phi_i, nlnl')$ Feshbach configuration be written as

$$TV_2 = -\frac{R(Z-\sigma)^2}{\left(n-\delta_{nl}\right)^2} - \frac{R(Z-\sigma)^2}{\left(n-\delta_{nl'}\right)^2}$$

where the δ's are the quantum defects appropriate to the (ϕ_i, nl) and (ϕ_i, nl') parent configurations, $Z = 1$ for our purpose, and σ is a parameter expressing the mutual screening of the two outermost electrons with respect to the core. For a large number of atoms and ions in $^2(\phi_i, n\text{s}^2)$ configurations, the experimental values of σ cluster very tightly about the average value of 0.257, which leads to a Feshbach decrement of $-0.104\ R/(n-\delta)^2$. This compares closely with the value of $-0.125\ R/(n-\delta)^2$ computed under the assumption that the electrons remain always diametrically opposite (*vide supra*). For completely uncorrelated motions of the two electrons, $\sigma = 0.50$, and there is zero binding energy of the incident electron for $\sigma = 0.293$. For the case where $\sigma = 0.25$, $r_1 = -r_2$ and the system is said to be at the Wannier ridge [R9].

In the case of the $^2(\phi_i, n\text{p}^2)$ configuration, the atomic data yields a wider range of σ values, $\sigma = 0.3$ being typical. In this case, the Feshbach decrement is zero or even positive, suggesting that such resonances will

† Actually, there is an ambiguity here since either $^3(\phi_i, n\text{s})$ or $^1(\phi_i, n\text{s})$ can function as the parent state for the $^2(\phi_i, n\text{s}^2)$ Feshbach configuration. Energetically, this is of little consequence since the singlet-triplet splits of molecular Rydberg states are only 1000–2000 cm^{-1}. Note however, that the splitting between $^3(1\text{s}, 2\text{s})$ and $^1(1\text{s}, 2\text{s})$ in helium is 12 900 cm^{-1}, and that the $^2(1\text{s}, 2\text{s}^2)$ Feshbach state is *ca.* 8000 cm^{-1} below $^3(1\text{s}, 2\text{s})$ in this system. In this work, we choose the singlet state as the parent of the doublet Feshbach configuration.

occur at or very close to the parent excitation frequencies. On the other hand, σ is approximately 0.22 for the $^2(\phi_i, nsnp)$ configuration in atoms and ions, placing the Feshbach resonance approximately 1000 cm^{-1} above the (ϕ_i, ns) parent frequency. Qualitatively, it is easy to see why Feshbach configurations in which the two Rydberg electrons have different n and/or l are less tightly bound, for in this case the differing radial and/or angular parts of the wavefunctions no longer allow the two electrons to remain always on opposite sides of the core. Similar angular and radial correlation effects appear as well in so-called "planetary atoms" in which two electrons occupy excited Rydberg orbitals in a neutral atom [F26]. To date, there are no molecular analogs of planetary atoms, but an example is sure to come out of the spate of recent two-color MPI experiments on molecules. The term values of molecular planetary states with two electrons in ns orbitals will follow the formula given above by Read with σ close to 0.25 and $Z = 2$ [R8], however the δ's will be smaller than those appropriate to the one-electron excitations.

As discussed in the many Chapters that follow, the $^2(\phi_i, ns^2)$ resonances in large-molecule systems have Feshbach decrements of 4000–5000 cm^{-1}, just as theorized and observed for atomic systems. It is most gratifying to see how nicely this concept carries over from atoms to polyatomic molecules. As in atoms, the Feshbach decrements for $^2(\phi_i, nsnp)$ and $^2(\phi_i, np^2)$ configurations in molecules are observed over a much wider range. Note however that in correlation-dominated systems such as TNI Feshbach resonances and planetary neutrals, the independent-particle quantum numbers are inappropriate even as a low-level approximation [R9]. Thus, one must understand that though there is an implied parental relationship in configurations written as $(\phi_i, 5s)$ and $^2(\phi_i, 5s^2)$, the single configuration designated $^2(\phi_i, 5s^2)$ is totally inadequate for describing the Feshbach state.

Define a two-electron term value TV_2 for a Feshbach state as the energy difference between the relevant negative (daughter) and positive (grandparent) ions. Clearly then, $TV_2 = TV_1 - \Delta F$, where TV_1 is the term value for the parent Rydberg state and ΔF is the Feshbach decrement. While ΔF is sensibly constant, TV_1 can vary widely for ns orbitals, and so then can TV_2. This is undoubtedly the explanation for the variability of the TV_2 values from 35 900 to 28 000 cm^{-1} on going from F_2 to I_2 in the halogen series [S61, S63]. This is contrary to Schulz's statement based on limited data that TV_2 is a constant (33 000 cm^{-1}) [S26]. Since both TV_1 [**I**, 65] and ΔF are readily estimated, it is now trivial to make an accurate estimate of Feshbach frequencies if the relevant ionization potential is known. Similarly, when a Rydberg assignment is uncertain (as in ketene, Section XVIII.A, for example) or the location of a molecular Rydberg state is unknown, the observation of the Feshbach resonance frequency can lead directly to an assignment. Another useful relationship in this regard is demonstrated by Spence [S65], who showed that within a group of closely related molecules there is a linear relationship between the Feshbach resonance frequency and the molecular ionization potential, Figure II.B-2, much like that between the Rydberg frequency in the parent and the ionization potential [M42]. Though the evidence is scant as yet, there is some evidence that the Feshbach decrement can increase rapidly as the hole descends into the inner-shell, reaching 26 600 cm^{-1} for the $^2(1s_C, 3sa_1^2)$ state of methane [M18]. Such large decrements are not explained by the simple theories discussed above.

It would appear that there is a phenomenological test for distinguishing between valence and Rydberg TNI resonances, just as there is for distinguishing between valence and Rydberg excitations in neutral-molecule spectra [**I**, 85]. If the target molecule is studied in both the vapor phase and the condensed phase, it is expected that the valence TNI resonances will survive in the condensed phase with only a slight perturbation compared to the vapor phase, whereas the Feshbach resonances of the vapor phase will be obliterated totally or very strongly broadened and shifted to higher frequencies in the condensed phase. As an example of this test in action, note that in methane (Section III.A), the valence TNI resonances below the $^2(1t_2, 3sa_1^2)$ Feshbach resonance are readily seen in the electron-transmission spectrum of a thin film of the material [H71], and though no experiments exist on the fate of Feshbach resonances in the condensed phase, they are most likely strongly perturbed in a manner paralleling that of their parent states. Inasmuch as such experiments are very infrequently reported, it is best to rely upon the difference of parent and daughter resonance frequencies when assigning TNI resonances.

II.C. Discrete Features in the Ionization Continuum

Molecular excitations of several types are possible beyond the outer-shell and inner-shell ionization limits. Thus, discrete peaks result from shake-up excitations, shape resonances and two-electron promotions. Moreover, the molecular ionization continuum itself may be modulated so as to have local peaks and valleys. Though these latter may appear on occasion to be discrete excitations between specific orbitals, this is a deception. Assignments of features within and upon ionization continua in general are not very secure.

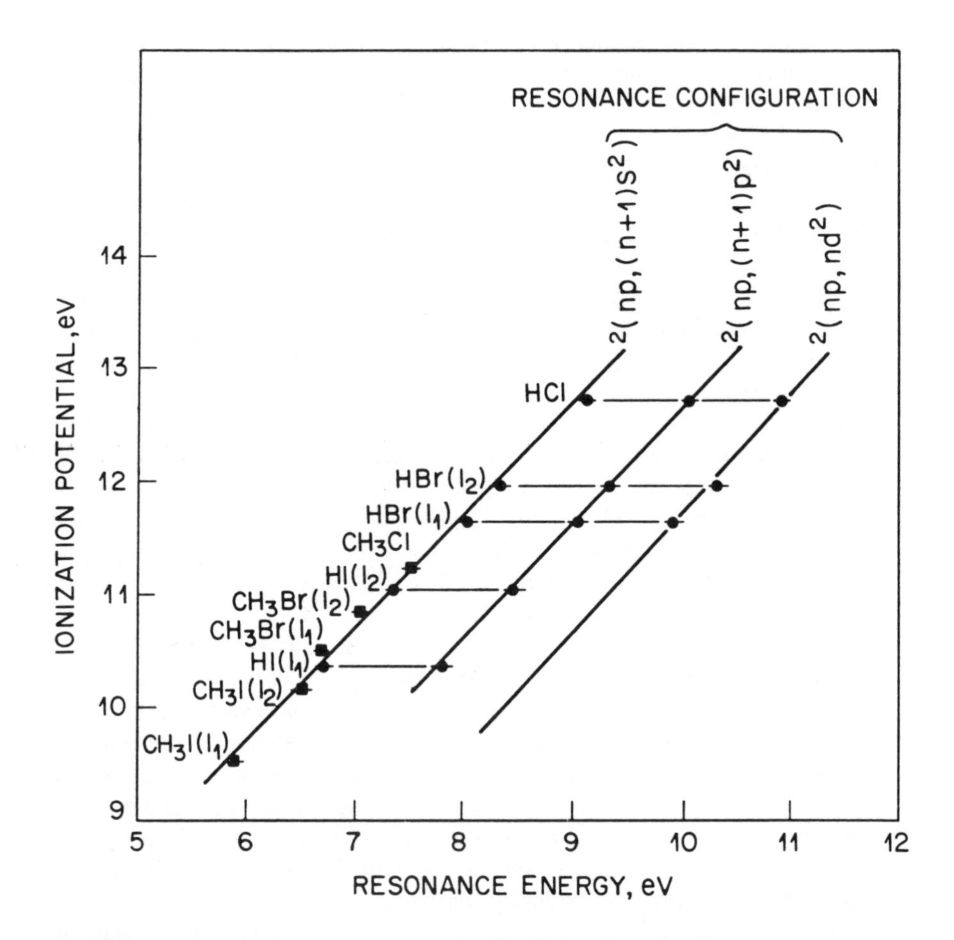

Fig. II.B-2. Linear relationships between the Feshbach resonance energies and corresponding ionization potentials (I_1 and I_2) in the hydrogen and methyl halides [S65].

Shake-Up Transitions

XPS and electron-impact energy loss spectroscopy are fruitful sources of information on bound excitations beyond the ionization threshold. In these experiments, an inner-shell electron is ionized and concomitant with this, one of the valence electrons in orbital ϕ_i of symmetry Γ_α is promoted to orbital ϕ_j, also of symmetry Γ_α. Since the inner-shell hole is localized, the point group in which Γ_α is a representation is that of the molecule with a localized hole. The monopole shake-up excitation $\phi_i \rightarrow \phi_j$ is induced by the "instantaneous" change of potential when an inner electron is ionized with appreciable excess energy. Frequencies of shake-up excitation will depend upon which atom is ionized and the densities which ϕ_i and ϕ_j have at the center bearing the hole. Valence excitation frequencies of shake-up excitations maybe higher, lower or equal to those in the neutral molecule [*cf.* the discussions on inner-shell spectra of formamide (Section XV.A) and benzene (Section XIX.A), for example]. Shake-up excitations of Rydberg transitions also are possible, however in such a case an electron in a Rydberg orbital is bound to a doubly-charged core, Section I.B. For Rydberg excitations, this increased effective nuclear charge acts to expand the frequency scale by a factor of 2−3 (Section I.B), so that a Rydberg excitation having a frequency of 56 000 cm^{-1} (*7 eV*) in the neutral molecule will fall between 120 000 and 160 000 cm^{-1} (*15 and 20 eV*) beyond the ionization limit in the shake-up spectrum; *cf.* [M64, S87] for examples of this in the $1s_O$ spectrum of water. As with neutral-molecule excitations, the Rydberg nature of a shake-up band could be probed by comparing vapor phase and condensed phase shake-up spectra.

Shape Resonances

A second type of excitation beyond each of the molecular ionization potentials involves the excitation $\phi_i^o \rightarrow \phi_j^*$ in the neutral molecule with an energy E_{ij} *larger* than the binding energy of an electron in ϕ_i^o. Depending upon the rates of autoionization and/or fragmentation, the excitation to (ϕ_i^o, ϕ_j^*) may be sharp or it may be broadened so as to be totally lost in the underlying ionization continuum. When such an excitation is observable as a discrete transition, it is here termed a "shape resonance".† By our definition, shape resonances are found in molecules but not in atoms, and are valence excitations since all of the Rydberg excitations have energies E_{ij} smaller than the binding energy of an electron in ϕ_i^o. This is an arbitrary restriction on the definition and is at odds with the definitions used by others, but has advantages in the context of this book. As examples of shape resonances, one may mention the $1s_B \rightarrow \pi^*$ inner-shell excitation in BF_3 (Fig. V.C-1) and the $\pi \rightarrow \pi^*$ excitation of

† The term derives from the idea that the excited state is stabilized against immediate autoionization due to a centrifugal term in the effective Hamiltonian causing a hump in the potential's shape, which in turn acts to trap the electron. The phrase "superexcited state" also is used to describe discrete absorptions above the ionization limit [L27].

formaldehyde computed to lie beyond the $(\pi)^{-1}$ ionization potential, Section XII.A. Shape resonances are observed both in direct absorption, in energy loss and in molecular photoionization experiments of various types. Viewed in terms of the effective potential discussed in Section I.D, a shape resonance can be thought of as arising when the inner well (Fig. I.D-2) is so narrow and/or so shallow that the lowest level for this well has a small but *positive* energy. In this case, the wavefunction is still localized in the vicinity of the inner well [C39], and is broadened energetically due to its autoionization-limited lifetime.

One often reads in the literature ([I, D6] or [B34], for example) of shape resonances described in terms of a centrifugal barrier resulting from high angular momentum components of the out-going electron which act to momentarily trap the electron in the region of the molecular core. Such an explanation at first sight is unrelated to the orbital approach familiar to chemical spectroscopists. However, Thiel and coworkers have shown convincingly that the terminating wavefunctions of molecular shape resonances indeed can be viewed as the antibonding virtual MO's of the ground state [K38, T9]. There is most likely a strong correspondence between the inner-well and outer-well state distinction which arises in the centrifugal-barrier picture and the valence and Rydberg states of the orbital picture. However, a recent paper by Natoli [N17] is sharply critical of the centrifugal-barrier picture of shape resonances in molecules. We will take the terminating levels in shape resonances to be antibonding MO's with the understanding that they more or less can be mixed with the underlying ionization continua.

In the outer-shell spectra of alkyl halides, the $\mathscr{A}(n\text{p}, \sigma^*)$ term values show an inverse correlation with the C—X bond strengths, with the strongest bonds placing σ^* highest and therefore with the smallest term values (Section I.C). A similar situation holds for inner-shell shape resonances in hydrocarbons, where the $1s_C \rightarrow \sigma^*(\text{C—C})$ frequency is shown to be strongly dependent upon the C—C bond length [H66]. As the C—C bond shortens, the shape resonance to $\sigma^*(\text{C—C})$ moves rapidly to higher frequency. In a molecule such as butadiene which has two different C—C bond lengths, two shape resonances are seen at frequencies correlating with the C—C and C=C bond lengths [H66]. If valid, this analysis leads one to the surprising picture of noninteracting σ^* MO's for the C—C and C=C groups in this molecule.

Two-Electron Excitations

The shake-up transition is a type of two-electron excitation which arises when the sudden perturbation of one electron leaving the molecule with high energy induces monopole excitations in the remaining electrons of the positive ion. Two-electron excitations also can be produced in neutral molecules with nonzero intensity through the agency of configuration interaction. In this case, the two-electron excited configuration cannot be reached from the ground state via electric-dipole radiation, however, if this configuration is mixed with a singly-excited configuration which can be reached from the ground state, then the two-electron excitation appears by virtue of its partial one-electron character. A clear example of a two-electron excitation of this type is reported by Shaw *et al.* [S45], who observe a vibronically structured band in CO which is 20 000−40 000 cm^{-1} (*3−6 eV*) beyond the $(1s_C)^{-1}$ ionization potential. This transition involves the simultaneous excitation of electrons in $1s_C$ and $2p\sigma$ orbitals into $2p\pi^*$ and 3p orbitals.

Photoionization Cross-Section Maxima and Minima

There can be no Rydberg excitations beyond the originating-orbital ionization potential, yet peaks which are closely related to the Rydberg excitations also can appear beyond the originating-orbital's ionization limit. The spectral properties of the Rydberg-series transitions continue smoothly through ionization and into the photoionization continuum, with cross sections which may be maximal either at threshold or considerably removed from threshold. In the latter case, the peaks result from the *delayed onset* of photoionization. These peaks in the various photoionization channels in the past have been mistaken for discrete MO excitations. Care must be taken in distinguishing shape resonances from the peaks in the photoionization cross section.

The maximizing of the photoionization cross section at a distance beyond threshold usually is explained in terms of the angular momentum of the ionizing electron. If the electron ionizes as a wave of large l, then there is an associated centrifugal barrier which limits the approach of the low-energy electron near threshold to the molecular core and thereby reduces the overlap with the originating orbital. This in turn reduces the spectral intensity. However, at higher energies of the ionizing electron the centrifugal barrier can be penetrated and the transition moment rises until the out-of-phase loop of the ionizing electron's wavefunction starts to diminish the overlap with the core orbital. Acting in combination, these effects lead to a maximum in the photoionization cross section which is either close to threshold or removed (delayed) depending upon the angular momentum of the out-going electron. The $4d_{Xe}$ ionizations of XeF_2 and XeF_4 are beautiful examples of the delayed onset of photoionization in molecules, having 4d $\rightarrow \epsilon$f channel maxima more than 180 000 cm^{-1} (*20 eV*) beyond the (4d)$^{-1}$ ionization limits [**II**, 342]. The variation of overlap density with electron wavelength which leads to a maximum in the photoionization cross section also can lead to "Cooper minima" in certain cases [B34].

In the case of an nf electron excitation, both ϵd and ϵg waves are available for ionization and two peaks will appear removed from threshold in the absorption cross section. It is also important to note that the $l \rightarrow l + 1$ propensity rule of atomic spectra also holds for ionization into the continuum [**I**, F3] so that in a molecule of O_h symmetry for example, the 2p $\rightarrow \epsilon$dt_{2g} and 2p $\rightarrow \epsilon$de_g transitions will be more intense than 2p $\rightarrow \epsilon$sa_{1g}. The interested reader is referred to Berkowitz's thorough discussion of molecular photoionization [B34].

In view of the strong perturbation wreaked on Rydberg excitations by condensed phases (Section II.D), it is a most interesting question as to the fate of the photoionization cross-section peaks of a molecule when it is incorporated into a solid. As seen from the electron-impact spectra of Killat [K16] on various organic substances, Fig. II.C-1, the prominent peaks at *ca.* 120 000 cm^{-1} (*15 eV*) in the vapor spectra which are presumed to be such cross-section maxima, are completely damped in the spectra of the solids! This phenomenon is worthy of more research, for it may yield a method by which shape resonances and photoionization cross-section maxima can be distinguished.†

Inner-shell ionization continua also can be modulated by the EXAFS phenomenon in which a damped sine-wave modulation of the continuum absorption results from internal diffraction of the photoejected electron by its surrounding atomic neighbors. The series of EXAFS peaks and troughs typically extend over many hundreds eV and are characterized by a regular pattern of undulations, which also can be calculated from the molecular structure.

† We note however that the shape of the photoionization continuum for the Ar atom out to 500 eV is identical to that of the solid within experimental error [S60]. The different responses of the ionization continua in the solid hydrocarbons and the solid rare gases may be due to differing electron mobilities in these phases [**I**, 85].

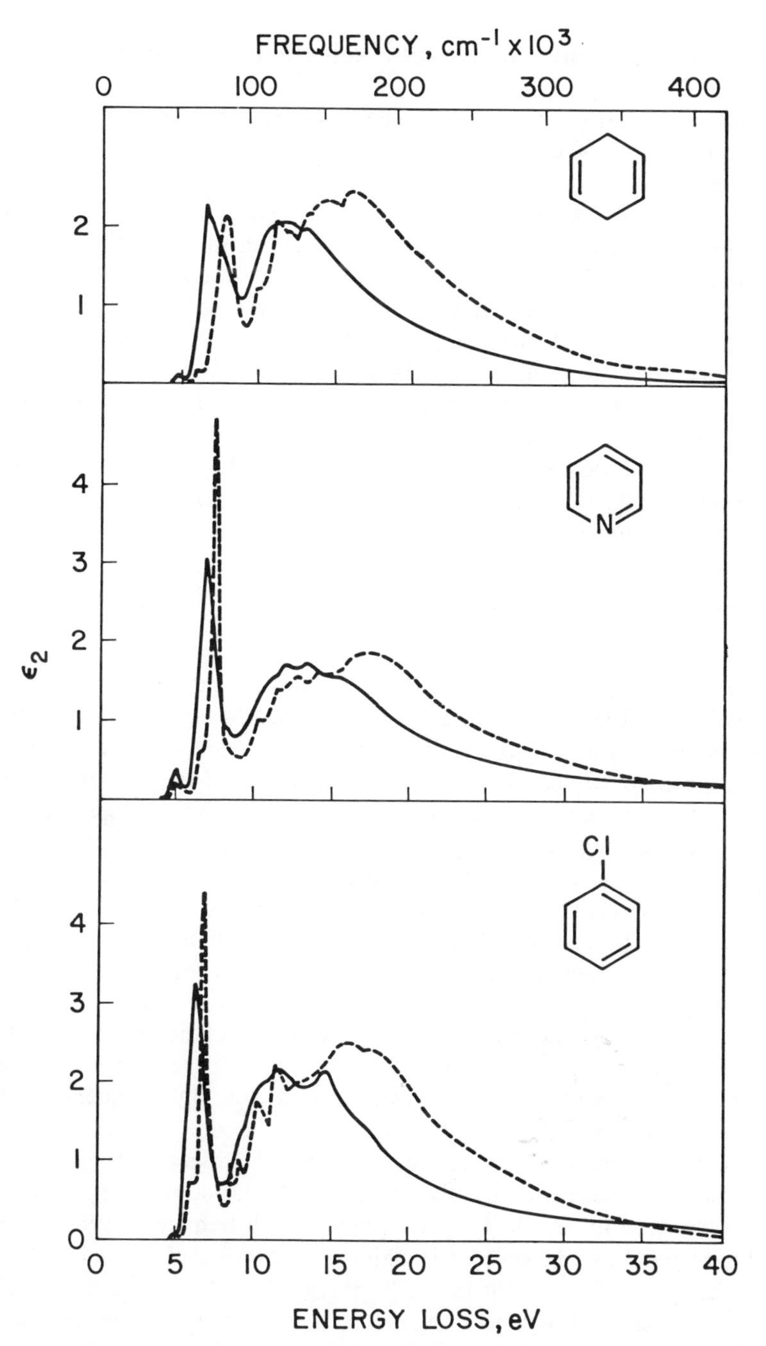

Fig. II.C-1. Energy loss spectra of 1,4-cyclohexadiene, pyridine and chlorobenzene in the vapor phase (—) and as thin films (- -) [K16].

II.D. Vacuum-Ultraviolet Excitations in Dense Phases

There is special interest in the response of vacuum-ultraviolet transitions to various forms of external perturbation, for this offers one of the few experimental opportunities for distinguishing between Rydberg and valence upper states [I, 76–91]. Such external perturbations include a high pressure of a second gas, placing the molecule in fluid solutions, lowering the temperature of such solutions, and placing the absorbing molecule in solid matrices or crystals. In all of the above, the common element is the extreme sensitivity of the energy of the big-orbit Rydberg state to the presence of foreign atoms within the Rydberg-orbital radius. Because this perturbation of the Rydberg state is independent of the originating orbital, the dense-phase perturbations also are expected in the inner-shell Rydberg spectra of molecules, and these have been observed in several cases. These dense-phase effects also are useful in freeing valence excitations of Rydberg contamination.

High-Pressure Gases and Fluid Solutions

The general qualitative effect of a high-pressure gas (50–150 atm) on a molecular Rydberg excitation is to asymmetrically broaden it (severalfold if it is otherwise sharp) while shifting the center-of-gravity of each vibronic component profile to higher frequencies. The same perturbation when applied to a valence excitation produces neither of these effects to any noticeable degree. The pressure-induced frequency shifts have been calculated for several molecular Rydberg excitations with interesting results. In the H_2 molecule symmetrically perturbed by six He atoms, the transition to the $2p\sigma$ Rydberg orbital is computed to be high-frequency shifted and the apparent Rydberg character of the orbital increases upon perturbation [E13]. However, when the He atoms are randomly disposed about the H_2 molecule, the calculated transition frequency to $2p\sigma$ returns nearly to that of the unperturbed system. It was concluded on this basis that solution in fluids will not necessarily lead to a high-frequency shift of a Rydberg excitation.

Ab-initio calculations on the lowest singlet Rydberg state of ammonia perturbed symmetrically by six He atoms [K33] and on the Rydberg-excited water dimer [V6] again predict high-frequency shifts due to exchange repulsion, however it is claimed in the case of ammonia that the perturbation does not result in the Rydberg orbital becoming more diffuse. However, increased Rydberg character is predicted for the case of the lowest triplet Rydberg state of ammonia perturbed by two H_2 molecules [K7]. Experimentally, the Rydberg bands of NH_3 between 48 000 and

72 000 cm^{-1} (*5.9 and 8.8 eV*) shift somewhat to lower frequencies when pressurized with up to 106 atm of Ar [M58].

The perturbation regime between that of the high-pressure gas and the normal-density solution is nicely bridged by the beautiful work of Messing *et al.* [M54, M55] in fluid rare gases in the density ranges 0−1.4 gm/cm^3 (Ar) and 0−2.3 gm/cm^3 (Kr). By way of example, the 5p → 6s Rydberg transition of CH_3I in liquid Ar solution is traced in Fig. II.D-1 as a function of solvent density, ρ. One sees that the perturbation effects up to $\rho = 0.43$ gm/cm^3 are much like those of N_2 gas up to $\rho = 0.15$ gm/cm^3 [I, 80], which is to say, the perturbation induces strongly asymmetric broadening to the high-frequency side. However, at Ar densities beyond 0.43 gm/cm^3, the broadening no longer increases but the center-of-gravity of the band moves rapidly to higher frequencies. Careful measurement of peak positions shows in fact that at low densities there is actually a slight shift to lower frequencies for the Rydberg transitions to 6s, 6p and 5d, but that these then rapidly shift in the opposite direction at higher densities.

Now it must be remembered that though the lower members of a vapor-phase Rydberg series may shift to higher frequencies upon solution, in fact the ionization limit of the solute in general is lower when in solution, which means that at a certain point in the Rydberg series, the perturbation shifts must be to *lower* frequencies instead. That point in the CH_3I/Ar system comes at $n = 7$ in the ns series, for the $n = 7-10$ members all shift to lower frequencies upon solution in Ar, with the shift of the highest-n component being the largest. Thus the general statement in regard Rydberg excitations being shifted to higher frequencies upon solution (*cf.* [Z4] for example) must be amended to include only those of lowest principal quantum number.

In parallel experiments on H_2CO in liquid Ar, Fig. II.D-2, the n_O → 3s Rydberg transition first broadens strongly at low Ar density and then shifts to higher frequency by approximately 2200 cm^{-1} as the density is run up to 1.4 gm/cm^3. Under the same conditions, the n_O → 3p and n_O → 3d transitions first shift to the low-frequency side and then reverse their direction. The core splittings of the (n_O, 3p) and (n_O, 3d) configurations are observed in these solutions, however by $n = 4$, the nonspherical core is sufficiently screened by the solvent so that the core splittings disappear.

The lowest Rydberg excitations of a molecule when dissolved in a liquified rare gas at reduced temperature and atmospheric pressure are shifted by 2000−3000 cm^{-1} to higher frequencies. Further lowering of the temperature has been found to strongly shift these molecular Rydberg excitations to yet higher frequencies [Z2, Z3]. Thus on going from 193 K to 119 K in liquid Kr, the 5p → 6s band of CH_3I shifts from 50 840 to

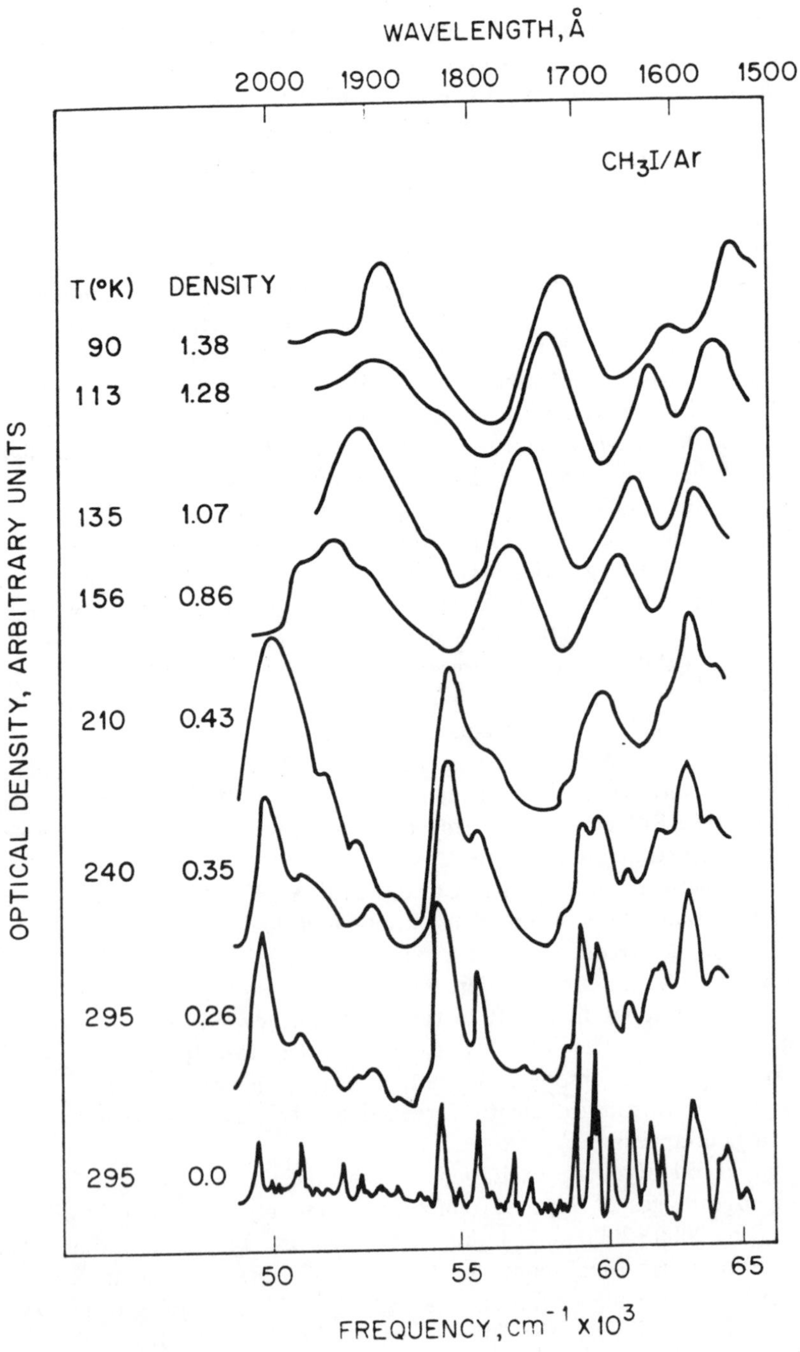

Fig. II.D-1. Absorption spectra of Ar solutions of CH_3I at various temperatures and densities [M55].

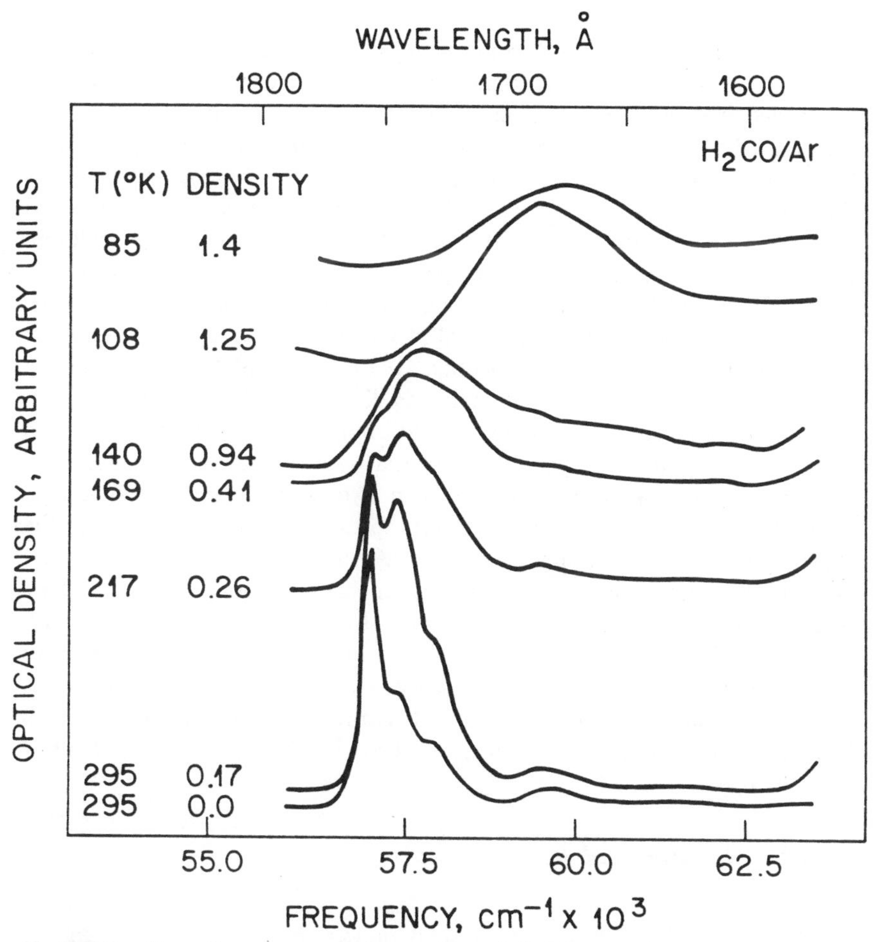

Fig. II.D-2. Absorption spectra of Ar solutions of H_2CO at various temperatures and densities [M55].

52 300 cm^{-1} (*vert., 6.303 to 6.48 eV*). The frequencies of the higher Rydberg bands of CH_3I also are shifted upward by very nearly the same amount in this temperature interval. The generality of this phenomenon is illustrated by high-frequency shifts of the Rydberg excitations of CH_3I, CF_3I and CS_2 in liquified Ar, Kr, Xe, CH_4 and CF_4 upon lowering the solvent temperature. In contrast, under the same conditions, valence excitations show a slight shift to lower frequencies as the temperature of the solution is reduced [Z4]. Thus the direction of frequency shift of a solute upon lowering the temperature of solution in a nonpolar solvent also can be of use in distinguishing between valence and Rydberg upper states.

Interestingly, the general conclusion in absorption spectroscopy that Rydberg excitations in solutions of the more common solvents are broadened beyond detection seems not to hold for circular dichroism spectra, where the signed nature of the differential absorption works to keep the Rydberg transition visible in spite of the broadening. This is illustrated in Fig. II.D-3 using the olefin Δ^5-cholestene, in which the broad absorption in the 45 000–60 000 cm^{-1} (*5.6–7.4 eV*) region in solution normally would be assigned as $\pi \rightarrow \pi^*$, but in fact is resolved into two oppositely signed bands in the CD spectrum, $\pi \rightarrow 3s$ and $\pi \rightarrow \pi^*$ [D45]. Several other examples can be quoted of Rydberg excitations which appear in the CD spectra of solutions (Sections VII.C, X.B and XVII.B). Though the Rydberg excitations are detectable as distinct bands in the CD spectra of solutions, the bands are nonetheless strongly shifted to high frequencies as compared with the vapor-phase spectra. As in the rare-gas solutions, lowering the temperature of solutions in hydrocarbon and other organic solvents increases the effective internal pressure and further shifts the Rydberg CD bands to higher frequency. A 100 K decrease of temperature typically produces a shift about half that incurred on going from the vapor to the solution at room temperature, Fig. II.D-3. This extreme sensitivity of Rydberg CD bands in solution to decreasing temperature can be used to test the Rydberg/valence nature of upper states in chiral molecules. It is anticipated that the Rydberg bands of achiral molecules in solution will behave as discussed above when probed by MCD rather than CD spectroscopy.

Solid Solutions and Pure Crystals

In the case of a molecule in a solid matrix, it has been customary to argue that the lowest transition of a Rydberg series in the solute will appear broadened and shifted in the matrix, but largely will retain its molecular character. Higher members of the series, however, are said to more closely resemble the Wannier excitons of the pure matrix, and to have little or no free-molecule character [J18]. This view is challenged by Hasnain *et al.* [H23], who studied the Rydberg spectrum of benzene in rare-gas matrices and concluded that the excitations up to high principal quantum number are those of the free molecule, albeit shifted from the vapor phase. In a similar vein, Resca and Resta [R10] point out that in many cases the Wannier formula for the matrix frequencies does poorly for low n, but that if a quantum defect is introduced as in the traditional free-molecule Rydberg formula, then all observed transition frequencies are fit much better. Though these best-fit δ's are often close to those of

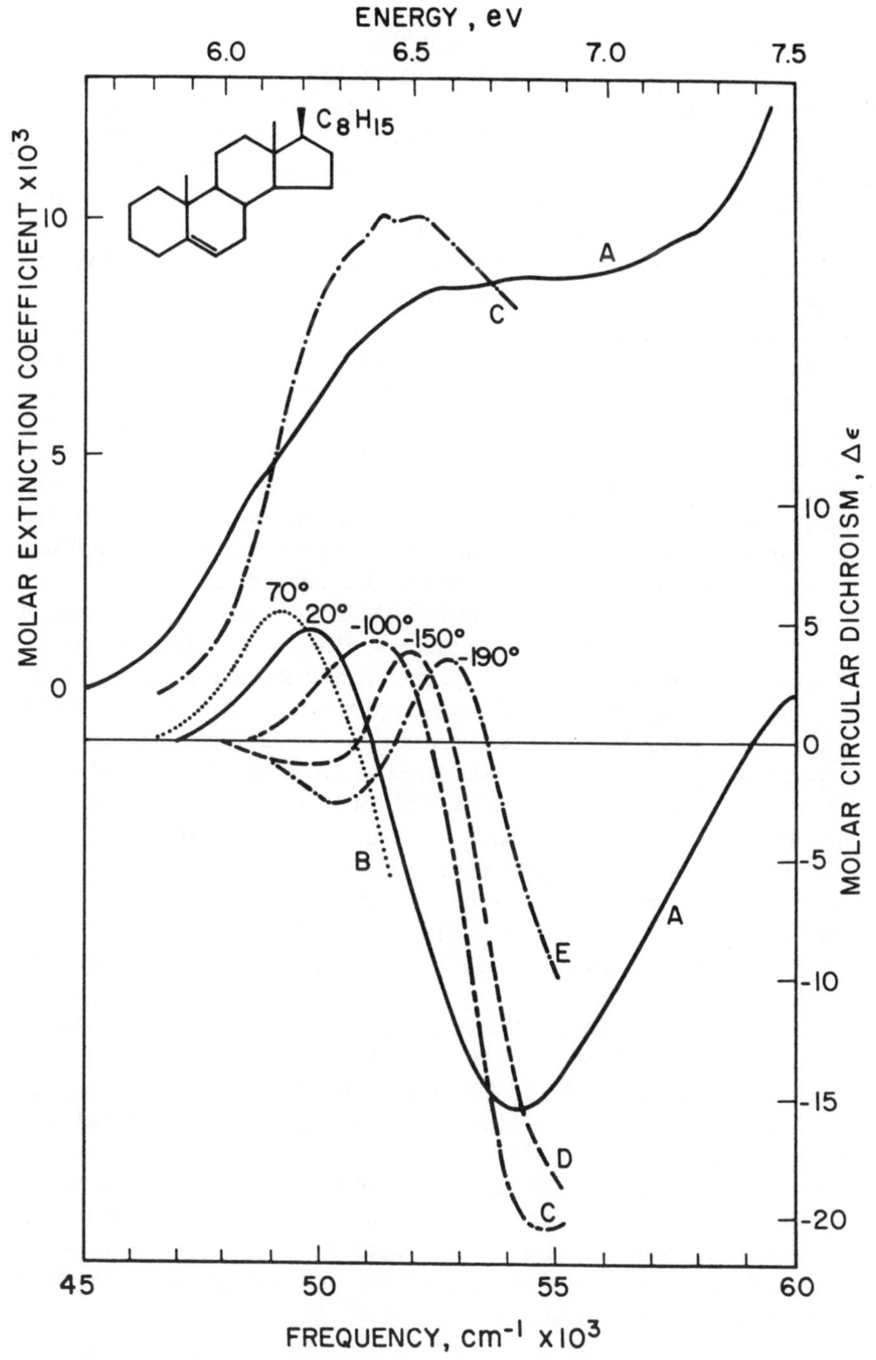

Fig. II.D-3. Absorption spectra (*upper curves*) and CD spectra (*lower curves*) of Δ^5-cholestene in pentane (**A**), in isooctane (**B**) and in 3-methyl pentane (**C**, **D**, **E**) [D45].

the vapor phase, it is nonetheless clear that the higher levels of the series overwhelmingly must involve the host-matrix orbitals.

In regard the fate of molecular Rydberg excited states in solid rare gases, Goodman and Brus [G24, G25] have put forward a most interesting idea. In excitation and fluorescence experiments on NO in rare gases, they find that the transition between the ground state and (π, 3s) level involves a broad phonon wing extending to high frequencies, implying a large lattice reorganization upon excitation. It is postulated that in the Rydberg upper state, the reorganized lattice is displaced outward so as to form a 10-Å cavity in which the excited molecular system forms a "bubble state". Order-of-magnitude calculations show that the bubble state is relevant only for the $n = 3$ Rydberg orbitals of NO and only for the lighter rare gases with their smaller lattice energies. Though it would seem from these considerations that bubble states would not be relevant for the more ordinary solvents of higher boiling point, there is large-molecule fluorescence data suggesting this may not be so (Section X.B). In the most general case, one should consider Rydberg states in condensed phases as linear combinations of Wannier and bubble-state character.

In the diatomic systems N_2 and NO [B60], it is observed that certain valence excitations in the vapor phase are strongly perturbed by nearby Rydberg excitations of the same symmetry, as witnessed by irregularities in vibrational frequencies and Franck-Condon factors. On the other hand, when investigated in condensed phases, the same transitions are now deperturbed, presumably due to the selective action of the condensed-phase effect on Rydberg upper states [I, 85]. The mutual deperturbation of interacting Rydberg and valence configurations also can be achieved through the use of high-pressure gas [M58, R42]. It is but a slight extrapolation from this situation to that of molecules such as H_2O (Section VII.A) or H_2S (Section VIII.A) and their derivatives in which the valence configuration (n_X, $a\sigma^*$) (n_X being the lone-pair orbital on atom X) is totally dispersed among the members of the conjugate Rydberg set (n_X, nsa_1) such that there is no specific excitation which can be labelled "$n_X \rightarrow a_1\sigma^*$", Section I.D. In such a situation, the largest part of $a_1\sigma^*$ appears mixed with the lowest nsa_1 Rydberg orbital, however the resulting state is still predominantly Rydberg though broadened and intensified by the mixing. Upon placing such a system in a condensed phase, the $a_1\sigma^*$ valence state again will be deperturbed, *i.e.*, freed of its Rydberg admixtures, and vice versa. In this case, the predominantly $n_X \rightarrow nsa_1$ Rydberg excitation of the vapor phase will seem to appear in the condensed phase as well, however in the latter it will have been transformed into the (n_X, $a_1\sigma^*$) valence state!

The argument given above explains in a natural way the otherwise puzzling presence of what appear to be the lowest $\phi_i \rightarrow nsa_1$ Rydberg excitations in the spectra of many molecules in condensed phases. The activity of such a deperturbation mechanism may be signalled by odd frequency shifts, a change of band shape and/or especially an increased oscillator strength.

A theoretical study of the $^3(3a_1, 3s)$ Rydberg state of ammonia perturbed by two molecules of hydrogen [K7] bears obliquely on the above discussion, for it is found that upon placing the two perturber molecules within the 3s sphere of the ammonia, the Rydberg character of the 3s orbital actually increases! This can be interpreted as an unmixing of the $(3a_1, 3s)$ Rydberg and $(3a_1, \sigma^*)$ valence-conjugate configurations by the external perturbation.

Note again however, that Rydberg excitations readily are found in solution spectra using CD, and so the observation of such a band could be interpreted as due to either a Rydberg excitation, or to its deperturbed conjugate valence excitation. The temperature sensitivity of the band frequency is then the important criterion for determining the upper orbital type, as per the discussion given above. With the current popularity of spectroscopy in molecular beams, it should be pointed out that the effect of clustering in the beam offers an excellent test of the Rydberg/valence character of a molecular transition [G9].

Plasmons and Collective-Electron Behavior

The question of molecular plasmon resonances [I, 35] properly belongs in this Section on condensed-phase effects, for Williams *et al.* [B45, W29] argue that such excitations can occur only in condensed phases. Further, they explain that many of the "plasmons" observed in molecular liquids in fact have only partial plasmon character (or collective behavior), depending upon the separation of the relevant peaks in the $\mathrm{Im}(-1/\epsilon)$ and ϵ_2 spectra. Thus, for example, a peak in the ϵ_2 spectrum of liquid glycerol at 97 000 cm^{-1} (*12 eV*) has its complement at 165 000 cm^{-1} (*20.4 eV*) in the spectrum of the energy-loss function $\mathrm{Im}(-1/\epsilon)$, Fig. II.D-4, indicating strong collective character for the absorption process. Similarly, the peak at *ca.* 177 000 cm^{-1} (*22 eV*) in the $\mathrm{Im}(-1/\epsilon)$ spectrum of solid benzene appears at approximately 121 000 cm^{-1} (*15 eV*) in the ϵ_2 spectrum [K14], again signalling a plasmon excitation. In contrast, the $^1A_g \rightarrow {}^1E_{1u}$ transition of liquid benzene has its ϵ_2 peak at 52 000 cm^{-1} (*6.4 eV*) and the corresponding peak in $\mathrm{Im}(-1/\epsilon)$ falls at 63 000 cm^{-1} (*7.8 eV*). In this case, there is very little collective behavior, as expected for a one-

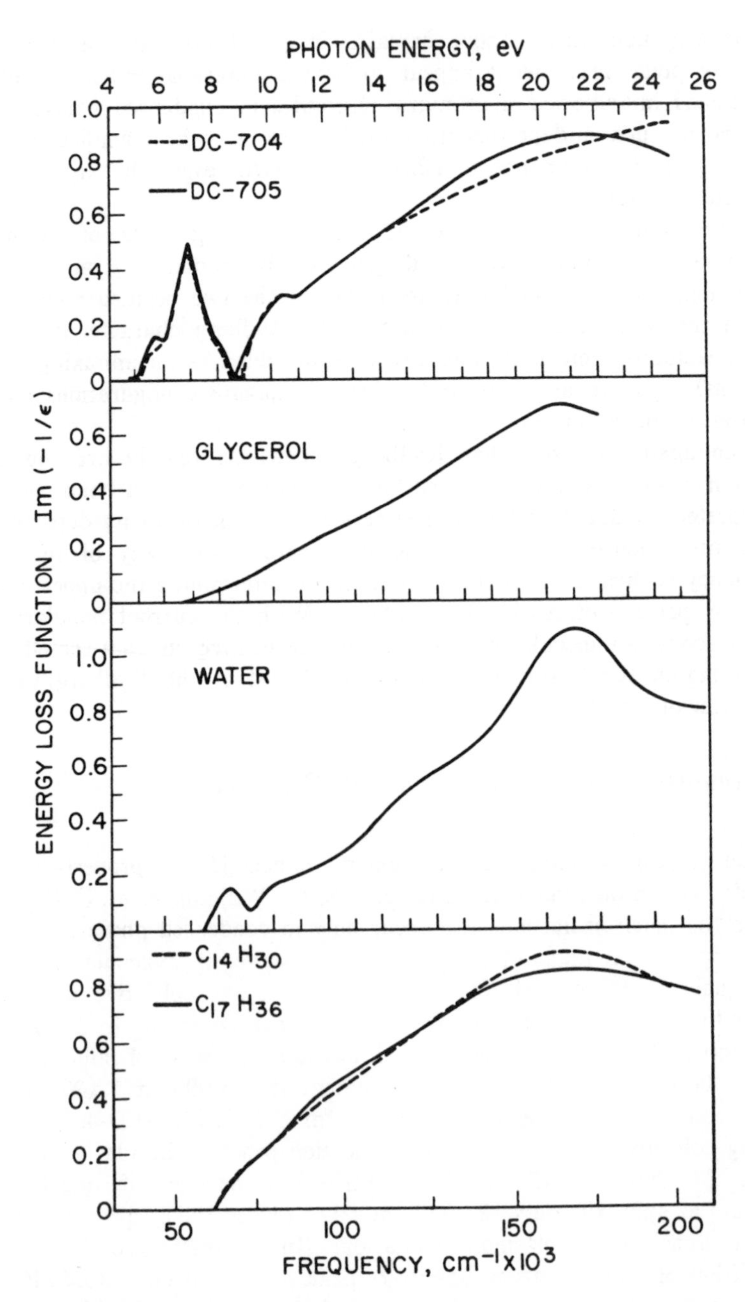

Fig. II.D-4. Plasmon excitations in a variety of liquids in the vicinity of 170 000 cm^{-1} (*20 eV*) [H37]. The pump oils DC-704 and DC-705 are phenyl derivatives of methylated trisiloxanes.

electron transition. Koch and Otto [K30] point out that the generally held opinion that plasmon excitation is forbidden optically (at normal incidence) but allowed by inelastic electron impact applies only to metals and not to insulators or free molecules. However, in contrast to Williams *et al.*, Koch and Otto argue that a plasmon-like resonance is possible in a free molecule, though no examples exist as yet. Yet a third view is that of Berg and Robinson [B31], who claim that collective oscillations simply will not be found in insulators.

The earlier assumptions that the seemingly ever-present broad transitions at 120 000–200 000 cm^{-1} (*15–25 eV*) in the spectra of most vapor-phase organic molecules are plasmon excitations (*cf.* [**I**, 109], for example) are fallacious. Rather, these peaks do not correspond to excitation to a specific "state," but to the maxima of one or more photoionization cross sections of the molecule, Section II.C. It would be most interesting to know how such photoionization maxima are affected by going into a condensed phase, and to see how the plasmon would develop as a function of density in the liquified rare gases. In regard the first point, energy-loss spectra of several organic molecules show that the broad intense peaks in the 121 000–242 000 cm^{-1} (*15–30 eV*) region of the vapor species (Fig. II.C-1) are almost totally damped in the spectra of the crystals [K16].

II.E. Addendum

F. Sette, J. Stöhr and A. P. Hitchcock, Correlation between intramolecular bond lengths and K-shell σ-shape resonances in gas-phase molecules, *Chem. Phys. Lett.* **110**, 517 (1984).

F. Sette, J. Stöhr and A. P. Hitchcock, Determination of intramolecular bond lengths in gas phase molecules from K shell shape resonances, *J. Chem. Phys.* **81**, 4906 (1984).

J. Hormes, On Rydberg transitions of matrix isolated atoms, *Chem. Phys. Lett.* **112**, 431 (1984).

S. P. McGlynn, G. L. Findley and R. H. Huebner, "Photophysics and Photochemistry in the Vacuum Ultraviolet", D. Reidel, Dordrecht, 1985.

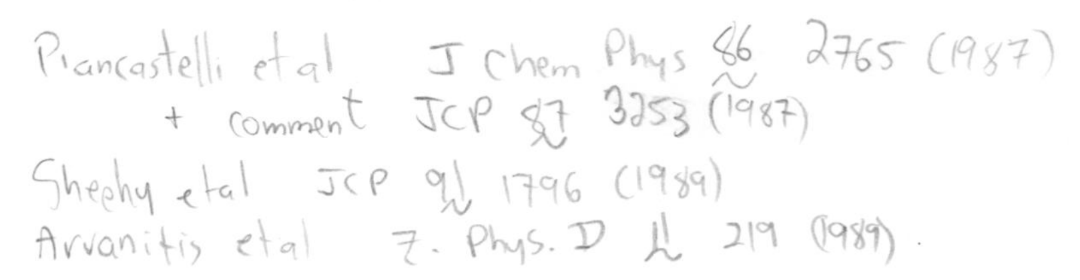

CHAPTER III

Alkanes

Prior to our studies published in Volumes **I** and **II**, the bulk of experimental and theoretical work on the electronic spectra of alkanes was discussed in terms of valence assignments. In contrast, it was our conclusion based on term-value considerations that most if not all the transitions observed in the spectra of alkanes are Rydberg excitations. Since that time, all *ab initio* theoretical studies in sufficiently large basis sets have come to the same conclusion. Along with the amines (Section VI.A) and oxo compounds (Chapter VII), the alkanes are among the most valence-poor chromophores to have been investigated. This lack of observable valence transitions in the alkanes could be due either to the complete miscibility of the σ^* antibonding MO's with the conjugate ns and np Rydberg manifolds (Section I.D), or it could be that the valence levels lie far above the relevant ionization potentials resulting in very broad $\sigma \rightarrow \sigma^*$ shape resonances as yet undiscovered (Section II.C). In two alkanes however (Section III.E), valence excitations appear without question below the first ionization potential and with enormous intensity! These are in fact the molecular analogs of the atomic giant resonances resulting from 4f wavefunction collapse (Section I.D). In stark contrast to the Rydberg picture of the spectra of the smaller acyclic alkanes, the most recent interpretations of the polyethylene spectrum rely totally on valence assignments. However, in Section III.D we point out that assignments based upon Rydberg term values also are tenable.

III.A. Methane and Its Fragments

Technical advances and understanding in the area of molecular Rydberg spectroscopy often follow progress made in atomic Rydberg spectroscopy by a decade or two. Thus it is interesting to read [W5] that a radio recombination line recently observed at 26 MHz in the carbon atom involves principal quantum numbers in the range 630–640! Inasmuch as the highest value of n so far identified in a polyatomic molecule is between 40 and 50 [F9, F45], molecular spectroscopists still have a long way to go.

New information on the Rydberg excitations of the methyl radical [**I**, 118] has come from the MPI spectrum of this transient species. Several of the $2p\pi \rightarrow ns$ and $2p\pi \rightarrow nd$ bands earlier observed by Herzberg as one-photon absorptions [**I**, H41] appear as two- and three-photon resonances in the MPI spectrum of the CH_3 radical generated by thermal decomposition of appropriate materials [D23, D25]. The special value of MPI spectroscopy lies in observing transitions such as $2p\pi \rightarrow np$ which are allowed only as two-photon absorptions. In fact, several new origins have been observed in the two-photon resonant MPI spectra of the methyl-h_3 and -d_3 radicals, corresponding to $2p\pi \rightarrow np\pi$ promotions [$^2A_2'' \rightarrow {}^2A_2''$, 59 972 cm^{-1} (*advert., 7.4354 eV*) for $n = 3$ in CH_3] [D24, H74]. The upper-state symmetry of this Rydberg series is specified in part by its quantum defect of 0.6 and by a rotational analysis. Yet another Rydberg series in the MPI spectrum of the methyl radical is identified as $2p\pi \rightarrow nf$ ($^2A_2'' \rightarrow {}^2E'$), again on the basis of a quantum defect of 0.0 and a rotational analysis. By photolyzing methyl iodide in the $\mathscr{A}(5p, \sigma^*)$-band region, a two-photon MPI resonance in the methyl radical photofragment ($2p\pi \rightarrow 3d$) is observed (67 820 cm^{-1} *advert., 8.408 eV*) [G11] which does not appear in either the one- or three-photon absorption spectra. As with the corresponding Rydberg states of H_2O, NH_3, C_2H_5 and CH_4, the $^2(2p\pi, 3s)$ Rydberg excited state of the methyl radical is unstable with respect to predissociation and simultaneous deRydbergization, here going to H and CH_2 fragments, each in a valence configuration at infinite separation [Y4]. The deRydbergization proceeds via the conjugate $a_1\sigma^*$ MO. The Rydberg spectrum of the methylene radical earlier was discussed from the term-value point of view [**I**, 119]; an *ab initio* calculation [R41] has since confirmed the assignments proposed in that work.

Sharp and Dowell [S43] report a double-humped feature in the electron-impact dissociative attachment spectrum of methane-h_4 which we interpret as involving the lowest Feshbach resonance in the molecule. The resonance peaks at 74 000 and 82 000 cm^{-1} (*vert., 9.2 and 10.2 eV*) no doubt have as parents the $^1(1t_2, 3s)$ configuration with Jahn-Teller components at 78 200 and 84 700 cm^{-1} (*vert., 9.70 and 10.5 eV*) [V31]. The frequency differences of 3000−4000 cm^{-1} between the $^2(1t_2, 3s^2)$ Feshbach states and the corresponding $^1(1t_2, 3s)$ parent Rydberg states are as expected (Section II.B). Carrying the analogy between Rydberg and Feshbach resonance band shapes one step further, we note that the corresponding TNI resonance in methane-d_4 [S43] shows only a single broad peak (76 000 cm^{-1} *vert., 9.4 eV*), as does the $1t_2 \rightarrow 3s$ absorption in CD_4 [D27]. The $^2(2p, 3s^2)$ Feshbach resonance states of methane correspond to the lowest resonance observed in neon (129 900 cm^{-1}, *16.11 eV*) [S25], the united atom of methane. A weak resonance at 96 800 cm^{-1} (*vert., 12.0 eV*) [M16] in methane has a 4000-cm^{-1} decrement with respect to the $(1t_2, 3d)$ Rydberg state expected at 100 900 cm^{-1} (*12.51 eV*), suggesting a $^2(1t_2, 3d^2)$ Feshbach configuration. As regards the $^2(1t_2, 3d^2)$ Feshbach decrement, the fit to expectation will not be very good owing to unknown multiplet splittings in both the Rydberg and Feshbach configurations.

It is significant that we have located the Feshbach resonance in methane having $^1(1t_2, 3s)$ as parent, for this means that all TNI resonances observed at lower frequencies must be valence resonances involving the σ^*(C−H) MO's. Three such valence TNI resonance peaks were reported at 52 700, 59 400 and 65 700 cm^{-1} (*6.53, 7.37 and 8.15 eV*), and were assigned to the three Jahn-Teller split components of the $^2(N, t_2\sigma^*)$ valence-state negative ion [B56]. If assigned correctly, the size of the Jahn-Teller splitting induced by placing an electron in the $t_2\sigma^*$ antibonding orbital set (13 000 cm^{-1}) is seen to be close to that observed by removing one from the $t_2\sigma$ bonding set (12 900 cm^{-1}) [I, B63]. In accord with this assignment, the angular pattern of electron scattering at 40 000−60 500 cm^{-1} (*5.0−7.50 eV*) incident energy shows a very strong $l = 2$ component, implying $^2T_2$ symmetry of the negative-ion resonance configuration [R40, T4, T5]. Note however, that though Mathur [M16] questions the reality of the three peaks mentioned above, he and several others [R40, T4] do report a single broad resonance at *ca.* 61 000 cm^{-1} (*vert., 7.6 eV*) in methane. Another valence TNI resonance is observed in methane at 19 400 cm^{-1} (*vert., 2.4 eV*) [B54], this one displaying several vibrational quanta of 1900 cm^{-1}. By default, this resonance is assigned to the $^2A_1$ state having the incident electron momentarily trapped in the $a_1\sigma^*$ MO. In studying electron transmission through solid methane films,

Huang and Hamill [H71] found a sharp negative-ion resonance at 20 000 cm^{-1} (*vert., 2.5 eV*) as well as broad features at 48 000–65 000 cm^{-1} (*6.0–8.0 eV*), all of which correlate with resonances observed in the vapor phase.[†] The correlation between vapor-phase and condensed-phase TNI transitions is most interesting, for the hypothesis that the TNI resonances below 72 000 cm^{-1} (*9 eV*) in methane have valence-shell configurations is strongly supported by their appearance in the solid-film spectrum.[§] An X-α calculation also assigns the TNI resonances at 48 000–65 000 cm^{-1} to capture of an electron in $t_2\sigma^*$, but places that to $a_1\sigma^*$ below 8000 cm^{-1} (*1 eV*) [T15]. See Section IX.A for a discussion of the related high-frequency TNI bands in silane.

It is at first sight queer that σ^* valence MO's are totally missing in the absorption and energy-loss spectra of methane, [**I**, 108; **II**, 297] yet are so obvious and indeed prominent at relatively low frequencies in its TNI spectrum. A likely explanation is that the σ^* levels are totally miscible with the Rydberg levels in the absorption case where the effective potential for the optical electron is that of the CH_4^+ core (Section I.D), whereas for TNI resonances, the effective potential is that of the neutral CH_4 core for which there is no Rydberg manifold in which the σ^* MO's can dissolve.

Excitation to the lowest triplet state in methane is observed by electron impact to fall at 71 000 cm^{-1} (*vert., 8.8 eV*), the upper state most likely being $^3(1t_2, 3s)$ [J7]. Since a $1t_2 \rightarrow 3s$ assignment of the 71 000-cm^{-1} band leads to a singlet-triplet splitting of 7000 cm^{-1}, whereas the more common value is 2000 cm^{-1} for Rydberg states terminating at 3s, one concludes that the $(1t_2, 3s)$ configuration of methane has a larger than usual complement of the valence conjugate configuration $(1t_2, a_1\sigma^*)$. If this is so, it will be reflected as well in a larger-than-normal value of the self-energy integral J_{3s3s} (Section I.D). In contrast, the singlet-triplet splittings of the lowest Rydberg configurations of ethane [B76] and propane [R18] are calculated to be less than 2000 cm^{-1}, while a value of 4000–5000 cm^{-1} is observed for ethane [J7].

[†] Note however that Mathur [M16] does not find the 19 400-cm^{-1} resonance in methane vapor and suggests that it is due instead to a nitrogen impurity. However, the clear presence of this band, slightly shifted, in other alkanes (Sections III.C and III.E) argues against Mathur's suggestion.

[§] Because Feshbach resonances involve configurations of strong Rydberg parentage, they will be severely perturbed in condensed phases (Section II.D), whereas valence TNI resonances will be much less susceptible to such perturbations.

Careful electron-impact work on the energy losses in CH_4 and CD_4 reveals a distinct but very weak band centered at 85 900 cm^{-1} (*10.65 eV*) which is either a Jahn-Teller component of the $1t_2 \rightarrow 3s$ transition or a multiplet component of $1t_2 \rightarrow 3p$ [D26]. The latter is more likely both in view of the observed low intensity, and upon comparison of the $1t_2 \rightarrow 3s$ band envelope with the $(1t_2)^{-1}$ band of the photoelectron spectrum [**I**, 111]. According to theory [M65], transitions are allowed to only the lowest of the four multiplets arising from the $(1t_2, 3p)$ configuration; a component of the $1t_2 \rightarrow 4s$ transition is peaked at 93 640 cm^{-1} (*11.61 eV*) in CH_4 and displays multiple quanta of 1150-cm^{-1} vibrations which are nontotally symmetric [D26, L15]. It is possible as well that this transition also is overlapped by one or more of the forbidden components of the $1t_2 \rightarrow 3p$ Rydberg complex. The inelastic electron-scattering cross sections up to 121 000 cm^{-1} (*15 eV*) in methane are reported in the work of Vuskovic and Trajmar [V31]. Integration of the energy loss in methane from 69 000 to 88 300 cm^{-1} (*8.55 to 10.95 eV*) yields an oscillator strength of 0.29 for the $1t_2 \rightarrow 3s$ excitation [C17].

Ab initio calculations on the lower states of methane again place the Rydberg excitations from $1t_2$ to 3s and 3p lowest, with the conjugate valence excitations predicted to be more than 80 000 cm^{-1} (*10 eV*) higher [K32].† For the $(1t_2, 3s)$ state at the ground-state equilibrium geometry, it is calculated that the effective nuclear charge for the 3s MO is 1.37 and that the maximum for the 3s orbital is at 4.8 a_o [M65]. This clearly shows the predominantly Rydberg nature of the $3sa_1$ MO of methane. As with the earlier theoretical studies on water and ammonia, calculations on the lowest singlet Rydberg state of methane $(1t_2, 3s)$ reveal that though it is strongly Rydberg at the ground-state equilibrium geometry, on dissociation along the coordinate leading to $CH_2 + H_2$ in C_{2v} symmetry, this state rapidly loses its Rydberg character to become fully valence once the C–H distance exceeds 3 a_o [G26, G27, G29]. Though the mixing of $a_1\sigma^*$ with $3sa_1$ in methane can be appreciable and is responsible for the dissociation of the $(1t_2, 3sa_1)$ Rydberg state, the $(1t_2)^{-1}$ ionic state is calculated to be distorted but bound, implying that the σ^* antibonding MO does not mix into the continuum orbitals of the ion. Thus one can

† Note however that the Rydberg basis set was not large enough (Section I.D) for an adequate test of the miscibility of the $a_1\sigma^*$ valence MO in the *n*s Rydberg sea. Further theoretical work on Rydberg/valence mixing in methane should be performed using very large Rydberg basis sets so as to provide the maximum Rydberg space for dissolution of the valence level.

rationalize why the $(1t_2)^{-1}$ band of the photoelectron spectrum is vibronically structured but the $1t_2 \rightarrow 3sa_1$ band of the electronic excitation spectrum is not, by considering the differential mixing with $a_1\sigma^*$. Similarly, we argue that the $(1t_2, 2t_2\sigma^*)$ valence level is not appreciably mixed into the conjugate $(1t_2, 3p)$ Rydberg configuration inasmuch as $1t_2 \rightarrow 3p$ is observed to be a vibrationally structured transition. That $a_1\sigma^*$ should mix with the nsa_1 manifold while $t_2\sigma^*$ does not mix with the npt_2 manifold follows from the fact that the ns orbitals are more penetrating than np and so have a larger overlap with the σ^* MO of the same symmetry (Section I.D). Perhaps the rapid deRydbergization of the $^1(1t_2, 3s)$ state of methane is in some way responsible for the large singlet-triplet splitting of the $(1t_2, 3s)$ configurations as measured from strongly nonvertical transitions.

Beyond the Rydberg transitions in methane originating at $1t_2$, the optical absorption shows a monotonic decrease from 150 000 to 555 000 cm^{-1} (*18.6 to 68.8 eV*) [L13]. In contrast to this featureless curve, measurement of the yield of Lyman-α emission beyond 150 000 cm^{-1} (*18.6 eV*) in methane reveals two nicely structured features at 164 900 and 176 100 cm^{-1} (*vert., 20.45 and 21.83 eV*); the vibrational intervals of *ca.* 2000 cm^{-1} correspond to the upper-state C—H stretch, as in the $(2sa_1)^{-1}$ photoelectron band [W38]. The first of the emission-yield peaks is visible as well in the electron-impact spectrum of methane taken at scattering angles of *ca.* 15° [D26], Fig. III.A-1. Measured with respect to the $(2sa_1)^{-1}$ ionization potential at 185 000 cm^{-1} (*vert., 22.94 eV*), the two peaks in question are seen to have $(2sa_1, 3p)$ and $(2sa_1, 4p)$ upper states by virtue of their respective 20 000- and 9000-cm^{-1} term values. These and other Rydberg states are revealed as well in the yield curves for various ionic fragments generated by irradiating methane in the extreme ultraviolet [M6, M16, M17]. Thus, peaks at 157 000, 167 000 and 172 000 cm^{-1} (*19.5, 20.7 and 21.3 eV*) in such yield curves are readily assigned as singlet excitations from $2sa_1$ to 3s (forbidden), to 3p (allowed), and to 3d (allowed), respectively. These features correlate with the electron impact energy-loss spectrum at 10–15° scattering angle, Fig. III.A-1, and in part with structure found in the measurement of the photoelectron asymmetry parameter β in the same region [M7]. Related features are observed in the spectra of silane and germane [D27], Section IX.A. The energy-loss spectrum of methane, Fig. III.A-1, shows yet more structure, with a clear transition $(2sa_1 \rightarrow 3dt_2)$ at 173 000 cm^{-1} (*advert., 21.50 eV*) and another following this (177 500 cm^{-1} *advert., 22.00 eV*) which is undoubtedly a member of a Rydberg series converging on $(2sa_1)^{-1}$, but which is otherwise difficult to assign. Using the electron-ion coincidence technique, Backx and van der Wiel [B2] have shown that

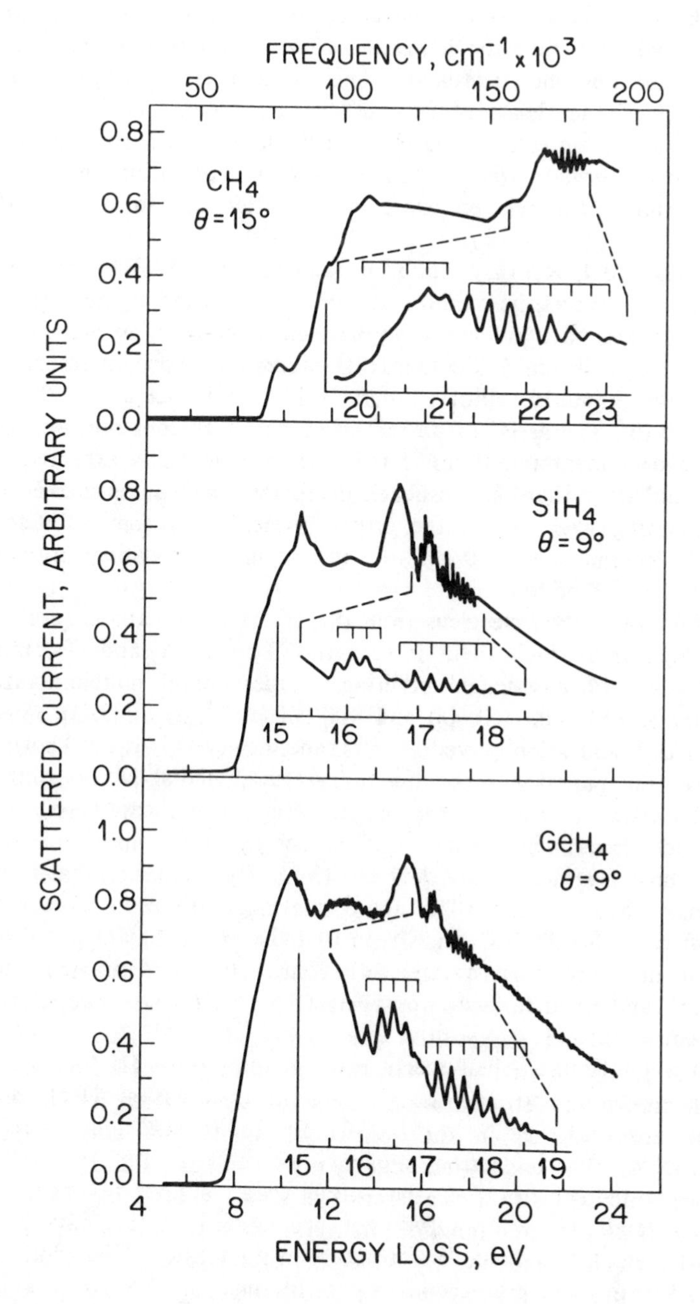

Fig. III.A-1. Electron impact energy-loss spectra of methane, silane and germane at impact energies of 200 eV [D27].

the peak of the $(1t_2)^{-1}$ ionization efficiency in methane is very close to threshold, whereas that for ionization from $2sa_1$ peaks at *ca.* 125 000 cm^{-1} *(15 eV)* beyond threshold. This pattern is consistent with the explanation of the delayed onset of the photoionization maximum as discussed in Section II.C, if the electron in the case of $(1t_2)^{-1}$ ionization radiates as an ϵs wave, whereas that for $(2sa_1)^{-1}$ ionization radiates as an ϵp wave and has a delayed onset in the latter due to the centrifugal barrier.

The $^1(2sa_1, 3s)$ Rydberg state of methane at 157 000 cm^{-1} (*vert., 19.5 eV*) can serve as the parent state for the excitation of the $^2(2sa_1, 3s^2)$ negative-ion Feshbach resonance on electron impact. A normal Feshbach decrement of 4000 cm^{-1} (Section II.B) places the expected resonance at 153 000 cm^{-1}. Indeed, Mathur reports a TNI resonance at 149 000 cm^{-1} (*vert., 18.5 eV*) and assigns it to the above Feshbach configuration [M16]. Here, the decrement (8000 cm^{-1}) is larger than normally expected (4000 cm^{-1}), suggesting that the Feshbach decrement may increase the deeper the originating MO lies in the target molecule. A somewhat smaller Feshbach decrement of 6400 cm^{-1} for this resonance is evident in the work of Marmet and Binette [M6].

The removal of two electrons from the $1t_2$ MO of methane results in a $1t_2^4$ configuration which yields in turn the $^3T_1$, $^1A_1$, ^1E and $^1T_2$ states of the di-cation, of unspecified ordering. Experimental double ionization potentials of 254 000, 310 000 and 400 000 cm^{-1} (*vert., 31.5, 38.4 and 50 eV*) [A20, R1] are observed for methane, however, it is not known how these are to be paired with the di-cation states quoted above. Nonetheless, these ionization potentials may be relevant to the results of (e, 2e) coincidence experiments which reveal energy losses at 225 000 and 250 000 cm^{-1} (*vert., 27.9 and 31.0 eV*) [V5]. Remembering that Rydberg term values in the cation will be approximately 2.8 times as large as those in the neutral (Section I.B), it is seen that the excitations at 225 000 and 250 000 cm^{-1} have term values with respect to the ionization limit at 310 000 cm^{-1} which are appropriate for $(1t_2, 3s)$ and $(1t_2, 3p)$ assignments in the $(1t_2)^{-1}$ positive ion.

The carbon 1s inner-shell spectrum of methane, Fig. III.A-2, has been investigated both optically using synchrotron radiation [E1] and by electron impact [H55, T19, W19]. Of these studies, the highest resolution was achieved in the electron impact work of Tronc *et al.* [T19]. As summarized in Table III.A-I, the energy losses are readily assigned as $1s_C \rightarrow ns$ and $1s_C \rightarrow np$ Rydberg promotions, with accompanying nontotally symmetric vibrational structure in the former. The width of the $1s_C \rightarrow 3p$ transition corresponds to a lifetime of 7×10^{-15} sec. In Section I.B, it was noted that the (ϕ_i, ns) Rydberg term value often was

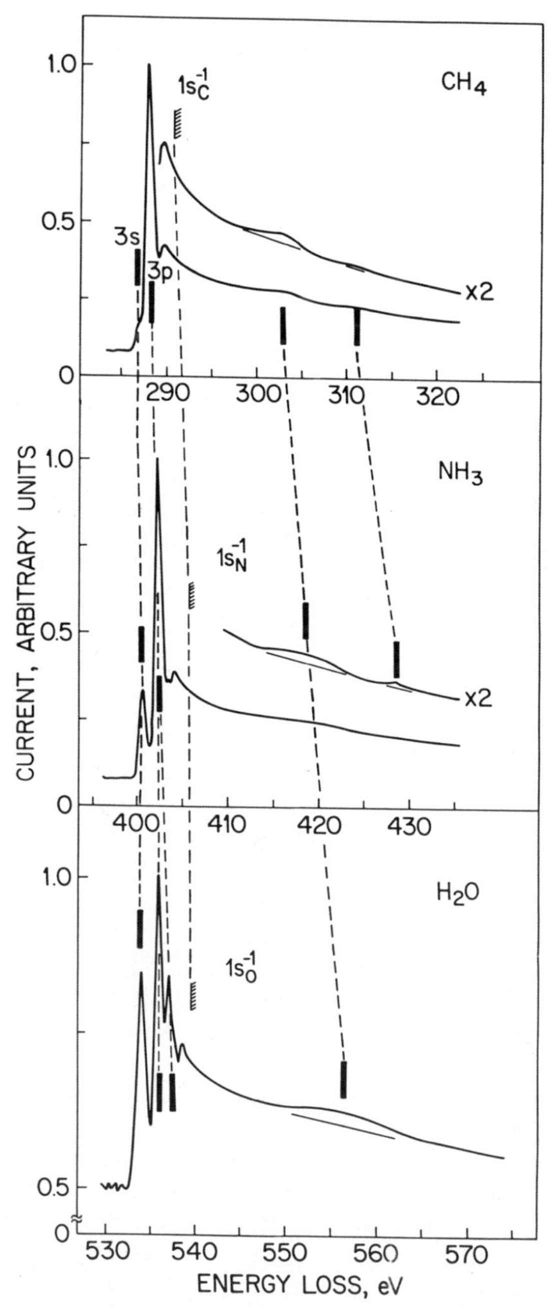

Fig. III.A-2. Inner shell energy-loss spectra of the first-row hydrides determined at an impact energy of 2500 eV and zero scattering angle [W19].

TABLE III.A-I

RYDBERG EXCITATIONS IN THE $1s_C$ SPECTRUM OF METHANE[a]

Energy Loss, eV	Term Value, cm^{-1}	Assignment
287.05	29 900	$(1s_C, 3s)$
288.00	22 260	$(1s_C, 3p)$
288.24	-	$(1s_C, 3p) + \nu'$[b]
288.37	-	$(1s_C, 3p) + \nu_1'$
288.70	-	$(1s_C, 3p) + 2\nu_1'$
289.14	13 060	$(1s_C, 3d)$, $(1s_C, 3p) + 3\nu_1'$
289.44	10 650	$(1s_C, 4p)$
289.83	7500	$(1s_C, 4d)$, $(1s_C, 5s)$, $(1s_C, 4p) + \nu_1'$
290.01	6050	$(1s_C, 5p)$
290.76	I.P.	$(1s_C)^{-1}$

[a] Reference [T19].

[b] Vibrational assignment uncertain.

increased by several thousand cm^{-1} when ϕ_i is changed from an outer-shell orbital to an inner-shell one. In methane, however, the term value actually decreases in the inner-shell spectrum, which is most peculiar. This negative "exhaltation" is seen again in ethane (Section III.B), however the corresponding $\phi_i^{-1} \rightarrow 4s$ band of silane has the expected positive exhaltation (Section IX.A). As mentioned in Section I.B, this unusual behavior of the term-value exhaltation may be due to a variation of the Rydberg/valence mixing as the hole descends from the valence to the inner shell.

Another puzzle in the inner-shell spectrum of methane is the band at 288.70 eV having a term value of 16 600 cm^{-1}. Tronc *et al.* [T19] suggest that it is a static Jahn-Teller component of the $(1s_C, 3p)$ configuration, however, since the degeneracy is in the "nonbonding" Rydberg orbital rather than in the bonding MO's as in $(1t_2, 3s)$, this assignment seems unlikely (Section I.B), and a $1s_C \rightarrow 4s$ assignment is similarly unsatisfactory on the basis of intensity. It is possible that this is a $1s_C \rightarrow 3dt_2$ transition with a large inner-shell term value exhaltation, but it is difficult to see why the transitions to 3s and 3p are not exhalted as well. A similar situation appears in ammonia (Section VI.A). Our suggestion here is that the methane band in question corresponds to the $1s_C \rightarrow 3p$ excitation accompanied by two quanta of ν_1', with the $v = 1$ and $v = 3$ bands coming at 288.37 and 289.14 eV, respectively, Table III.A-I. If assigned correctly as vibrational structure, these excitation frequencies will show large isotope effects in CD_4. Going beyond the $(1s_C)^{-1}$ ionization potential of methane, Wight and Brion [W19] report two broad bands which are 95 000 and 160 000 cm^{-1} (*vert., 11.8 and 19.8 eV*) beyond the ionization limit. The second of these corresponds to the $1t_2 \rightarrow 3p$ shake-up excitation in the $(1s_C)^{-1}$ positive ion [M64].

The appearance of the symmetry-forbidden $1s \rightarrow 3s$ excitation in the methane inner-shell spectrum must be due to either an electric quadrupole absorption, to vibronically induced absorption, or to a combination of these. If the first mechanism is operative, the intensity ratio of the transitions to $^1(1s_C, 3s)$ in CD_4 and CH_4 will be 1.0, whereas it is estimated that the ratio will be 0.75 if the Herzberg-Teller vibronic mechanism is operative instead [H55]. The experimental intensity ratio of 0.81 ± 0.08 shows clearly that the absorption mechanism is essentially vibronic and electric dipole in nature.

Elastic electron scattering experiments on CH_4 near the $(1s_C)^{-1}$ ionization limit reveal a negative-ion resonance at 283.75 eV which is postulated to have the $^2(1s_C, 3sa_1^2)$ Feshbach configuration [M18]. If the assignment is correct, then the Feshbach decrement for this excitation amounts to 26 600 cm^{-1}! This is an extreme example of the trend noted in Section II.B of increasing Feshbach decrement as the excitation approaches the inner-shell. The angular scattering pattern of this resonance is consistent with the $^2(1s_C, 3sa_1^2)$ Feshbach assignment [M19].†

III.B. Ethane

Recent experimental and theoretical work on ethane [I, 120] has focussed on the structured absorption at 75 800 cm^{-1} (*vert., 9.40 eV*), with relatively less attention being paid to the other bands in the spectrum. Custer and Simpson [C48] studied the structured bands of ethane-h_6 and

† Mathur's later objection [M19] to the assignment of this band to the $^2(1s_C, 3sa_1^2)$ configuration on the grounds that $1s_C \rightarrow 3sa_1$ is forbidden optically is irrelevant for an electron impact excitation at threshold with a large scattering angle.

-d_6 as a function of temperature, and fit the rovibronic band shapes on the assumption that in the upper state, methyl-group rotation is hindered by a 600-cm^{-1} barrier. Our earlier conclusion that the absorption in the 70 000–80 000 cm^{-1} (*8.7–10.0 eV*) region consists of both $3a_g \rightarrow 3p$ and $1e_g \rightarrow 3p$ promotions [I, 124] has been confirmed repeatedly by theory [B76, C1, C2, C3, R15, R16]. In the most detailed calculation [B76], the Rydberg excitations from the $3a_{1g}$ and $1e_g$ MO's to $3p\sigma$ and $3p\pi$ are spread over a region of only 1500 cm^{-1}. Of these, the excitation to ($3a_{1g}$, $3p\sigma$) is calculated to be especially intense since it is mixed with an intense valence $\sigma \rightarrow \sigma^*$ excitation of the same symmetry, however, this is not necessarily the excitation responsible for the structured band.

A very interesting calculation by Richartz *et al.* [R15, R16] on the lowest photoelectron band of ethane (97 500 cm^{-1} *vert., 12.09 eV*) bears on the assignment of the 75 800-cm^{-1} transition in the absorption spectrum. These authors calculate that in the ^2E_g ionic state there is a Jahn-Teller splitting into ^2A_g and ^2B_g components, and that the vibrational fine structure observed in the first photoelectron band is compatible only with ionization terminating at ^2B_g. This overlaps structureless ionization leading to the ^2A_g ions. Applying the above conclusion and the standard argument relating photoelectron and Rydberg band envelopes [I, 73] to the 75 800-cm^{-1} absorption, one is led immediately to a $1e_g \rightarrow 3p\sigma$ assignment for the structured band at 75 800 cm^{-1} in ethane.

In the Rydberg transitions of both methane (Section III.A) and cyclopropane (Section III.E) we note that when the hole is in a degenerate valence orbital, the vibronic structure appears to be chaotic, whereas if the degeneracy is in the nonbonding Rydberg orbital, the vibrational progression is much more regular. Accordingly, the $1e_g \rightarrow 3p\sigma$ transition in ethane should have a very complicated and irregular vibronic envelope. At low resolution in ethane-h_6, the absorption in question appears to be a regular progression of *ca.* 1200 cm^{-1}, however at higher resolution the envelope is seen to consist of many more vibrations [I, P9]. Since the photoelectron band was analyzed using only two active vibrations (ν_3' and ν_{11}', accidentally degenerate) [R16] the analysis may be faulty if indeed the photoelectron band is related to the band observed optically at 75 800 cm^{-1}.

Using electron impact, Johnson *et al.* [J7] observe clearly the shoulder at 69 000 cm^{-1} (*vert., 8.55 eV*) in ethane which is reported only occasionally in the optical spectra. Measurement of the energy and angular dependence of the 69 000-cm^{-1} band intensity with respect to that at 75 800 cm^{-1} identifies the former as electric-quadrupole allowed. Our earlier assignment for this band [I, 126] was a combination of $3a_{1g} \rightarrow 3s$

and $1e_g \rightarrow 3s$, both of which are electric-quadrupole allowed. Theory [B76, C3] also supports such an assignment. Johnson *et al.* [J7] further point out that intervals of 5600 cm^{-1} appear not only in the photoelectron spectrum of ethane but repeatedly in its energy-loss spectrum [91 900 and 97 600 cm^{-1} (*11.4 and 12.1 eV*); 104 800 and 110 500 cm^{-1} (*12.99 and 13.70 eV*)], and they propose corresponding Rydberg assignments based upon promotions to 3p and 4p orbitals.

Several attempts have been made to relate the Rydberg spectroscopy of ethane to its vacuum-ultraviolet photochemistry [C3, G29, S13], and these results are relevant to our work as well. Thus Caldwell and Gordon [C3] point out that the $^1(e_g, 3s)$ configuration of ethane is dissociative with respect to the formation of CH_3CH and H_2 and that the 3s Rydberg orbital shrinks into a valence one as the H_2 fragment moves away. This dissociative behavior explains the lack of vibronic structure in the $e_g \rightarrow 3s$ transition. On the other hand, the ^2E_g cation is stable and the $(e_g)^{-1}$ band of the photoelectron spectrum is nicely structured. Closely similar structure is seen in the transition from e_g to 3p, and presumably this state also is bound. Thus the situation in ethane parallels that of methane closely:- the lowest excitation to 3s is strongly mixed with the conjugate excitation to σ^*, leading to a dissociative upper state, featureless absorption and a rapid deRydbergization along the dissociation coordinate. Excitations to 3p by contrast are bound as is the cation toward which the lowest Rydbergs converge, implying that $(\phi_i^o, 3p)$ and the $(\phi_i)^{-1}$ cationic state are not mixed appreciably with σ^*. Note however, that the *ab initio* calculation of Buenker and Peyerimhoff [B76] argues for "substantial" contributions of σ^* to 3p in ethane.

A valence TNI resonance (Section II.B) at 18 100 cm^{-1} (*vert., 2.25 eV*) in ethane [C23, H52] is quite close to that of methane at 20 000 cm^{-1} (*2.48 eV*), implying that the σ^* MO in both cases involves C—H orbitals; the lowest unoccupied valence MO in ethane is calculated to be $3a_{2u}$ [I, S37]. Feshbach resonances are expected in ethane at 65 000 cm^{-1} [$^2(1e_g, 3s^2)$] and at 71 000 cm^{-1} [$^2(1e_g, 3p^2)$], with a continuous profile in the former and a complex vibronic envelope expected in the latter, as in the corresponding parent excitations.

At higher frequencies in ethane (150 000 to 500 000 cm^{-1}, *18.6 to 62.0 eV*), the optical absorption intensity shows a monotonic decrease which quantitatively matches that for ethylene [L13, P22]. As was the case with methane (Section III.A), many transitions will be found in this "continuum" in the region of 160 000 cm^{-1} (*20 eV*) when probed using techniques other than straightforward absorption.

In the region of $1s_C$ ethane absorption (*ca. 290 eV*), several energy losses are observed in the electron-impact spectrum [H56], the lowest two of which are seen optically as well [E1]. The inner-shell energy loss spectrum of ethane is much like that of methane except that the $1s_C \rightarrow 3s$ transition in ethane (28 900-cm^{-1} term value) is more intense relative to the transition to 3p (21 800-cm^{-1} term value) than it is in methane. As in methane, the inner-shell "exhaltation" of the ($1s_C$, 3s) term value of ethane is somewhat negative. Bands in the inner-shell spectrum of ethane with term values of 10 500 and 7260 cm^{-1} are assigned as $1s_C \rightarrow 4p$ and 5p, respectively. In the outer-shell region, the corresponding term values are 29 500 (3s), 21 700 (3p) and 9500 cm^{-1} (4p) [**I**, 125]. As with methane, there are no readily identifiable valence excitations appearing on either side of the $(1s_C)^{-1}$ ionization potential of ethane, however, Hitchcock *et al.* [H66] compare the electron-energy-loss spectra of methane and ethane and conclude that there is a $1s_C \rightarrow \sigma^*$(C—C) shape resonance in the latter 4000 cm^{-1} above the $(1s_C)^{-1}$ ionization limit. If correct, this would be some of the first evidence we have concerning the fate of σ^* levels in alkanes.

III.C. Acyclic Alkanes

Absorption spectra of the smaller acyclic alkanes ($CH_4 - C_4H_{10}$) in the vapor and solid phases have been reviewed [K30]. Calculations on the excited states of such alkanes have proceeded in two orthogonal directions. On the one hand, some calculations ignore Rydberg excited states and correlate low-lying absorptions and consequent photochemistry with excited valence configurations (see, for example, [S1]). In contrast to these, calculations which include Rydberg AO's in the basis set invariably predict that all the low-lying excitations in alkanes are Rydberg rather than valence.

An *ab initio* calculation including Rydberg AO's is reported by Richartz *et al.* [R18] for the excited states of propane [**I**, 129]. The calculation supports completely our initial suggestion [**I**, 133] that the absorption features in this molecule correspond to excitations to the few lowest-lying Rydberg orbitals (3s, 3p, 3d) from each of the occupied MO's. However, the detailed assignments differ considerably because the first three ionization potentials are calculated to be much closer together than is observed in the photoelectron spectrum. The theorists argue that the calculated ionization potentials are truly vertical whereas the peaks observed in the photoelectron spectrum are not. Even were this the case, the relevant Rydberg excitations would show the same nonverticality and one then would have to compare theory with nonvertical absorption frequencies rather than with the observed absorption peaks, as they did.

As a consequence of this approach, the lowest ionization potential of propane is taken by Richartz *et al.* as 96 550 cm^{-1} (*11.97 eV*), leaving them no reasonable assignment for the absorption observed at 64 000 cm^{-1} (*vert., 7.93 eV*). On the other hand, we contend that the lowest vertical ionization potential corresponds to the first peak in the photoelectron spectrum at 91 900 cm^{-1} (*vert., 11.39 eV*), thereby giving the 64 000-cm^{-1} band of propane a respectable (ϕ_i^o, 3s) term value of 27 900 cm^{-1} [**I**, 133]. The photoelectron energies and orbital assignments previously assumed in assigning the Rydberg spectrum of propane [**I**, 133] are supported by the *ab initio* calculations of Müller *et al.* [M78]. The calculated singlet-triplet splits of the propane Rydberg states scale with the quantum defect δ (Section I.B), and the transitions to 3s are predicted to be much weaker than those to 3p in general. Another *ab initio* calculation of the propane spectrum [C1] also claims success while ignoring the transition to 3s at 64 000 cm^{-1}, however the same type of calculation does predict correctly the transition to 3s in *n*-butane at 64 000 cm^{-1} (*vert., 7.93 eV*). Peaks in the threshold photoionization cross section spectrum of propane in the 88 000−160 000-cm^{-1} (*11−20-eV*) region correspond closely with those observed in absorption [S74], implying that all the higher Rydberg excitations are more or less autoionized.

The optical spectrum of neopentane [**I**, 138] shows a broad alkane-like profile with a major peak at 129 000 cm^{-1} (*vert., 16.0 eV*) and subsidiary shoulders at lower frequencies, Fig. III.C-1 [K29]. Just at the peak, there is a set of sharp vibronic features having Fano antiresonance profiles [K29, P22], which are assigned as 2s → 3p, 4p, 5p on the basis of their term values.† Each origin is accompanied by totally symmetric vibrations, as is the origin of the $(2s)^{-1}$ photoelectron band. These sharp features, which are so unusual in acyclic alkane spectra, are not observed in the electron-impact spectrum of neopentane [J7], which otherwise shows a large number of energy-loss bands assigned as Rydberg excitations. The lowest triplet state of neopentane is but 2000 cm^{-1} below the complementary ($4t_2$, 3s) singlet state, resulting in a singlet-triplet splitting which is much more normal than those observed for methane (7000 cm^{-1}) or ethane (4500 cm^{-1}).

† Theoretical work by Druger [D46] on antiresonance line shapes in molecules which is based upon a more general set of assumptions than in the past suggests that the observation of the antiresonance profile in molecules will be infrequent, arising only when there is a fortuitous combination of coupling matrix elements.

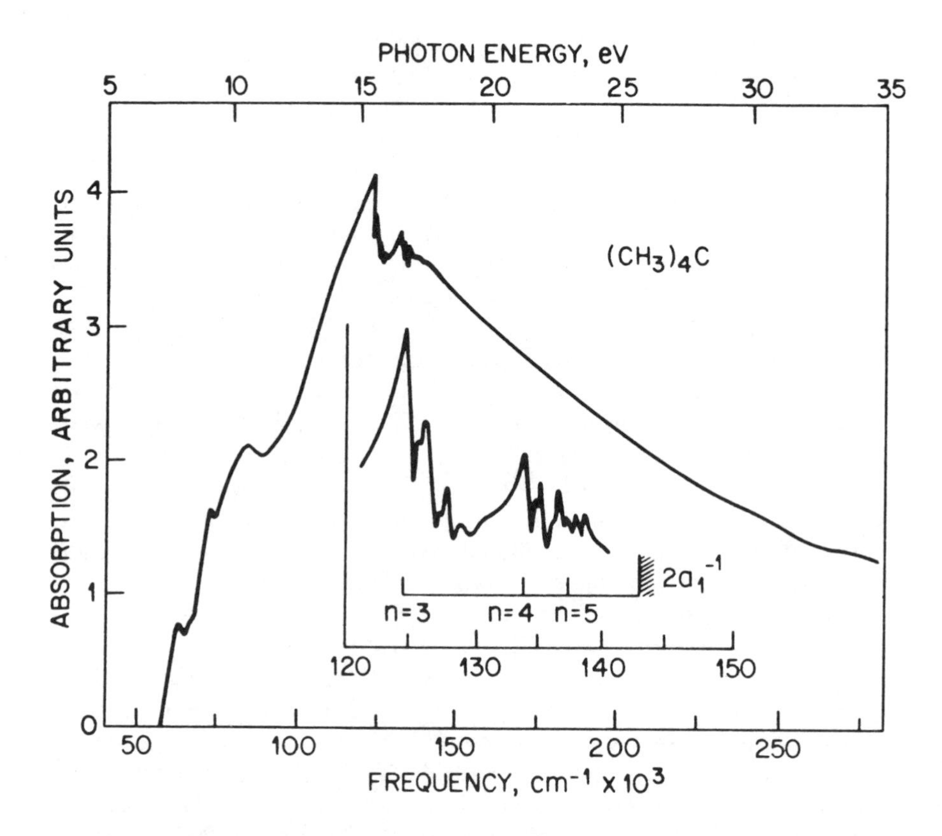

Fig. III.C-1. Optical absorption spectrum of neopentane, showing the Fano-resonance-like structure at *ca.* 130 000 cm^{-1} (*inset*) [K29].

The spectrum of neopentane is remarkable not only for the unique example it offers of antiresonances in the absorption spectrum of an alkane, but also for the continuous background against which the antiresonance is impressed. At its peak, the broad 129 000-cm^{-1} (*16 eV*) band of neopentane has an absorption cross section far larger than 150 Mb, which implies that this may be another example of a d → f giant resonance in a highly symmetric alkane (Section I.D). Indeed, in neopentane there is a $2t_1\sigma^*$(C–H) virtual valence MO for which the only conjugate Rydberg is 4f. Similarly, among the occupied orbitals, the $1e$(C–H) MO alone transforms as a pure d orbital in the T_d point group [K18]. As a consequence of these group theoretical facts, we are led to assign the intense 129 000-cm^{-1} band of neopentane as a d → f giant resonance [$1e\sigma$(C–H) → $2t_1\sigma^*$(C–H)]. The excitation is also a shape resonance since the $(1e)^{-1}$ ionization potential is only 114 000 cm^{-1} (*vert., 14.1 eV*). The proof of this valence assignment awaits a determination of the spectrum of solid neopentane. Similar features should appear in tetramethyl silane and tetrasilyl methane.

A number of what appear to be valence TNI resonances (Section II.B) are observed in thin films of the acyclic alkanes between 24 000 and 48 000 cm^{-1} (*3 and 6 eV*) in addition to others at approximately 5000 cm^{-1} (*0.6 eV*) [H52, L23]. Though assignments of such features await high-level calculations, it is clear that these are not Feshbach resonances based both on their appearance in solid phases and on their term values.

The circular dichroism spectrum of the optically active alkane (−)(3*S*:5*S*)-2,2,3,5-tetramethyl heptane reveals a single broad negative CD band (67 600 cm^{-1} *vert., 8.38 eV*) in the region 59 000−71 000 cm^{-1} (*7.3−8.8 eV*), in which region the absorption rises monotonically [A14]. Comparison with the absorption spectra of similarly large *n*-alkanes [**I**, 117] shows that the 67 600-cm^{-1} band is not the lowest excitation in the molecule, though no lower CD bands are resolved. Photoionization yields, cross sections and oscillator strength distributions have been determined for *n*-hexane [H75, P21].

The absorption spectra of alkyl radicals closely follow the pattern of absorption in the alkanes, which is to say, all the lower excitations are assignable as Rydberg transitions to 3s, 3p and 3d [W12] with regular term values. Thus, on going from $H_3C\cdot$ to $(CH_3)_3C\cdot$, the $(\phi_i^o$, 3s) term value decreases in a regular way from 33 100 to 25 800 cm^{-1} while the $(\phi_i^o$, 3p) and $(\phi_i^o$, 3d) term values hold at 18 000 and 13 000 cm^{-1}, respectively. In each case, the originating orbital ϕ_i^o is the 2p orbital on carbon carrying the unpaired electron. Calculations which do a fine job of explaining these Rydberg excitations also predict distinct valence excitations at frequencies just within the Rydberg manifold [L17]. Unlike the situation with the alkanes themselves, there is no mixing here because such valence transitions involve promotions to ϕ_i^o from the lower MO's, and so are not conjugate with the overlapping Rydberg levels. As appropriate for Rydberg excitations, these bands shift to higher frequency by *ca.* 2500 cm^{-1} upon solution. Calculations on the ethyl radical excited to the lowest 3s Rydberg level predict that it is deRydbergized along the coordinate leading to ground-state C_2H_4 plus H [E14], as with many other hydrides.

III.D. The Longer Alkanes, Polyethylene and Diamond

Our understanding of the excitations in polyethylene [**I**, 135] is very much in flux for several reasons. Experimentally, variability in density, crystallinity, methods of film preparation, radiation damage and solvent impurities all lead to spectral uncertainties. Theoretically, the electronic band structure of polyethylene has been formulated in terms of valence AO's only, and the spectrum interpreted in terms of interband transitions. In contrast, our approach [**I**, 105] has been to consider the low-lying transitions of the smaller alkanes as involving Rydberg excitations which would appear as excitons in a solid such as polyethylene rather than as interband transitions. To make a start, we focus on the polarized absorption data of Hashimoto *et al.* [H21] on low-density polyethylene, Figure III.D-1, though the results of other studies differ somewhat from this and from one another. The range of the optical absorption work on polyethylene is extended by the electron-impact spectra of Ritsko [R25], Fig. III.D-1.

Many studies report a weak excitation in polyethylene at 52 000 cm^{-1} (*vert., 6.4 eV*) which is due to absorption by radiation-induced unsaturation. Hashimoto *et al.* quote the intrinsic absorption edge in low-density polyethylene as 64 900 cm^{-1} (*8.05 eV*), whereas using electron impact, Ritsko deduces the far lower value of 58 000 cm^{-1} (*7.2 eV*). A shoulder with electric-dipole polarization parallel to the polyethylene chain axis appears at 68 000 cm^{-1} (*vert., 8.4 eV*, Fig. III.D-1), followed immediately by a perpendicularly-polarized band at 72 600 cm^{-1} (*vert., 9.00 eV*). The first of these correlates with the sharp spike previously found at 67 300 cm^{-1} (*vert., 8.34 eV*) in polyethylene at 4.2 K [**I**, 135]. Beyond this, the parallel-polarized absorption then mounts to peaks at 92 800 and 129 000 cm^{-1} (*vert., 11.5 and 16.0 eV*). The electron-impact energy loss spectrum of polyethylene is dominated by a broad structureless band centered at 158 000 cm^{-1} (*19.6 eV*) and a shoulder at 174 000 cm^{-1} (*vert., 21.6 eV*). Working at apparently lower resolution, Painter *et al.* [P3] investigated the optical absorption of polyethylene from 4000 to 613 000 cm^{-1} (*0.5 to 76.0 eV*), but report peaks only at 81 000 and 166 000 cm^{-1} (*10.0 and 20.6 eV*). Appropriate integration of their absorption curve yields an f-sum value of 16, as expected for a $-CH_2-CH_2-$ monomeric unit.

The peak at *ca.* 160 000–170 000 cm^{-1} (*19.8–21.1 eV*) in polyethylene is most interesting. As seen in Fig. II.D-4, peaks much like it appear in a wide diversity of materials, including carbon and polystyrene as well as those in the Figure [B11]. It is generally accepted that this feature corresponds to a collective excitation of the valence electrons, called a plasmon, Section II.D. Thus it is predicted for a plasmon that there will be a peak in the energy loss function Im $(-1/\epsilon)$ at the plasmon frequency, with a corresponding minimum in $|\epsilon|$ at the same frequency, as observed for polyethylene [R25]. Further, variations of the amplitude and

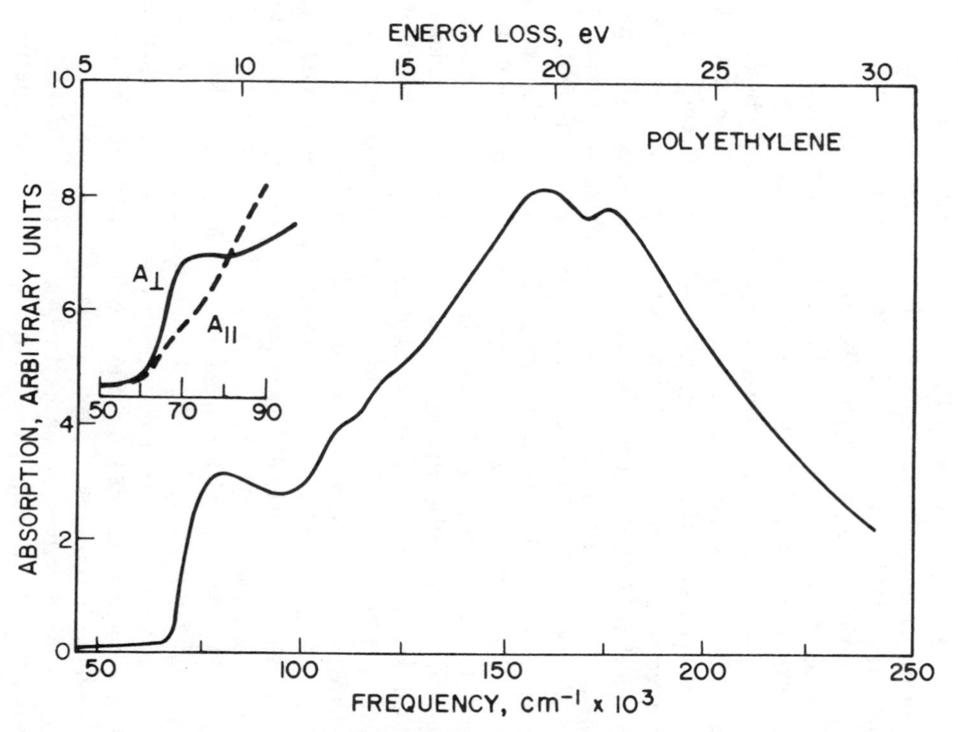

Fig. III.D-1. Electron-impact energy loss spectrum of pristine polyethylene at 80 KeV impact energy and zero scattering angle [R25]. The inset shows the optical absorption spectrum recorded with the light polarized along the polymer chain (- -) and perpendicular thereto (—) [H21].

frequency of the 160 000-cm^{-1} energy loss of polyethylene with scattering angle agree with those expected for plasmon excitation. In contrast, for an energy loss around 58 000 cm^{-1} (*7.19 eV*), Ritsko finds no shift of the threshold with increasing momentum transfer, indicating a highly localized, excitonic one-electron excitation. However, it is difficult to say just what feature is being excited at this frequency; possibly it is the co-excitation of the band edges of the 68 000- and 72 600-cm^{-1} features.

Hashimoto *et al.* [H21] have summarized the excitation data on several different polyethylenes and polyethylene-like alkanes, from which one can recognize certain constancies. Thus these spectra show an absorption tail in the 58 000–65 000 cm^{-1} (*7.2–8.1 eV*) region with perpendicular polarization, a band with the same perpendicular polarization centered at 72 000 cm^{-1} (*9.0 eV*), a parallel-polarized band at 93 600 cm^{-1} (*vert., 11.6 eV*), and possibly a parallel-polarized band at 66 000 cm^{-1} (*vert., 8.2 eV*). Spectra of several polyethylene-like systems also have been reported. Heller *et al.* [H37] studied the reflective properties of the liquid alkanes n-$C_{14}H_{30}$ and n-$C_{17}H_{36}$. In each, a transition is observed at 70 000 cm^{-1} (*vert., 8.7 eV*), corresponding to the 72 000-cm^{-1} band of polyethylene. Bands at 79 000 cm^{-1} (*vert., 9.8 eV*) in n-$C_{14}H_{30}$ and at 83 000 cm^{-1} (*vert., 10.3 eV*) in n-$C_{17}H_{36}$ are closest to the band at 93 000 cm^{-1} (*11.5 eV*) in polyethylene, but the discrepancies in frequency are surprisingly large. The reflectivity data show that all the above bands in the liquified alkanes are one-electron excitations, whereas the bands centered at *ca.* 170 000 cm^{-1} (*21 eV*) in the Im $(-1/\epsilon)$ spectra of these systems have an "appreciable degree of collective behavior" [H37]. Spectra of n-$C_{26}H_{54}$ [H37, O8] and of n-$C_{36}H_{74}$ [H21, O8] are appropriately polyethylene-like. Polypropylene [C8] exhibits monotonically decreasing absorption in the region $0.0208-1.25 \times 10^7$ cm^{-1} (*25.8–1550 eV*); this is the high-frequency tail of the plasmon excitation peaked at *ca.* 160 000 cm^{-1} (*20 eV*) and overlapping the various photoionization continua. The reflection spectrum of liquid squalane ($C_{30}H_{62}$) has a peak in the absorption coefficient at 107 000 cm^{-1} (*13.3 eV*) and one in Im $(-1/\epsilon)$ at 169 000 cm^{-1} (*21 eV*) [P4].

In assigning the excitations in polyethylene, our first inclination, as with other alkanes, is to turn to its photoelectron spectrum in an attempt to deduce reasonable term values. This will be somewhat uncertain here because the 3s and 3p term values will lie close in polyethylene and also because the peak absorption and ionization frequencies are not well established in this molecule. Further, we necessarily will be dealing with the difficult problem of Rydberg excitations in a condensed phase of low electronic mobility [**I**, 85].

The polyethylene ionization potentials have been determined by Delhalle *et al.* [D13] and by Seki *et al.* [S38], and can be assigned symmetries in accord with the results of valence-band calculations [A23, D18, F1, K5, M23]. As shown in Table III.D-I, the photoelectron work reveals four low-lying features which have been assigned as involving C—H bonds and two deeper ones involving C—C bonds. Though the symmetries of these orbitals at $k = 0$ are given in the Table in the idealized point group D_{2h}, the selection rules based on this point group will be relaxed by chain folding, branching, interchain interactions, *etc.*

TABLE III.D-I

RYDBERG EXCITATIONS IN POLYETHYLENE

Transition Frequency, cm^{-1} (*eV*)[a]		Ionization Potential, cm^{-1} (*eV*)[b,c]		Term Value, cm^{-1}
(54 000)[d]	(*6.70*)	76 000	(*9.4*, b_{2g})	22 000
68 000	(*8.4*)	87 900	(*10.9*, b_{1g})	20 000
72 600	(*9.00*)	99 200	(*12.3*, a_g)	26 600
92 800	(*11.5*)	109 000	(*13.5*, b_{2u})	16 200
129 000	(*16.0*)	143 000	(*17.7*, b_{3u})	14 000
170 000[e]	(*21.1*)	188 000	(*23.3*, a_g)	18 000

[a] Reference [H21].

[b] Symmetry of ionized orbital in idealized D_{2h} point group also given in parenthesis.

[c] References [D13, F1].

[d] Predicted but not yet observed.

[e] This is the average value for the peak reported between 158 000 and 174 000 cm^{-1} (*19.6 and 21.6 eV*) by various workers.

Since neither the appropriate symmetry nor the excitation and ionization frequencies are known with any certainty in polyethylene, we work at the lowest level, ignoring selection rules and taking transitions to 3s and 3p as degenerate for term-value purposes. As is clear from the data in Table III.D-II, (ϕ_i^o, 3s) Rydberg term values of approximately 21 000 cm^{-1} are to be expected in an alkane as large as polyethylene. The "$(b_{2g})^{-1}$" ionization at 76 000 cm^{-1} (*vert., 9.4 eV*) will lead to (b_{2g}, 3s/3p) excitons in the 54 000—58 000 cm^{-1} (*6.7—7.2 eV*) region.

TABLE III.D-II

CONVERGENCE OF THE TERM VALUES OF THE LOWEST RYDBERG STATES IN THE ALKANES[a]

Compound	Term Value, cm^{-1}
Methyl radical	33 187
Methane	31 600
Ethane	29 500
Cyclopropane	27 580
Butane	25 500
Cyclohexane	22 920
Norbornane	23 200
Bicyclo[2.2.2]octane	23 600
trans-Decalin	21 800
Adamantane	22 700

[a] Reference [H29].

Weak bands are reported just in this frequency range, but have been assigned as due to −C=C− absorption induced by radiation damage. Possibly, some of this absorption is intrinsic. The weak 68 000-cm^{-1} (*vert., 8.4-eV*) absorption band of polyethylene has a term value of 20 000 cm^{-1} with respect to the "$(b_{1g})^{-1}$" ionization at 87 900 cm^{-1} (*vert., 10.9 eV*) and so is readily assigned as (b_{1g}, 3s/3p) in the upper state. Transitions from b_{1g} to both 3s and 3p would be forbidden in the idealized D_{2h} geometry at $k = 0$. The intense absorption feature at 72 600 cm^{-1} (*vert., 9.00 eV*) is best explained in the Rydberg exciton picture as having an (a_g, 3s/3p) upper state, even though its term value with respect to the "$(a_g)^{-1}$" ionization potential at 99 200 cm^{-1} (*vert., 12.3 eV*) is somewhat high at 26 600 cm^{-1}. The possibility also exists that this is a valence $\sigma \rightarrow \sigma^*$ shape resonance, however it must be pointed out that such an excitation never has been documented in a long-chain alkane. The remainder of the absorption and energy-loss features in polyethylene similarly can be assigned to transitions to 3s/3p as shown in Table III.D-I. Note that this scheme works no worse for the band at 170 000 cm^{-1}, even though previous claims are that it is a collective excitation rather than a one-electron promotion.

It would be of great interest to detect the Feshbach resonances to $^2(\phi_i^o, 3s^2)$ in the longer alkanes in the vapor phase (and in polyethylene itself) so as to gauge the true positions of the neutral-molecule Rydberg excitations in these systems. MCD experiments on polyethylene also might prove to be of value in interpreting its spectrum.

Start with neopentane and replace all its H atoms by CH_3 groups; if one continues to replace all H atoms with CH_3 groups in cycle after cycle with appropriate ring-closing, the diamond lattice results. Viewed in this way, diamond is the limiting form of highly alkylated methane. As studied by electron impact [E4], diamond shows a one-electron excitation as a shoulder at 72 000 cm^{-1} (*vert., 9 eV*). Though at the same frequency as the 72 000-cm^{-1} band of polyethylene, Fig. III.D-1, this band originates with the σ(C—H) MO's of polyethylene and with the σ(C—C) MO's of diamond. The C—C bond orbitals of polyethylene lead to transitions at much higher frequencies [129 000 cm^{-1} (*16.0 eV*) and above]. The 72 000-cm^{-1} band of diamond is followed by two broad, intense energy losses centered at 190 000 and 270 000 cm^{-1} (*23.5 and 33.3 eV*), which were assigned as a possible surface plasmon and as a bulk plasmon, respectively. Plasmon losses observed in graphite [C4] at 56 000 and 218 000 cm^{-1} (*vert., 7 and 27 eV*) were assigned as π and σ plasmons, though we are tempted to assign the first of these as a one-electron $\pi \rightarrow \pi^*$ excitation. The regular progression of the bulk plasmon frequency from 270 000 to 218 000 to 170 000 cm^{-1} on going from diamond to graphite to polyethylene is understandable if the only electrons involved in the collective oscillations are those in the σ(C—C) bonds. In this case, the plasmon frequency will vary as the square root of the number of C—C σ bonds per C atom, *i.e.*, in the ratio $2 : \sqrt{3} : \sqrt{2}$, in fair agreement with the observed ratio.

III.E. Cyclic Alkanes

Considerable new data and *ab initio* calculations allow us to synthesize a more coherent picture of the excited states of cyclopropane [**I**, 141]. The weak bands observed previously in the 52 000—55 000 cm^{-1} (*6.4—6.8 eV*) region of the one-photon absorption spectrum [**I**, 141] appear strongly in the MPI spectrum as two-photon resonances at the one-photon frequencies [R32]. The coincidence of one-photon and two-photon frequencies proves the transition to be electronically allowed in both cases, which leads directly to a $^1A_1' \rightarrow {}^1E'$ assignment in the D_{3h} point group. The origin at 51 060 cm^{-1} (*6.330 eV*) has a term value of 27 580 cm^{-1} as appropriate for a $3e' \rightarrow 3s$ Rydberg promotion in a molecule containing three carbon atoms. A strong Jahn-Teller effect in the ($3e'$, 3s) state results in a highly irregular vibronic envelope in the MPI spectrum, as confirmed theoretically [D51]. According to calculation [G22], the allowed $3e' \rightarrow 3s$ transition in cyclopropane has only about 10 percent of the oscillator strength normally observed for such allowed transitions.

The region of $3e' \rightarrow 3p$ absorption in cyclopropane (60 000–66 000 cm^{-1}, *7.4–8.2 eV*) displays three MCD peaks:- a structureless peak of positive rotation at 61 500 cm^{-1} (*7.62 eV*) and a structureless negative rotation peaking at 65 400 cm^{-1} (*8.11 eV*) upon which is weakly impressed the structured, positive MCD band of the $3e' \rightarrow 3p$ transition at *ca.* 62 000 cm^{-1} (*7.7 eV*) [G7]. The ($3e'$, $3p\sigma$) configuration has three components ($^1A_2'$, $^1A_1'$ and $2^1E'$), with one-photon transitions allowed only to $2^1E'$. The structured feature (61 000–63 000 cm^{-1}, *7.6–7.8 eV*) is a vibronically allowed transition either to $^1A_2'$ or $^1A_1'$, while the MCD band peaking at 65 400 cm^{-1} is the transition to $2^1E'$. The above explanation, given by Goldstein *et al.* [G22] on the basis of *ab initio* calculation, reinterprets the experimental work of Gedanken and Schnepp [G7], who otherwise assigned the two MCD peaks at 61 500 and 65 400 cm^{-1} as a valence A term rather than two Rydberg B terms. In *trans*-1,2-dimethyl cyclopropane [G7], two CD peaks of opposite sign are recorded at 50 000 and 52 600 cm^{-1} (*vert., 6.2 and 6.52 eV*) and probably represent transitions to 3s from the MO's which form the degenerate $3e'$ set in cyclopropane. This can be checked by comparing the splitting in the first band of the photoelectron spectrum of the dimethyl derivative with that in the CD spectrum.

The absorption feature at 69 000 cm^{-1} (*vert., 8.55 eV*) in cyclopropane has been reassigned by Gedanken and Schnepp as valence $1e'' \rightarrow 4e'$, by Fridh [F32] as Rydberg (unspecified), and by Goldstein *et al.* [G22] as a Rydberg terminating at 3d. Even the assignment of the valence band at 83 100 cm^{-1} (*vert., 10.3 eV*) with its huge oscillator strength of 0.7, Fig. III.E-1, has been called into question by Goldstein *et al.*, who compute that this region should be home to intense $1e'' \rightarrow 3p$ Rydberg excitations. A highly structured absorption in cyclopropane at 107 000 cm^{-1} (*vert., 13.3 eV*) [F32, K36] is the $a_2''\pi \rightarrow 3s$ Rydberg transition associated with the highly structured $(a_2''\pi)^{-1}$ PES band at 134 000 cm^{-1} (*vert., 16.6 eV*).

Even if the Rydberg excitations predicted by Goldstein *et al.* [G22] did fall at 83 000 cm^{-1} in cyclopropane, the observed oscillator strength is just too large for them alone and so there still must be an intense valence excitation at the same frequency. The spectrum of a solid film of cyclopropane [I, 141] convincingly shows that the 83 000-cm^{-1} peak indeed is valence. How is it that evidence for valence excitations is almost

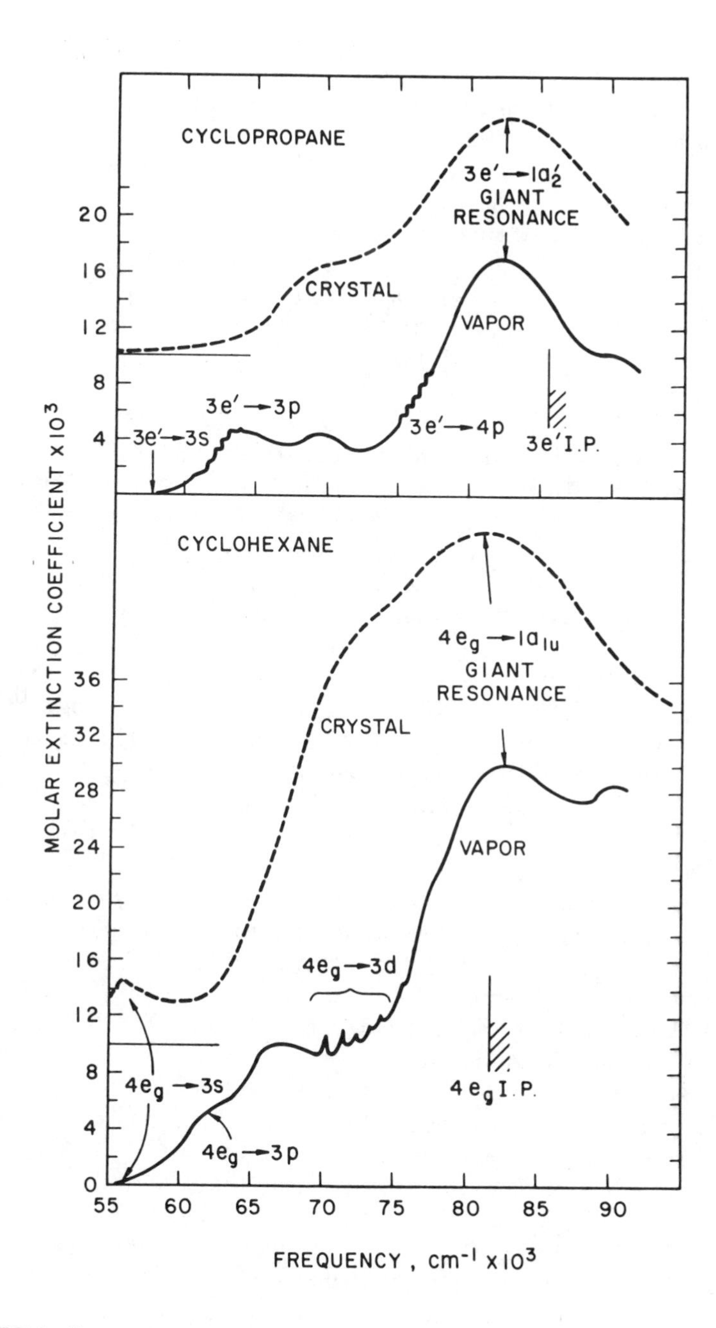

Fig. III.E-1. Spectra in the vapor and solid phases illustrating d → f giant resonances in cyclopropane and cyclohexane [R36].

nonexistent in alkanes, yet in cyclopropane we have a transition as undeniably valence as the N → V band of ethylene, if not even more so?

The answer lies in the high symmetry of cyclopropane. According to the orbital calculations, the valence excitation at 83 000 cm^{-1} is $3e'\sigma \rightarrow 2a_2'\sigma^*$ in the D_{3h} point group. Because the $2a_2'\sigma^*$ antibonding valence orbital has three σ_v nodal planes, it has the symmetry of an f atomic orbital and its conjugate Rydberg configuration is therefore $(3e', nfa_2')$. However, as discussed in Section I.D, the near-zero penetration of nf wavefunctions will keep the conjugate $1a_2'\sigma^*$ and nfa_2' functions spatially separate and thereby unmixed. Consequently the $3e'\sigma \rightarrow 1a_2'\sigma^*$ transition maintains its valence character in spite of the near degeneracy with its conjugate Rydberg sea.

When considered from the point of view of the double-well effective potential of Section I.D, we see immediately that the 83 000-cm^{-1} band of cyclopropane, Fig. III.E-1, represents the molecular analog of upper-state 4f wavefunction collapse, and because the $3e'$ MO has d-orbital symmetry, the concomitant d → f giant resonance [R36]. Apparently the modulating effect of the molecular geometry is able to pull the "nonRydberg" phenomenon from the large-Z end of the periodic table into the first row!

The frequency of a temporary-negative-ion resonance in cyclopropane observed at 40 300 cm^{-1} (*vert., 5.0 eV*) [F32] is too low to be a $^2(3e', 3s^2)$ Feshbach resonance, Section II.B, and so the transition is assigned instead as a valence TNI resonance. Because 40 300 cm^{-1} is a high frequency for the lowest valence TNI resonance in an alkane, one guesses that at least one resonance at *ca.* 15 000 cm^{-1} (*2 eV*) and probably several others between 15 000 and 40 000 cm^{-1} (*2 and 5 eV*) have yet to be uncovered in cyclopropane.

The lowest Rydberg excitations in several high-symmetry cyclic alkanes are electronically forbidden in one-photon absorption, but allowed for two-photon absorption. Because of the molecular rigidity, these two-photon resonances in the MPI spectra are expected to be spectrally well defined. In cyclohexane [**I**, 147], the lowest Rydberg excitation ($e_g \rightarrow$ 3s) appears weakly in the one-photon absorption spectrum via a false origin at 57 130 cm^{-1}. The MPI spectrum of cyclohexane first was said to show the $e_g \rightarrow$ 3s transition as a two-photon resonance with the origin at 55 730 cm^{-1} (*6.909 eV*) [H29], however by using a jet-cooled molecular beam, this feature has been shown to be a hot band instead, with the true origin at 56 766 cm^{-1} (*advert., 7.0379 eV*) in C_6H_{12}, and at 57 146 cm^{-1} (*advert., 7.0850 eV*) in C_6D_{12} [W14]. The vibronics of this transition are complicated by nonlinear vibronic coupling and by what Whetten and Grant [W14] call "Jahn-Teller vibronic pseudorotation." In liquid cyclohexane, there is a sharp one-photon absorption edge at 57 500 cm^{-1}

(*7.13 eV*) peaking at 60 000 cm^{-1} (*7.44 eV*) which corresponds to the externally perturbed $e_g \rightarrow$ 3s promotion [K19]. Emission from the (e_g, 3s) state of cyclohexane competes with intersystem crossing on the nanosecond time scale [F11].

The electron-impact spectrum of cyclohexane [K16] resembles the one-photon optical spectrum, and though it is of poorer resolution, it also is of greater extent. As with all alkanes, there is a broad and intense peak centered at 125 000 cm^{-1} (*15.5 eV*), which is most likely due to a maximum in the photoionization cross section. In the solid phase, the electron-impact spectrum of cyclohexane [K16] still shows considerable structure, however this cannot be placed in one-to-one correspondence with the vapor-phase spectrum with any confidence. Most interestingly, the 125 000-cm^{-1} prominence in the vapor phase is quenched totally in the solid, whereas the intense feature at 84 000 cm^{-1} (*vert., 10.4 eV*) in the vapor spectrum [**I**, 148] appears at 81 000 cm^{-1} (*vert., 10.0 eV*) in the solid and at 78 000 cm^{-1} (*vert., 9.7 eV*) in the liquid [K19]. The quenching of the 125 000-cm^{-1} "peak" in the solid suggests that it might be due to the delayed onset of the photoionization maximum (Section II.C), while the persistence of the peak at 84 000 cm^{-1} (Fig. III.E-1) strongly suggests that it is a valence excitation, perhaps related to that in the same region of cyclopropane. In fact, in the allowed valence excitation $e_g \rightarrow a_{2u}$, the e_g and a_{2u} orbitals transform as atomic d and f orbitals, respectively. Thus in the 84 000-cm^{-1} band of cyclohexane we have a possible third example of 4f wavefunction collapse and a consequent d $\rightarrow$ f giant resonance in an alkane (Section I.D) [R36].

In *trans*-decalin, norbornane and bicyclo[2.2.2]octane [**I**, 151] the lowest excitations to 3s are clearly revealed as two-photon resonances in the MPI spectra, while otherwise being one-photon forbidden [H29]. The $4a_2 \rightarrow$ 3s MPI excitation in norbornane (Fig. II.A-2) is accompanied by considerable excitation of ν_{15}, the wing-flapping motion, and in this regard closely resembles the lowest Rydberg excitations in norbornadiene and Dewar benzene (Section XVII.C).

The one- and two-photon spectra of adamantane [H29; **I**, 150] both commence with a sharp-line origin at 52 345 cm^{-1} (*6.4898 eV*) having a typical 3s term value (22 700 cm^{-1}). In the T_d symmetry of adamantane, only excitation to a T_2 upper state can be both one- and two-photon allowed, and so the originating orbital must have t_2 symmetry if the transition is to terminate at 3s. This is confirmed by MO calculations. All assignable hot bands in the adamantane MPI spectrum are nontotally symmetric, signaling significant Jahn-Teller effects in the T_2 upper state.

Observation of the two-photon excitations to the 3s levels of the larger cyclic alkanes allows one to demonstrate that the depressing effects of alkyl groups on the 3s term values occur as well for alkanes themselves as for other alkylated chromophores, Table III.D-II. In the largest alkanes, the 3s term value converges upon *ca.* 22 000 cm^{-1}, just as it does in all other heavily alkylated chromophores [**I**, 51].

CHAPTER IV

Alkyl Halides

Spectra of the alkyl halides are distinguished from those of the alkanes in large part by the low-lying Rydberg excitations which originate with lone-pair (np) halogen electrons; because such transitions are nonbonding → nonbonding, they are unusually sharp and Franck-Condon vertical. Moreover, in the alkyl halides, the σ^*(C–X) valence orbitals are accessible at low frequencies, in contrast to the σ^*(C–H) and σ^*(C–C) MO's of the alkanes. Thus most alkyl halides show low-lying lone pair → σ^*(C–X) transitions preceding the lowest Rydberg excitation. Valence transitions of this sort, here classified as $\mathscr{A}(n\mathrm{p}, \sigma^*)$ bands (Section I.C), are listed in Table I.C-III for a number of halide-containing systems. As discussed in Section I.C, even though the $\mathscr{A}$ bands are valence excitations, there is an obvious regularity in their term values, and so we shall discuss their frequencies from this point of view. Several general trends are apparent in the $\mathscr{A}(n\mathrm{p}, \sigma^*)$ term values of the alkyl halides listed in the Table. Thus it is seen that in the series $CH_{4-y}X_y$, the $\mathscr{A}$-band term values increase as y goes from 1 to 4, that the term values increase as H is replaced by F, and that in general for a given degree of halogenation, the term values increase as one goes down a column of the periodic table. In addition to the $\mathscr{A}(n\mathrm{p}, \sigma^*)$ bands, vibronically structured σ(H–X) → σ^*(H–X) bands have been identified in the hydrogen halides

[J1, S70] at relatively low frequencies (76 210 cm^{-1} *adiabat., 9.448 eV* in HCl) and the corresponding σ(C–X) → σ^*(C–X) transitions are expected in the alkyl halides at approximately the same frequencies. Note as well the earlier assignment [**I**, 190] of intense $\sigma \rightarrow \sigma^*$ bands in the longer perfluoroalkanes.

In a prototypical heavy-atom alkyl halide such as CH_3X, with X = Cl, Br or I, the $\mathscr{A}$(np, σ^*) band is followed by the np → (n + 1)s Rydberg excitation. Through spin-orbit coupling and exchange effects, the (np, (n + 1)s) configuration is split into four states:- $^3E_{(2)}$, $^3E_{(1)}$, $^3E_{(0\pm)}$ and $^1E_{(1)}$, in order of increasing energy [G35, M44].† The splitting between the $^3E_{(1)}$ and $^1E_{(1)}$ states is equal to that between the corresponding $^2E_{3/2}$ and $^2E_{1/2}$ ionic states. Transitions to all four components are detectable when X = Br or I, whereas for X = Cl only the $^3E_{(1)} - {}^1E_{(1)}$ splitting is observed and for X = F, one observes only the transitions to configurations ^{3}(2p, 3s) and ^{1}(2p, 3s). The above generalities hold for inner-shell and outer-shell spectra alike, except for the case of X = F, where the orbital ordering may depend on the originating MO. Note that the np → (n + 1)s Rydberg transition and the $\mathscr{A}$(np, σ^*(C–X)) band transition are Rydberg/valence conjugates and that the four-fold spin-orbit splitting of the Rydberg configuration discussed above also will occur for the conjugate valence configuration; however, because of their broadness, these components of the $\mathscr{A}$(np, σ^*) band are rarely resolved.

IV.A. Alkyl Chlorides

In methyl chloride [**I**, 158], the $\mathscr{A}$(3p, σ^*) band in the MCD spectrum shows a B term centered at 57 400 cm^{-1} (*7.12 eV*); a second peak of opposite sign at 61 500 cm^{-1} (*7.62 eV*) is due to the ν_2'' hot band of the 3p → 4s origin [F2]. The $\mathscr{A}$(3p, σ^*)-band term value of 31 000 cm^{-1} in methyl chloride is meaningless by itself, but is valuable as a standard against which to measure the term values of other halogen-containing molecules.

Unlike the situation in the other methyl halides, the 3p → 4s Rydberg transition in methyl chloride is vibronically poorly defined, and even

† Actually, the $^3E_{(0\pm)}$ level consists of two nearly degenerate levels of A_1 and A_2 symmetry which never have been resolved.

though deuterium substitution has been used, the vibronic assignments within this band remain uncertain [F2]. Though an anomalous $E_{3/2} - E_{1/2}$ splitting of 954 cm^{-1} has been identified in the 3p → 4s band, this interval achieves a more normal value of *ca.* 625 cm^{-1} in the higher-*n* Rydberg members [F2, H67]. Felps *et al.* [F2] similarly deduce the $E_{3/2} - E_{1/2}$ splitting in the higher members of the 3p → *n*p series of methyl chloride ($n \geqslant 4$) and Truch *et al.* [T21] report that the $n = 4$ and 5 members of each of the $E_{3/2}$ and $E_{1/2}$ core configurations of (3p, *n*p) are split into resolvable npe and npa_1 components. Strangely, it is found that $4pe$ is below $4pa_1$, but $5pe$ is above $5pa_1$. Perhaps this reflects the elevation of (3p, $4pa_1$) by interaction with (3p, $4sa_1$) and with its valence conjugate (3p, σ^*), whereas (3p, $5pa_1$) is not mixed in this way.

A quantitative optical study of methyl chloride [W37] in the deep ultraviolet reveals four new excited states. Broad, structureless transitions at 139 000 and 150 000 cm^{-1} (*vert., 17.2 and 18.6 eV*) originate at the $2sa_1$ MO of the methyl group (ionization potential of 173 000 cm^{-1} *vert., 21.5 eV*) and terminate at 3s and 3p Rydberg orbitals, respectively.† The second of these correlates with the prominent $2sa_1$ → 3p excitation at 165 000 cm^{-1} (*vert., 20.40 eV*) in methane, whereas the first is symmetry forbidden in methane and appears only as a weak shoulder (157 000 cm^{-1} *vert., 19.5 eV*) in that molecule (Section III.A). Assignments of a vibronically structured band at *ca.* 166 000 cm^{-1} (*vert., 20.6 eV*) and a smooth band at 225 000 cm^{-1} (*vert., 27.9 eV*) in methyl chloride are uncertain, however the latter is very close to the $(3s_{Cl})^{-1}$ ionization limit and may correspond to a maximum in the photoionization cross section for the 3s → εp channel.

Reflectivity measurements on liquid CCl_4 (see [I, 175]) place the peak of the $\mathscr{A}$(3p, σ^*) band at 61 300 cm^{-1} (*7.60 eV*) and the subsequent $\sigma \rightarrow \sigma^*$ valence excitation at 72 000 cm^{-1} (*vert., 8.9 eV*) [A19]. The $\mathscr{A}$(3p, σ^*) band at 57 100 cm^{-1} (*vert., 7.08 eV*) in vapor-phase carbon tetrachloride has an *A* term in the MCD spectrum appropriate to a degenerate upper state, and is assigned as $3t_2 \rightarrow 3a_1\sigma^*$ [R47]. Intense energy losses at 81 000 and 113 000 cm^{-1} (*vert., 10 and 14 eV*) in the electron impact spectrum of CCl_4 have been shown to be electric-dipole allowed, $^1A_1 \rightarrow {}^1T_2$ [L12]; the first of these has been assigned as $\sigma \rightarrow \sigma^*$ [I, 176]. Since only three $^1A_1 \rightarrow {}^1T_2(\sigma \rightarrow \sigma^*)$ excitations are possible in

† The lowest Rydberg orbital is here taken to have $n = 3$ since it is the CH_3 portion of the molecule which is excited.

CCl_4, all three are accounted for provided the band at 72 000 cm^{-1} in the liquid is distinct from that at 81 000 cm^{-1} in the vapor. Between 121 000 and 240 000 cm^{-1} (*15 and 30 eV*), there are several unassigned, sharp energy-loss features in carbon tetrachloride, including a nice Fano antiresonance at 181 000 cm^{-1} (*22.5 eV*). The bands at 129 000 and 141 000 cm^{-1} (*vert., 16.0 and 17.5 eV*) are affiliated with the MO ionizing at 160 000 cm^{-1} (*vert., 19.9 eV*), and terminate at 4s and 4p Rydberg orbitals, respectively, as judged by their term values.

Unfortunately, in conventional spectroscopy, the manifold of σ^*(C–X) levels in most alkyl halides is in part overlaid by Rydberg excitations and so is partially masked. Though the most direct indication of just where the σ^*(C–Cl) orbitals are located in the chloromethanes comes from the electron-transmission spectroscopic work of Burrow *et al.* [B85], Fig. IV.A-1, interpretations of these valence temporary-negative-ion (TNI) resonances (Section II.B) as involving electron capture in σ^*(C–Cl) MO's (see, for example, [P25]) are in some cases questionable. Especially suspicious are the claims for very large splittings between pairs of σ^*(C–Cl) group orbitals. It is these large splittings which contrast with our interpretation of the neutral-molecule valence excitations in the chloromethanes (*vide infra*).

In addition to valence TNI resonances, several other types of electron-impact resonance are observed in the alkyl halides. In the chloromethanes, neutral-molecule resonances are observed 2400–4000 cm^{-1} (*0.3–0.5 eV*) below the lowest singlet $\mathscr{A}$(3p, σ^*) frequencies, and are thought to involve ^{3}(3p, σ^*) valence configurations [V21]. Temporary negative-ion Feshbach resonances are observed in methyl chloride [60 000 cm^{-1} *vert., 7.5 eV*; $^2(3pe, 4s^2)$] [S65, V21] and carbon tetrachloride [64 000 cm^{-1} *vert., 8.0 eV*; $^2(3pt_1, 4s^2)$] [V21]. Related excitations also are reported for carbon tetrafluoride (Section IV.D). Each of these Feshbach resonances appropriately is *ca.* 4000 cm^{-1} below the corresponding neutral-molecule Rydberg configuration (Section II.B).

Going beyond the lowest Feshbach resonances to $^2(np, (n + 1)s^2)$ negative-ion configurations, other higher resonances have been recorded but left unassigned in CH_3I, CH_3Br and CH_3Cl [S65]. Spence has shown that among the hydrogen halides and methyl halides, and also among the rare gases, there is a linear relationship between ground-state ionization potentials and the Feshbach resonance frequencies [S64, S65], Fig. II.B-2. The unassigned higher resonances fit nicely onto this chart. Thus the resonance at 77 800 cm^{-1} (*vert., 9.65 eV*) in CH_3Cl has the Feshbach configuration $^2(3p, 3d^2)$. A second resonance in CH_3Cl at 72 300 cm^{-1}

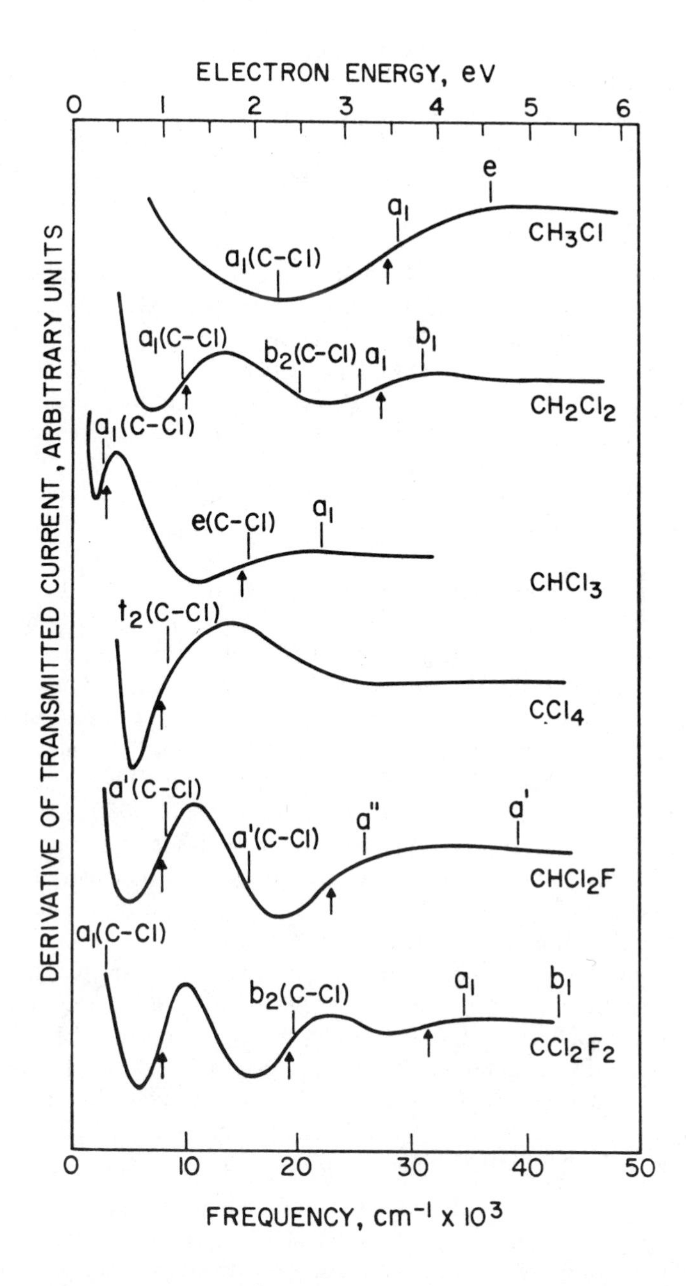

Fig. IV.A-1. Valence temporary-negative-ion spectra of the chloromethanes, $CHCl_2F$ and CCl_2F_2 [B85]. The arrows indicate the midpoints of the resonance frequencies while the vertical bars represent the calculated resonance frequencies. Those resonances not otherwise specified involve either σ^*(C–H) or σ^*(C–F) orbitals.

(*vert., 8.96 eV*) lies neatly between those for $^2(3p, 4p^2)$ and $^2(3p, 3d^2)$ and so is assigned as $^2(3p, 4p3d)$.†

The $1s_C$ inner-shell spectra of the chloromethanes are a fascinating blend of transitions terminating at Rydberg and σ^* valence orbitals, Fig. IV.A-2 [H59]. In CCl_4, no traces of the forbidden $1s_C \rightarrow 4s$ or the allowed $1s_C \rightarrow 4p$ transitions appear; the spectrum is dominated by the forbidden $1s_C \rightarrow a_1\sigma^*$(C–Cl) and allowed $1s_C \rightarrow t_2\sigma^*$(C–Cl) valence excitations. These $\mathscr{A}(1s_C, \sigma^*)$ bands are observed to be split by 7200 cm^{-1} in the $1s_C$ inner-shell spectrum, while in the TNI spectrum [B85] of CCl_4 the $a_1\sigma^*$ and $t_2\sigma^*$ MO's are separated by 7600 cm^{-1}. This close correspondence is very encouraging. In MCD work on the carbon tetrahalides, Rowe and Gedanken [R47] raise the question of the ordering of the $a_1\sigma^*$ and $t_2\sigma^*$ MO's in these systems; it is abundantly clear that in carbon tetrachloride $a_1\sigma^*$ is below $t_2\sigma^*$ and that the σ^* ordering in CBr_4 and CI_4 can be ascertained by assigning their $1s_C$ spectra. As can be seen from the term values, the orbital ordering in CCl_4 is in Regime Z, Fig. I.D-3.

In each of the chloromethanes CH_2Cl_2, $CHCl_3$ and CCl_4, the σ^*(C–Cl) manifold splits into two components, and we argue on general grounds that this splitting should be largest in CCl_4 (7600 cm^{-1}) and smallest in CH_2Cl_2 (if the C–H and C–Cl bond orbitals are relatively unmixed). In fact, the data of Fig. IV.A-2 show this to be the case among the $\mathscr{A}(1s_C, \sigma^*)$ transitions, broad and overlapped though they may be. Note however, that in the TNI spectra of these compounds, Fig. IV.A-1, the σ^* splitting is assigned as 17 300 cm^{-1} in CH_2Cl_2 and 11 900 cm^{-1} in $CHCl_3$. Because of this discordance with the inner-shell spectral splittings, it is suggested that some of the higher-frequency TNI resonances in these systems involve σ^*(C–H) rather than σ^*(C–Cl) MO's and that certain of the σ^*(C–Cl) resonances are too broad to be seen.

Measurements of the photoelectron angular distribution parameter β in the chloromethanes as a function of frequency showed local minima in the region of 160 000 cm^{-1} (*20 eV*) which were tentatively attributed to σ^* shape resonances originating at the $3p_{Cl}$ lone-pair orbitals [K11]. With the deepest $3p_{Cl}$ level in CCl_4 at 108 400 cm^{-1} (*vert., 13.44 eV*), clearly the highest frequency $\mathscr{A}(3p_{Cl}, \sigma^*)$ band will come at 70 000 cm^{-1}. Consequently, the perturbations to β at 160 000 cm^{-1} cannot involve

† Hybrid molecular resonances of this sort in which the two Rydberg electrons are in orbitals of different angular momentum are discussed for the hydrogen halides by Spence and Noguchi [S62].

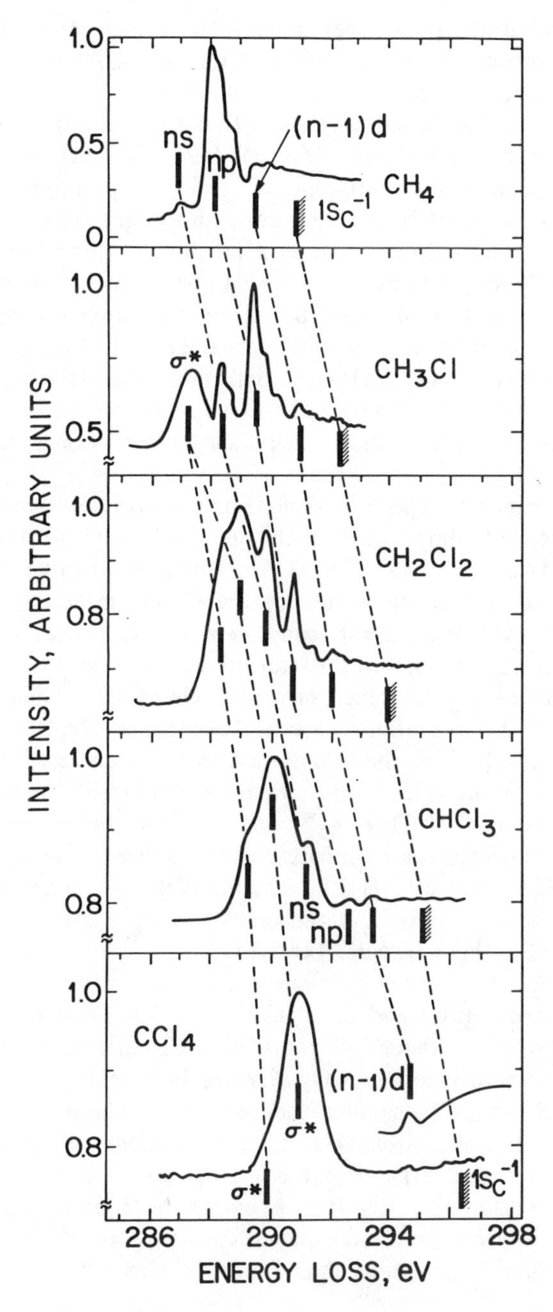

Fig. IV.A-2. Correlation of the energy-loss bands of methane and the chloromethanes (2500 eV impact energy and 0° scattering) originating at $1s_C$ [H59].

$(3p_{Cl}, \sigma^*)$ configurations. Cooper minima [M48] are reported in the lone pair photoionization cross sections of CCl_4 at 320 000–400 000 cm^{-1} (*40–50 eV*) [C6].

In addition to the broad $\mathscr{A}(1s_C, \sigma^*)$ valence bands, the inner-shell spectra of the chloromethanes (Fig. IV.A-2) also display many sharp features having term values appropriate for Rydberg assignments. In the series from CH_4 to $CHCl_3$, the intensity ratios of the transitions from $1s_C$ to 3s and 3p increase strongly but in a regular way, and core splitting of the $(1s_C, 3p)$ configurations is evident in the spectra of CH_2Cl_2 and CH_3Cl. The $1s_C$ inner-shell spectrum of ethyl chloride can be interpreted in terms of the inner-shell spectra of CH_4 and CH_3Cl, again with considerable overlap. {The $(1s_C)^{-1}$ ionization potentials in C_2H_5Cl are separated by 8000 cm^{-1} [H59]}. The first few Rydberg excitations in the $1s_C$ spectrum of CH_3Cl show well-resolved vibrational bands of a_1 symmetry [H61].

In the chlorine 2p spectral region, the absorption patterns of the chloromethanes are complicated by the $^2P_{3/2} - {}^2P_{1/2}$ spin-orbit splitting of 13 000 cm^{-1} (*1.6 eV*) [C36, H59, O15]. However, in this and in the $2s_{Cl}$ spectra, the lowest-frequency valence transitions to σ^* are still easily detected, and the $\mathscr{A}(2p_{Cl}, \sigma^*)$ bands are found to have term values somewhat larger than when the transition originates at $1s_C$, whereas $(2s_{Cl}, \sigma^*)$ term values are much larger, Table I.C-II. There are modest modulations of the ionization continua beyond the $(2p_{Cl})^{-1}$ and $(1s_C)^{-1}$ ionization thresholds in the chloromethanes which are attributed to EXAFS interferences [H59], as in carbon tetrafluoride [B66]. The $1s_{Cl}$ spectra of CH_3Cl, C_2H_5Cl, CF_2Cl_2 and C_2H_3Cl are too broad to show much detail, but what are apparently the valence $\mathscr{A}(1s_{Cl}, \sigma^*)$ bands appear at 2823–2825 eV [H10]. The 46 000-cm^{-1} term value of the $\mathscr{A}(1s_{Cl}, \sigma^*)$ band of CH_3Cl is consonant with the other $\mathscr{A}$-band term values observed in this molecule, Table I.C-II.

The trends of the various $\mathscr{A}$-band term values in the chloromethanes, Table I.C-II, are regular and most instructive. One sees first that as the hydrogen atoms of methane are replaced with chlorine atoms, the term value for a given originating MO increases in a regular way. We have argued as well that the smaller the principal quantum number of the originating orbital, the stronger is the antishielding effect (Section I.C) and consequently, the larger is the σ^* term value. This effect is clearly seen as an increasing (ϕ_i^o, σ^*) term value as the ϕ_i level descends in the chlorine atom. The largest value should occur in CCl_4 for $1s_{Cl}$ excitation, where the $(1s_{Cl}, \sigma^*)$ term value should be approximately 60 000 cm^{-1}.

Spectra of the fluorinated chloromethanes are discussed in Section IV.E.

IV.B. Alkyl Bromides

The MCD spectrum of CH_3Br in the $\mathscr{A}$(4p, σ^*) region shows negative *B* terms at 44 000 and 49 000 cm^{-1} (*vert., 5.4 and 6.1 eV*) [G5]; these are two of the four spin-orbit components of the 4p $\rightarrow \sigma^*$ transition in methyl bromide. Of the four components of the conjugate 4p $\rightarrow$ 5s transition in methyl bromide, transitions to $^3E_1$ and $^1E_1$ are expected to show *A* terms in the MCD spectrum, with equal upper-state magnetic moments [M43]. This is as observed, proving that the origins at 56 030 and 59 165 cm^{-1} (*6.9467 and 7.3354 eV*) have $^3E_1$ and $^1E_1$ upper-state symmetries, respectively.

In a molecule containing a single Br atom, the transition $4p\pi^4 \rightarrow 4p\pi^3 5s^1$ will split into four components under spin-orbit (ζ) interaction and into two components under the exchange (K) interaction. The relative importance of these two factors have been tabulated by Felps *et al.* [F6] for a series of compounds, with interesting results. Analysis of the spectra [F2, H67] show that HBr and the alkyl bromides are in the intermediate coupling regime where $K/\zeta = 0.3-1.0$. In the series from HBr to C_3H_7Br, the ζ value is constant at 1280 $\pm$ 50 cm^{-1}, as deduced from the splitting of the (4p, 5s) configuration. Since this involves spin-orbit coupling in the core, it involves only the originating orbital of the 4p $\rightarrow$ 5s transition, and therefore is insensitive to the hydrogen or alkyl-group substitution. On the other hand, the exchange integral K(4p, 5s) drops monotonically from 1146 cm^{-1} in HBr to 438 cm^{-1} in C_3H_7Br. With the 4p orbital pinned to the bromine atom, the decreasing value of the K integral $\langle 4p5s|4p5s \rangle$ indicates the dramatically smaller 4p $-$ 5s overlap in the presence of larger alkyl groups. This is totally in accord with our earlier conclusions [I, 51] about the delocalization of Rydberg orbitals over alkyl groups and the convergence of the term values in highly alkylated systems, and with the proposal (Section I.B) that the singlet-triplet exchange splitting will decrease as the term value decreases.

Rydberg series are observed converging on the $^2E_{3/2}$ and $^2E_{1/2}$ ionization limits of methyl bromide [I, 162]. Those transitions converging on $^2E_{3/2}$ are sharp (assigned as 4p $\rightarrow$ nd; $n \leqslant 32$) whereas those which are above the $^2E_{3/2}$ limit and converge on $^2E_{1/2}$ are broad and show definite signs of antiresonance profiles [C41]; the members of this second series (also assigned as 4p $\rightarrow$ nd) below the $^2E_{3/2}$ limit show broad profiles and intensity anomalies according to Baig *et al.* [B6]. There is clearly an interaction here between the two channels leading to $^2E_{3/2}$ and $^2E_{1/2}$ ions. Because of its poorer definition, the autoionizing 4p → *n*d series going to $^2E_{1/2}$ can be delineated only up to $n = 15$. The observation of transitions terminating at such high *n* values is due in large part to the highly vertical nature of the 4p → *n*d Franck-Condon factors [C9]. However, see below for an alternate assignment of these bands.

In addition to the 4p → *n*s series having $E_{3/2}$ and $E_{1/2}$ core configurations earlier reported by Price for CH_3Br [I, P33], Causley and Russell [C9] report the corresponding 4p → *n*p series to high *n*, but report transitions to 4d with $E_{3/2}$ and $E_{1/2}$ cores without any higher members of the series discernible. This is odd, for the long series at high frequencies reported by Baig *et al.* [B6] is said by them to involve 4p → *n*d excitations. In the region of common overlap, the supposed 4p → *n*d transition frequencies of Baig *et al.* closely match those assigned as 4p → *n*s by Price and by Causley and Russell. The assignments as 4p → *n*s seem secure, for the series in question extrapolates backward to what is clearly the 4p → 5s transition rather than to 4p → 4d. The 4p → *n*s assignments moreover are energetically consistent with those of the 4p → *n*p series [C9] as laid out by Causley and Russell, whereas this is not so for 4p → *n*d assignments. Thus it is concluded that the nth members of the 4p → *n*s series have been misassigned [B6] as the $(n - 1)$ members of the 4p → *n*d series. The transitions to the (4p, 5s) and (4p, 5p) configurations with $E_{3/2}$ cores involve the excitation of single vibrational quanta of ν_4' and ν_6' of *e* symmetry, suggesting a Jahn-Teller interaction in these states [C9, F2]. Strangely, the transitions to the corresponding states having the $^2E_{1/2}$ core do not show these *e* vibrations [C9].

The $1s_C$ inner-shell spectrum of methyl bromide, Fig. IV.B-1, displays the usual broad $1s_C \rightarrow \sigma^*$ valence excitation, followed by Rydberg excitations to 4s and 4p showing vibrational structure. In accord with the dictum that the term value will increase as the hole descends, Section I.C, the 34 000-cm^{-1} term value of the $\mathscr{A}(4p, \sigma^*)$ band of methyl bromide becomes 42 700 cm^{-1} for the $\mathscr{A}(1s_C, \sigma^*)$ band and 44 400 cm^{-1} for the $\mathscr{A}(3d_{Br}, \sigma^*)$ band, Table I.C-II [H57, H61]. Similarly the (4p, 5s) Rydberg term value of 28 940 cm^{-1} becomes 29 800 cm^{-1} in the $(1s_C, 5s)$ state of CH_3Br, however the $3d_{Br} \rightarrow 5s$ transition is not observed, largely for symmetry reasons. The transitions to 5p are split into *e* and a_1 components separated by 5200 cm^{-1} in the $1s_C$ spectrum and 4000 cm^{-1} in the $3d_{Br}$ spectrum. As regards orbital ordering, the alkyl bromides are securely in Regime Z (Section I.D). The 3d → ϵf ionization continuum in CH_3Br shows a delayed onset with the peak approximately 30 eV beyond threshold [H57].

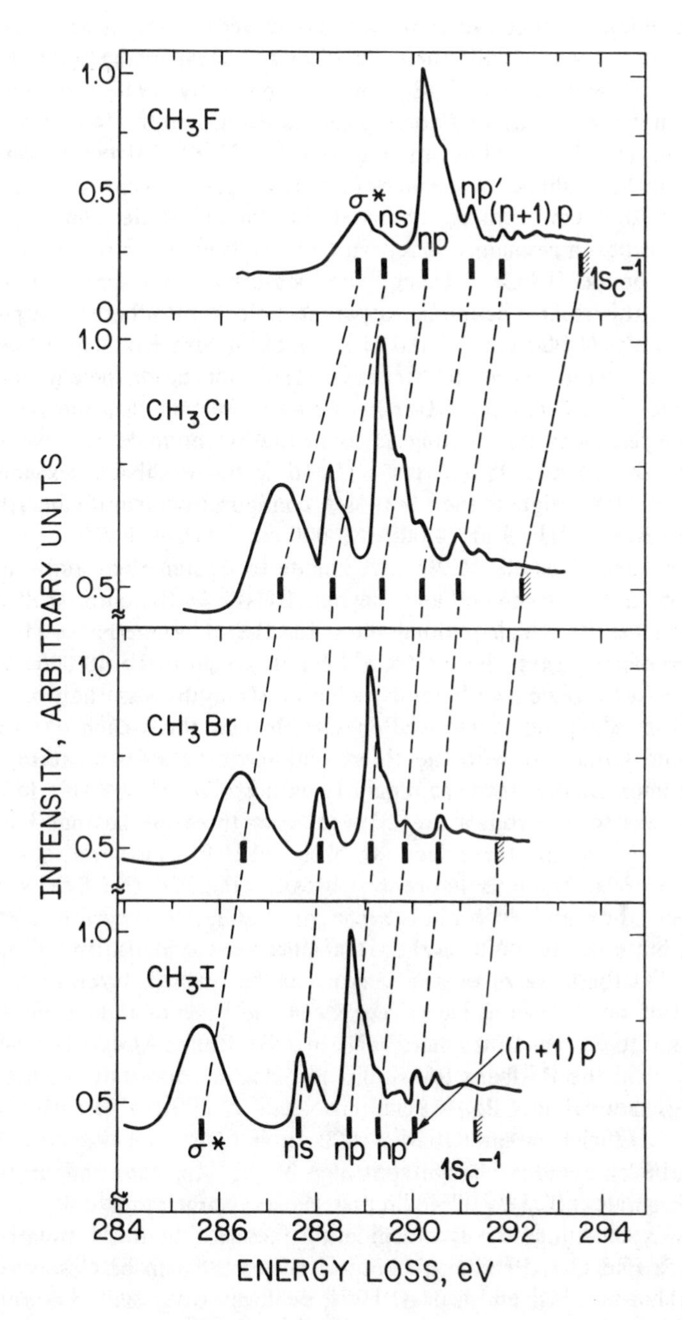

Fig. IV.B-1. Electron impact energy-loss spectra of the methyl halides originating at the $1s_C$ orbitals at 2500 eV impact energy and 0° scattering angle [H57, H61].

Negative-ion Feshbach resonances are observed in CH_3Br at 53 840 and 56 660 cm^{-1} (*vert., 6.675 and 7.025 eV*) corresponding to $^2(4p, 5s^2)$ configurations with $^2E_{3/2}$ and $^2E_{1/2}$ cores, respectively, while a resonance at 71 100 cm^{-1} (*vert., 8.81 eV*) corresponds to excitation to $^2(4p, 3d^2)$ with a $^2E_{1/2}$ core [S65] according to Fig. II.B-2. The 2600-cm^{-1} spin-orbit splitting of the Feshbach resonances in methyl bromide compares favorably with the 2800-cm^{-1} value observed in the photoelectron spectrum. Another Feshbach resonance is reported at 76 800 cm^{-1} (*vert., 9.52 eV*) in methyl bromide [S65]. Using the standard 4000-cm^{-1} Feshbach decrement, this resonance implies a parent-molecule Rydberg absorption at approximately 81 000 cm^{-1} (*10.0 eV*), which in turn has a term value of 28 000 cm^{-1} with respect to the $e\sigma(C{-}H)^{-1}$ ionization potential of the CH_3 group at 109 000 cm^{-1} (*vert., 13.5 eV*). Noting that the two lower Rydberg excitations terminating at 5s in methyl bromide also have term values of 28 940 and 28 810 cm^{-1} [**I**, 166], the Feshbach resonance at 76 800 cm^{-1} is assigned the $^2(e\sigma, 3s^2)$ configuration, paralleling that in methane (Section III.A) at 74 000–82 000 cm^{-1} (*9.2–10 eV*).

Causley and Russell [C9] have made a systematic study of the absorption in the bromomethanes beyond CH_3Br in the outer-shell region and report results which parallel those for the chloromethanes [**II**, 305]. In the bromomethanes, the $\mathscr{A}(4p, \sigma^*)$ bands come in the 40 000–55 000 cm^{-1} (*5–7 eV*) region with total oscillator strengths less than 0.1. One such $\mathscr{A}(4p, \sigma^*)$ band is resolved in CH_3Br and three each are seen in $CHBr_3$ and CBr_4. As with the chloromethanes, the $\mathscr{A}$-band term values in the bromomethanes increase regularly with bromination, Table I.C-III.

Transitions to *n*s orbitals are allowed from three of the four lone-pair MO's of CH_2Br_2, and three such are observed [C9]. However, absorption in the 4p → 5s region is noticeably broader than in CH_3Br. Whereas excitations to 5p and 4d orbitals are weak in CH_3Br, they are intense in CH_2Br_2, but no series built on these transitions are evident. In $CHBr_3$ and CBr_4, no Rydberg series of any length can be divined, however the first members of such series can be assigned on the basis of their term values. Transitions to 5s are broad here, and in CBr_4 the $\mathscr{A}(4p, \sigma^*)$ bands are narrower than the Rydberg transitions to 5s! An amazingly intense band (4p → 5p) appears in CBr_4 (65 340 cm^{-1} *advert., 8.101 eV*) with a molar extinction coefficient greater than 50 000 ($f = 0.27$).

The spin-orbit and exchange splittings of the (4p, 5s) configuration of methyl bromide will be altered in the lower symmetry of ethyl bromide provided hyperconjugation is a competing factor. Detailed comparison of the CH_3Br and C_2H_5Br vibronic spectra shows them to be closely related (as are those of CH_3I and C_2H_5I) [B7], demonstrating that spin-orbit and exchange effects are still dominant compared to the effect of hyperconjugation. An interesting vibronic effect appears in C_2H_5Br where the vibronic transition to $^3E_1 + 1\nu_1'$ is zero-order degenerate with the electronic transition to $^1E_1$. The consequent vibronic mixing intensifies the former and splits the two bands by 270 cm^{-1} [F4]. In C_2D_5Br, the ν_1' frequency is much lower and the vibronic resonance is destroyed. Vibronic splittings similar to that in C_2H_5Br also are seen in methyl iodide, Section IV.C.

The spectra of the fluorinated alkyl bromides are discussed in Section IV.E.

IV.C. Alkyl Iodides

The smooth, symmetric contour of the $\mathscr{A}(5p, \sigma^*)$ band of methyl iodide [**I**, 157; **II**, 303] has been resolved into three distinct electronic states in the MCD spectrum [G5]. In a detailed vibronic analysis of the 5p → 6s Rydberg transition, Felps *et al.* [F2] readily assign the vibrations in the $^3E_{(1)}$ and $^1E_{(1)}$ states and then from the unassigned bands assemble progressions built upon deduced origins for transitions assumed to terminate at $^3E_{(2)}$ and $^3E_{(0\pm)}$. The electronic origins so deduced are actually within 40 cm^{-1} of the electronic origins observed as two-photon resonances in the MPI spectrum of methyl iodide, Fig. II.A-1. More or less the same vibrational frequencies appear in each of the four 5p → 6s transitions of methyl iodide. A similar assignment strategy based upon two unobserved origins was successful in assigning the four *n*p → (*n* + 1)s states of CH_3Br, but was not practical for CH_3Cl. The (5p, 6s) Rydberg term values of the alkyl iodides CH_3I through $(CH_3)_3CI$ have been evaluated [K8], and found to decrease smoothly as the alkyl groups become more bulky, as reported in [**I**, 166]. In agreement with the general claim of Hochmann *et al.* [H67], a plot of ionization potential *versus* term value for these compounds in linear, as are those for the rare gases isoelectronic with H–X.

Abundant confirmation of the detailed 5p → 6s assignments in CH_3I is found in the MCD spectrum of this molecule [M47, S32, S33]. The origin of the transition to $^3E_{(2)}$ and all its vibronic members (*e* symmetry) have positive *A* terms and the same upper-state magnetic moment. A similar statement holds for the transition to $^3E_{(1)}$, except that nondegenerate vibrations appear. In the transition to $^3E_{(0\pm)}$ there is no angular momentum in the upper state and so *B* terms appear instead at the MCD origin and for all vibronic (*e*) attachments. The *A* terms in the transition to $^1E_{(1)}$ are much like those to $^3E_{(1)}$, and the relative magnetic moments observed for each of these states is close to that calculated theoretically. Though the transitions to $^3E_{(2)}$ and $^3E_{(0\pm)}$ are said to get their intensities vibronically from those to $^3E_{(1)}$ and $^1E_{(1)}$, there are noticeable intensities at the origins, and so other (nonvibronic) mechanisms also must be at work.

In CH_3X, the np → (n + 1)s promotion leads to only four distinct levels under spin-orbit coupling, whereas the np → (n + 1)p transition yields 10 [G35]. One of these 10 5p → 6p components is seen in the MPI spectrum of methyl iodide with an origin at 58 330 cm^{-1} (*advert., 7.2318 eV*) [G11]. This origin and its associated vibronics (totally symmetric and nontotally symmetric) are observed as two-photon resonances. By comparison, the lowest 5p → 6p one-photon origin is at 59 000 cm^{-1} (*advert., 7.315 eV*) [R35], demonstrating that the MPI origin is two-photon allowed and one-photon forbidden, and therefore has A_2 symmetry and a $^2E_{3/2}$ core. A second origin in the MPI spectrum, at 63 505 cm^{-1} (*advert., 7.8734 eV*) is 5200 cm^{-1} above the first MPI origin and so involves the $^2E_{1/2}$ core configuration, 5200 cm^{-1} being the $^2E_{1/2} - {}^2E_{3/2}$ spin-orbit interval of the ionic core. Using a spectrograph of high resolution, Baig *et al.* [B4] have recorded the (5pπ, nd) Rydberg series of methyl iodide up to principal quantum numbers of more than 30, converging on both $^2E_{3/2}$ and $^2E_{1/2}$.

The effects of external perturbation on Rydberg transitions have been illuminated using dilute solutions of methyl iodide in liquified rare gases. Working in liquid argon, Messing *et al.* [M54, M55] recorded the 5p → 6s spectrum of CH_3I in a density range of 0.0−1.38 gm/cm^3, Fig. II.D-1. At the low-density end, each vibronic band broadens to high frequency, as when perturbed by high-pressure nitrogen gas [**I**, 80]. At densities beyond 0.86, the 5p → 6s and 5p → 6p bands then shift bodily to high frequency, maintaining the unperturbed $E_{1/2} - E_{3/2}$ split, and retaining in part their molecular character as appropriate for "intermediate excitons." On the other hand, the higher Rydberg states of CH_3I in high-density Ar are transformed into $n \geqslant 2$ Wannier excitons. See Section II.D for a further discussion of the spectrum of methyl iodide in liquid argon solution.

Zelikina *et al.* [Z2, Z3] report extreme broadening of the Rydberg transitions of CH_3I and CF_3I in a variety of liquified gases, as well as a strong dependence of the absorption frequencies on temperature. With respect to Rydberg absorption frequencies in the vapor phase, CF_4 as solvent produces the largest shift to high frequency and xenon, the least. On the other hand, the depression of the ionization potential on solution is largest in Xe (*ca.* 16 000 cm^{-1}, *2 eV*) and least in Ar (9500 cm^{-1}, *1.18 eV*) [Z3].

Three negative-ion excited states of CH_3I are reported as Feshbach resonances at 47 600, 52 630 and 60 500 cm^{-1} (*vert., 5.90, 6.525 and 7.50 eV*) [S65]. Remembering that such resonances to Feshbach negative-ion configurations are about 4000 cm^{-1} below the corresponding $^1(\phi_i^o, \phi_j^{+1})$ neutral-molecule configurations (Section II.B), clearly the first of these Feshbach resonances has a $^2(5p, 6s^2)$ configuration with a $^2E_{3/2}$ core, while the other two are $^2(5p, 6s^2)$ and $^2(5p, 6p^2)$ with $^2E_{1/2}$ core configurations. The transition to $^2(5p, 6p^2)$ with a $^2E_{3/2}$ core is expected at 55 500 cm^{-1} (*6.88 eV*). These assignments fit nicely among the data points of Fig. II.B-2.

The so-called "vibronic doubling" seen in the ethyl bromide spectrum (Section IV.B) also appears in the spectra of methyl iodide-h_3 and -d_3 [M44]. In the former, a vibronic band of the $^3E_{(2)}$ state is coincident with a level of the $^3E_{(1)}$ state, resulting in a large redistribution of intensity, but a relatively small shift of frequency (15 cm^{-1}). In the perdeutero compound, it is components of the $^3E_{(0\pm)}$ and $^1E_{(1)}$ states which interact with the same consequences.

Tsai and Baer [T23] have studied the photoionization spectrum of CH_3I in the region between the $^2E_{3/2}$ and $^2E_{1/2}$ ionization potentials, and observe that each of the transitions from $5p(^2E_{1/2})$ to 7d and 8d are split into two components. This splitting is also evident in the absorption spectrum in the same region [B4]. Though Tsai and Baer attribute this to the splitting of the *n*d levels under the influence of the quadrupole moment of the CH_3I^+ core, it resembles the vibronic doubling seen at lower frequencies (*vide supra*), and should be checked by studying either the photoionization or absorption spectrum of CD_3I. The doubling does not appear in the corresponding transitions of HI. The very interesting Fano line shapes appearing between the $^2P_{3/2}$ and $^2P_{1/2}$ ionization limits of Xe do not appear in the equivalent region of the CH_3I spectrum. Application of multichannel quantum defect theory to the long Rydberg series of the methyl iodide spectrum reveals interactions between channels converging on the $^2E_{3/2}$ and $^2E_{1/2}$ ionic states and best-fit parameters which are close to those derived for Xe [D1].

Since the $\sigma \rightarrow \sigma^*$ valence bands of HF, HCl and of HBr are observed at 83 275, 76 245 and 74 900 cm^{-1} (*adiabat., 10.324, 9.4530, and 9.29 eV*) [J1, S70], one is tempted to look for corresponding $\sigma \rightarrow \sigma^*$(C–X) excitations in the methyl halides in approximately the same spectral region. In fact, possible $\sigma \rightarrow \sigma^*$ candidates are seen in the electron-impact spectrum of methyl iodide [H57], Fig. IV.C-1, as broad peaks at 68 000, 79 000 and 89 000 cm^{-1} (*vert., 8.4, 9.8 and 11.0 eV*). We note however

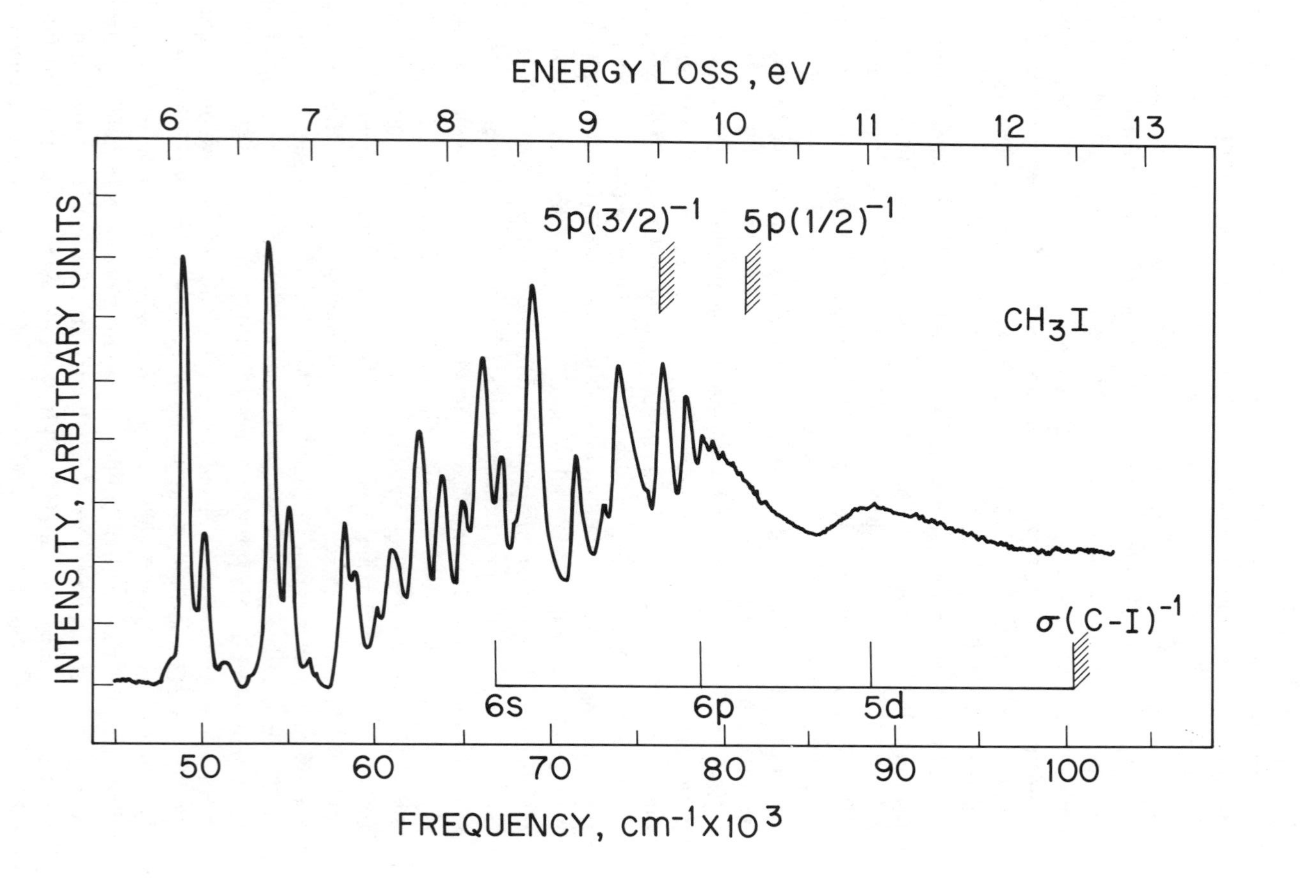
ENERGY LOSS, eV
6
7
8
9
10
11
12
13
5p(3/2)$^{-1}$
5p(1/2)$^{-1}$
CH_3I
INTENSITY, ARBITRARY UNITS
σ(C-I)$^{-1}$
6s
6p
5d
50
60
70
80
90
100
FREQUENCY, cm^{-1}×10^3

that these three broad bands have appropriate term values with respect to the $\sigma(C-I)^{-1}$ ionization potential at 100 800 cm^{-1} (*vert., 12.50 eV*) for assignment as $\sigma \rightarrow 6s$, 6p and 5d Rydberg excitations. Possibly, these Rydberg excitations are preceded by a $\sigma \rightarrow \sigma^*$ excitation centered at 63 000 cm^{-1} (*7.8 eV*).

The $1s_C$ inner-shell spectrum of methyl iodide [H61] holds no surprises, Fig. IV.B-1. The broad $\mathscr{A}(1s_C, \sigma^*)$ band has a term value of 46 000 cm^{-1}, and is the most intense band in the spectrum. As with methyl chloride and bromide, the lowest $1s_C \rightarrow ns$ transition in methyl iodide shows a reasonable term value (29 400 cm^{-1}) and the excitation of a quantum of ν_1' (2400 cm^{-1}), while the transition to the lowest np level is split into an intense transition to an e component and a weak one to an a_1 component. Additionally, the corresponding $4d_I$ spectrum of methyl iodide in the 400 000–480 000 cm^{-1} (*50–60 eV*) region has been recorded optically [O14] and by electron impact [H57]. Excitation of an electron from $4d_I$ results in $4d_{5/2}$ and $4d_{3/2}$ core configurations split by 13 700 cm^{-1}. Thus the $4d_I$ spectrum of methyl iodide begins with two broad $\mathscr{A}(4d_I, \sigma^*)$ bands separated by 13 700 cm^{-1}. Transitions from $4d_I$ to 6s are not observed, in accord with the atomic selection rule, whereas transitions from both core components to $6pe$ and $6pa_1$ are observed. This latter band has been assigned as $4d_I \rightarrow 4f$ [O14], but its term value of 15 730 cm^{-1} suggests a $6pa_1$ terminating orbital instead. The $\sigma^* < ns < np$ upper-level ordering for methyl iodide closely parallels that given for I_2 in the 4d region, and both molecules also show shake-up excitations in the 480 000–600 000 cm^{-1} (*60–75 eV*) span.

An exceedingly broad absorption feature having a width at half-height of *ca.* 200 000 cm^{-1} (*25 eV*) and centered approximately 260 000 cm^{-1} (*32 eV*) beyond the $(4d_I)^{-1}$ ionization threshold [H57] has been assigned as a giant resonance [O14]. However, by its frequency and its width, we feel it is better described simply as the delayed onset of photoionization in the ϵf channel (Section II.C). Closely similar delayed-onset peaks are observed in the $4d_{Xe}$ spectra of XeF_2 and XeF_4 [**II**, 342] and the $3d_I$ spectrum of CH_3I.

The effects of lowering the ϕ_i level on the $\mathscr{A}(\phi_i, \sigma^*)$ band and Rydberg term values are clearly evident in methyl iodide. For the $\mathscr{A}(\phi_i, \sigma^*)$ band, we note term values of 38 530 cm^{-1} for $\mathscr{A}(5p, \sigma^*)$, 46 000 cm^{-1} for $\mathscr{A}(4d_I, \sigma^*)$, and 45 600 cm^{-1} for $\mathscr{A}(1s_C, \sigma^*)$, while for the conjugate transitions to 6s, the term values are 27 320 cm^{-1} for (5p, 6s) and 29 400 cm^{-1} for $(1s_C, 6s)$. The splitting of 6p into e and a_1 components is 4400 cm^{-1} in the $(1s_C, 6p)$ state, 3200 cm^{-1} in the $(4d_{5/2}, 6p)$ state and 4800 cm^{-1} in the $(4d_{3/2}, 6p)$ state. A splitting of 5200 cm^{-1} has been noted for the $5p \rightarrow 6p$ outer-shell transition [G11], but this is more likely $E_{3/2} - E_{1/2}$ splitting than (5p, 6p) multiplet splitting.

The spectrum of CH_2I_2 (67 000−91 000 cm^{-1}, *8.3−11.3 eV*) [O6] is said to show a $5p \rightarrow ns$ Rydberg series converging on the third ionization potential at 82 990 cm^{-1} (*advert., 10.29 eV*), however the series is not convincing. Spectra of the fluorinated alkyl iodides are discussed in Section IV.E.

IV.D. Alkyl Fluorides

We have seen that the term values for the $\mathscr{A}(np, \sigma^*)$ bands of the methyl halides increase in the order fluoride to iodide (Section I.C). On the other hand, the $(np, (n + 1)s)$ term values of the conjugate Rydberg levels decrease in the same order. These trends of opposite slope result in a Rydberg-above-$\mathscr{A}$-band pattern for excitations from np in the chloride, bromide and iodide, but a Rydberg-below-$\mathscr{A}$-band pattern for the fluoride. This inversion in the alkyl fluorides has been the source of confusion and uncertainty in the past, but we now have a somewhat clearer picture of its origin and consequences.

Our initial interpretation of the spectra of the alkyl fluorides as consisting solely of low-lying Rydberg excitations [**I**, 178] has survived intact in the outer-shell region, but has been questioned with good reason in the inner-shell region. The older experimental work on the outer-shell spectrum of carbon tetrafluoride has been confirmed [H61, W20], and the Rydberg assignments of [**I**, 186] have been expanded by new experimental work. In CF_4, vibronically structured Rydberg transitions are newly reported in absorption at 170 400 (*advert., 21.13 eV*), at 165 800 and at 169 000 cm^{-1} (*vert., 20.56 and 20.95 eV*) [L14]. These bands display vibronic progressions of 600−700 cm^{-1} and can be readily assigned on the basis of their absorption frequencies as $4a_1 \rightarrow 3s$ (32 300-cm^{-1} term value), $3t_2 \rightarrow 3d$ (12 600-cm^{-1} term value) and $3t_2 \rightarrow 4p$ (9400-cm^{-1} term value), respectively. The $4a_1$ and $3t_2$ MO's of CF_4 encompass the σ(C−F) bonds. Transitions which are energetically and vibrationally similar to the $3t_2 \rightarrow 3d$ band of CF_4 are observed in CF_3H and CF_3Cl (Section IV.E).

A Rydberg series of purported window resonances in carbon tetrafluoride originating at $2s_C$ has been assigned as $2sa_1 \rightarrow np$, but the series starts at $n = 2$ [L14] and involves several doubtful features. A few of these "window resonances" also are observed in the electron-impact spectrum of CF_4 [H61], but the authors of the electron impact work do not assign them as such. The forbidden Rydberg transition $1e \rightarrow 3s$ in carbon tetrafluoride is uncovered in the trapped-electron spectrum at 120 200 cm^{-1} (*14.9 eV*) with a 29 000-cm^{-1} term value [V21]. The total dissociation cross sections for CF_4, CF_3H, C_2F_6 and C_3F_8 all peak at *ca.* 800 000 cm^{-1} (*100 eV*) as do the cross sections for total electronic excitation [W31]. At these maxima, the cross sections for CF_4, CF_3H and CH_4 are equal to one another as are those of C_2H_6 and C_2F_6.

Corresponding to the CF_4 neutral-molecule parent transitions $1t_1 \rightarrow 3s$ and $1t_1 \rightarrow 3p$ at 102 000 (*vert., 12.7 eV*) and 110 000 cm^{-1} (*vert., 13.6 eV*) [I, 179], negative-ion Feshbach resonances to $^2(1t_1, 3s^2)$ and $^2(1t_1, 3p^2)$ configurations are observed at 96 800 and 105 000 cm^{-1} (*vert., 12.0 and 13.0 eV*) in this molecule [V21]. It is common that the negative-ion resonance to a doubly-occupied Rydberg orbital is about 4000 cm^{-1} below the corresponding neutral-molecule transition (Section II.B). Another negative-ion resonance in CF_4 in the region of 0−8000 cm^{-1} (*0−1 eV*) [V21] is said to involve placing the odd electron in a nonbonding 3s orbital since the vibronic structure in this band (1280-cm^{-1} progression) matches a neutral-molecule ground state frequency. However, we do not believe such orbitals exist for anions, and moreover, the length of the progression is totally at odds with the claim that the half-filled orbital in the anion is nonbonding. Yet other negative-ion resonances in CF_4 are located at 53 700 and 56 900 cm^{-1} (*vert., 6.7 and 7.1 eV*) [H78, S66]; these and the one mentioned above presumably are related to the methane valence TNI resonances observed in the same region and assigned as terminating at Jahn-Teller components of the $t_2\sigma^*$ MO [B56].

The low-lying temporary-negative-ion resonance at *ca.* 4000 cm^{-1} (*vert., 0.5 eV*) in CF_4 is a most intriguing item. Low-frequency valence TNI resonances such as this are known in the chloromethanes to involve excitations into the σ^*(C−Cl) orbitals [B85], as do the neutral-molecule $\mathscr{A}(\phi_i, \sigma^*)$ bands. However, such $\mathscr{A}(2p, \sigma^*)$ bands never have been identified in the outer-shell spectra of the fluoromethanes, and thus excitations involving σ^* are presumed to be high in frequency. However, it is known from the perfluoro effect [I, B67] that fluorination drastically lowers σ MO's, and we now expand that to include σ^* MO's. According to Tossell and Davenport [T15], the lowest unoccupied orbital in CF_4 is an $a_1\sigma^*$ MO, and so the 4000-cm^{-1} valence TNI resonance configuration is assigned as $^2(N, a_1\sigma^*)$. The vibrational interval of 1280 cm^{-1} presumably corresponds to the excitation of ν_3', a t_2 deformation mode.

A quantitative study of the absorption spectra of CH_3F and CHF_3 in the region 131 000−571 000 cm^{-1} (*16.3−70.8 eV*) has uncovered several new bands [W37]. In CHF_3, the previously identified $3e\sigma, 5a_1\sigma \rightarrow 3d$ transition (160 500 cm^{-1} *vert., 19.90 eV*) is observed optically to consist of a 910 cm^{-1} vibrational progression (ν_2'), while the $4a_1 \rightarrow 3d$ band at 186 850 cm^{-1} (*vert., 23.16 eV*) shows vibrational intervals of 1080 cm^{-1}. The $3e\sigma$, $5a_1\sigma$ orbitals of CF_3H are derived directly from the $3t_2\sigma$ MO's of CF_4 and consequently the 160 500-cm^{-1} band of CF_3H with its 910-cm^{-1} vibrations ($3e\sigma$, $5a_1\sigma \rightarrow 3d$) correlates directly with the 165 800-cm^{-1} band of CF_4 with its 600−700 cm^{-1} vibrations ($3t_2\sigma \rightarrow 3d$). Bands peaking at *ca.* 320 000 cm^{-1} (*40 eV*) in CF_4, CF_3Cl [L14], CF_3H and CH_3F [W37] but absent in CH_4 would appear to be Rydberg excitations originating at the 2s orbitals of the fluorine atoms.

Photofragmentation of CF_3X (X = H, Cl, Br) produces CF_3 in an excited state from which fluorescence is emitted with an origin at 51 700 cm^{-1} (*6.41 eV*) [S80, S81]. The excited state has a term value of 22 800 cm^{-1} with respect to the $(2p_C)^{-1}$ ionization threshold at 74 500 cm^{-1} (*adiabat., 9.24 eV*), suggesting a $3p \rightarrow 2p_C$ fluorescent transition rather than $4s \rightarrow 2p_C$ as calculated [W7].

The inner-shell spectra of methyl fluoride and carbon tetrafluoride have been studied by electron impact [H57, H61, W20] and the full series of compounds from methane to carbon tetrafluoride has been studied optically [B66], Fig. IV.D-1. Looking at the $1s_C$ spectrum of CH_4, recall that the $1s_C \rightarrow 3s$ transition is very weak because it is only vibronically allowed, whereas $1s_C \rightarrow 3p$ is intense and though the upper state is triply degenerate ($^1T_2$), it shows no Jahn-Teller splitting in the optical spectrum [B66]. Note however, that the $1s_C \rightarrow 3p$ band of methane [H59] in the electron-impact spectrum does show a splitting of *ca.* 5500 cm^{-1}, Fig. IV.A-2. As expected, the $1s_C \rightarrow 4p$ transition also is discernible in methane. Inner-shell spectra of the fluoromethanes originating at $1s_C$ will relate to this pattern in methane, with due allowance for the complications of lower symmetry and possible valence $\mathscr{A}(1s_C, \sigma^*)$-band transitions. Inner-shell Rydberg excitations from $1s_F$ in CH_3F and CF_4 are predicted to be very broad due to dissociation of the $(1s_F)^{-1}$ cation [G31], as observed [W20].

Before going further with the alkyl fluorides, we pause to consider the nature of transitions to σ^*(C−X) orbitals in alkyl halides in general. In methyl chloride, bromide and iodide, the photoelectron spectra show that the halogen np lone pair orbitals are uppermost. Thus the $\mathscr{A}(np, \sigma^*)$ band is a low-lying excitation in these molecules, even lower than the $np \rightarrow (n+1)s$ Rydberg excitations. However, in methyl fluoride, the fluorine $2p\pi e$ lone-pair orbitals are *not* uppermost, but lie below the $2e$ methyl-group orbitals by 32 000 cm^{-1} (*4 eV*). Thus in CH_3F, the $\mathscr{A}(2p\pi e, \sigma^*(C-F))$ band must be at frequencies far above those for the first few Rydberg excitations originating at $2e\sigma$(C−H), as also must be true for the $2e\sigma \rightarrow \sigma^*(C-F)$ valence excitation. Though

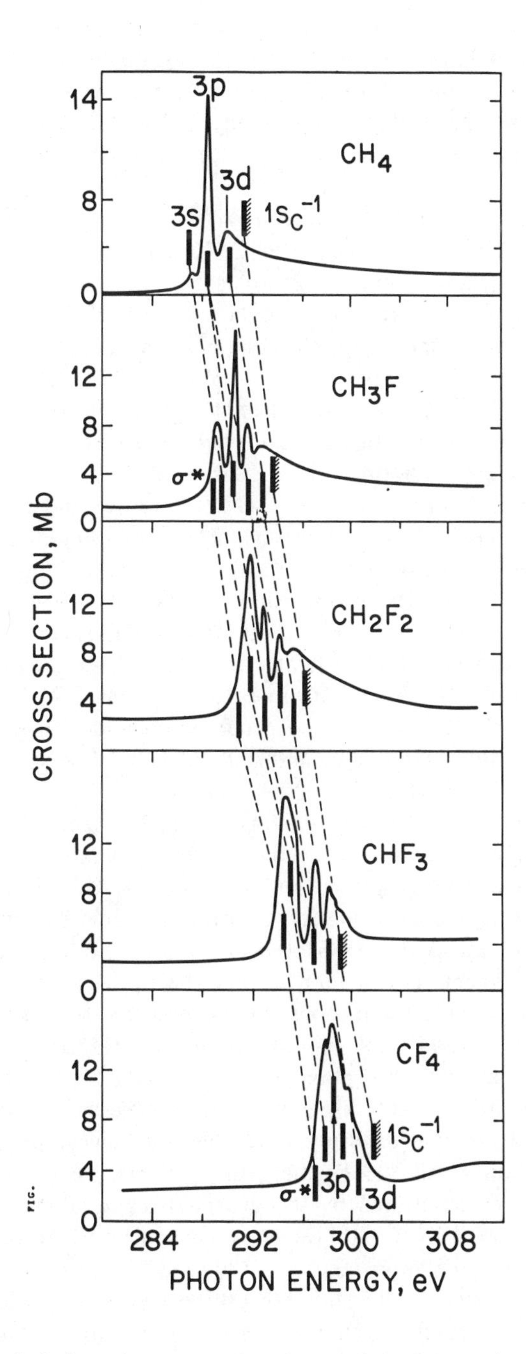

Fig. IV.D-1. Optical absorption spectra of methane and the fluoromethanes in the region of the $(1s_C)^{-1}$ ionization limits [B66].

$\mathscr{A}(2p\pi, \sigma^*(C-F))$ bands are not going to be the lowest outer-shell excitations in CH_3F, CH_2F_2 and CHF_3, the argument given above would not apply to CF_4, and $\mathscr{A}(2p\pi, \sigma^*)$ bands in the lowest Rydberg region might be expected here. Nonetheless, no $\mathscr{A}(2p\pi, \sigma^*)$ band in the outer-shell region ever has been reported either for this molecule or for any other alkyl fluoride.

The systematic variation of the $\mathscr{A}(np, \sigma^*)$ and $\mathscr{A}(1s_C, \sigma^*)$ band outer-shell and inner-shell term values with C–X bond strength in the methyl halides, Fig. I.C-3, allows one to extrapolate values of these quantities for methyl fluoride. Thus, with a C–F bond strength of 110 kcal/mole, one estimates that the $(2p\pi e, \sigma^*)$ and $(1s_C, \sigma^*)$ term values for CH_3F are 27 500 and 35 000 cm^{-1}, respectively. With a $(2e, 3s)$ term value of 33 000 cm^{-1} in the outer-shell spectrum [**I**, 180], one sees the likelihood that the $\mathscr{A}(np, \sigma^*)$ band will not be a noticeable factor in the outer-shell spectrum of the alkyl fluorides since it is weak and will be overlaid by Rydberg transitions to 3s and 3p. On the other hand, antishielding exhaltation (Section I.C) places the $(1s_C, \sigma^*)$ term value at approximately 35 000 cm^{-1} in the inner-shell spectrum, thereby making it essentially degenerate with the Rydberg transition to $(1s_C, 3s)$. Thus, $\mathscr{A}(\phi_i^{-1}, \sigma^*)$ bands will be important factors in the inner-shell spectra of the alkyl fluorides, for in this case the $\mathscr{A}$ bands are intense and having larger term values, will precede the lowest Rydberg absorptions. We are clearly in Regime Y, Fig. I.D-3, and must be prepared to find strongly overlapped and possibly strongly mixed Rydberg and valence σ^* levels in the inner-shell spectra of alkyl fluorides.

If we now apply the above arguments to the $1s_C$ spectra of the methyl halides, we expect $\mathscr{A}(1s_C, \sigma^*(C-X))$ valence bands to precede the Rydberg excitations originating at $1s_C$ in CH_3Cl, CH_3Br and CH_3I, whereas in CH_3F the two lowest transitions from $1s_C$ will be near-degenerate, terminating at both 3s and σ^*. Given the high penetration of ns orbitals, the expectation in fact, is that the excitations to $a\sigma^*$ may be totally dissolved in the excitations to ns and so not appear as such (Section I.D), depending upon their location with respect to the Rydberg manifold. In the $1s_C$ absorption region of CH_3Cl, CH_3Br and CH_3I, Fig. IV.B-1, the expected $\mathscr{A}(1s_C, \sigma^*(C-X))$ valence bands, broad and low frequency, do lead the absorption parades, followed by the sharper Rydberg transitions. All the Rydberg and valence assignments listed in the Figure are well supported by the observed term values, Tables I.B-II and I.C-II. However, in CH_3F the lowest-frequency feature has a term value appropriate to $1s_C \rightarrow 3s$ (35 500 cm^{-1}), but a width more appropriate to $1s_C \rightarrow \sigma^*$. Though the entire CH_3F inner-shell spectrum can be assigned as Rydberg on the basis of term values, Hitchcock and

Brion [H61] feel that the first band consists of distinct but overlapped $1s_C \rightarrow 3s$ and $\mathscr{A}(1s_C, \sigma^*)$ band transitions. That this is indeed the case is more evident in the CH_3F spectrum of Fig. IV.B-1, which is of higher resolution than that of Fig. IV.D-1. Thus, if a distinct $1s_C \rightarrow \sigma^*$ band is anywhere in the inner-shell spectrum of CH_3F, it is most logical to assign it to the broad band preceding the transitions to 3s and 3p; this is in accord with the expectation that the inner-shell antishielding will raise the valence term values much more than the Rydberg term values (Sections I.B and I.C). It also follows from this analysis that the outer-shell $\mathscr{A}(np, \sigma^*)$ band of CH_3F falls approximately midway between the Rydberg excitations to 3s (74 200 cm^{-1} *vert., 9.20 eV*) and 3p (84 000 cm^{-1} *vert., 10.4 eV*).

From this point, we follow the $1s_C$ transitions along the road from CH_3F to CF_4 [B66]. Going straight to CF_4, if one were to argue that σ^* orbitals are unimportant in the alkyl-fluoride spectra, then the $1s_C$ spectrum of CF_4 should closely resemble that of CH_4. Experimentally, this is not the case at all, Fig. IV.D-1, for there are far too many bands in CF_4 and they are all relatively intense.† It is possible that $1s_C \rightarrow 3p$ in CF_4 has a large Jahn-Teller split (but unlikely since the degeneracy is in a nonbonding orbital, Section I.B) while there is none in CH_4, and that the vibronic borrowing is much more effective in intensifying $1s_C \rightarrow 3s$ in CF_4 owing to the nearness of $1s_C \rightarrow 3p$. It is nonetheless difficult to assign bands such as those at 298.0 and 298.5 eV in CF_4 (Fig. IV.D-1) as Rydberg excitations with term values of 29 800 and 27 400 cm^{-1} while maintaining the band at 297.6 eV as $1s_C \rightarrow 3s$ with a term value of 32 300 cm^{-1}. Similar problems exist for the other alkyl fluorides in the regions between $1s_C \rightarrow 3s$ and $1s_C \rightarrow 3p$ excitations. The inescapable conclusion is that the fluoromethanes are in Regime Y, Fig. I.D-3, so that the Rydberg and σ^* orbitals indeed are near-degenerate and strongly overlapped in the inner-shell spectra, while the Rydberg orbitals are far below σ^* in the outer-shell spectra. Also see Table I.C-V, which shows the large $\mathscr{A}$-band term values found in perfluorinated systems.

The conclusion that the inner-shell spectra of the alkyl fluorides are increasingly dominated by valence transitions to σ^*(C—F) MO's with increasing fluorination is supported as well by the close resemblance of the $1s_C$ inner-shell spectra of the alkyl fluorides (Fig. IV.D-1) and those of the

† Note also that the optical [B66] and electron impact spectra [W20] in the $1s_C$ region of CF_4 are not in very good quantitative agreement.

corresponding chloromethanes (Figs. IV.A-2 and IV.B-1). This similarity is lacking totally in the outer-shell spectra owing to a radically different ordering of both originating and terminating MO's in these two systems. The emergence of significant transitions to σ^* on going from CH_4 to CHF_3 and CF_4 is not without parallels, for this is also the case on going from H_2O to F_2O (Section VII.A), from NH_3 to NF_3 (Section VI.A), from PH_3 to PF_3 [**I**, 239] and from the larger alkanes to their perfluoro derivatives [**I**, 190].

A scattered-wave calculation on the $1s_C$ spectrum of CF_4 yields a set of σ^* MO's distinct from but interspersed among their Rydberg conjugates [H6], whereas in an *ab initio* MO calculation, the σ^* MO's are so thoroughly mixed with their Rydberg conjugates that no valence levels are discernible as such [S30]. According to this latter work, the first four features in the $1s_C$ spectrum of CF_4 are a heavy mixture of 3s and σ^* orbitals complicated by vibronic interactions. Broad features beyond the $(1s_F)^{-1}$ photoionization threshold of CF_4 have been assigned as EXAFS interferences [B66], however this interpretation has been called into question [H59].

At the next level upward in complexity from the fluoromethanes are the fluoroethanes, on which considerable spectroscopic work has been done recently. The lowest ionization potential moves from 96 800 to 118 000 cm^{-1} (*vert., 12.0 to 14.6 eV*) on going from ethane to hexafluoroethane, and the vacuum-ultraviolet absorption edge correspondingly shifts to higher frequency in the series [S21]. As in ethane itself, the lowest Rydberg transition to 3s is forbidden in hexafluoroethane [**I**, 188], while that to 3p is observed to be allowed. In ethyl fluoride and 1,1-difluoroethane, the lowest transitions to 3s remain weak in spite of the formally lower symmetry, and it is not until 1,1,1-trifluoroethane is reached that one sees clear transitions to both 3s (84 900 cm^{-1} *vert., 10.5 eV*) and 3p orbitals (92 000 cm^{-1} *vert., 11.4 eV*) [S21]. The former has a term value of 29 500 cm^{-1}. Extension of the ethyl fluoride spectrum beyond that in [**II**, 306] reveals other bands at 112 000 and 128 000 cm^{-1} (*vert., 13.9 and 15.9 eV*) [S21]. Quantitative absorption studies of C_2F_6 in the 130 000–570 000 cm^{-1} (*16–71 eV*) and 800 000–2 200 000 cm^{-1} (*100–275 eV*) regions are reported [C36, L14, W31].

The valence TNI resonance at *ca.* 56 000 cm^{-1} (*vert., 7 eV*) in CF_4 appears between 32 000 and 40 000 cm^{-1} (*4 and 5 eV*) in C_2F_6 and moves downward from there in the larger perfluoroalkanes [H78, S66] reaching 9700 cm^{-1} (*vert., 1.2 eV*) in n-C_6F_{14}. Interestingly, in n-C_4F_{10}, i-C_4F_{10} and n-C_5F_{12}, valence TNI resonances also appear at 4800 cm^{-1} (*0.6 eV*), and at 4400 cm^{-1} (*0.55 eV*) in n-C_6F_{14}; these frequencies are close to that observed in CF_4 (*ca.* 4000 cm^{-1}, *0.5 eV*) [V21], and as in that molecule, must involve $^2(N, \sigma^*(C-F))$ resonance configurations. A similar excitation is reported in C_2F_6 [M70] followed by valence TNI resonances at 30 200 and 36 400 cm^{-1} (*vert., 3.75 and 4.51 eV*) [S66]. The F^- dissociative attachment spectrum of perfluorocyclobutane shows peaks at 64 000 and 83 100 cm^{-1} (*vert., 8.0 and 10.3 eV*) [C23] which have the proper term values with respect to the PES bands at 98 600 and 115 300 cm^{-1} (*vert., 12.23 and 14.30 eV*) [R35] for assignment as $^2(\phi_i^o, 3s^2)$ Feshbach resonances.

IV.E. Fluoroalkyl Halides

The recent recognition of the deleterious effects of various halocarbons on the earth's ozone layer has focused considerable attention on the spectra of the fluorinated alkyl halides. We illustrate the general effects of fluorination on alkyl-halide spectra with a specific discussion of the spectrum of CF_2Cl_2 [I, 177]. One notices first that the electron impact spectrum of CF_2Cl_2, Fig. IV.E-1 [H76, K20], is much sharper than the optical spectrum of CH_2Cl_2 [II, 305]. The spectrum of CF_2Cl_2 commences with two clearly resolved $\mathscr{A}(3p, \sigma^*)$ bands at 56 300 and 65 600 cm^{-1} (*vert., 6.98 and 8.13 eV*). The splitting here amounts to 9300 cm^{-1}, and though the σ^* MO's in the valence TNI spectrum of CF_2Cl_2 are separated by 11 000 cm^{-1} [B85, V21], one is not certain that the 9300 cm^{-1} represents σ^* splitting, for the difference between the fourth and first ionization potentials of CF_2Cl_2 is 9700 cm^{-1}. Indeed, in CH_2Cl_2, the first two outer-shell $\mathscr{A}(3p, \sigma^*)$ band absorptions are split by only 2600 cm^{-1}, whereas the apparent σ^* split in the valence TNI spectrum is 17 300 cm^{-1} [B85]. Multiple $\mathscr{A}(3p, \sigma^*)$ bands are observed as well in $CFCl_3$ and theory explains their separation as owing to the difference in originating-MO ionization potentials [P25]. (However, see the discussion on this point in Section IV.A). The lowest-frequency $\mathscr{A}(3p, \sigma^*)$ band in CF_2Cl_2 has a term value of 43 100 cm^{-1}, whereas in CH_2Cl_2 it is only 35 900 cm^{-1}. Sufficient data has not been collected yet to say whether or not it is generally true that fluorination decreases the σ^* splitting (CH_2Cl_2, 17 300 cm^{-1}; $CHFCl_2$, 14 900 cm^{-1}; CF_2Cl_2, 11 000 cm^{-1}) [B85], however, it is clear from Table I.C-IV that fluorination does increase $\mathscr{A}(3p, \sigma^*)$-band term values significantly. In fact, one sees that in the halomethanes, addition of fluorine stabilizes σ^* much more than it stabilizes the lowest *n*s Rydberg orbital.

Consonant with the better resolution in CF_2Cl_2, transitions from three of the four chlorine 3p lone-pair orbitals to 4s and all the transitions from 3p to 4p are observed [J4, K20]. Yet another dozen or so transitions in

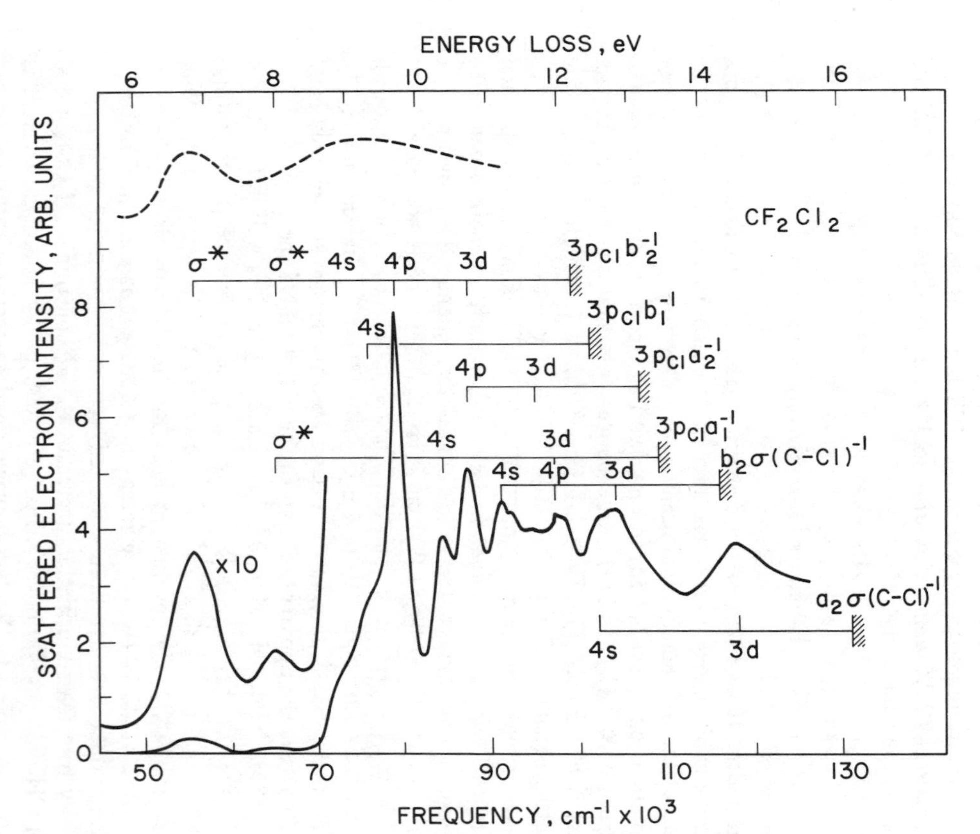

Fig. IV.E-1. The electron-impact energy loss spectrum of CF_2Cl_2 recorded at zero scattering angle and an impact energy of 500 eV [K20]. The dashed curve is the optical absorption spectrum of a solid film of CF_2Cl_2 [M63], while the ionization potentials are taken from [C49].

CF_2Cl_2 are assigned to low-lying Rydberg excitations involving the lowest seven ionization potentials. The average (ϕ_i^o, 4s), (ϕ_i^o, 4p) and (ϕ_i^o, 3d) term values are 27 400, 22 000 and 13 600 cm^{-1}, respectively. In the solid phase, the CF_2Cl_2 spectrum shows a clear $\mathscr{A}$(3p, σ^*) band at 56 100 cm^{-1} (*vert., 6.96 eV*), whereas the weaker $\mathscr{A}$(3p, σ^*) band at 65 600 cm^{-1} (*vert., 8.13 eV*) in the vapor seems not to be present, Fig. IV.E-1 [M63]. The prominent Rydberg excitations in the range 72 000−88 000 cm^{-1} (*9−11 eV*) in the vapor phase also are missing in the neat spectrum. Being 3400 cm^{-1} below the 3p → 4s Rydberg transition at 75 400 cm^{-1} (*vert., 9.35 eV*), the TNI resonance at 72 000 cm^{-1} (*vert., 8.9 eV*) in CF_2Cl_2 vapor is confidently assigned to the ^{2}(3p, $4s^2$) Feshbach configuration [V21]. The corresponding Feshbach resonance in CF_3Cl is observed at 74 000 cm^{-1} (*vert., 9.2 eV*).

Because the energy-loss spectrum of CF_2Cl_2 beyond 97 000 cm^{-1} (*12 eV*) matches exactly the ionization yield spectrum [K20], every absorption beyond this frequency must eventually ionize. In the region between 129 000 and 290 000 cm^{-1} (*16 and 36 eV*) there is only a broad peak at 157 000 cm^{-1} (*19.5 eV*), which is most likely a Rydberg transition originating at $2s_C$.

In an optical study of the mixed systems $CH_xF_yCl_{4-x-y}$ Gilbert *et al.* [G16] summarize the common characteristics of the outer-shell absorption in this class of fluoroalkyl halides:

i) Transitions from the chlorine 3p lone pairs to 4s and 4p orbitals occur between 70 000 and 95 000 cm^{-1} (*8.7 and 12 eV*).

ii) Rydberg transitions in the range 90 000 to 105 000 cm^{-1} (*11 to 13 eV*) originate with σ(C−H) and/or σ(C−Cl) MO's.

iii) Beyond 105 000 cm^{-1} (*13 eV*), the Rydberg excitations originate with fluorine lone pairs and/or σ(C−F) orbitals.

iv) Absorption up to 130 000 cm^{-1} (*16 eV*) is discrete, but there are no further distinct bands of high or even modest intensity between 130 000 and 160 000 cm^{-1} (*16 and 20 eV*).

Studies on $CFCl_3$ and CF_3Cl by King and McConkey [K20] yield spectral results much like those described above for CF_2Cl_2. Threshold electron impact spectra of these systems are presented in [V21] and the optical absorption spectra taken with synchrotron radiation are displayed in [J4]. A high-quality calculation on $CFCl_3$ by Peyerimhoff and Buenker [P25] confirms that in the lowest-frequency valence TNI resonance, the incident electron occupies the $12a_1\sigma^*$(C−Cl) MO, leading eventually to Cl^- and $CFCl_2$. The first two singlet excited states of $CFCl_3$ are calculated to involve $\mathscr{A}$(3p, σ^*) band promotions from the $2a_2$ and $10e$

lone pair orbitals on the chlorine atoms to the $12a_1\sigma^*$(C–Cl) MO, in agreement with the term-value arguments used in constructing Table I.C-III. In each of these upper states, the singlet-triplet splitting is calculated to be *ca.* 6500 cm^{-1}, in consonance with the valence nature of $\mathscr{A}$(3p, σ^*) band promotions, but approximately twice as large as is observed in the $\mathscr{A}$(3p, σ^*) bands of the chloromethanes (Section IV.A).

Particularly noteworthy is the 148 000–164 000-cm^{-1} (*18.3–20.3-eV*) region of CF_3Cl in which weak peaks are found at 151 400 and 156 800 cm^{-1} (*18.77 and 19.44 eV*) displaying vibrations of 530–700 cm^{-1}, the CF_3 deformation [L14]. The first of these has an 11 100-cm^{-1} term value with respect to the $(4a_1)^{-1}$ ionization potential peaked at 162 500 cm^{-1} (*20.15 eV*) which in turn shows a long vibrational progression of 640 cm^{-1} [C49]. The second band may relate to the structured ionization of the $2e$ orbital centered at *ca.* 169 000 cm^{-1} (*21.0 eV*). The term values for both excitations suggest 3d terminating orbitals. These excitations of CF_3Cl correlate with similar features in CF_4 in the 160 000–180 000-cm^{-1} (*19.8–22.3-eV*) region which originate with the $3t_2$ MO, and to vibronically structured features in CF_3H at 160 500 cm^{-1} assigned as $3e, 5a_1 \rightarrow$ 3d (*cf.* Section IV.D). $\mathscr{A}(n\text{p}, \sigma^*)$ band profiles in the fluorinated halomethanes can be found in [G34, H72, M63] and the corresponding term values are listed in Table I.C-III.

In the inner-shell absorption region, only chlorine 2p spectra are available for the chlorofluoromethanes [C36, O15]. The resolution here is not too high, the spectra are doubled by $^2P_{3/2} - {}^2P_{1/2}$ spin-orbit splitting and no $(2p_{Cl})^{-1}$ ionization potentials are known. Still, for CF_2Cl_2, we note a split of 24 000 cm^{-1} between the first two bands, assumed to be $2p_{Cl} \rightarrow \sigma^*$ and $2p_{Cl} \rightarrow 4s$. If correct, and assuming a ($2p_{Cl}$, 4s) term value of 27 000 cm^{-1}, this leads to a ($2p_{Cl}$, σ^*) term value of 51 000 cm^{-1}. That this is significantly larger than the outer-shell (3p, σ^*) term value of 43 100 cm^{-1} is in accord with our experience that the $\mathscr{A}(\phi_i, \sigma^*)$-band term value increases as the principal quantum number of the hole decreases (Section I.C). Absorption spectra in the $2p_{Cl}$ region of the substituted ethanes C_2F_5Cl and $C_2Cl_2F_4$ are reported at low resolution [C36]. Ionization potentials are lacking and so term values cannot be calculated.

All aspects of the fluorochloromethanes discussed above appear in the spectra of the fluorobromomethanes as well, but displaced to lower frequencies [D44, R48, S12, S82]. In the mixed halide CF_2ClBr, $\mathscr{A}(n\text{p}, \sigma^*)$ bands are observed at 48 800 cm^{-1} (*6.05 eV*, $4p_{Br} \rightarrow \sigma^*$) and 61 000 cm^{-1} (*7.56 eV*, $3p_{Cl} \rightarrow \sigma^*$). The remainder of its spectrum is similarly a composite of Rydberg excitations originating with Cl and Br lone-pair orbitals [D44]. There is no evidence for mixed-bond transitions such as $4p_{Br} \rightarrow \sigma^*(C-Cl)$ in the mixed halomethanes.

Vacuum-ultraviolet absorption spectra of the fluorochloroethanes are in line with those of the fluorochloromethanes. In fact, Doucet *et al.* [D43] state that the spectrum of C_2F_5Cl up to 84 000 cm^{-1} (*10.4 eV*) is so similar to that of CF_3Cl that they could be mistaken for one another without "a close look at the numerical data." Fluorination of C_2H_5Cl to first form CH_3CF_2Cl and then C_2F_5Cl results in the progressive high-frequency shift of the $3p \rightarrow 4s$ excitation from 63 400 cm^{-1} (*vert., 7.86 eV*) to 73 900 cm^{-1} (*vert., 9.16 eV*) and thence to 78 300 cm^{-1} (*vert., 9.71 eV*). Throughout this series, the (3p, 4s) Rydberg term values remain surprising constant at 26 000 ± 1000 cm^{-1}. A transition to 4p is observed in CH_3CF_2Cl (81 300 cm^{-1} *vert., 10.1 eV*), while in C_2F_5Cl the equivalent transition lies beyond the range investigated, but is predicted to fall at 85 700 cm^{-1} (*10.6 eV*) on the basis of a 21 000-cm^{-1} term value. Fluorination has a much larger effect on the term value of the $\mathscr{A}(3p, \sigma^*)$ band of ethyl chloride, increasing it from 30 000 cm^{-1} (C_2H_5Cl) to 36 900 cm^{-1} (CH_3CF_2Cl), and finally to 37 800 cm^{-1} (C_2F_5Cl). In the fluorinated dichloroethanes ClF_2C-CF_2Cl and $ClF_2C-CFHCl$ there is a $3p_{Cl} - 3p_{Cl}$ splitting of *ca.* 4500 cm^{-1} which appears in the photoelectron spectrum and in the Rydberg transitions to 4s, but not in the $\mathscr{A}(3p, \sigma^*)$ bands, while in $CF_2Cl-CFCl_2$, two $\mathscr{A}(3p, \sigma^*)$ bands appear (55 000 and 62 500 cm^{-1} *vert., 6.82 and 7.75 eV*), along with two transitions each to 4s and 4p [D43].

Absorption spectra of CF_3I, CF_3CFICF_3 and $CF_3OCF_2CF_2I$ each show $\mathscr{A}(5p, \sigma^*)$ bands at *ca.* 37 000 cm^{-1} (*vert., 4.6 eV*), transitions to (5p, 6s) ($E_{3/2}$) at 59 000 cm^{-1} (*vert., 7.31 eV*) and to (5p, 6s) ($E_{1/2}$) at 64 500 cm^{-1} (*vert., 8.00 eV*) [P35]. Though the spectra extend to 87 000 cm^{-1} (*10.8 eV*), no assignments have been offered for the bands beyond 65 000 cm^{-1} (*8.1 eV*); the spectrum of the iodo-ether looks much like those of the alkyl iodides (Section IV.C). A brief discussion is given in [C20] of the Jahn-Teller effect on the ν_6 hot-band sequences within the $5p \rightarrow 6s$ Rydberg band of CF_3I.

IV.F. Addendum

J.-M. Dumas, P. Dupuis, G. Pfister-Guillouzo and C. Sandorfy, Ionization potentials and ultraviolet absorption spectra of fluorocarbon anesthetics, *Can. J. Spectrosc.* **26**, 102 (1981).

A. P. Hitchcock, G. R. J. Williams, C. E. Brion and P. W. Langhoff, Experimental and theoretical studies of the valence-shell dipole excitation spectrum and absolute photoabsorption cross section of hydrogen fluoride, *Chem. Phys.* **88**, 65 (1984).

T. A. Carlson, A. Fahlman, W. A. Svensson, M. O. Krause, T. A. Whitley, F. A. Grimm, M. N. Piancastelli and J. W. Taylor, Angle-resolved photoelectron cross section of CF_4, *J. Chem. Phys.* **81**, 3828 (1984).

R. Reininger, A. M. Köhler, V. Saile and G. L. Findley, Pressure effects on photoionization in CH_3I, to be published.

CHAPTER V

Boron Compounds

V.A. Boron Hydrides

With time, the diborane spectrum [I, 192] proves to be more intractable and therefore more interesting. By now, it is a trivial exercise to match the features of the electronic spectrum of a substance with its photoelectron spectrum and extract a few Rydberg assignments on the basis of the bands' term values. In diborane, even this was not possible at first. On taking a second look, it is now a little more clear as to where the Rydberg excitations might be, especially with the calculations of Elbert *et al.* [E7] now in hand. A strong sense of uncertainty nonetheless pervades the assignments in diborane.

Note first that since the uppermost filled MO of diborane has b_{2g} symmetry, Rydberg transitions from it to ns and nd will be parity forbidden, and that transitions will be allowed from it to only two of the three np components [E7]. It is calculated that transitions to the two allowed components of the (b_{2g}, 3p) complex are separated by only 500 cm^{-1} [E7], and it is expected that their term values will be approximately 20 000 cm^{-1}. On this basis, with a $(b_{2g})^{-1}$ ionization potential of 95 260 cm^{-1} (*vert., 11.81 eV*) it is natural to assign the intense feature at 75 000 cm^{-1} (*vert., 9.3 eV*) in the optical spectrum of diborane [I, 192] to the unresolved $b_{2g} \rightarrow$ 3p promotions. Note however, that since its oscillator strength of 0.3 is more than twice as large as that ordinarily found for Rydberg excitations to 3p, there must be another excitation overlapping this band. The next higher peak in diborane, at 86 200 cm^{-1} (*vert., 10.7 eV*), has a term value (9060 cm^{-1}) appropriate for the $b_{2g} \rightarrow$ 4p excitation.

It is tempting to assign the transition from b_{2g} to 3s to the broad, weak feature at 62 000–68 000 cm^{-1} (*vert., 7.7–8.4 eV*), however, this leads to an unreasonably large term value of approximately 30 000 cm^{-1}, whereas *ca.* 25 000 cm^{-1} is expected and the *ab initio* calculation [E7] predicts a (b_{2g}, 3s) term value of only 21 600 cm^{-1}. The exact frequency of the $b_{2g} \rightarrow$ 3s transition in diborane could be more easily determined by observing it as an allowed two-photon resonance in the MPI spectrum, Section II.A. The valence transition to $^1(\pi, \pi^*)$ in diborane is calculated to come at 86 700 cm^{-1} (*10.75 eV*) and to have an oscillator strength of 0.33; this latter value equals that predicted for the $\pi \rightarrow \pi^*$ transition in the isoelectronic molecule ethylene. Most reasonably, the $\pi \rightarrow \pi^*$ label is assigned to the intense feature at *ca.* 90 000 cm^{-1} (*vert., 11 eV*) in diborane. The earlier assignment of the weak band at 55 000 cm^{-1} (*vert., 6.8 eV*) as $b_{2g} \rightarrow b_{3g}$ (valence) [I, 193] is confirmed by the calculation of Elbert *et al.* [E7].

The energy-loss spectrum of diborane taken with 30 keV electrons at low resolution [H54] reproduces the optical spectrum [I, 192] fairly well and extends it considerably. Tentative Rydberg assignments are presented in Table V.A-I, with due consideration being given to term values and symmetry. If these assignments are correct, it is odd that the Rydberg excitations associated with the higher ionization potentials are so readily identified whereas those associated with the lowest ionization potential are still in doubt. The final excitation in the energy-loss spectrum, at 134 000 cm^{-1} (*vert., 16.6 eV*), is assigned as a valence shape resonance terminating at π^*. The average values of the (ϕ_i^o, 3s) (25 000 cm^{-1}) and (ϕ_i^o, 3d) (14 000 cm^{-1}) term values derived in the Table are noticeably larger than those calculated:- 21 600 and 9200 cm^{-1} [E7].

The inner-shell absorption spectrum of diborane [Z7] is interesting with regard to the outer-shell discussion above. The $1s_B$ spectrum of diborane begins with an intense band having a 50 800-cm^{-1} term value [R31] and a dipole-allowed $1s_B(b_{1u}) \rightarrow b_{3g}\pi^*$ assignment. This is followed by a second band having a term value of 19 400 cm^{-1} which is taken to be an unresolved blend of $1s_B \rightarrow$ 3s and $1s_B \rightarrow$ 3p. Two other bands, 4000 and 39 000 cm^{-1} (*0.5 and 4.8 eV*) beyond the $(1s_B)^{-1}$ ionization potential possibly are shape resonances terminating at σ^* [E7].

TABLE V.A-I.

EXCITATIONS AND ASSIGNMENTS IN DIBORANE[a]

Frequency, cm^{-1} (*eV*)		Ionization Potential, cm^{-1} (*eV*)		Term Value, cm^{-1}	Upper State
97 600	(*12.1*)	112 000	(*13.9*)	14 400	(b_{3u}, 3d)
104 000	(*12.9*)	118 000	(*14.7*)	14 000	(b_{2u}, 3d)
104 000	(*12.9*)	130 000	(*16.1*)	26 000	(b_{1u}, 3s)
109 000	(*13.5*)	118 000	(*14.7*)	9000	(b_{2u}, 4d)
115 000	(*14.3*)	130 000	(*16.1*)	15 000	(b_{1u}, 3d)
121 000	(*15.0*)	130 000	(*16.1*)	9000	(b_{1u}, 4d)
125 000	(*15.5*)	130 000	(*16.1*)	5000	(b_{1u}, 5d)
129 000	(*16.0*)	-	-	-	-
134 000	(*16.6*)	130 000	(16.1)	– 4000	(b_{1u}, π^*)

[a] Reference [H54].

V.B. Boron-Nitrogen Compounds

Dimethylamino-, *bis*-dimethylamino- and *tris*-dimethylaminoboranes, methylated or halogenated, have been studied in the vacuum ultraviolet by Fuss and Bock [F51, F52]. In the dimethylaminoboranes $[(CH_3)_2\ddot{N}-BR_2]$, there is a prominent $\pi \rightarrow \pi^*$ transition at *ca.* 53 000 cm^{-1} (*vert., 6.6 eV*) having considerable N → B charge-transfer character. This is followed by what are undoubtedly Rydberg excitations originating at π_N. In the *bis*- and *tris*-dimethylaminoboranes, one sees a surprising sharpening of the spectra, with many vibrational intervals clearly visible. In these systems, the $\pi_N \rightarrow 3s$ transition falls at 40 000–45 000 cm^{-1} (*vert., 5.0–5.6 eV*), followed by an intense $\pi_N \rightarrow \pi_B^*$ transition at 44 000–47 000 cm^{-1} (*vert., 5.5–5.8 eV*). As expected, this valence transition is overlapped by $\pi_N \rightarrow 3p$ Rydberg transitions and followed by several members of a $\pi_N \rightarrow nd$ Rydberg series ($\delta = 0.20$). The absorption spectrum of *tris*-dimethylaminoborane in *n*-hexane solution loses all Rydberg fine structure, revealing a peak at 46 000 cm^{-1} (*5.70 eV*), which is in part $\pi_N \rightarrow \pi_B^*$ and in part the remnants of the $\pi_N \rightarrow 3p$ excitations.

Assignments in the electronic spectrum of borazine (*"inorganic benzene"*) remain uncertain; most workers feel that the structured bands in the vicinity of 55 000 cm^{-1} (*vert., 6.8 eV*), Fig. V.B-1, are analogous to the benzene transitions $^1A_{1g} \rightarrow {}^1B_{2u}$ ($^1A_1' \rightarrow {}^1A_2'$, vibronically induced in borazine) and/or $^1A_{1g} \rightarrow {}^1B_{1u}$ ($^1A_1' \rightarrow {}^1A_1'$, vibronically induced in borazine) [**I**, 202]. The MPI spectrum of borazine two-photon resonant in the 54 000–56 000 cm^{-1} (*6.7–6.9 eV*) region [R33] shows vibronic features at frequencies identical to those observed in the one-photon spectrum. Since the only upper-state symmetry which is both one-photon and two-photon allowed from the ground state is $^1E'$, which correlates with both $^1E_{2g}$ and $^1E_{1u}$ in benzene, it is seen that neither of the previous $^1A_2'$ or $^1A_1'$ upper-state assignments is correct in the 54 000–56 000 cm^{-1} region. It is still an open question as to the assignment of the borazine vibronic structure in the 50 000–54 000 cm^{-1} (*6.2–6.7 eV*) region.

A more direct look at the Rydberg spectrum of borazine is contained in the electron-impact work of Doering *et al.* [D38]. As shown in Fig. V.B-1, the energy-loss spectrum of borazine contains a large number of bands most of which can be assigned as Rydberg excitations on the basis of their term values and their relative intensities. The average term values in Fig. V.B-1 are 25 500 cm^{-1} for (ϕ_i^o, 3s), 19 000 cm^{-1} for (ϕ_i^o, 3p) and 11 700 cm^{-1} for (ϕ_i^o, 3d), each of which is perfectly normal. The Rydberg term-value approach leaves unassigned the two transitions at 52 000 and 61 700 cm^{-1} (*vert., 6.45 and 7.65 eV*), in agreement with earlier work which assigned these to $\pi \rightarrow \pi^*$ excitations. No singlet → triplet excitations were found below 50 000 cm^{-1} (*6.2 eV*) in the electron-impact energy loss spectra.

The lowest triplet state in the trapped-electron spectrum of borazine is observed at 64 000 cm^{-1} (*vert., 7.9 eV*) [D39]. This was assigned as an overlapping transition to the three $\pi \rightarrow \pi^*$ configurations $^3A_1'$, $^3A_2'$ and $^3E'$, however since $^1E'$ peaks at 61 700 cm^{-1} and the other two singlets are most likely below it, these assignments are doubtful. As always, the situation is further complicated by the possibility of triplet Rydberg assignments, which are always prominent in the trapped-electron spectra of molecular systems. In order to test this possibility, the trapped-electron peak frequencies have been inserted in Fig. V.B-1. Allowing for singlet-triplet splits of about 2000 cm^{-1} [**I**, 22], it is seen from the Figure that the four triplet peaks of borazine readily can be assigned Rydberg configurations (Table V.B-I). Similar Rydberg assignments based upon term values also are listed in the Table for the triplet levels of N-trimethyl borazine [D39, **I**, 204].

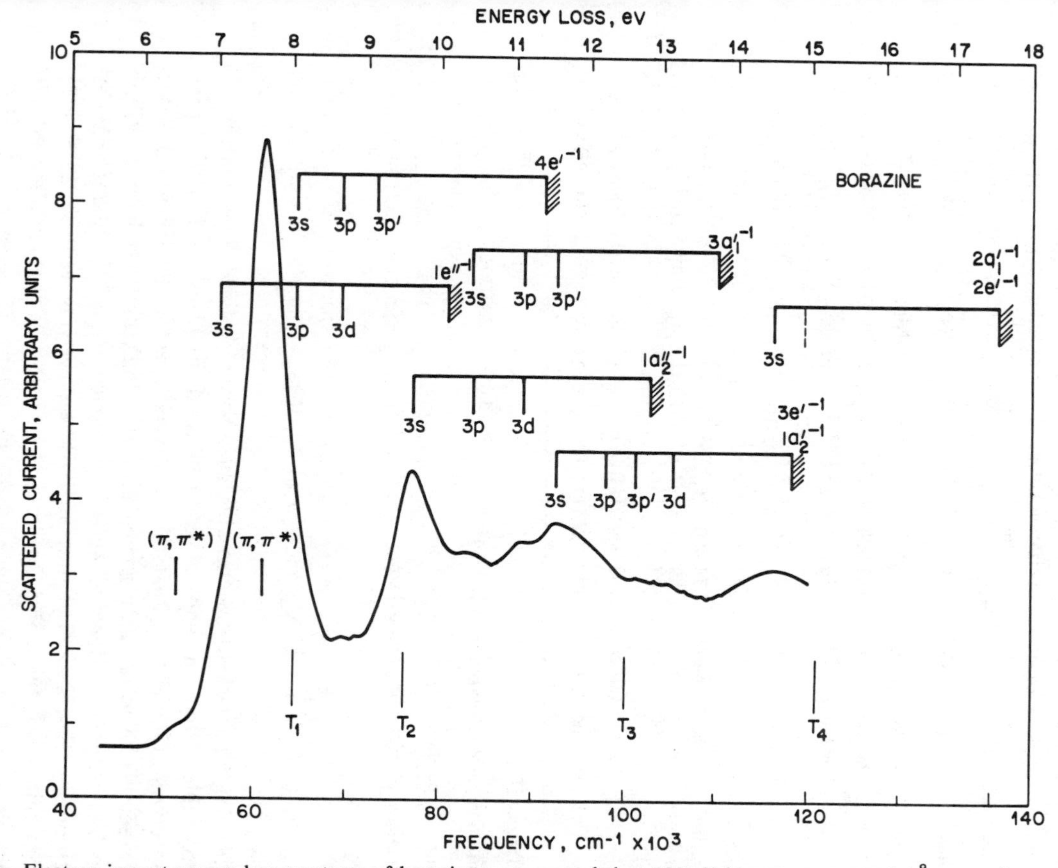

Fig. V.B-1. Electron impact energy-loss spectrum of borazine vapor recorded at 100 eV impact energy and 1 ° scattering angle [D38]. With these impact and scattering parameters, the spectrum is dominated by electric-dipole transitions. The vertical ionization potentials are taken from [**I**, B39], and the triplet frequencies from [D39]. The dashed line represents a predicted level not yet observed.

TABLE V.B-I.

TRIPLET EXCITATIONS AND TERM VALUES IN BORAZINES[a]

	Triplet Frequency, cm^{-1} (*eV*)	Ionization Potential, cm^{-1} (*eV*)	Term Value, cm^{-1}	Upper State
Borazine[b]	64 000 (*7.9*)	81 380 (*10.09*)	17 400	$^3(1e'', 3p)$
	64 000 (*7.9*)	91 950 (*11.40*)	28 000	$^3(4e', 3s)$
	76 000 (*9.4*)	103 500 (*12.83*)	27 500	$^3(1a_2'', 3s)$
	100 000 (*12.4*)	119 000 (*14.75*)	19 000	$^3(3e', 3p'); (1a_2', 3p')$
	121 000 (*15.0*)	141 000 (*17.5*)	20 000	$^3(2e', 3p); (2a_1', 3p)$
N-Trimethyl borazine[c]	53 000 (*6.6*)	74 800 (*9.27*)	21 800	$^3(\phi_i, 3p)$[d]
	65 000 (*8.1*)	89 500 (*11.1*)	24 500	$^3(\phi_i, 3s)$[d]
	82 300 (*10.2*)	100 000 (*12.4*)	17 700	$^3(\phi_i, 3p)$[d]
	104 000 (*12.9*)	128 000 (*15.9*)	24 000	$^3(\phi_i, 3s)$[d]

[a] Reference [D39].

[b] Ionization potentials taken from [I, B39].

[c] Ionization potentials taken from [I, L30].

[d] Symmetry of originating MO unknown.

V.C. Boron Halides

The $1s_B$ inner-shell spectra of the boron halides have attracted considerable attention and have been studied from several different perspectives. One expects that the inner-shell spectra of the boron halides will resemble those of the corresponding trihalomethanes (Chapter IV) in regard transitions and term values to σ^* valence and ϕ_j^{+1} Rydberg orbitals. Additionally, the boron trihalides will show low-lying valence transitions terminating at π^*, the empty $2p\pi$ orbital on boron. These expectations are born out in the excellent inner-shell spectra of Ishiguro *et al.* [I6], Fig. V.C-1. We focus first on the inner-shell spectrum of boron trifluoride. The intense band in the $1s_B$ spectrum at 195.5 eV is universally agreed to be a valence excitation, $1s_B \rightarrow a_2'' 2p\pi^*$. This assignment has been deduced using term-value arguments [R31], semiempirical MO theory [N13], *ab initio* MO theory [I6, S30] and the X-α scattered-wave approach [S88].

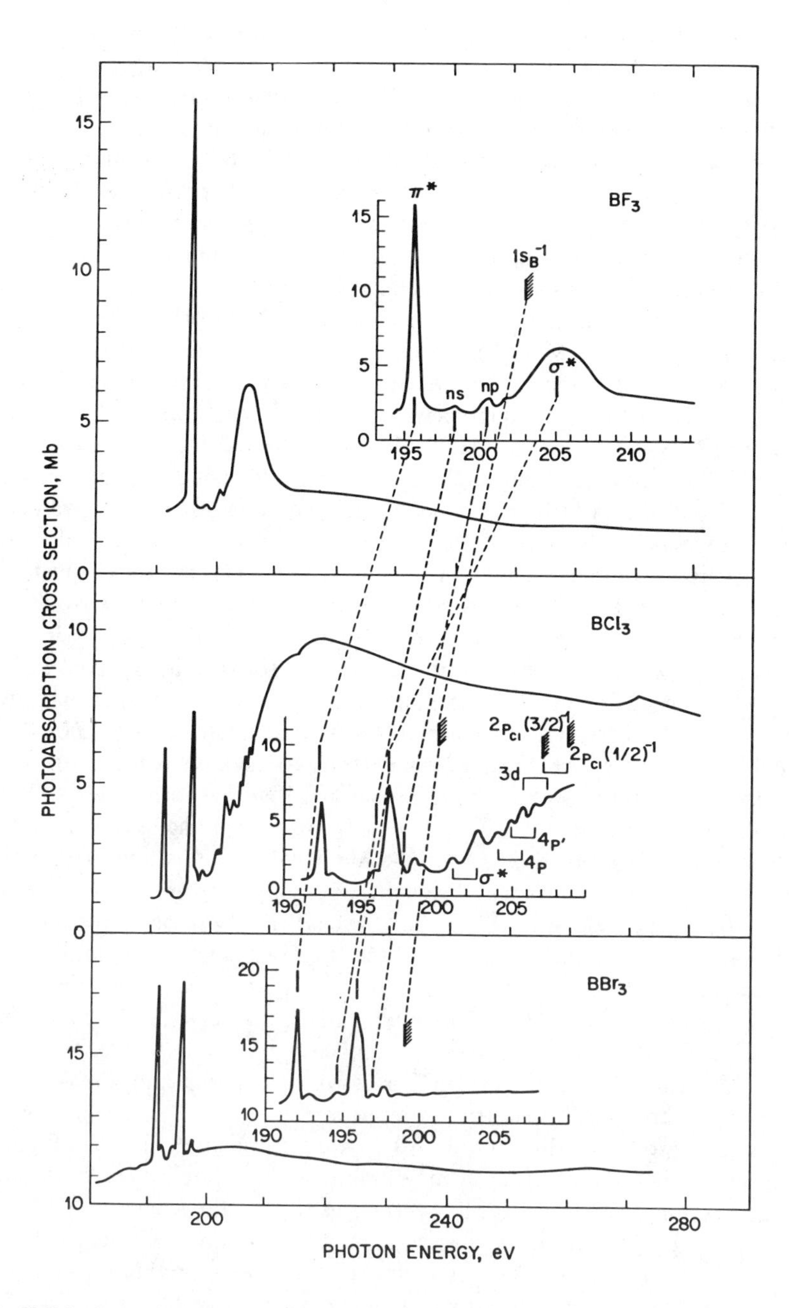

Fig. V.C-1. Inner-shell absorption spectra of the boron halides [I6]. Insets are on expanded scales.

The first Rydberg excitation, $1s_B \rightarrow 3s$, is assigned to the weak peak at 198.2 eV. Its term value of 37 100 cm^{-1} is rather larger than the 33 000 cm^{-1} expected in the outer shell, and therefore must be exhalted by antishielding (Section I.B). *Ab initio* calculations in a basis including Rydberg orbitals [I6] show that the $3sa_1$ MO in BF_3 is strongly mixed with its $a_1 2p\sigma^*$ valence conjugate [**I**, 24] and has an orbital extent about half-way between those of 2p valence and 3p Rydberg orbitals. The remaining band in the $1s_B$ spectrum is assigned to unresolved $1s_B \rightarrow 3p$ Rydberg excitations as indicated in Fig. V.C-1.

The valence excitation from $1s_B$ to $e'2p\sigma^*$ involves the strongly antibonding σ^*(B—F) MO's and so is considerably higher in frequency; it is assigned to the shape resonance 18 500 cm^{-1} beyond the $(1s_B)^{-1}$ ionization potential. In agreement with the $e'2p\sigma^*$ molecular orbital assignment, Swanson *et al.* [S88] interpret this shape resonance as a transition to an inner-well level of e' symmetry using the double-well model. That the $\mathscr{A}(1s_B, \sigma^*)$ band of BF_3 has a negative term value is totally anomalous in view of the positive (ϕ_i, σ^*) term values of other perfluorinated systems listed in Table I.C-V. However, it must be recognized that the B—F σ bond of BF_3 is an extremely strong one [E(B—F) = 153 kcal/mole] and so σ^* will be very high lying in this compound, Section I.C [I6]. A similar shape resonance is expected in the outer-shell spectrum of BF_3, though with a somewhat smaller negative term value. Interestingly, even for this shape resonance of BF_3 there is a calculated mixing of the valence level with its conjugate channel [I6].

Inasmuch as the $1s_B \rightarrow \sigma^*$ band of boron trifluoride terminates at a σ^* orbital, it is related to the $\mathscr{A}(1s_C, \sigma^*)$ bands of halogen-containing molecules [**I**, 156]. In the case of the alkyl halides, it is known that the $\mathscr{A}(1s_C, \sigma^*)$ band excitation frequencies drop sharply in the series F, Cl, Br, I, as the R—X bond strengths decrease, Section I.C. Consequently, it is no surprise to see that in boron trichloride and tribromide with their much lower B—X σ bond strengths, the corresponding $\mathscr{A}(1s_B, \sigma^*(\text{B—X}))$ bands fall *below* the $(1s_B)^{-1}$ ionization potential, Fig. V.C-1, in contrast to the case in boron trifluoride. The $(1s_B, \sigma^*)$ term value is 23 000 cm^{-1} in boron trichloride and 26 000 cm^{-1} in boron tribromide. The transitions from $1s_B$ to π^* and the higher Rydberg orbitals in BCl_3 and BBr_3 are traced out in the Figure, considering both the order-of-magnitude higher intensity of allowed valence excitations in inner-shell spectra and term-value trends.

Considerable structure is observed in BCl_3 immediately beyond the $(1s_B)^{-1}$ ionization limit. This is owing to transitions originating at $2p_{Cl}$, each of which is split into two components by exactly the spin-orbit interval in the $(2p_{Cl})^{-1}$ (^2E) positive ion. The first band in the $2p_{Cl}$

spectrum (201.0 eV) has a term value of 47 600 cm^{-1}, which compares well with the 49 600-cm^{-1} term value for the $2p_{Cl} \rightarrow \sigma^*$ inner-shell valence transition of $HCCl_3$ (Table I.C-II). Apparently $2p_{Cl} \rightarrow \pi_B^*$ is too weak to appear in BCl_3 for lack of spatial overlap. Transitions from $2p_{Cl}$ to components of 4p and 3d as indicated in the Figure have term values which lead to unambiguous assignments.

The division of the BF_3 inner-shell spectrum into valence excitations and Rydberg excitations as described above is supported by the X-ray spectra of the alkali-metal tetrafluoroborate salts [S30]. In forming BF_4^- from BF_3, the low-lying $a_2''2p\pi^*$ nonbonding MO of BF_3 is converted into a bonding σ/antibonding σ^* pair. Consequently, the $1s_B \rightarrow a_2''2p\pi^*$ band is absent in BF_4^-, while the $1s_B \rightarrow a_1\sigma^*$ and $1s_B \rightarrow t_2\sigma^*$ transitions are observed, with the latter Jahn-Teller split and appearing as a shape resonance above the $(1s_B)^{-1}$ ionization potential, as with the $1s_B \rightarrow \sigma^*$ band of BF_3 in Fig. V.C-1. Of course, the Rydberg excitations of the $1s_B$ spectrum do not appear in the spectra of solids such as these (Section I.B).

Theoretical work on vibronic profiles in the photoelectron spectrum of boron trifluoride raises some interesting possibilities regarding the Rydberg bands associated with the ionic states of the valence shell. Haller *et al.* [H5] calculate very strong vibronic mixing between the various outer-shell ionic states via Jahn-Teller active modes which can mix states of different symmetry. The corresponding Rydberg states converging upon the $^2A_1'$ and $^2E'$ ionic states also may be mixed strongly in this way.

V.D. Addendum

F. A. Gianturco, E. Semprini and F. Stefani, Final-state effects on the assignment of boron K-shell spectra in BF_3 molecules, *Nuovo Cimento* **2D**, 687 (1983).

CHAPTER VI

Amines, Phosphines, Arsine

VI.A. Ammonia and Amine Derivatives

A reassessment of the spectroscopic data on ammonia [**I**, 208; **II**, 312] and its interpretation has been carried out by Colson and coworkers [G18, G19, N24, N25] based upon the generation of considerable new MPI and one-photon absorption data on static and expansion-cooled samples of NH_3 and ND_3. The assignments of the Rydberg states of NH_3 originating at the nitrogen lone-pair orbital $3a_1$ as deduced from the one-photon absorption and MPI studies are summarized in Table VI.A-I. Suto and Lee [S84] present a very clear one-photon spectrum of ammonia from 50 000 to 100 000 cm^{-1} (*6.2 to 12.4 eV*) showing several of the transitions earlier identified in the MPI work.

The $3a_1 \rightarrow 3sa_1'$ two-photon excitation of ammonia is observed in the MPI spectrum at exactly the frequencies previously reported for one-

TABLE VI.A-I.

RYDBERG STATE ASSIGNMENTS[a] IN AMMONIA-h_3 FOR TRANSITIONS ORIGINATING AT $3a_1$

Upper Orbital[b]	Origin Frequency, cm^{-1} (*eV*)	Spectrum[c]	Term Value
$3sa_1'$	46 130 (*5.719*)	MPI, $1h\nu$	35 820
$3pe'$	59 200 (*7.34*) [d]	MPI, $1h\nu$	22 800
	62 010 (*7.6880*)	$1h\nu$	19 940
$3pa_2''$	63 880 (*7.920*)	MPI	18 070
$3de'$	69 764 (*8.6494*)	MPI, $1h\nu$	12 190
$3da_1'$	70 067 (*8.6870*)	$1h\nu$	11 880
$4sa_1'$	70 100 (*8.69*)	$1h\nu$	11 800
$4pe'$	71 315 (*8.8417*)	MPI	10 635
$4pa_2''$	72 900 (*9.0382*)	MPI	9050
$4de'$	75 205 (*9.3240*)	MPI, $1h\nu$	6740

[a] Reference [G19].

[b] Symmetry appropriate to the D_{3h} point group of the planar upper state.

[c] Multiphonon ionization (MPI); one-photon absorption ($1h\nu$).

[d] Assigned as a dynamic Jahn-Teller splitting [G19].

photon absorption.† The vibronic profile of this transition in electron impact shows a variation with scattering angle which has been explained as due to a pseudo-Jahn-Teller interaction involving the ground state via an a_2'' vibration [O2]. Resonance Raman spectroscopy within the $3a_1 \rightarrow 3sa_1'$ band documents the previously suspected occurrence of the symmetric N–H stretch (ν_1') along with the ν_2' bending in the Franck-

† In ammonia and the amines, the originating orbital symmetry is given in the C_{3v} point group appropriate to the pyramidal structure of the ground state, whereas the terminating orbital symmetry is given in the D_{3h} point group appropriate to the planar structure of the Rydberg states.

Condon envelope [Z6], as does the theoretical band-shape calculation of Avouris *et al.* [A36]. Two transitions earlier assigned by Walsh and Warsop [I, W10] as terminating at 4s and 5s also have been observed in the jet-cooled MPI spectrum, but rotational analyses lead instead to 3de' and 4de' terminating-orbital assignments. The $3a_1 \rightarrow 3pe'$ promotion appears in the MPI spectrum as a three-photon resonance in the 59 000–70 000 cm^{-1} (*7.3–8.7 eV*) region, with frequencies identical to those in the one-photon spectrum.

A new excited state in ammonia, observed as a three-photon resonance in the MPI spectrum, has an origin at 63 880 vacuum cm^{-1} (*7.9199 eV*) followed by a long vibrational progression in ν_2' (840–1040 cm^{-1}). This new transition is assigned as $3a_1 \rightarrow 3pa_2''$ on the basis of a rotational analysis. Following this, Glownia *et al.* [G19] then assign the 62 010-cm^{-1} (*7.6680-eV*) origin and its ν_2' progression as a part of the $3a_1 \rightarrow 3pe'$ system induced by a dynamic Jahn-Teller interaction involving ν_3' (e'). This transition in turn is built upon the $3a_1 \rightarrow 3pe'$ origin component at 59 200 cm^{-1} (*7.34 eV*). According to this analysis, the core splitting between $3pa_2''$ and the center of gravity of the 3pe' complex in ammonia amounts to approximately 2500 cm^{-1}. Various calculations [R12, R49] place the transition from $3a_1$ to 3pe' below that to $3pa_2''$ by 1000–3000 cm^{-1}. The corresponding transitions terminating within the $3a_1 \rightarrow 4p$ manifold do not appear in the one-photon spectrum, but are observed as three-photon resonances in the MPI spectrum at 71 315 vacuum cm^{-1} (*advert., 8.8417 eV*, $3a_1 \rightarrow 4pe'$) and 72 900 vacuum cm^{-1} (*advert., 9.0382 eV*, $3a_1 \rightarrow 4pa_2''$); the 4p core splitting has decreased to 1585 cm^{-1}. In accord with the $\Delta l = +1$ propensity rule on atomic intensities, the transitions from $3a_1$ to nd in ammonia are observed to be much more intense than those to ns or np. Excitations from $3a_1$ to 3de' and 3da_1' have origins at 69 764 and 70 067 cm^{-1} (*8.6494 and 8.6870 eV*) in the one-photon spectrum, and the corresponding bands involving 4d orbitals occur near 75 200 cm^{-1} (*9.32 eV*).

As with H_2O, calculations on the excited states of NH_3 show that all the lower states are predominantly Rydberg, having some σ^* valence character of the appropriate symmetry, but with no orbital predominantly valence in character [R12, R49]. This fits nicely with the earlier suggestion [I, 215] that the apparent continuum underlying the structured $3a_1 \rightarrow 3sa_1'$ transition is *not* due to the conjugate valence excitation $3a_1 \rightarrow a_1\sigma^*$. Modeling of the absorption band shape confirms that even though the continuous absorption underlying the vibronic components of the $3a_1 \rightarrow 3sa_1'$ transition amounts to 1/3 to 1/2 of the total oscillator strength in this region [G37, S84], this absorption is simply the result of overlapping vibronic-band tails [A36]. Note however that the mixing of

$a\sigma^*$ into 3s does not proceed nearly as far in NH_3 as in H_2O (Section VII.A), as evidenced by the close match of PES and Rydberg absorption profiles in the former (Section I.D).

The $3a_1 \rightarrow 3sa_1'$ and $3a_1 \rightarrow 3pe'$ transitions in ammonia have measured oscillator strengths of 0.066 and 0.00734 [S84], respectively, in accord with the upper limit of 0.08 per degree of degeneracy for Rydberg transitions in large molecules [I, 29]. The trend of $n_N \rightarrow 3s$ oscillator strength values† in the series of alkyl amines RNH_2 closely tracks that measured for the $n_O \rightarrow 3s$ transitions in the corresponding alcohols ROH [O1]; Berkowitz discusses the oscillator-strength distribution in ammonia [B34]. The $3sa_1 \rightarrow 3a_1'$ fluorescence emission and absorption of ND_3 have been studied and yield an oscillator strength for this transition of 0.080 [G37], which is at the upper limit of those found for molecular Rydberg transitions [I, 30]. In spite of the appreciable oscillator strength, the fluorescence quantum yield is only 8.3×10^{-5} owing to a competing predissociation to NH_2 + H. In this fragmentation, the 3s Rydberg orbital is transformed as the H atom moves outward, becoming more of a σ^* valence orbital at intermediate distances and eventually becoming a 1s orbital on the H atom at large N–H distance [E12, M76, R49]. The excited configurations $^3(3a_1, 3sa_1')$ and $^1(3a_1, 3pa_2'')$ also show such a "deRydbergization" along the N–H coordinate, as do the lower Rydberg states of H_2O (Section VII.A), CH_4, CH_3 and C_2H_5 (Chapter III), SiH_4 (Section IX.A) and PH_3 (Section VI.B).

It has been noted that the $3a_1 \rightarrow 3sa_1'$ transition of ammonia in the vapor phase is relatively insensitive to high-pressure perturbation [I, 83]. Careful measurements on the $3a_1 \rightarrow 3sa_1'$ and $3a_1 \rightarrow 3p$ systems pressurized with up to 140 atm of argon show that the bands noticeably broaden, but rather than shift to higher frequencies as normally happens, the bands shift in the opposite direction: -40 cm^{-1} for the transition to $3sa_1'$ and -200 cm^{-1} for the transition to $3pe'$ [M58]. As discussed in Section II.D, several systems are known in which the high-pressure perturbation induces low-frequency shifts at low density which then reverse rapidly at higher densities; it is likely that the NH_3/Ar system is another of this type.

Kassab *et al.* [K7] find theoretically that while placing two H_2 molecules within the space of the $3sa_1'$ Rydberg orbital of ammonia acts to increase the Rydberg character of that orbital significantly, it has only a

† The $3a_1$ orbital of ammonia is written as n_N in its derivatives.

small effect on the energy. On the other hand, calculations on NH_3 with He atoms as perturbers [K33] show that the $3sa_1'$ Rydberg orbital distorts significantly in its efforts to become orthogonal to the perturber orbitals, but there is no increase in Rydberg orbital size. For a perturbation consisting of six He atoms, the $(3a_1, 3sa_1')$, $(3a_1, 3pe')$ and $(3a_1, 3pa_2'')$ levels are calculated to be shifted to higher frequencies by 3800, 4400 and 4000 cm^{-1}, respectively. In a solid-argon matrix [C44], the $3a_1 \rightarrow 3sa_1'$ transition of ammonia-h_3 displays a vibrational progression of 1830 cm^{-1} $(2\nu_2')$, becoming 1330 cm^{-1} in ammonia-d_3. The temperature dependence of the band intensities suggests that there is no nuclear spin interconversion in the matrix and that the molecules are rotating freely.

The transition to the $^3(3a_1, 3sa_1')$ level in ammonia is centered at *ca.* 48 000 cm^{-1} (*5.9 eV*), and is thought to be the lowest triplet state of this molecule; the "triplet" reported at 36 000 cm^{-1} (*vert., 4.5 eV*) would seem to be an artifact [A3, J6]. As is the case with the band at 36 000 cm^{-1} (*vert., 4.5 eV*) in water [E3], that in ammonia at 36 000 cm^{-1} may be due to inelastic scattering from ammonia adsorbed on the spectrometer slits. The singlet-triplet split of 3500 cm^{-1} for the $(3a_1, 3sa_1')$ configuration of ammonia is somewhat larger than expected for a $(\phi_i^o, 3s)$ Rydberg configuration, but much smaller than that observed for the corresponding configuration of methane (Section III.A).

Two peaks in the dissociative-electron-attachment spectrum of ammonia are thought to coincide with the formation of temporary negative ions [S44]. Coming at 45 600 and 84 700 cm^{-1} (*vert., 5.65 and 10.5 eV*), the first of the attachment peaks is just below the $3a_1 \rightarrow 3sa_1'$ transition in the neutral molecule (51 500 cm^{-1} *vert., 6.38 eV*). According to the discussion in Section II.B, the 5900-cm^{-1} difference between the neutral-molecule and negative-ion frequencies is a clear sign that the 45 600-cm^{-1} transition is a Feshbach resonance having the $^2(3a_1, 3sa_1'^2)$ configuration. It is natural then to assume that the TNI resonance at 84 700 cm^{-1} represents a second Feshbach resonance, $^2(1e, 3sa_1'^2)$, in which case the $1e \rightarrow 3sa_1'$ neutral-molecule transition is expected at about 5000 cm^{-1} higher frequency, *i.e.*, at 90 000 cm^{-1} (*11.2 eV*). This agrees exactly with the broad peak in the electron-impact spectrum of ammonia at 91 100 cm^{-1} (*vert., 11.3 eV*) [I, 209] having a term value of 37 100 cm^{-1} with respect to the $(1e)^{-1}$ ionization potential. The geometric structure calculated for the ^2E core of NH_3^+ [K34, R45] no doubt is applicable as well to the $(1e, 3sa_1')$ Rydberg and $^2(1e, 3sa_1'^2)$ Feshbach resonance states. Still missing for ammonia are the valence TNI resonances to $a_1\sigma^*$ and $e\sigma^*$ MO's, transitions which already have been reported for phosphine and arsine [E2] in the 16 000–48 000 cm^{-1} (*2–6 eV*) region (Section VI.B).[†]

By virtue of its unit positive charge, the ammonium ion can bind an electron even though it is a closed-shell system. The odd electron however, must be accommodated in a Rydberg orbital, even in the ground state. Recall that for the CH_4 molecule, the (ϕ_i^o, 3s) and (ϕ_i^o, 3p) outer-shell term values are 31 600 and 21 000 cm^{-1}, respectively. In the NH_4 radical, the nuclear charge is larger by +1 than in CH_4, while this is shielded by the sixth electron in the $1t_2$ MO. Because this shielding is imperfect, the Rydberg term values for NH_4 will be close to but somewhat larger than those in CH_4. By a similar argument, the Rydberg term values for the H_3O radical will be somewhat larger than those for NH_3. Several high-quality calculations on Rydberg radicals strongly support the term-value approach for these systems [B65, H24, M50, R7]. Thus in the calculation of Raynor and Herschbach [R7], the electron in the $3sa_1$ MO of NH_4 in its ground state is calculated to be bound by 32 500 cm^{-1}, whereas an electron in the 3p orbital is bound by 20 700 cm^{-1} and an electron in the e and t_2 components of 3d is bound by 12 400 and 13 200 cm^{-1}, respectively. The very close fit of these calculated values to our general expectations regarding polyatomic Rydberg term values [**I**, 65] suggests strongly that the ground-state Rydberg radicals are very little different (if at all) from their excited-state relatives, and that the penetration of the Rydberg orbital into the core has very little to do with whether the core has an open- or closed-shell electronic arrangement.[§] As with the other first-row hydrides in (ϕ_i^o, $3sa_1$) configurations, the ground state of NH_4 is calculated to predissociate to H + NH_3 by tunneling through a low barrier [M50].

Experimentally, Rydberg-to-Rydberg emissions [H45, H46] and absorptions [W16] of NH_4 and ND_4 radicals have been identified in the visible region. A transition having an origin at 14 828 cm^{-1} (*1.8384 eV*) in ND_4 is assigned as one between 3s and 3p Rydberg orbitals [W10, W16]. This assignment is supported by the term-value perspective, for the

[†] Note that there are no virtual orbitals of predominant σ^* character in the MO manifold of the neutral NH_3 molecule because the σ^* orbitals are totally dissolved in the sea of surrounding conjugate Rydberg orbitals. However, in the negative ion, the optical electron does not experience a $-1/r$ potential and so does not have a set of Rydberg orbitals in which the σ^* MO's can be dissolved (Section I.B).

[§] Friedrich *et al.* however have a contrary view, claiming that exchange effects are significant in Rydberg term-value measurements [F33].

$3sa_1 \rightarrow 3pt_2$ Rydberg-to-Rydberg excitation is expected to have a frequency equal to the difference of (ϕ_i^o, 3s) and (ϕ_i^o, 3p) term values, *i.e.*, 33 000 − 18 000 = 15 000 cm^{-1}. The suggestion that the 14 828-cm^{-1} transition corresponds to $3de \rightarrow 3pt_2$ [H45] does not agree with the term-value argument, which would place this transition at *ca.* 6000 cm^{-1}. Assignment of a second transition at 17 600 cm^{-1} (*2.18 eV*) in NH_4 is more problematic, for this value is a few thousand cm^{-1} smaller than the frequency difference expected between 3s and 3d orbitals and several thousand cm^{-1} smaller than the frequency difference expected between 3s and 4s. Since both of these options are only allowed vibronically, the transition could appear shifted to low frequency by one quantum of the ground-state t_2 vibration. Herzberg and Hougen [H46] offer a tentative $3dt_2 \rightarrow 3sa_1$ assignment for the band at 17 600 cm^{-1}.

High-quality inner-shell excitation spectra are available for ammonia and methyl amine, Figs. III.A-2 and VI.A-1 [W19]. In ammonia, the $1s_N \rightarrow 3sa_1$ excitation in the 400 eV region has a term value of 40 300 cm^{-1}; this is larger than the outer-shell term values of 35 760 ($3a_1$, $3sa_1'$) and 37 100 cm^{-1} ($1e$, $3sa_1'$) due to the antishielding of the $1s_N$ hole (Section I.B). This transition in ammonia is more intense than the corresponding bands of methane and neon, but weaker than that of water [A7]. In ammonia there next follows an inner-shell Rydberg transition to 3p with a very large term value of 27 400 cm^{-1}. This is reminiscent of the situation in water (Section VII.A) where the (n_O, σ^*) valence conjugate is strongly mixed into the (n_O, 3p) manifold to push one 3p component strongly downward, leading to a seemingly anomalous term value. The second transition to 3p has a 21 000-cm^{-1} term value in the optical spectrum [A7], but is missing from the electron impact spectrum. Thus it is seen that the 3p splitting in the $1s_N$ spectrum amounts to 6400 cm^{-1} (4600 cm^{-1} if only the optical values are used), while the splitting of the 3p manifold in the outer-shell spectrum of ammonia is known from the $3a_1 \rightarrow 3p$ transitions to amount to 4730 cm^{-1} (*vide supra*), with extreme term values of 22 800 and 18 070 cm^{-1} [G19]. The assignment of an inner-shell transition in ammonia with a term value of 16 900 cm^{-1} (Fig. VI.A-1) is in limbo, for it is several thousand cm^{-1} too low for assignment as $1s_N \rightarrow 3p$ (Wight and Brion [W19] choose this assignment) and a few thousand cm^{-1} too high for assignment as $1s_N \rightarrow 4s$ [A7]. Of the two, we prefer the latter choice. In fact, the 27 400 and 16 900-cm^{-1} term values both might be anomalous due to an unusually strong interaction between them induced by the antishielding. A similar peak with an anomalous 16 600-cm^{-1} term value also appears in the inner-shell spectrum of methane (Section III.A). An inner-shell transition in ammonia with a 12 100-cm^{-1} term value is unambiguously assigned as

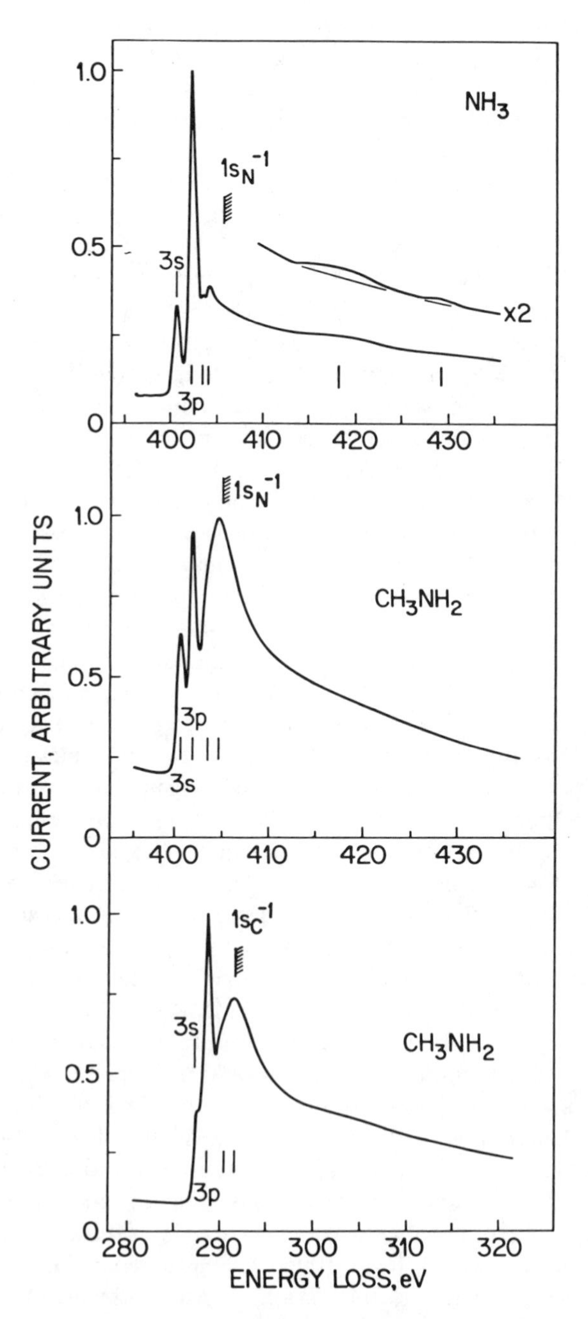

Fig. VI.A-1. Inelastic electron-scattering spectra of ammonia and methyl amine in the 1s regions at an impact energy of 2500 eV [W19].

$1s_N \rightarrow 3d$. Though it may be that the σ^* MO's are playing an indirect role in the Rydberg core splitting, it is clear from the observed band widths that none of the observed inner-shell spectral bands of ammonia below threshold directly involve σ^* MO's. Two broad bands beyond the $1s_N$ threshold of ammonia are either shake-up bands involving σ^* valence MO's, shape resonances, or EXAFS interferences, Section II.C. A calculation would be most welcome on the extreme splitting of the ($1s_N$, np) and ($1s_N$, 3d) configurations of ammonia by the core asymmetry and the possible roles of the σ^* MO's in this.

Methylation of ammonia reduces the $1s_N \rightarrow 3s$ term value from 40 300 cm^{-1} to 36 200 cm^{-1}, Fig. VI.A-1 [W19]. The corresponding $1s_C \rightarrow 3s$ term value, 33 100 cm^{-1}, is lower still. Predictably (Section I.B), the outer-shell (n_N, 3s) term value of methyl amine is smaller than both of these, being 31 200 cm^{-1} [I, 210]. Thus one sees that the 1s hole antishielding is operating in these systems so as to increase the inner-shell (ϕ_i^{-1}, 3s) term values while at the same time alkylation is operating to reduce them. In methyl amine, the first transition to 3p has a term value of 25 800 cm^{-1} in the $1s_N$ spectrum and 25 000 cm^{-1} in the $1s_C$ spectrum, thus mirroring the anomalously high value found for ammonia.

In the outer-shell spectrum of nitrogen trifluoride, $n_N \rightarrow \sigma^*$ and $n_N \rightarrow 3s$ transitions were observed with term values of 40 100 and 34 900 cm^{-1}, respectively [I, 228; **II**, 315]. Thanks to the $1s_N$ antishielding effect (Section I.C), one expects an approximately 8000 cm^{-1} exhaltation of the ($1s_N$, σ^*) term value compared with that for (n_N, σ^*), raising it to 48 000 cm^{-1}. This fits nicely with the intense, broad features observed in the $1s_N$ and $1s_F$ absorption spectra of NF_3 having term values of 52 000 cm^{-1} [B13, B63, V26]. On the same basis, the relatively weak $1s_N \rightarrow 3s$ Rydberg transition is expected to have a term value of approximately 39 000 cm^{-1}, and is probably hidden in the high-frequency wing of the conjugate $\mathscr{A}(1s_N, \sigma^*)$ band. The appearance of low-lying $\mathscr{A}(1s_N, \sigma^*)$ bands in NF_3 contrasts strongly with the situation in BF_3, where the $\mathscr{A}(1s_B, \sigma^*)$ band is observed as a shape resonance (Section V.C). This is a consequence of the B—F bond strength (153 kcal/mole) being so much larger than that for the N—F bond (67 kcal/mole) [K47]. In the intermediate case of CF_4 with a bond strength of 115 kcal/mole, the $\mathscr{A}(2p, \sigma^*)$ bands are quasi-degenerate with the Rydberg excitations originating at the same MO's. A sharp transition in the $1s_N$ spectrum of NF_3 has a 17 000-cm^{-1} term value indicating a $1s_N \rightarrow 3p$ assignment. Between this sharp band and the $\mathscr{A}(1s_N, \sigma^*)$ transition, a very weak feature is reported with a term value of 27 000 cm^{-1}; this possibly is the second component of $1s_N \rightarrow 3p$. Though the inner-shell 3p term values of

NF_3 (27 000 and 17 000 cm^{-1}) are very close to those of the second and third bands of the inner-shell spectrum of NH_3 (27 400 and 16 900 cm^{-1}), the intensity ratios in the two situations are reversed.

In the (n_N, 3s) upper state of methyl amine-d_2 [T25, **I**, 215] in the vicinity of 42 000 cm^{-1} (*5.2 eV*), detailed analysis reveals that the methyl group is in free rotation. The peak of this transition (47 000 cm^{-1}, *5.8 eV*) is 4000 cm^{-1} above a TNI resonance at 43 000 cm^{-1} (*vert., 5.3 eV*) [A3], thus identifying the resonance state as $^2(n_N, 3s^2)$. The singlet-triplet split in the (n_N, 3s) configuration of methyl amine is 2700 cm^{-1} [A3]. Two transitions in trimethyl amine at 72 500 and 83 000 cm^{-1} (*vert., 8.99 and 10.3 eV*) are beyond the first ionization limit [G41], but are assignable as Rydberg transitions associated with the Jahn-Teller components of the $(4e')^{-1}$ ionizations at 99 200−102 700 cm^{-1} (*vert., 12.3−12.73 eV*) [K18]. Trialkyl amines larger than $(CH_3)_3N$ have continuous absorption in the 60 000−85 000 cm^{-1} (*7.4−10.5 eV*) region [G41].

Two transitions are observed in the MPI spectrum of ethylenimine [**I**, 220], both of the Rydberg type [R32]. The first is the $n_N \rightarrow 3s$ transition, with broad vibronic structure in the 47 000−53 000 cm^{-1} (*5.8−6.6 eV*) region. Though the one-photon $n_N \rightarrow 3s$ transition shows no vibronic structure, the MPI band of the same transition (two-photon resonant) prominently displays quanta of 920 (N−H wag) and 1340 cm^{-1}. The second MPI transition has its origin at 53 250 cm^{-1} (*6.602 eV*), whereas the nearest origin in the one-photon spectrum is at 55 550 cm^{-1} (*6.887 eV*). These are two of the three $n_N \rightarrow 3p$ components, with adiabatic term values of 19 860 and 22 160 cm^{-1}. An *ab initio* calculation [R6] confirms the Rydberg assignments given above. Using electron-impact spectroscopy, Fridh [F32] has extended the ethylenimine spectrum so as to newly uncover intense transitions at 105 000 and 115 000 cm^{-1} (*vert., 13.0 and 14.3 eV*), which he assigns as valence excitations. These may relate to the giant resonance (Section I.D) discovered for cyclopropane at *ca.* 84 000 cm^{-1} (*vert., 10.5 eV*, Section III.E).

The CD spectrum of the simplest optically-active amine, propylenimine in the *anti* conformation, has been recorded in the vapor phase and in solution [G32]. The $n_N \rightarrow 3s$ transition, a clear shoulder at 50 000 cm^{-1} (*vert., 6.2 eV*) in the vapor phase, has a term value of 27 200 cm^{-1} [Y2], to be compared with 29 700 cm^{-1} in ethylenimine. Note however that ethylenimine has quite regular 3p term values of 19 860 and 22 160 cm^{-1} [**I**, 210] whereas the structured bands corresponding to $n_N \rightarrow 3p$ excitations in propylenimine have term values of 22 900−26 000 cm^{-1}. These latter bands indeed are Rydberg, in spite of their anomalous term values, as witnessed by their disappearance in *n*-hexane solution, however they may not correspond to the two excitations to 3p observed in ethylenimine. The $n_N \rightarrow 3p$ transition in trimethylenimine, isomeric with propylenimine, has a normal (n_N, 3p) term value of 20 700 cm^{-1} [C29]. In piperidine the imino hydrogen is either axial or equatorial [A6] and the Rydberg transitions to 3p from n_N are separated by 4000 cm^{-1} in the two conformers.

MPI spectra of the first Rydberg excitation of diazabicyclooctane (DABCO) in the 36 000−38 000 cm^{-1} (*4.5−4.7 eV*) region [G23, P14] convincingly show the transition to be one-photon forbidden/two-photon allowed, as required by the $n_+ \rightarrow 3sa_1'$ assignment [**I**, 226]. A component of the $n_+ \rightarrow 3p$ Rydberg excitation in DABCO at 40 000−43 500 cm^{-1} (*4.96−5.39 eV*)† displays a clear *A* term in the MCD spectrum [H1] and therefore can be assigned as $n_+ \rightarrow 3pe'$ [P11, P14]. Absorption in ABCU, a relative of ABCO (the monoaza derivative of DABCO) in which each of the three dimethylene bridges is replaced with a trimethylene bridge, displays peaks at 35 932, 38 700 and 43 197 cm^{-1} (*advert., 4.4549, 4.80 and 5.3556 eV*) [H8]. These transitions have term values of 20 700, 17 900 and 13 400 cm^{-1}, respectively, with respect to the $(n_N)^{-1}$ ionization potential at 56 600 cm^{-1} (*adiabat., 7.02 eV*) [**I**, R20] and clearly terminate at 3s, 3p and 3d Rydberg orbitals, respectively. Being only of C_{3v} symmetry, ABCO has a two-photon resonant MPI spectrum identical to its one-photon absorption spectrum [P11].

Rather novel experiments on DABCO and on its monoaza parent ABCO [**I**, 223; **II**, 313] lead to the observation of exceedingly high Rydberg levels in these molecules [F9, F45]. Using two-photon excitation, the DABCO molecule is first placed in a particular vibronic sublevel of the (n_+, $3sa_1'$) Rydberg state. A second laser then excites the electron in the 3s orbital into higher *n*p and *n*d Rydberg orbitals. Such $3s \rightarrow n,l,\lambda$ Rydberg-to-Rydberg transitions are pseudo-atomic in regard the Franck-Condon envelopes, Fig. VI.A-2, and so can be followed to *n* values beyond 40! Being pseudo-atomic, there are no vibrational sub-bands in the Rydberg-to-Rydberg spectrum. In the D_{3h} point group of DABCO, excitations from $3sa_1'$ are allowed to npe', npa_2'' and to nde' orbitals. Indeed, three Rydberg series can be constructed from the peaks in Fig. VI.A-2, with

† In DABCO, the n_+ orbital is the higher-energy lone pair MO, consisting of the in-phase combination of nitrogen lone pair AO's coupled through the σ framework of the cage.

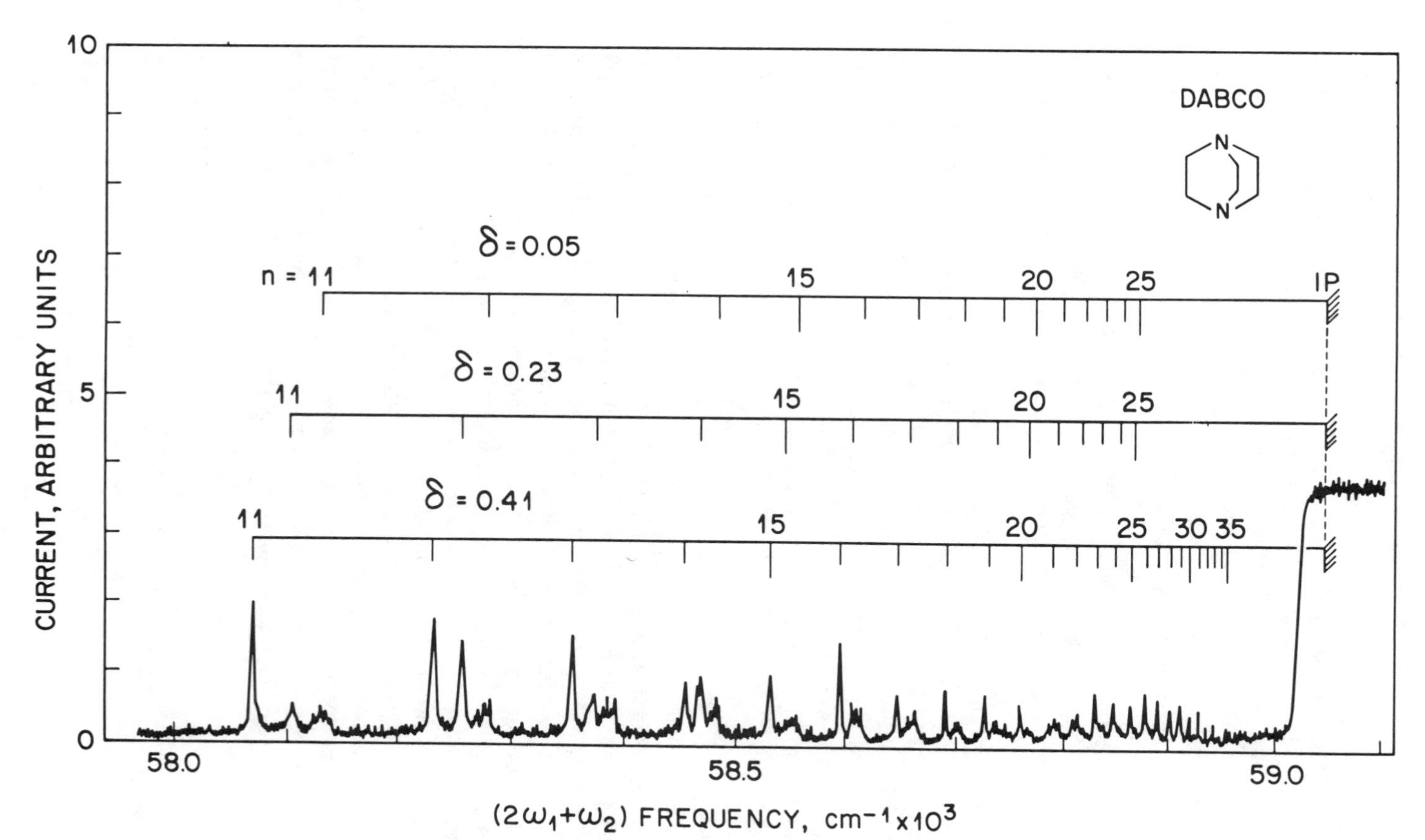

Fig. VI.A-2. Two color Rydberg-to-Rydberg MPI spectrum of DABCO cooled in a free-jet expansion [F45].

quantum defects of approximately 0.5, 0.2 and 0.0. While the δ values of 0.5 and 0.0 are commonplace for series terminating at np and nd orbitals, respectively, the 0.2 value is assigned to the second np series with reluctance. The fate of molecules in such high Rydberg states is reviewed by Freund [F27].

VI.B. Phosphines and Arsine

According to calculations, the continuous nature of the $n_P \rightarrow 4s$ absorption band of phosphine† (55 700 cm^{-1} *vert., 6.90 eV*) [**I**, 231] is due to predissociation along the P—H coordinate in the upper state [M77]. As with ammonia, the Rydberg orbital in the lowest singlet Rydberg state of phosphine is centered upon the central atom when in the ground-state geometry, but as the H and PH_2 fragments separate, the 4s Rydberg orbital on the phosphorus atom rapidly collapses into the 1s orbital of the departing H atom via the σ^*(P—H) valence MO.

The $2p_P$ inner-shell spectrum of phosphine (Fig. VI.B-1) [F33] looks much like that of silane and has been assigned in a parallel manner using term values [R31]. Thus the first broad region of absorption is resolved into a $2p_P \rightarrow \sigma^*$ valence excitation (39 200-cm^{-1} term value) and a $2p_P \rightarrow 4s$ Rydberg excitation (34 400-cm^{-1} term value); each of the bands is split by the spin-orbit coupling of the $2p^5$ core configuration in the upper states. The ($2p_P$, 4s) term value of 34 400 cm^{-1} is reasonable in view of the outer-shell (n_P, 4s) term value of 31 000 cm^{-1} and the exhaltation of the inner-shell term value expected due to antishielding (Section I.B). In solid PH_3 [F33], Fig. VI.B-1, the inner-shell $\mathscr{A}(2p_P, \sigma^*)$ band is shifted to lower frequency, whereas the conjugate Rydberg excitation has shifted strongly to higher frequency, as is characteristic of a Rydberg excitation in a matrix of moderate electron mobility. These excitations of phosphine in the vapor phase are followed by transitions to higher np and nd levels as shown in the Figure. The pattern of levels in phosphine is very much like that deduced for silane, Fig. IX.A-1. The theoretical work of Friedrich *et al.* [F33] reverses the assignments of the two features in phosphine here assigned as the $\mathscr{A}(2p_P, \sigma^*)$ band and the ($2p_P$, 4s) Rydberg band,

† The 3p and 4p lone-pair orbitals of PH_3 and AsH_3 are written here as n_P and n_{As}, respectively.

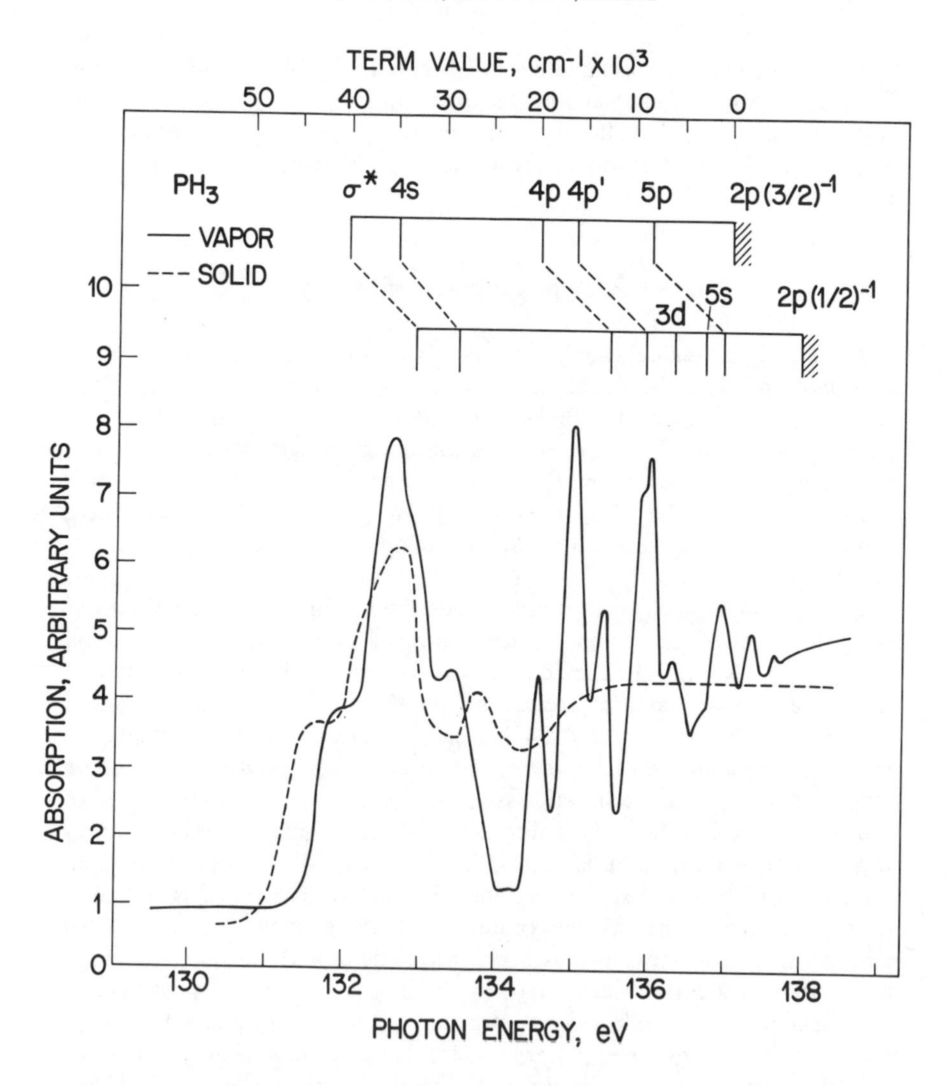

Fig. VI.B-1. Inner-shell absorption spectra of phosphine in the vapor phase (—) and as a neat polycrystalline film at 50 K (- -) [F33].

however this reversal leads to the Rydberg and valence excitations in the solid each shifting in a manner opposite to that normally assigned to its orbital type, Section II.D. These broad bands are followed by sharper structures, of which the lowest two have been assigned as $2p_P \rightarrow 4p$ and $2p_P \rightarrow 3d$ [R31]. On the other hand, Schwarz *et al.* [F33, S27] prefer assignments in which 4p is replaced with 5s, thereby setting 5s *below* 3d.

The appearance of low-lying excitations to σ^* in the inner-shell spectrum of phosphine but not in its outer-shell spectrum is understandable from the far larger σ^* inner-shell term value exhaltation produced by antishielding compared to that for inner-shell Rydberg excitations (Sections I.B and I.C). With inner-shell σ^* and 4s term-value exhaltations of 10 000 and 3000 cm^{-1}, respectively (Section I.C), one is then led to the conclusion that the outer-shell transitions to σ^* and 4s in PH_3 are nearly degenerate and strongly mixed. This may account for the continuous nature of the $n_P \rightarrow 4s$ transition of PH_3. It is especially intriguing to note that though there is clear evidence for a low-lying transition to σ^* in the inner-shell spectrum of phosphine while a corresponding outer-shell transition cannot be seen, a similar situation holds for silane (Section IX.A) but not ammonia. Such behavior places phosphine and silane in Regime Y, Fig. I.D-3.

Electron impact on both phosphine and arsine yield negative-ion fragments with peak efficiencies at or close to energies corresponding to temporary-negative-ion (TNI) formation [E2]. Focusing on PH_3, the TNI resonances are observed at 18 000, 44 000, 51 000, 60 000 and 66 000 cm^{-1} (*vert., 2.2, 5.4, 6.3, 7.5 and 8.2 eV*). Using the general rule that Feshbach resonances fall approximately 4000 cm^{-1} below the corresponding neutral-molecule parent Rydberg states (Section II.B), one sees that the (n_P, 4s), (n_P, 4p) and (n_P, 3d) parent states of phosphine at 55 500, 68 000 and 72 200 cm^{-1} (*vert., 6.88, 8.43 and 8.95 eV*) [**I**, 234] reasonably correlate with $^2(n_P, 4s^2)$, $^2(n_P, 4p^2)$ and $^2(n_P, 3d^2)$ Feshbach resonances at 51 000, 60 000 and 66 000 cm^{-1}. Most interestingly, the existence of a $^2(n_P, 3d^2)$ Feshbach configuration in phosphine strongly supports the idea [**I**, 229] that transitions of the sort $3p \rightarrow 3d$ in phosphorus compounds are Rydberg even though the principal quantum number in the upper level does not increase beyond that of the outer shell. Parallel Feshbach assignments [$^2(n_{As}, 5s^2)$ and $^2(n_{As}, 5p^2)$] [**I**, 234] hold for the arsine TNI resonances observed at 50 000 and 61 000 cm^{-1} (*vert., 6.2 and 7.6 eV*) [E2]. The two valence TNI resonances at *ca.* 17 000 and 43 000 cm^{-1} (*vert., 2.1 and 5.3 eV*) in both phosphine and arsine involve occupation of the $a_1\sigma^*$ and $e\sigma^*$ valence MO's by the incident electron.

The inner-shell spectrum of PCl_3 [T13] is a most interesting affair. In excitations from the 2p level of the P atom, very high term values of 52 400 and 48 400 cm^{-1} are observed, clearly indicating $\mathscr{A}(2p_P, \sigma^*(P{-}Cl))$ band transitions. When originating with the 2p levels of the Cl atoms, the corresponding term values are 46 800 and 36 800 cm^{-1} in PCl_3, and the equivalent inner-shell $\mathscr{A}(2p_{Si}, \sigma^*(Si{-}Cl))$-band transitions of $SiCl_4$ have term values of 51 000 and 37 500 cm^{-1}. The splittings between the pairs of $\mathscr{A}(n\text{p}, \sigma^*)$ bands in the spectra represent the e/a_1 splitting (PCl_3) or t_2/a_1 splitting ($SiCl_4$) among the σ^* local MO's. The experimental spectra also are doubled due to the $^2P_{3/2} - {}^2P_{1/2}$ spin-orbit splitting of the $2p^5$ core configuration. It is most peculiar that the $2p \rightarrow ns$ transitions do not appear in either the $2p_P$ or $2p_{Cl}$ spectra of PCl_3, but that $2p \rightarrow 4p$ (21 000 cm^{-1} term value in the P spectrum and 23 100 cm^{-1} in the Cl spectrum) and $2p \rightarrow 5p$ (9600 cm^{-1} term value in the P spectrum and 10 600 cm^{-1} in the Cl spectrum) are clearly evident. What appears to be a shake-up transition in the $2p_{Cl}$ spectrum is spin-orbit split by 12 500 cm^{-1}.

The outer-shell spectrum of PCl_3 [I, 241] has a pair of $\mathscr{A}$-band-like transitions at 46 100 and 48 600 cm^{-1} (*vert., 5.72 and 6.02 eV*) and another at 57 200 cm^{-1} (*vert., 7.09 eV*) which most likely terminate at σ^*(P—Cl) orbitals, but which may originate at either n_P [ionization potential 84 850 cm^{-1} (*vert., 10.52 eV*)] or $3p_{Cl}$ levels [ionization potential 94 450 cm^{-1} (*vert., 11.71 eV*)]. Comparison with the $\mathscr{A}(3\text{p}, \sigma^*)$-band term values of $HCCl_3$ (Table I.C-II) shows that the first two transitions originate at n_P (with term values of 38 750 and 36 250 cm^{-1}) and the third at $3p_{Cl}$ (with a term value of 37 250 cm^{-1}).

The reflection spectrum of liquid hexamethyl phosphoramide, $[(CH_3)_2N]_3P{=}O$, reveals three one-electron valence excitations centered at 48 000, 60 000 and 93 000 cm^{-1} (*6.0, 7.5 and 11.5 eV*) [B47]. In addition to these, the energy-loss function shows an intense peak at 170 000 cm^{-1} (*vert., 21 eV*) attributed to a many-electron collective excitation (plasmon), Section II.D. See Fig. II.D-4 for plasmon excitations in other liquids.

Electronic excitations in PF_6^-, AsF_6^-, AsF_3 and AsF_5 are discussed in Section XXI.A.

VI.C. Addendum

F. Alberti, K. P. Huber and J. K. G. Watson, Absorption spectrum and analysis of the ND_4 Schüler band, *J. Mol. Spectrosc.* **107**, 133 (1984).

M. Fujii, T. Ebata, N. Mikami and M. Ito, Two-color multiphoton ionization and fluorescence dip spectra of diazabicyclo[2.2.2]octane in a supersonic free jet. Rydberg states ($n = 5-39$) and autoionization, *J. Phys. Chem.* **88**, 4265 (1984).

V. Vaida, W. Hess and J. L. Roebber, The direct ultraviolet absorption spectrum of the $\tilde{A}\ ^1A_2'' \leftarrow \tilde{X}\ ^1A_1$ transition of jet-cooled ammonia, *J. Phys. Chem.* **88**, 3397 (1984).

R. N. S. Sodhi, C. E. Brion and R. G. Cavell, Inner shell excitation, valence excitation and core ionization in NF_3 studied by electron energy loss and X-ray photoelectron spectroscopies, *J. Electron Spectrosc. Related Phenomena* **34**, 373 (1984).

J. Kaspar and V. H. Smith, Jr., Rydberg transitions in the ammonium radical; scattered-wave local-spin-density calculations, *Chem. Phys.* **90**, 47 (1984).

K. Hirao, SAC-CI calculations for Rydberg levels of the ammonium radical, *J. Am. Chem. Soc.* **106**, 6283 (1984).

A. M. Weber, A. Acharya and D. H. Parker, Resonance-enhanced multiphoton ionization spectroscopy of ABCO and ABCU: core splitting of the 3p Rydberg orbitals, *J. Phys. Chem.* **88**, 6087 (1984).

M. Furlan, M. J. Hubin-Fraskin, J. Delwiche, D. Roy and J. E. Collin, High resolution electron energy loss spectroscopy of NH_3 in the 5.5 to 11 eV energy range, submitted to *J. Chem. Phys.*

R. N. S. Sodhi and C. E. Brion, Electronic excitation in phosphorus containing molecules. I. Inner shell electron energy loss spectra of PH_3, $P(CH_3)_3$, PF_3 and PCl_3, submitted to *J. Electron Spectrosc. Related Phenomena.*

R. N. S. Sodhi and C. E. Brion, Electronic excitation in phosphorus containing molecules. II. Inner shell electron energy loss spectra of PF_5, OPF_3 and $OPCl_3$, submitted to *J. Electron Spectrosc. Related Phenomena.*

R. N. S. Sodhi and C. E. Brion, Electronic excitation in phosphorus containing molecules. III. Valence shell electron energy loss spectra of $P(CH_3)_3$, PCl_3, PF_3, $OPCl_3$ and PF_5, submitted to *J. Electron Spectrosc. Related Phenomena.*

R. N. S. Sodhi and C. E. Brion, Inner shell electron energy loss spectra of the methyl amines and ammonia, submitted to *J. Electron Spectrosc. Related Phenomena.*

CHAPTER VII

Oxo Compounds

What follows below on the spectroscopic properties of water and its alkyl derivatives can be summarized in three words: Rydberg, Rydberg, Rydberg. In this sense, the spectra of the oxo compounds closely resemble those of the alkanes and amines. Though there is considerable evidence for the mixing of the Rydberg states of oxo compounds with their conjugate valence configurations, questions remain as to the quantitative extent of the mixing throughout a given Rydberg series, and the dependence of this mixing on the originating orbital. Further, after mixing, do distinct levels of predominantly σ^* valence character still exist? If so, where are they? These are the most important unanswered questions one can ask in regard the electronic structures of the first-row hydrides and their alkyl derivatives.

VII.A. Water and Hydrogen Peroxide

Water Vapor

Theory and experiment on the water molecule [**I**, 245; **II**, 316] are converging upon a common description of the electronic excitations originating at $1b_1$, the uppermost lone-pair orbital. Calculations in large basis sets uniformly predict that the lowest 20–30 excited states of water are Rydberg in nature [B74, D21, D57, G20, S68, W2], in agreement with the conclusions derived from considerations of term values and condensed-phase perturbation effects [I, 245]. The agreement is not as harmonious in the realm of valence excited states of water. Thus, in one calculation the $1b_1 \rightarrow a_1\sigma^*$ transition is said to fall at 177 000 cm^{-1} (*vert., 22 eV*) [W28], Wang *et al.* place it at 60 000 cm^{-1} (*vert., 7.4 eV*) [W6], while in other calculations, the $(1b_1, a_1\sigma^*)$ valence configuration conjugate to $(1b_1, 3sa_1)$ is calculated to be mixed into all the $(1b_1, nsa_1)$ Rydberg levels to such an extent that there is no final level containing enough of this configuration to be called "$a_1\sigma^*$" [D21, S68]. Since the latter calculations are the most comprehensive, we take the view that the $(1b_1, a_1\sigma^*)$ valence configuration of water is totally dissolved in the conjugate Rydberg sea. The $(1b_1, a_1\sigma^*)$ valence configuration appears in highest concentration when mixed with $(1b_1, 3sa_1)$ and accounts for the large oscillator strength calculated for the $1b_1 \rightarrow 3sa_1$ transition compared to those calculated for the higher $1b_1 \rightarrow nsa_1$ excitations.

Though the excitation to $^1(1b_1, \sigma^*)$ in water cannot be seen directly, the effect of its dissolution in the Rydberg sea is evidenced by comparison of the $(1b_1)^{-1}$ PES band shape and that of the optical excitation to $^1(1b_1, 3sa_1)$. If the Rydberg excitation is not mixed with its valence conjugate, then its Franck-Condon envelope and that of the corresponding PES band are identical, whereas mixing of the valence conjugate into the Rydberg state but not the ionic state results in significantly different band shapes. As shown in Fig. VII.A-1, these differences are extremely large in the case of excitations from the $1b_1$ MO in water as expected for strong $\sigma^*/3s$ mixing. There is no feature of the H_2O spectrum which can be assigned a valence upper state with any confidence, regardless of the originating MO. (However, see [W6] for a contrary view.)

The presence in study after study of an energy loss at 32 000–48 000 cm^{-1} (*4–6 eV*) in the electron impact spectrum of water vapor has perplexed the spectroscopic community for years. Most recently, Edmonson *et al.* [E3] have concluded that the excitation is artifactual, being related in some way to stray electrons inelastically scattered off water adsorbed onto the metal slits of the spectrometer. If correct, this is a most welcome resolution of the problem, for there is no way to rationalize such a low-lying band in water other than to reassign the 60 000-cm^{-1} (*7.4-eV*) band as largely valence in character with a

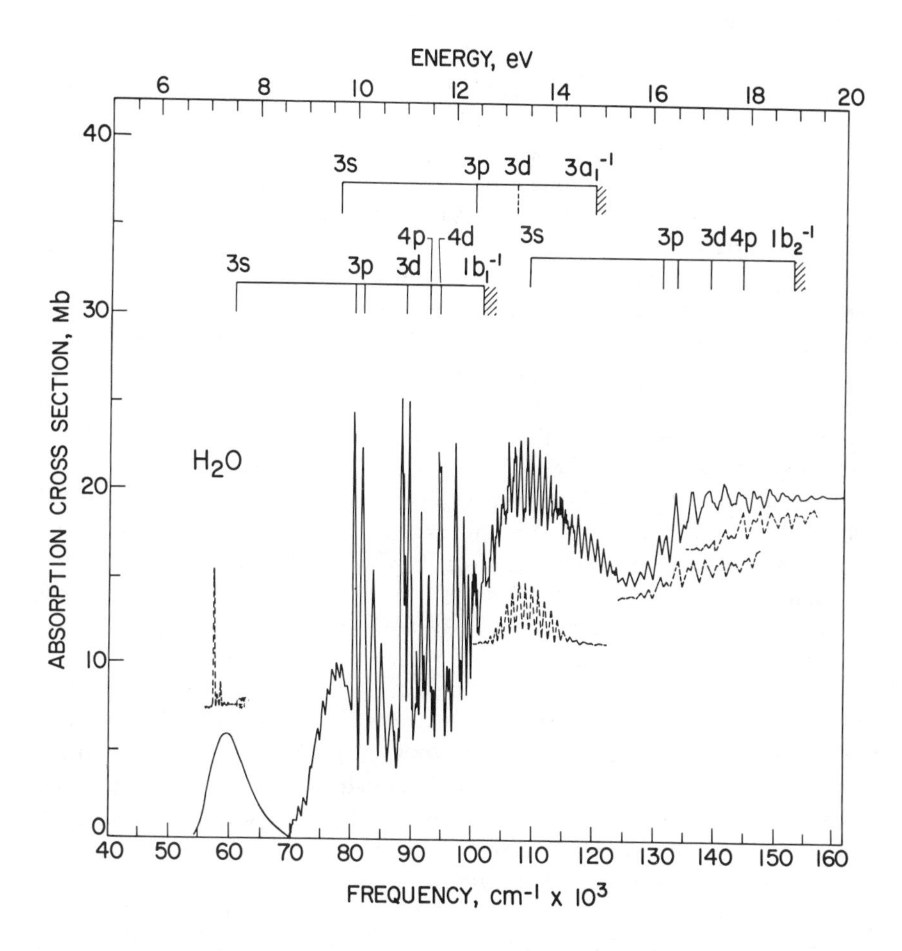

Fig. VII.A-1. Absorption spectrum of water vapor (—) and profiles of the $(1b_1)^{-1}$, $(3a_1)^{-1}$ and $(1b_2)^{-1}$ photoelectron bands (- -) aligned with various absorption features [G44].

correspondingly large (20 000 cm^{-1}, *2.5 eV*) singlet-triplet split.† Such an assignment contradicts the conclusions of all calculations of the last 10

† Actually, one can imagine that the $^3(1b_1, \sigma^*)$ and $^1(1b_1, \sigma^*)$ valence configurations are strongly split by the exchange interaction and that $^1(1b_1, \sigma^*)$ subsequently is dissolved totally in the Rydberg sea (Section I.D). This leads to a $(1b_1, \sigma^*)$ triplet lying far below the Rydberg manifold, but lacking the distinct presence of the complementary singlet state.

years on water. A relevant experiment in this regard is that of Hunter *et al.* [H77], who studied the electron reflection spectrum of 400-Å thick ice films on Kovar and observed an intense loss at 34 000 cm^{-1} (*vert., 4.2 eV*).

As in the past [I, 245], we continue to view the 60 000-cm^{-1} band of water as $1b_1 \rightarrow 3sa_1$ strongly perturbed by the conjugate $1b_1 \rightarrow a_1\sigma^*$ valence promotion. An accurate measurement of the $1b_1 \rightarrow 3sa_1$ oscillator strength [L8] yields 0.060, to be compared with calculated values of 0.06 [B74], 0.03 [S68] and 0.02 [D21]. The relative generalized oscillator strengths as measured by electron impact for the $3a_1 \rightarrow 3sa_1$ (78 200 cm^{-1}, *9.70 eV*) and $1b_1 \rightarrow 3pb_1$ (82 110 cm^{-1}, *10.18 eV*) transitions of H_2O show maxima and minima very similar to those of the $1b_1 \rightarrow 3sa_1$ transition [M56]. The dipole oscillator strength distribution for water over a very large spectral region has been considered critically by Berkowitz [B34] and by Zeiss *et al.* [Z1].

Due to correlation effects, the value of $<r^2>^{1/2}$ for the $^1(1b_1, 3sa_1)$ configuration of water is approximately 10 % larger than that for $^3(1b_1, 3sa_1)$ [W2]. Frequencies of the higher $1b_1 \rightarrow nsa_1$ and $3a_1 \rightarrow nsa_1$ transitions in water have been calculated with high accuracy, but none have been observed beyond $n = 4$. The calculations of Wadt and Goddard nicely illustrate how the $(1b_1, 3sa_1)$ term value of water drops upon methylating it to first form methanol and then dimethyl ether [W2].

According to Buenker and Peyerimhoff's calculation [B74], the HOH angle will open somewhat in the $^1(1b_1, 3s)$ state of water, whereas experimentally, Wang *et al.* [W6] actually deduce the opposite. They find clear evidence for a vibrational progression of ν_2' in the $1b_1 \rightarrow 3sa_1$ bands of H_2O and D_2O with frequency separations 20 % larger than in the respective ground states; this is rationalized in terms of a $1b_1$ orbital which is H—H nonbonding, and a $3sa_1$ orbital which is appreciably H—H bonding in nature, leading to a *closing* of the HOH angle in the upper state.

Theory has been especially useful as a guide in picking out the three components of the $1b_1 \rightarrow 3p$ transition in H_2O [B74, D21, G20, W2]. It is computed that the configurations $(1b_1, 3pa_1)$ and $(1b_1, 3pb_1)$ lie very close to one another and removed from the $(1b_1)^{-1}$ ionization potential by the normal $(\phi_i^o, 3p)$ term value, *i.e.*, approximately 20 000 cm^{-1}. However, the third component, $(1b_1, 3pb_2)$, is calculated to be 5000—10 000 cm^{-1} lower in frequency, placing it anomalously close to the $(1b_1, 3sa_1)$ state and on top of $(3a_1, 3sa_1)$. In the C_{2v} point group of H_2O, the $3pb_1$ orbital has its axis (x) perpendicular to the plane, the axis of the $3pa_1$ orbital is in-plane and parallels the C_2 axis (z) of the molecule, whereas the $3pb_2$ orbital has its axis in-plane and perpendicular to the molecular C_2 axis (y). Note that there are only two virtual valence MO's in water, $a_1\sigma^*$ and $b_2\sigma^*$, and therefore only transitions terminating at a_1 and b_2 Rydberg orbitals can have valence conjugates. It is the mixing of the conjugate ($1b_1$, $b_2\sigma^*$) valence and ($1b_1$, $3pb_2$) Rydberg configurations which shifts the latter out of its normal 3p position and increases its term value by almost 50 % while reducing its value of $<r^2>^{1/2}$ to only 68 % of that of the other 3p MO's [G20]. Nonetheless, compared to the MO's having $n = 2$, the $3pb_2/b_2\sigma^*$ hybrid MO is still quite diffuse and Rydberg-like. In a calculation on F_2O, the transition to ($2b_1$, $3pb_2$) again is calculated to lie quite low with respect to the two other 3p components [V4], while in hydrogen peroxide the 3p core splitting is observed to be *ca.* 5000 cm^{-1} (*vide infra*).

Experimentally, transitions from $1b_1$ to the three components of 3p in water are observed in electron impact at 73 400 ($3pb_2$), 80 700 ($3pa_1$) and 82 040 cm^{-1} ($3pb_1$) (*9.1, 10.01 and 10.171 eV*) [C25, **II**, K16]. Comparable 3p core splittings are observed in the transitions from $1s_O$ to 3p in the inner-shell spectrum of water [A7, W19]. Experimental proof of such a large 3p core splitting in the outer-shell was slow in coming because the out-lying $1b_1 \rightarrow 3pb_2$ transition is forbidden and so has a very low oscillator strength in absorption. The related $3a_1 \rightarrow 3p$ Rydberg core splittings have been calculated [B74, D21, W2] and again show the $3pb_2$ component significantly below the other two. The three-photon resonant MPI transitions to the ($1b_1$, $3pa_1$) levels of H_2O and D_2O have been recorded and rotationally analyzed to reveal two distinct predissociation mechanisms [A34]. An unusually large number of Rydberg excitations terminating at triplet configurations involving ns, np and nd orbitals are cataloged by Chutjian *et al.* [C25], Table I.B-I.† See Section I.B for a discussion of Rydberg singlet-triplet splitting and its relationship to term values. Good agreement with the experimental singlet-triplet splittings in water are obtained in calculations by Durand and Volatron [D57] using orbitals which have been carefully optimized.

Transitions in the 87 000–90 000 cm^{-1} (*10.8–11.2 eV*) span corresponding to the five components of the $1b_1 \rightarrow 3d$ promotions in H_2O and D_2O have been identified [C25, C40, G44, I5, W6] and several have been rotationally resolved. The calculated frequencies of these bands are in good agreement with their observed positions. Singlet Rydberg series built

† In fact, more triplets have been cataloged for water (21) than singlets for most other molecules!

upon three of the $1b_1 \rightarrow 3d$ transitions have quantum defects of 0.109, 0.054 and 0.026 [I5], and several of the corresponding triplet transitions have been identified by low-energy/large-angle electron impact spectroscopy [C25]. Connerade *et al.* [C40] have shown that several higher members of the $1b_1 \rightarrow nd$ series can be rotationally resolved. Complete vibronic analyses of members of the $3a_1 \rightarrow nd$ series are offered by Ishiguro *et al.* [I5].

The straightforward assignment of the five $1b_1 \rightarrow 3d$ excitations in water seemingly is upset by the MPI work of Ashfold *et al.* [A33], who find a three-photon resonant electronic origin in H_2O at 84 434 cm^{-1} (*adiabat., 10.468 eV*; 84 646 cm^{-1} *adiabat., 10.494 eV* in D_2O). Simulation of the rotational envelope of this band identifies it as being $^1A_1 \rightarrow {}^1B_1$, and Ashfold *et al.* assign it as $1b_1 \rightarrow 3da_1$. With a term value of 17 300 cm^{-1} with respect to the $(1b_1)^{-1}$ ionization potential, the MPI band is out of range for assignment as $1b_1 \rightarrow 3d$, all components of which are otherwise assigned anyway. We note however, that this band is close to the $1b_1 \rightarrow 4sa_1$ excitation of water suspected by Chutjian *et al.* [C25] to lie among several other vibronic bands at approximately 84 850 cm^{-1} (*advert., 10.52 eV*), and we assign it accordingly. What is apparently the same transition is seen optically as a one-photon promotion in H_2O at 84 395 cm^{-1} (*10.463 eV*) [**I**, B18]. Theory concurs in placing the $1b_1 \rightarrow 4sa_1$ excitation in the vicinity of 85 000 cm^{-1} (*10.5 eV*) [D21, W2]. If the assignment to $4sa_1$ is correct, it is most interesting to note the large differences in Franck-Condon verticality in the transitions to $3sa_1$ and $4sa_1$ implying much stronger mixing of $3sa_1$ than $4sa_1$ with the $a_1\sigma^*$ valence conjugate.

An alternate interpretation of the outer- and inner-shell spectra of water is given by Wang *et al.* [W6]. In brief, they take the 60 000-cm^{-1} band as essentially valence, $1b_1 \rightarrow a_1\sigma^*$, and in turn place the $1b_1 \rightarrow 3sa_1$ origin at 73 270 cm^{-1} (*9.084 eV*), thereby giving it a higher frequency than the origin of the $3a_1 \rightarrow 3sa_1$ band! Their σ^*-below-3s ordering is applied as well to the interpretation of the inner-shell spectrum. Though the limited set of arguments used by Wang *et al.* to support these conclusions is internally consistent, the conclusions run counter to our empirically determined scheme of term values and to the results of many *ab initio* calculations [G20, S68, W2] which support the more conventional assignments of [**I**, 259].

At the higher end of the outer-shell spectrum of water vapor, a very broad continuum peaks at 156 000 cm^{-1} (*19.4 eV*) in both H_2O and D_2O [P30]. The peak is due to the maximum in the $(1b_1)^{-1}$ photoionization cross section, in which the most intense partial channel is $1b_1 \rightarrow \epsilon da_1$

[D21, W28].† Upon this continuum rests a highly structured feature at 126 000–139 100 cm^{-1} (*15.62–17.25 eV*) with vibronic spacings identical to those in the $(1b_2)^{-1}$ photoelectron band 15 700 cm^{-1} higher in frequency [G44, P30], Fig. VII.A-1. The 15 700-cm^{-1} term value identifies the transition as $1b_2 \rightarrow 4sa_1$; due to overlapping absorption, it is impossible to tell if the corresponding $1b_2 \rightarrow 3sa_1$ transition is structured or not. That part of the $1b_2 \rightarrow 4sa_1$ vibronic band below the (0,0) of the $(1b_2)^{-1}$ ionization is much sharper than that part beyond this point, owing to autoionization in the latter. A broad shoulder, seen clearly in water vapor at 218 000 cm^{-1} (*vert., 27.1 eV*), has a term value of 42 000 cm^{-1} with respect to the $(2a_1)^{-1}$ ionization potential at 260 000 cm^{-1} (*vert., 32.2 eV*). The transition is undoubtedly $2a_1 \rightarrow 3sa_1$. The term values of the four outer-shell transitions to $3sa_1$ in water thus are seen to be 41 800, 39 500, 43 300 and 42 000 cm^{-1}, whereas for excitation from the inner shell $1s_O$ orbital, the $(1s_O, 3sa_1)$ term value is 46 000 cm^{-1}. As predicted, Section I.B, all the orbitals composed of central atom 2s and 2p AO's have essentially the same 3s term value, but upon decreasing the principal quantum number of the originating orbital by 1 as for excitations originating at $1s_O$, there is a noticeable term-value exhaltation due to antishielding.

Water exhibits several high-lying negative-ion resonances in addition to those discussed in [**II**, 317]. Experimentally, Mathur and Hasted [M12] report a negative-ion resonance at 64 200 ± 1600 cm^{-1} (*vert., 7.96 ± 0.2 eV*) having a full width of 24 000 cm^{-1}. This resonance is confirmed by Seng and Linder [S41], who conclude that there are two resonances in the 48 000–64 000 cm^{-1} (*6–8 eV*) region. Additionally, the electron-impact dissociative attachment work of Jungen *et al.* [J22] on H_2O and D_2O shows peaks at 56 000, 73 000 and 95 200 cm^{-1} (*vert., 7.0, 9.1 and 11.8 eV*). Jungen *et al.* argue that the three states observed by them correspond to $^2(\phi_i, 3sa_1^2)$ Feshbach resonances with excitations from the $1b_1$, $3a_1$ and $1b_2$ valence MO's respectively, Table VII.A-I. With

† It is interesting to note that in the calculation of Diercksen *et al.* [D21], a broad autoionizing $1b_2 \rightarrow b_2\sigma^*$ valence excitation is predicted (*ca.* 135 000 cm^{-1}, *17 eV*) whereas the complementary valence excitations terminating at $a_1\sigma^*$ exist only as minority components associated with Rydberg or continuum functions. This is reminiscent of the situation in H_2S, Section VIII.A. The reluctance of the $b_2\sigma^*$ valence MO to dissolve in the Rydberg sea may relate to the fact that its nodal pattern [W6] is much more like that of $3db_2$ than $3pb_2$, which is to say, its prime Rydberg conjugate is only weakly penetrating, Section I.D. Nonetheless, it is clear from the core splitting of the 3p orbitals, that $b_2\sigma^*$ does mix somewhat with $3pb_2$ and stabilize it.

TABLE VII.A-I

TNI RESONANCES IN WATER AND THE SMALLER ALCOHOLS

Compound	TNI Frequency, cm^{-1} (*eV*)[a]	Parent Frequency, cm^{-1} (*eV*)[b]	Feshbach Decrement,[c] cm^{-1}	TNI Configuration
H_2O	56 000 (*7.0*)	60 000 (*7.44*)	4000	$^2(1b_1, 3s^2)$
	64 200 (*7.96*)	-	-	$^2(N, b_1\sigma^*)$
	73 000 (*9.10*)	79 500 (*9.86*)	6500	$^2(3a_1, 3s^2)$
	79 500 (*9.86*)	82 000 (*10.16*)	2500	$^2(1b_1, 3p^2)$
	88 700 (*11.0*)	91 700 (*11.37*)	3000	$^2(1b_1, 4p^2)$
	95 200 (*11.8*)	98 080 (*12.16*)[d]	2900	$^2(3a_1, 3p^2)$
CH_3OH	63 600 (*7.88*)	67 100 (*8.32*)	3500	$^2(2a'', 3p'^2)$
C_2H_5OH	60 400 (*7.49*)	65 570 (*8.129*)	5200	$^2(n_O, 3p'^2)$
C_3H_7OH	13 400 (*1.66*)	-	-	$^2(N, \sigma^*)$
	57 900 (*7.18*)	62 000 (*7.69*)	4100	$^2(n_O, 3p^2)$

[a] Reference [M12].

[b] Reference [I, 259].

[c] The difference between the TNI resonance frequency and the frequency of the neutral-molecule parent excitation (Section II.B).

[d] Calculated value from [D21].

these assignments, the Feshbach peaks are respectively 3500, 6100 and 16 000 cm^{-1} below the corresponding neutral-molecule excitations to the parent (ϕ_i^o, 3s) states. The first two of these Feshbach decrements are in the normal range for Feshbach resonances, but the third is much larger than is normally seen and merits further consideration. We note, Table VII.A-I, that the 95 200-cm^{-1} resonance is displaced from that at 73 000 cm^{-1} $^2(3a_1, 3sa_1^2)$ by just the interval which separates the $^2(1b_1, 3sa_1^2)$ and $^2(1b_1, 3p^2)$ resonances. Consequently it is natural to assign the 95 200-cm^{-1} feature as having the $^2(3a_1, 3p^2)$ configuration. With respect to the ($3a_1$, 3p) parent configuration calculated to lie at 98 080 cm^{-1} (*12.16 eV*) [D21], the resonance at 95 200 cm^{-1} has a more normal Feshbach decrement of 2900 cm^{-1}.

Interestingly, the negative-ion Feshbach states having the nominal $3sa_1^2$ occupation are calculated to involve not only the $^2(\phi_i^o, 3sa_1^2)/^2(\phi_i^o, a_1\sigma^{*2})$

Rydberg/valence conjugate pair configuration, but also the $^2(\phi_i^o, 3pb_2^2)/^2(\phi_i^o, b_2\sigma^{*2})$ conjugate pair configuration, so as to form a "semi-valence" orbital [J22]. This strong configuration mixing is not unexpected, because considerable correlation is involved in keeping the two Rydberg electrons away from one another. However, though the incorporated orbitals are σ^*, implying that the Feshbach orbital is *smaller* than the parent Rydberg orbital, calculations on atoms predict the opposite [W11].

One readily estimates that the excitation to the $^2(1b_1, 3pb_2^2)$ Feshbach state of water will come at 70 200 cm^{-1} (*8.7 eV*), which is far above the 64 200-cm^{-1} resonance observed by Mathur and Hasted [M12]. The frequency of their state, however, is just between those expected of the $^2(1b_1, 3sa_1^2)$ and $^2(1b_1, 3pb_2^2)$ states, and so may prove to have the hybrid configuration $^2(1b_1, 3sa_1 3pb_2)$. More likely, the extreme width of the 64 200-cm^{-1} resonance suggests that this is a valence TNI resonance having the $^2(N, b_1\sigma^*)$ configuration which is derived from the $^2(N, t_2\sigma^*)$ configuration of methane at approximately 60 000 cm^{-1} (*7.4 eV.*, Section III.A). The transition to $^2(N, a_1\sigma^*)$ in H_2O is expected at much lower frequency. The presence of these valence TNI resonances can be checked by determining the electron transmission spectrum of ice, for in this phase, the Feshbach resonances should be obliterated (Section II.D).

The $1s_O$ inner-shell spectrum of water vapor, as determined optically [A7] and by electron impact [W19], begins with a sharp, intense feature terminating at $3sa_1$ with a term value of 46 000 cm^{-1}, Fig. III.A-2. Appropriately, this value is larger than the average term value of 41 600 cm^{-1} of the outer-shell spectrum owing to the antishielding of the $1s_O$ hole (Section I.B). The most intense band in the $1s_O$ absorption spectrum has a term value of 30 600 cm^{-1}, and corresponds to the $1s \rightarrow 3pb_2$ excitation. The b_2 level is the lowest in the 3p manifold of water [W2] and is calculated to be 5000–10 000 cm^{-1} lower than the adjacent components $3pa_1$ and $3pb_1$. Indeed, the outer shell transition $1b_1 \rightarrow 3pb_2$ has a term value of 28 380 cm^{-1}, and the antishielding effect is expected to increase this somewhat in the $(1s_O, 3pb_2)$ configuration. A calculation on the $(1s_O, 3pb_2)$ state of water assigns considerable valence character to the $3pb_2$ orbital [D21], which is to say that there is strong mixing of Rydberg and conjugate valence configurations. A composite peak with a term value of 21 000 cm^{-1} corresponds to the $1s_O \rightarrow 3pa_1$ and $1s_O \rightarrow 3pb_1$ transitions (21 156 and 19 742 cm^{-1} term values in the outer-shell spectrum). There is surprisingly little $1s_O$ hole exhaltation of the term value here. A fourth peak in the inner-shell spectrum has a term value of 9600 cm^{-1} and terminates at 4p.

A somewhat different interpretation of the inner-shell spectrum of water follows from the theory of Diercksen *et al.* [D21]. These authors conclude that the band assigned here as Rydberg ($1s_O \rightarrow 3pb_2$) instead involves the $b_2\sigma^*$ valence MO as the terminating orbital. Since the ($1s_O$, $3pb_2$) and ($1s_O$, $b_2\sigma^*$) configurations are Rydberg-valence conjugates, they no doubt are mixed, which is why $3pb_2$ is separated from the other two components of 3p. However, though mixed, we prefer to argue on the basis of the narrow width of the transition, Fig. III.A-2, that the state in question remains predominantly Rydberg. A proper assignment here would be aided by determining the behavior of this band in the inner-shell absorption spectrum of ice (Section II.D). Diercksen *et al.* also assign the absorption peak at *ca.* 555 eV as the maximum in the $1s_O \rightarrow \epsilon pb_2$ photoionization cross section.

Using the equivalent-core analogy, Wight and Brion [W22] deduce that the $H_3O\cdot$ radical will have a set of outer-shell Rydberg term values equal to those of NH_3 in which the $1s_N$ inner-shell is excited so as to resemble the core of $H_3O\cdot$. If correct, the hydronium radical will have a 3s Rydberg ground state with an ionization potential of 40 000 cm^{-1} (*5.0 eV*). Indeed, a binding energy (term value) of 40 000 cm^{-1} for the 3s electron bound to the H_3O^+ core is little different from the average value of 41 600 cm^{-1} binding the 3s electron to the H_2O^+ core, as expected [**I**, 65]. Raynor and Herschbach [R7] calculate hydronium radical term values of 37 600 cm^{-1} (3s), 19 500 and 22 500 cm^{-1} ($3pa_1$ and $3pe$) and 11 700, 11 300 and 13 500 cm^{-1} ($3da_1$, $3de$ and $3de$). Thus by analogy with the Rydberg-to-Rydberg spectrum of NH_4 (Section VI.A), one expects a $3p \leftrightarrow 3s$ transition in H_3O at *ca.* 18 500 cm^{-1}, however the lifetime of the radical in the ^{2}(N, 3s) state will be shortened by predissociation into ground-state H and H_2O [G2], as in the (ϕ_i^o, 3s) Rydberg states of the other first-row hydrides.

It is interesting to note that the binding energy of a positron to OH^- is calculated to be 38 700 cm^{-1} [K3], while the term value of an electron in 3s bound to H_3O^+ is calculated to be 37 600 cm^{-1} [W22]. The near-equality of these binding energies suggests that the positron bound to OH^- is in a 3s-like orbital resembling that of the electron bound to H_3O^+. This conclusion parallels that deduced theoretically for the (Na^+, e^-)/(F^-, e^+) and (NO^+, e^-)/(CN^-, e^+) pairs, wherein the anion/positron complexes were found to have positron level spacings closely following the Rydberg pattern [K45].

Ice

The absorption spectra of ice [I, 251] in its various crystallographic forms are shifted from, but surprisingly like the vapor spectrum in its broadest features. Thus the $1b_1 \rightarrow 3sa_1$ band at 60 000 cm^{-1} (*vert., 7.4 eV*) in the vapor shifts upward to 66 000 cm^{-1} (*8.2 eV*) in the liquid and to 70 000 cm^{-1} (*8.7 eV*) in hexagonal, cubic and amorphous ices [H36, S39, S46, W8]. It has been calculated that the largest part of the high-frequency shift of the transition to 3s in the water dimer is due to the exchange repulsion between the 3s orbital and the neighboring ground-state molecule [V6]. This probably applies to ice as well. The reflection spectrum of a single crystal of hexagonal ice cleaved under vacuum [S39] shows fine structure on the $1b_1 \rightarrow 3sa_1$ peak which is similar to that observed and assigned as ν_2' vibrational structure in the vapor-phase spectrum [W6]. It is most interesting to note that though the very sharp $1b_1 \rightarrow np$ and $1b_1 \rightarrow nd$ Rydberg transitions of the vapor are so broadened in the ices that they cannot be seen, the half widths of the $1b_1 \rightarrow 3sa_1$ excitations in the vapor, liquid and solid phases remain equal! This particular aspect of the ($1b_1$, $3sa_1$) excited state reflects its strong mixing with the conjugate ($1b_1$, $a_1\sigma^*$) valence configuration. Though the Rydberg transitions of molecules in fluid solvents shift strongly to higher frequencies upon lowering the temperature, the $1b_1 \rightarrow 3sa_1$ band of ice does not shift with changing temperature [S46]. Perhaps this relates to the near-zero thermal-expansion coefficient of ice compared to the very large values for most liquids. The frequency of the $1b_1 \rightarrow 3sa_1$ band of solid D_2O is 400 cm^{-1} higher than that of solid H_2O [S46]. The x-ray fluorescence spectra of amorphous and crystalline ice [G14] show the $1b_1$ level split into two prominent components separated by 8000 cm^{-1}; this splitting (possibly due to the differing extents of hydrogen bonding in different molecules) should contribute to the width of any ultraviolet transition originating at $1b_1$.

An interesting but unproved possibility is that on going from the vapor phase of H_2O to the liquid or to ice, the condensed-phase effect acts to deperturb the $a_1\sigma^*$ MO (Section II.D) so as to resurrect it as a stand-alone orbital of the virtual manifold. Were this the case, then the 60 000-cm^{-1} band of the vapor would be classified as $1b_1 \rightarrow 3sa_1$ Rydberg, whereas the bands at 66 000 and 70 000 cm^{-1} in liquid water and ice would be the conjugate valence excitations $1b_1 \rightarrow a_1\sigma^*$.

The molecular-crystal nature of ice for spectroscopic purposes also is evident from its photoelectron spectrum [A1, S46] in which the $(1b_1)^{-1}$, $(3a_1)^{-1}$, $(1b_2)^{-1}$ and $(2a_1)^{-1}$ free-molecule ionizations are clearly seen in the solid with very nearly the free-molecule separations.

Unfortunately, charging of the ice film does not allow the direct determination of the absolute bulk ionization potentials, however combining the photoelectron data of [B14] and [A1] leads to a $(1b_1)^{-1}$ ionization potential of 89 000 cm^{-1} (*vert., 11 eV*) for ice. From this it is seen that the 41 800-cm^{-1} term value of the $(1b_1, 3s)$ level in the vapor phase is reduced to only 19 000 cm^{-1} in ice! The $3a_1 \rightarrow 3s$ and $1b_2 \rightarrow 3s$ excitations in water vapor at 79 500 and 106 200 cm^{-1} (*vert., 9.86 and 13.18 eV*), respectively, are shifted in hexagonal ice to 84 000 and 117 000 cm^{-1} (*vert., 10.4 and 14.5 eV*). The latter peak is not observed in amorphous ice, however. These excitations to 3s in ice have term values closely similar to that of the $1b_1 \rightarrow 3s$ transition.

Features still unassigned in the ice spectrum fall at 101 000, 141 000 and 201 000 cm^{-1} (*vert., 12.5, 17.5 and 25 eV*). Though it is tempting to try to assign these features to low-lying Rydberg states, the attempt fails quantitatively. For example, the band at 201 000 cm^{-1} in ice, if a Rydberg, most naturally would relate to the $(2a_1)^{-1}$ ionization potential at 260 000 cm^{-1} (*32.2 eV*) in the vapor. However, the $2a_1 \rightarrow 3s$ transition in the free molecule is observed at 218 000 cm^{-1} (*vert., 27.1 eV*) [P30] and is expected at *higher* frequencies in the solid, not lower. Possibly, the band at 141 000 cm^{-1} in ice is derived from the maximum in the $(1b_1)^{-1}$ photoionization cross section observed in the free molecule at 156 000 cm^{-1} (*vert., 19.4 eV*), and said to involve the $1b_1 \rightarrow 4a_1\sigma^*$ valence excitation [W28]. Assignments as two-electron excitations seem unlikely.

Liquid Water

Work on the reflection spectrum of liquid water [**I**, 251] shows optical constants characteristic of a volume plasmon at 173 000 cm^{-1} (*vert., 21.4 eV*), Fig. II.D-4. (However, for a contrary view, see [B31]). The transition has an oscillator strength of 2.0, a half width of 28 000 cm^{-1} (*3.5 eV*) and an upper-state lifetime of *ca.* 10^{-16} sec before it decays into the manifold of single-particle excitations [H36, K46]. This collective excitation is smeared out and of reduced intensity in ice, and does not appear in the energy-loss function of the vapor [H36]. Comparison of the optical properties of liquid H_2O and D_2O in the $1b_1 \rightarrow 3s$ region [H38] shows that the band of D_2O is narrower, with a larger cross section, and shifted 1300 cm^{-1} to higher frequency. The higher bands of liquid water are at frequencies intermediate between those of water vapor and ice [H36].

Hydrogen Peroxide

Interpretation of the electronic excitations in H_2O_2 closely follows that given above for water. Absorption commences with a weak, broad feature centered at 55 200 cm^{-1} (*vert., 6.84 eV*) [S83] having a term value of 37 600 cm^{-1} with respect to the $(4b)^{-1}$ ionization potential at 92 840 cm^{-1} (*vert., 11.51 eV*) [B68]. The $4b$ originating orbital is the out-of-phase combination of what otherwise are the $1b_1$ lone-pair orbitals of water, and the excitation clearly terminates at 3s strongly perturbed by its σ^* conjugate as in water. The 10 % decrease of the $(\phi_i^o, 3s)$ term value of hydrogen peroxide with respect to that in water is observed in several other M_2H_{2x-2} systems when compared with those in MH_x. Two vibronically structured excitations to components of 3p fall at 59 890 and 65 600 cm^{-1} (*adiabat., 7.425 and 8.133 eV*) in hydrogen peroxide. The first of these has a term value (25 100 cm^{-1}) which is beyond the range of term values normally found for excitations to 3p, whereas the second has a term value (19 400 cm^{-1}) which is quite normal. However, it must be remembered that in the isoelectronic molecule CH_3OH, the $(\phi_i^o, 3p)$ term values are 26 150 and 21 310 cm^{-1} [I, 259] and in H_2O the $(1b_1, 3pb_2)$ component has a term value of 28 380 cm^{-1} (*vide supra*). Thus, in a sense, the $(\phi_i^o, 3p)$ term values of H_2O_2 are not expected.

The transition to the lower-lying component of. 3p in H_2O_2 has a vibronic progression of 1340 cm^{-1} (ν_6', asymmetric OH bending), compared to values of 1140 cm^{-1} in both the $(4b)^{-1}$ ion [B68] and in the upper component of 3p. This suggests that the conjugate $(b\sigma^*)$ configuration which pushes the $(4b, 3pb)$ level about 5000 cm^{-1} to lower frequency also stiffens the force constant along the ν_6 coordinate. A similar mixing of $(4b, 3s)$ with $(4b, a\sigma^*)$ is responsible for the broad, featureless Franck-Condon profile of the transition to the former. Vibronic structure in the H_2O_2 spectrum curiously breaks off at 72 500 cm^{-1} (*8.98 eV*), while broad peaks are observed at 72 000 and 87 000 cm^{-1} (*vert., 8.9 and 10.8 eV*). These excitations probably correspond to Rydberg promotions converging upon the second ionization potential at 101 300 cm^{-1} (*vert., 12.56 eV*).

In view of the O–O bond strength of only 37 kcal/mole in hydrogen peroxide and the inverse relationship between bond strength and $\mathscr{A}$-band term value, one expects a low-lying $\mathscr{A}(4b_1, \sigma^*)$ valence excitation in H_2O_2, in strong contrast to the situation in H_2O. No doubt, this is the assignment for the broad, weak excitation in H_2O_2 extending far to the low-frequency side of the transition to $(4b_1, 3s)$. An $\mathscr{A}(1s_O, \sigma^*)$ band also is expected to be low lying in the inner-shell spectrum of H_2O_2.

VII.B. Alcohols

Details in the overall picture of the excited states of alcohols have been filled in over the past ten years, especially in the areas of inner-shell spectra, TNI resonances and *ab initio* calculations. The earlier suggestion that the spectra of alcohols (and ethers) begin with transitions from n_O[†] to 3s, 3p, 3p′ and 3d Rydberg orbitals [**I**, 254] has been confirmed repeatedly. Thus, for example, our initial assignments of the first three bands in the CD spectrum of (+)-*S*-2-butanol as terminating at 3s, 3p and 3p′ Rydberg orbitals [**II**, 318] since has been confirmed by an *ab initio* calculation [S35].

In the realm of outer-shell absorption spectra, data is now available for cyclohexanol [D20] and for *l*-borneol [S56]. In both of these substances, weak $n_O \rightarrow 3s$ excitations at 53 000−54 000 cm^{-1} (*vert., 6.6−6.7 eV*) are followed by far more intense $n_O \rightarrow 3p$ transitions at *ca.* 62 000 cm^{-1} (*vert., 7.7 eV*), and very intense and broad absorptions centered at *ca.* 72 000 cm^{-1} (*9.0 eV*) which contain both the $n_O \rightarrow 3d$ promotions and transitions originating at the second-lowest ionization potential and terminating at 3p. Now in water and the smaller alcohols, the core splitting of the (n_O, 3p) manifold was quite obvious in the absorption spectra (water, 7300 cm^{-1}; methanol, 4840 cm^{-1}; ethanol, 2580 cm^{-1}; *n*-propanol, 2000 cm^{-1} [**I**, 259]) but was unresolved in the absorption spectra of the larger systems. This absence of 3p core splitting holds as well for the absorption spectra of 2-butanol and *l*-borneol, however the core splittings instead are visible in the CD spectra of these systems. Thus in (+)-*S*-2-butanol [**II**, AD171], the $n_O \rightarrow 3p$ region of the CD spectrum shows two oppositely rotating bands, peaked at 61 500 and 67 000 cm^{-1} (*7.62 and 8.31 eV*), while in *l*-borneol, the peaks are at 59 500 and 63 500 cm^{-1} [S56]. Note however, that the two components of (n_O, 3p) observed in the CD spectra are not necessarily the same two resolved in the absorption spectra of the other alcohols, and also that the CD peaks may be more widely separated than the absorption peaks. The CD bands of *l*-borneol are high-frequency shifted upon solution in hexafluoroisopropanol, as appropriate for Rydberg upper states, Section II.D. Combining this new data on alcohols with that in [**I**, 259] reveals that in those cases where 3p core splitting can be resolved, there is always an $n_O \rightarrow 3p$ component at 62 000−63 000 cm^{-1} (*7.7−7.8 eV*) of variable

[†] The $1b_1$ orbital of H_2O is given the generic label n_O in the alcohols and ethers.

term value, while a higher component follows the ionization potential so as to yield a constant term value of 21 000 cm^{-1}. This latter transition is most likely $n_O \rightarrow 3p\pi$, while the first is $n_O \rightarrow 3p\sigma$ with $3p\sigma$ aligned along the C—O line [W2]. The CD bands of ten alcohols between 50 700 and 53 500 cm^{-1} (*6.3 and 6.6 eV*) are discussed in [T8], and it is concluded that the terminating orbital is predominantly σ^* based on a highly approximate calculation of the sign of rotation. Ogata *et al.* [O1] have measured the $n_O \rightarrow 3s$ oscillator strengths in a series of alcohols and find that of methanol the smallest and that of *t*-butanol the largest with isopropanol as a local maximum and isobutanol as a local minimum. The oscillator strength function df/dE has been measured for methanol and ethanol out to 130 000 cm^{-1} (*16 eV*) [C17].

The reflection spectrum of liquid glycerol has been redetermined using a new technique [B46]; as expected, the Rydberg absorptions do not appear, but a band is observed at 105 000 cm^{-1} (*vert., 13 eV*) which is assigned as $\sigma \rightarrow \sigma^*$, and the ubiquitous collective excitation peaks at 169 000 cm^{-1} (*21 eV*) in the energy-loss function, Fig. II.D-4.

Excitations in which an electron is attached to an otherwise neutral molecule via Rydberg or antibonding valence MO's for times longer than *ca.* 10^{-16} sec are called temporary-negative-ion (TNI) resonances (Section II.B). The excitation frequencies to such states and to the related neutral-molecule states of oxo compounds are gathered in Table VII.A-I; the TNI data for the alcohols is that of Mathur and Hasted [M12]. Using the standard Feshbach decrement of *ca.* 4000 cm^{-1} (Section II.B), one sees that many of these resonances correspond to excitations to $^2(n_O, 3p^2)$ configurations. The corresponding resonance in water is observed at 79 500 cm^{-1} (*vert., 9.86 eV*). The electron-transmission spectrum of propanol shows a resonance at 13 400 cm^{-1} (*1.66 eV*) which also appears in several alkanes; the specific upper-state electronic configurations in these states are unknown, but most likely involve occupation of σ^* MO's.

Interpretation of the inner-shell spectrum of methanol, Fig. VII.B-1 [W19], is straightforward. For excitations out of $1s_C$, the weak transition to 3s has a term value of 33 900 cm^{-1}, increasing to 38 700 cm^{-1} when the originating orbital is $1s_O$. On the basis of the antishielding argument, Section I.B, the latter value is quite reasonable when compared with an outer-shell (n_Oa'', 3s) term value of 34 070 cm^{-1}. However, on the same basis, we would expect the ($1s_C$, 3s) term value to be *ca.* 36 000 cm^{-1}. The ($1s_C$, 3p) and ($1s_C$, 3p$'$) term values, 23 400 and 16 100 cm^{-1}, are similarly about 5000 cm^{-1} less than expected in view of the outer-shell values, 26 150 and 21 310 cm^{-1}. Perhaps there is a calibration error in either the ionization potential or excitation potential data originating at $1s_C$ in methanol. Alternatively, the transition with the 16 100-cm^{-1} term value might be the $1s_C \rightarrow 4s$ transition rather than $1s_C \rightarrow 3p'$.

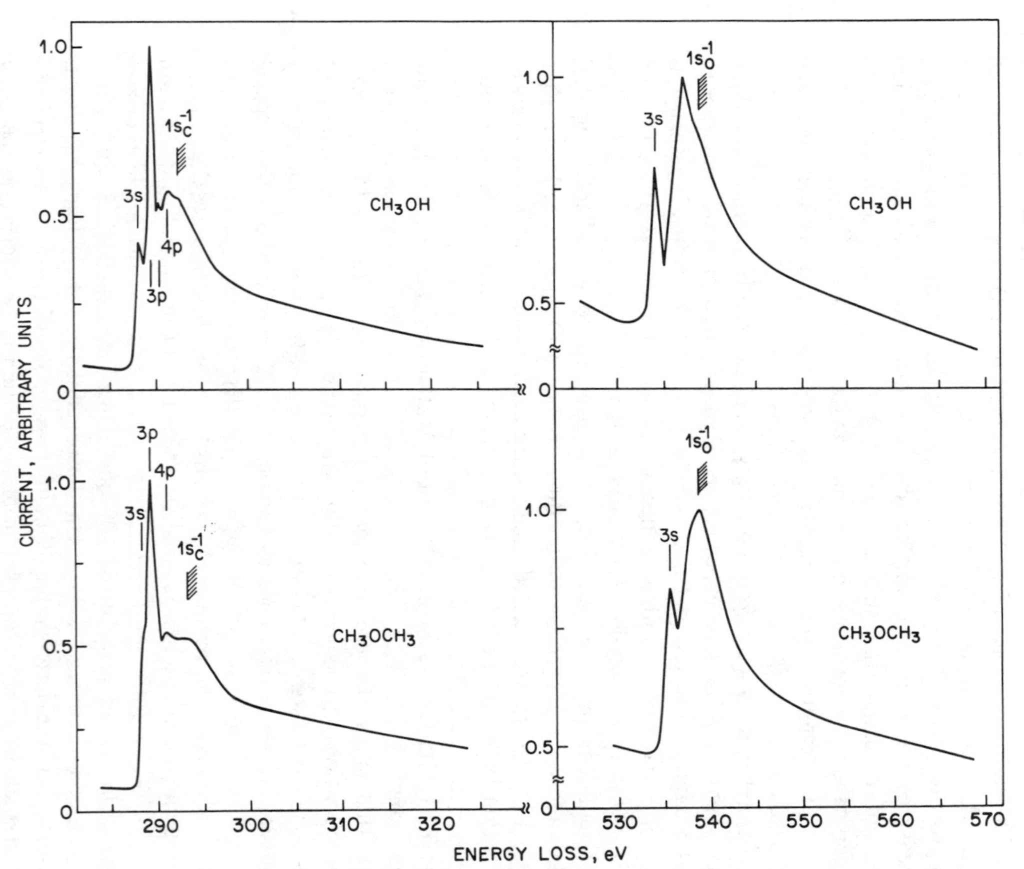

Fig. VII.B-1. Inner-shell energy loss spectra of methanol and dimethyl ether at 2500 eV incident energy and zero scattering angle [W19].

VII.C. Ethers

As with the alcohols, the alkyl ethers show an undeviating pattern of $n_O \rightarrow 3s$, 3p, 3d bands in the vacuum ultraviolet [**I**, 258; **II**, 319]. For ethers bearing alkyl groups of the size C_2H_5- or larger, these bands will come at approximately 53 000, 57 000 and 62 000 cm^{-1} (*vert., 6.6, 7.1 and 7.7 eV*). Besides dimethyl ether, there is only one other ether in which the core splitting of the (n_O, 3p) manifold has been resolved, and as in many of the alcohols (Section VII.B), this is found only in the CD spectrum (*vide infra*).

Studying the $n_O \rightarrow 3s$ transitions of diethyl ether-h_{10} and -d_{10} vapors [**I**, 266] at low temperature, McDiarmid identifies sequence bands corresponding to a vibration of very low frequency (at or below 100 cm^{-1}) and assigns it to torsional motion around the C—O bond [M24]. The prominent interval of 415 cm^{-1} in the $n_O \rightarrow 3s$ band of diethyl ether also appears in the same band of *trans*-dimethyl tetrahydrofuran as a long progression of characteristic frequency 390 cm^{-1} [K24]. McDiarmid has argued that since the dimethyl-ether bands at 58 820 and 61 390 cm^{-1} (*7.292 and 7.611 eV*) which were earlier assigned as components of $n_O \rightarrow 3p$ [**I**, 267] do not resemble the first two $n_O \rightarrow 3p$ bands of water, the first of these must be valence instead [M25]. Note however that the comparison is faulty because the *ab initio* calculation of Wadt and Goddard [W2] shows that the ordering of the two lowest 3p levels in water are reversed with respect to those in dimethyl ether as a consequence of the nonplanar structure of the latter (Section I.B).

The inner-shell $1s_C$ and $1s_O$ excitation spectra of dimethyl ether, Fig. VII.B-1 [W19], are revealing. Because of the antishielding effect, we expect the inner-shell term values for Rydberg excitations to be noticeably larger than their outer-shell counterparts (Section I.B). This is seen in the $1s_C$ spectrum, where the ($1s_C$, 3s) and ($1s_C$, 3p) term values of 30 200 and 23 000 cm^{-1} are indeed far larger than the outer-shell values for the ($2b_1$, 3s) and ($2b_1$, 3p) states, 25 730 and 18 470 cm^{-1}. Only one discrete excitation is observed in the $1s_O$ spectrum of dimethyl ether, with a term value of 25 000 cm^{-1}. Wight and Brion [W19] assign this as $1s_O \rightarrow 3s$, and we are forced to go along with this since there is no sign of a transition at lower frequency. However, on the basis of term values we would much prefer to assign it to $1s_O \rightarrow 3p$ with an exhalted term value due to the antishielding of the $1s_O$ hole. A $1s_O \rightarrow 3p$ assignment is favored over $1s_O \rightarrow 3s$ also on the basis of intensity, and from the fact that the ($1s_O$, 3s) term value in methanol is 38 500 cm^{-1} [W19] and would be expected to be only a few thousand cm^{-1} smaller in dimethyl ether.

The CD spectra of optically active dialkyl ethers in the vapor and solution phases reveal no transitions beyond those to 3s, 3p and 3d which one otherwise is accustomed to seeing in absorption. So, for example, $C_2H_5OCHCH_3C(CH_3)_3$ has CD peaks at 52 400, 56 500 and 61 000 cm^{-1} (*6.50, 7.00 and 7.56 eV*) in the vapor phase [B37]. In perfluoro-*n*-hexane solution, each of these CD peaks moves *ca.* 2000 cm^{-1} to higher frequency, and in hexafluoroisopropanol solution the upward shift from the vapor spectrum is 7000−10 000 cm^{-1}. Further, as discussed by Drake and Mason for Rydberg excitations [D45], these CD peaks shift to higher frequency as the solvent is cooled. Similar effects on the CD spectra of related aliphatic ethers are reported in [S42].

Fridh [F32] has extended the electron impact spectrum of ethylene oxide [**I**, 270] to 193 000 cm^{-1} (*24 eV*) and reports new features centered at 111 000, 125 000 and 148 000 cm^{-1} (*vert., 13.8, 15.5 and 18.4 eV*). The first of these broad bands Fridh assigns as a $5a_1 \rightarrow 7a_1$ valence transition, however as always, alternate Rydberg assignments are possible. Thus the 111 000-cm^{-1} band has a 23 000 cm^{-1} term value with respect to the $(5a_1)^{-1}$ ionization potential (134 000 cm^{-1} *vert., 16.6 eV*), suggesting a ($5a_1$, 3p) configuration, and a 29 000 cm^{-1} term value with respect to the $(1b_2)^{-1}$ ionization potential (140 000 cm^{-1} *vert., 17.4 eV*), suggesting a ($1b_2$, 3s) configuration. Similarly, the band at 125 000 cm^{-1} has a frequency appropriate to ($1b_2$, 4s), which in turn makes the $1b_2 \rightarrow 3s$ assignment a more logical choice for the 111 000-cm^{-1} band. Inasmuch as the 2s → 3s excitation is at 157 000 cm^{-1} (*vert., 19.5 eV*) in methane (Section III.A), the 148 000-cm^{-1} band of ethylene oxide probably involves Rydberg excitations originating with the 2s MO's of the $-CH_2CH_2-$ group. The electrodichroism spectrum of ethylene oxide [A18] in the $n_O \rightarrow 3s$ region reveals a transition-moment polarization perpendicular to that of the ground-state dipole moment, as expected.

The results of spectral experiments and calculations on derivatives of ethylene oxide follow the ether pattern discussed above. In the optically-active ether (−)(*S*,*S*)-2,3-dimethyl ethylene oxide [C35, L22], the CD spectrum reveals the transitions to 3s and 3d at 57 000 and 66 220 cm^{-1} (*vert., 7.064 and 8.211 eV*) and transitions to two components of 3p at 60 420 and 62 420 cm^{-1} (*advert., 7.491 and vert., 7.739 eV*). Once again it is CD which resolves the transitions to 3p and 3p′, for these two bands are not resolvable in absorption. Peaks in the CD spectrum of the related polymer poly[(*R*)-oxypropylene $[-H_2C\overset{O}{-}CHCH_3-]_n$ in hexafluoroisopropanol solution are at 3000 cm^{-1} higher frequency than in cyclohexane solution [H50], and also follow the ether pattern, as do those in several poly(alkyl ether) polymers investigated by Bertucci *et al.* [B38]. The appearance of Rydberg excitations in the CD spectra of solutions is well documented (Section II.D).

Dickinson and Johnson [D20] studied the absorption spectra of several cyclic alkyl ethers [**I**, 272; **II**, 319] as models for sugars. When free of pendant OH groups, the cyclic-ether spectra are highly structured vibronically, but no vibrational analyses are given. The band centers correspond to the standard ether pattern of Rydberg excitations. The spectra of tetrahydrofurans methylated in the 2 and 5 positions [K24] also are more or less vibronically structured in the $n_O \rightarrow 3s$ region (50 000–54 000 cm^{-1}, *6.2–6.7 eV*).

As expected, heavy fluorination of an ether shifts its lower ionization potentials to larger values, pulling the Rydberg spectra with them to higher frequencies. Thus in $CF_3CF_2CF_2OCHFCF_3$ [H11] the lowest ionization limit is shifted to 112 000 cm^{-1} (*vert., 13.89 eV*), and the lowest excitation peaks at 81 000 cm^{-1} (*10.0 eV*). The 31 000-cm^{-1} term value clearly signals a transition terminating at 3s, however the originating orbital may be either the oxygen lone pair n_O or a σ(C–C) orbital. The corresponding transition to 3p peaks strongly at 92 000 cm^{-1} (*11.41 eV*), with a 20 000 cm^{-1} term value. A puzzling pair of bands in this molecule which appear at 84 500 and 87 500 cm^{-1} (*vert., 10.5 and 10.8 eV*) possibly are related to other ionization processes overlapping that at 112 000 cm^{-1}. The spectrum of the anesthetic HCF_2OCF_2CHFCl [T22] consists of alkyl-chloride-type bands in the 55 000–75 000 cm^{-1} (*6.8–9.3 eV*) region and a peak at 79 000 cm^{-1} (*9.8 eV*) possibly due to the ether group.

Mixed-chromophore spectra also are found in derivatives of vinyl ethers. The first consequence of the dual presence of conjugated vinyl and ether groups in methyl vinyl ether is that the vinyl π orbitals push the oxygen lone-pair ionization potential upward to 97 600 cm^{-1} (*vert., 12.1 eV*) from *ca.* 78 000 cm^{-1} (*9.67 eV*) where it would be in methyl ethyl ether. In response to this, the vinyl ionization potential drops to 73 700 cm^{-1} (*vert., 9.14 eV*) [P32]. The ether group thereby surrenders the 50 000–60 000 cm^{-1} (*6.2–7.4 eV*) region to the vinyl-group absorptions ($\pi \rightarrow 3s$, 48 000 cm^{-1} *vert., 5.9 eV*; $\pi \rightarrow \pi^*$, 53 250 cm^{-1} *vert., 6.602 eV*; $\pi \rightarrow 3p$, 55 400 cm^{-1} *vert., 6.87 eV*; $\pi \rightarrow 3d$, 60 000 cm^{-1} *vert., 7.4 eV*). Following these discrete absorptions, a block of continuous absorption runs from 65 000 to 85 000 cm^{-1} (*8.1 to 10.5 eV*) representing both the $(\pi)^{-1}$ ionization continuum and the lower Rydberg excitations originating with the oxygen lone-pair orbital. The relatively low frequency of the $\pi \rightarrow \pi^*$ band of vinyl ethers is consonant with the low-frequency shift that occurs on going from ketones to carboxylic acids (Section XV.B) and from imines to oximes (Section XI.B). The $\pi \rightarrow 3s$ transition of methyl vinyl ether shows a number of 200–300 cm^{-1} intervals assigned as C–O–C stretching. In 2,3-dihydropyran, the positions of the $\pi \rightarrow 3p$ and $\pi \rightarrow \pi^*$ bands are reversed compared with that in methyl vinyl ether [P32]. A related ether, (*S*)-(–)-1,2,2-trimethylpropyl vinyl ether (t-C_4H_9–$CHCH_3OCH{=}CH_2$) in the vapor phase shows CD bands which correlate closely with the pattern described above for methyl vinyl ether [B39]. In a variety of solvents, the $\pi \rightarrow 3s$ CD band of this vinyl ether disappears, while the $\pi \rightarrow \pi^*$, $\pi \rightarrow 3p$ and $\pi \rightarrow 3d$ CD bands are shifted strongly to higher frequencies but remain clearly visible. Thus, one sees again that CD bands of Rydberg excitations tend to persist in condensed phases (Section II.D).

With an O–F bond strength of only 44.9 Kcal/mole in F_2O compared to an O–H bond strength of 110.8 Kcal/mole in H_2O, one strongly expects a low-lying σ^*(O–F) valence MO in F_2O, Section I.C. In fact, the (b_1, σ^*) level in F_2O may well precede the Rydberg levels. The spectrum of F_2O has not been recorded, however, Valenta *et al.* [V4] predict the $\mathscr{A}(2p, \sigma^*)$ band to be the lowest excited singlet state in F_2O and the related molecule CF_3OF does have such a band at 52 000 cm^{-1} (*vert., 6.4 eV*) [I, R18]. In regard having an $\mathscr{A}(2p, \sigma^*)$ band below the lowest Rydberg excitation, F_2O stands in the same relation to H_2O as does NF_3 to NH_3 (Section VI.A), PF_3 to PH_3 [I, 239] and the perfluoroalkanes to their alkane parents [I, 190].

VII.D. Addendum

R. A. Rosenberg, P. R. La Roe, V. Rehn, J. Stöhr, R. Jaeger and C. C. Parks, K-shell excitation of D_2O and H_2O ice: photoion and photoelectron yields, *Phys. Rev.* **28B**, 3026 (1983).

R. J. Buenker, G. Olbrich, H.-P. Schuchmann, B. L. Schürmann and C. von Sonntag, Photolysis of methanol at 185 nm. Quantum mechanical calculations and product study, *J. Amer. Chem. Soc.* **106**, 4362 (1984).

C. Bertucci, R. Lazzaroni and W. C. Johnson, Jr., Far-UV circular dichroism spectra, at 145-220 nm, of some cyclic ethers as model compounds for carbohydrates, *Carbohydr. Res.* **133**, 152 (1984).

H. A. Michelsen, R. P. Giugliano and J. J. BelBruno, Photochemistry and photophysics of small heterocyclic molecules: I. Multiphoton ionization and dissociation of N-isopropyl dimethyl oxaziridine, to be published in *J. Phys. Chem.*

CHAPTER VIII

Compounds of Sulfur, Selenium and Tellurium

VIII.A. Hydrogen Sulfide and Hydrogen Selenide

In large part, recent assignments in the vacuum-ultraviolet spectrum of hydrogen sulfide are in accord with earlier work [**I**, 278], though a few surprises also have appeared. In the region of 50 000 cm^{-1} (*6 eV*) [**I**, 27], the absorption in H_2S is an obvious overlap of at least two transitions, which we earlier had assigned as Rydberg $2b_1 \rightarrow 4sa_1$ and valence $2b_1 \rightarrow a_1\sigma^*$; in H_2S, the axes are taken as x (out-of-plane, b_1), y (in-plane, b_2) and z (in-plane C_2 axis, a_1), with the $2b_1$ orbital being the uppermost lone pair orbital on sulfur. Molecular orbital [R5, S47] and scattered-wave [R29] calculations confirm the simultaneous presence of Rydberg and valence excitations in this region, but assign the σ^* MO symmetry as b_2 rather than a_1. Note however that in a recent calculation [D22], it is claimed that the 50 000-cm^{-1} continuum of H_2S consists of but a single transition, $2b_1 \rightarrow 4s$, where 4s is strongly admixed with $a_1\sigma^*$ valence character. This calculation however, does not consider the forbidden $\mathscr{A}(2b_1, b_2\sigma^*)$ band. Though the calculation of Rauk and

Collins [R5] yields a distinct $2b_1 \rightarrow a_1\sigma^*$ valence excitation (76 800 cm^{-1}, *9.52 eV*), this is achieved in a basis having only one ns Rydberg type orbital; in fact, in studies with larger bases [D22, R29] there is no distinct $a_1\sigma^*$ level reported, and one assumes that as in H_2O (Section VII.A), the $a_1\sigma^*$ MO of H_2S is dispersed among all the nsa_1 Rydberg orbitals such that no orbital has a preponderance of $a_1\sigma^*$ character. Note however, that this does not apply to the $(2b_1, b_2\sigma^*)$ and $(2b_1, 4pb_2)$ conjugate configurations, for both are observed and calculated as distinct states. It is not at all clear why the $a_1\sigma^*$ orbital is so much more miscible with the nsa_1 manifold than is $b_2\sigma^*$ with the npb_2 manifold. A possible clue to this puzzle appears in the theoretical work of Roberge and Salahub [R29], who show that $a_1\sigma^*$ bears a strong resemblance to 4s as expected. On the other hand, $b_2\sigma^*$ does not look at all like $4pb_2$, but instead closely resembles $3db_2$! In this case, $b_2\sigma^*$ and the prime conjugate ndb_2 orbitals have a small spatial overlap and are well-separated energetically, and so will tend not to mix, Section I.D.

As mentioned above, the forbidden $\mathscr{A}(2b_1, b_2\sigma^*)$ valence excitation of H_2S is the broad feature centered at *ca.* 55 000 cm^{-1} (*6.9 eV*) while its "conjugate" transition $2b_1 \rightarrow 4pb_2$ (*vide supra*) is observed at 63 327 cm^{-1} (*advert., 7.851 eV*) [M11]. As in H_2O, the np manifold of H_2S is split with npb_2 lowest, however the $(2b_1, 4pb_2)$ term value is unexceptional (21 090 cm^{-1}), unlike the situation in H_2O where it is unusually large (28 380 cm^{-1}). This result is consonant with the $b_2\sigma^*$ nodal pattern in H_2S being much more like that of an ndb_2 orbital than an npb_2 orbital such that the $b_2\sigma^*$ level mixes only weakly with the npb_2 manifold. Presumably, the $b_2\sigma^*$ orbital in H_2O is much more compatible with and mixes more strongly with $3pb_2$. At the $n = 4$ level, the $2b_1 \rightarrow npa_1$ transition is split into four components by j-j coupling with term values in the range 17 280−18 437 cm^{-1}, while higher members are represented by a single component; the resulting Rydberg series is quite irregular. Masuko *et al.* [M11] identify not only the three $2b_1 \rightarrow n$p series, but an ns series and all five components of the $2b_1 \rightarrow n$d series as well. Only seven Rydberg series originating at $2b_1$ are allowed in H_2S, and the mechanisms whereby the forbidden $2b_1 \rightarrow npb_2$ and $2b_1 \rightarrow ndb_2$ series appear is not understood. Several of the bands in the 64 000−71 000-cm^{-1} (*8−9-eV*) region of the H_2S spectrum recorded by Masuko *et al.* also are found in the MPI spectrum as three-photon resonances [A4]. As with the $(2b_1)^{-1}$ band of the photoelectron spectrum [K4], the Rydberg transitions originating from the $2b_1$ MO are extremely vertical (except for $2b_1 \rightarrow 4s$), allowing series to be observed up to $n = 20$, and in one case, up to $n = 26$ [B5]. Paralleling the situation in H_2O, the transition to 4s in H_2S is broad whereas those to higher ns members are much more vertical, the implication being that the ($2b_1$, $a_1\sigma^*$) valence conjugate preferentially has been mixed into ($2b_1$, $4sa_1$) thereby broadening the transition to it. In general, the best calculations vary widely in their detailed assignments of the Rydberg series of H_2S originating at $2b_1$ (*cf.* [D22], for example) and little can be said now as to which assignments are correct.

The Rydberg transition at 71 897.3 cm^{-1} (*advert., 8.9139 eV*) is among the most intense in the H_2S spectrum [M11]. Rotational analyses of this band in the one-photon absorption spectrum [G1] and as a three-photon resonance in the MPI spectrum [A32] show that it has perpendicular polarization and a lifetime of *ca.* 2×10^{-12} sec. The out-of-plane polarization is compatible with $2b_1 \rightarrow nsa_1$, npa_1 and nda_1 assignments, while its 12 520-cm^{-1} term value is most compatible with $3da_1$ as the terminating orbital. Indeed, Masuko *et al.* [M11] assign it as the leading member (n = 3) of an nd series having $\delta = 0.037$; the propensity rule on Δl [I, 29] also favors a p → d assignment for this intense feature [D22]. Baig *et al.* [B5] track the $2b_1 \rightarrow nda_1$ series out to n = 26 and report clear perturbations at n = 10 and 11. We note that this is just the region (83 500 cm^{-1}, *10.3 eV*) in which one expects the $5a_1 \rightarrow 4pb_1$ transition, and so the perturbation may come from the interaction between these two $^1B_1$ channels. Another such interchannel interaction is predicted to occur between the members of the $2b_1 \rightarrow npb_1$ series and the $5a_1 \rightarrow 4sa_1$ configuration, where $4sa_1$ has significant σ^* content [D22]. However the ($5a_1$, $4sa_1$) configuration does not mix with the members of ($2b_1$, ndb_1).

In hydrogen selenide [I, 281], the transition corresponding to that at 71 897.3 cm^{-1} in H_2S falls at 67 430 cm^{-1} (*advert., 8.360 eV*), however Hollas *et al.* [H69] prefer to assign it as terminating at $6sa_1$ rather than $4da_1$ after comparing its spectrum with that of the selenium atom. In support of the assignment of the 67 430-cm^{-1} band as terminating at $4da_1$, Mayhew *et al.* [M20] place it as the n = 4 member of an nda_1 series extending to n = 31. In contrast, Rauk and Collins [R5] calculate that the 67 430-cm^{-1} feature is a valence $\mathscr{A}$($4pb_1$, $a_1\sigma^*$) band, however this erroneous conclusion is the result of using such a small Rydberg basis set that the ($4pb_1$, $a_1\sigma^*$) valence configuration does not have enough conjugate Rydberg levels for dissolution (Section I.D). Though the three other allowed $4b_1 \rightarrow n$d Rydberg series of H_2Se are observed [M20], along with the two allowed series to np, members of the allowed series to ns are strangely missing. It is reported [M20] that the Rydberg excitations of H_2Se do not involve excitation of the bending mode, as in H_2S, but contrary to the situation in H_2O.

A Feshbach resonance in H_2S centered at 63 400 cm^{-1} (*vert., 7.86 eV*) [II, AD163] has a Feshbach decrement (Section II.B) of 2600 cm^{-1} with respect to the parent ($2b_1$, 4p) Rydberg state at 65 985 cm^{-1} (*advert., 8.181 eV*). As with the related Feshbach resonance at 79 500 cm^{-1} (*advert., 9.86 eV*) in H_2O, that in H_2S at 63 400 cm^{-1} is assigned the $^2(2b_1, 4p^2)$ configuration. This resonance also is seen in the H^- photodetachment spectrum with an angular dependence which suggests that the negative ion has $^2A_1$ symmetry [A39]. Azria *et al.* [A39] also report H^- photodetachment peaks in H_2S at 49 000 and 79 000 cm^{-1} (*vert., 5.6 and 9.8 eV*), the first of which has a 2000-cm^{-1} Feshbach decrement with respect to the ($2b_1$, 4s) parent state at 51 000 cm^{-1}, and an angular dependence indicating a $^2B_1$ resonance state as appropriate for the $^2(2b_1, 4s^2)$ negative ion. The 79 000-cm^{-1} feature is most likely another Feshbach resonance associated with the $(3a_1)^{-1}$ ionization potential at 106 900 cm^{-1} (*vert., 13.25 eV*) and involving double occupation of the 4p Rydberg orbitals. In agreement with this, it is observed to have $^2A_1$ symmetry [A39, J2]. The TNI resonance observed at 18 600 cm^{-1} (*vert., 2.3 eV*) in H_2S [A39, J2, R39] is clearly a σ^* valence resonance, however its orbital assignment is still in question. According to the scattering studies of Rohr [R39], the angular scattering from this resonance indicates that the electron is captured in the $a_1\sigma^*$ MO. The transition to the $b_2\sigma^*$ component is calculated to be 17 000 cm^{-1} (*2.1 eV*) higher [G13]. In contrast, the theoretical work of Jain and Thompson [J2] leads them to conclude that the resonance involving $a_1\sigma^*$ comes very close to zero energy, whereas that at 18 600 cm^{-1} involves the $b_2\sigma^*$ MO.

The inner-shell spectrum of H_2S fits the pattern noted for the other second-row hydrides. In the region of 2475 eV ($1s_S$), the absorption commences with a relatively intense and asymmetric feature having a 48 400-cm^{-1} term value and a width so broad as to suggest an overlap with a second transition on the high-frequency side [L11]. This large term value is appropriate only for a $1s_S \rightarrow \sigma^*$ valence excitation. From $1s_S$, the valence excitations to $a_1\sigma^*$ and $b_2\sigma^*$ are allowed, however transposing the outer-shell argument to the inner-shell spectrum, one concludes that the $1s \rightarrow a_1\sigma^*$ transition is dissolved in the Rydberg manifold (*vide supra*), and that it is $1s \rightarrow b_2\sigma^*$ which appears in the inner-shell spectrum. Two other peaks in the $1s_S$ spectrum of H_2S are observed to have term values of 24 200 and 12 100 cm^{-1}, and are assigned as $1s_S \rightarrow 4p$ and $1s_S \rightarrow 5p$, respectively [M21]. The $1s_S \rightarrow 4s$ Rydberg excitation will not be an intense one, and is probably folded into the high-frequency wing of the $1s_S \rightarrow \sigma^*$ band.

Many of the features of the $1s_S$ spectrum also appear in the $2p_S$ spectrum of H_2S, though doubled by the spin-orbit splitting of the $2p^5$ core configuration, Fig. I.C-2 [S27]. Ignoring this splitting, we recognize a

$(2p_S, \sigma^*)$ term value of 46 000 cm^{-1}, while the outer-shell $\mathscr{A}(2b_1, \sigma^*)$ band has a term value of 37 000 cm^{-1}, and the $\mathscr{A}(1s_S, \sigma^*)$ band has a term value of 48 400 cm^{-1}. This trend in the term values of H_2S is that expected on the basis of antishielding arguments (Section I.C). We interpret the peak at 166.5 eV as being $2p_S \rightarrow 4s$ (31 500 cm^{-1} term value), followed by a sharp Rydberg excitation to 4p with a term value of 18 800 cm^{-1} [R31]. Schwarz [S28], on the other hand, assigns the first of these to a second $2p_S \rightarrow \sigma^*$ promotion, and places 4s and 4p as nearly degenerate (actually with 4p somewhat *below* 4s) with term values of *ca.* 20 000 cm^{-1}. A band with a term value of 12 700 cm^{-1} is a blend of $2p_S \rightarrow 5p$ and $2p_S \rightarrow 3d$ excitations.

VIII.B. Sulfide Derivatives

Alkyl Sulfides

In spite of the growing variety of sulfur-containing molecules investigated in the vacuum ultraviolet, the pattern of $n_S \rightarrow \sigma^*$, 4s, 4p, 3d transitions remains immutable.† (See, for example, the calculations on CH_3SH [R5]). Dozens of Rydberg excitations in dimethyl sulfide-h_6 and -d_6 [I, 248; II, 321] have been placed into ns (δ = 2.00), np (δ = 1.55) and nd (δ = 1.15) series [M26]. In this work, McDiarmid has chosen the quantum numbers and the quantum defect of the nd series such that it appears to begin at n = 4, thereby placing the 3d orbitals in the valence shell. We prefer to assign the $n_S \rightarrow 3d$ transition (55 600 cm^{-1} *advert., 6.89 eV*) [I, 285] as the leading member of the Rydberg series with δ = 0.15, however, as McDiarmid points out, it is anomalously broad for a Rydberg excitation in dimethyl sulfide. Our approach normally is to assume that a Rydberg excitation is broadened through mixing with its valence conjugate. Because such mixing is out of character for a transition terminating at 3d owing to its low penetration, this band of dimethyl sulfide deserves further study. A very broad absorption also appears in the vapor spectrum of dimethyl sulfide at 49 500 cm^{-1} (*vert., 6.14 eV*), a frequency most appropriate for an $n_S \rightarrow 4p$ assignment. However, the

† The $2b_1$ orbital of H_2S is written as n_S in its derivatives, while σ^* is a combination of σ^*(S–C) group orbitals of unspecified symmetry surrounding the sulfur atom.

band persists at very nearly this frequency in trifluoroethanol solution [M26], and so must terminate at σ^*. Under the same conditions, the $n_S \rightarrow 4s$ Rydberg transition at 44 000 cm^{-1} (*advert., 5.455 eV*) in the vapor phase is obliterated. A more specific assignment of the band at 49 500 cm^{-1} in dimethyl sulfide cannot be given since it is not known if it is the lowest transition to σ^*. A polarization determination by electric dichroism [A16] would be most welcome here.

As anticipated from the spectra of many other silyl compounds, that of disilyl sulfide is shifted to higher frequency compared to that of dimethyl sulfide; its lowest ionization potential is about 8000 cm^{-1} higher than that of dimethyl sulfide [M32]. The spectrum of digermyl sulfide is broad and ill-defined [M32].

Using the technique of electric-field dichroism, Altenloh and Russell [A16, A17] have investigated the transitions from n_S to 4s, 4p and 3d in the series of cyclic sulfides [**I**, 288; **II**, 321] from ethylene sulfide (thiirane) to pentamethylene sulfide. Among other quantities, they measured the angle θ between the local C_{2v} axis at the S atom and the transition-moment direction for each excitation. For transitions from n_S to 4s, the angle θ is measured to be 90°, as expected for a $b_1 \rightarrow a_1$ transition. In the case of transitions to 4p, core splitting is generally not observed in the cyclic sulfides, however Altenloh and Russell find a prominent transition with an (n_S, 4p) term value of 20 000 ± 1000 cm^{-1} and $\theta = 0°$. The transition therefore must be $n_S \rightarrow 4pb_1$. Similarly, only one transition to 3d is observed, with $\theta = 90°$, suggesting either $n_S \rightarrow 3da_1$ or $n_S \rightarrow 3da_2$ assignments. For a given sulfide, the upper-state polarizability increases in the order 4s < 4p << 3d. With increasing ring size, the (n_S, 4s) polarizability also increases, whereas the reverse trend is claimed for the (n_S, 4p) and (n_S, 3d) configurations. The dichroism results for dimethyl sulfide [A16] parallel those for ethylene sulfide.

Electronic-state assignments in the ethylene sulfide system also have been checked through a CD study of (+)-(*R*,*R*)-2,3-dimethyl thiirane in tandem with calculations of its absorption and CD spectra [C34]. In this work, it is concluded that $n_S \rightarrow 4s$ and $n_S \rightarrow 4p$ transitions appear as an unresolved complex in both absorption and CD in the 46 500–52 600 cm^{-1} (*5.77–6.52 eV*) region, and that the $n_S \rightarrow 3d$ band comes at 58 100 cm^{-1} (*vert., 7.21 eV*), in agreement with the earlier interpretation [**I**, 284]. Additionally, the Rydberg absorption in this system is preceded by a weak valence excitation, which is calculated to be $n_S \rightarrow b_2\sigma^*(C–S)$. This transition is closely analogous to the $\mathscr{A}(np, \sigma^*)$ band of alkyl halides (Section I.C), and correlates with the lowest valence excitation of H_2S (Section VIII.A). However, it is not known if this band of ethylene sulfide correlates with that at 49 500 cm^{-1} in dimethyl sulfide (*vide supra*).

In passing, we note that $CH_3SCH_2CH_2CH_3$ [O9] shows a pattern of $n_S \rightarrow$ 4s, 4p and 3d transitions in which the 4p level has a core splitting (*ca.* 800 cm^{-1}) and the transition to 4s rests upon a valence transition to σ^*. In the disulfide $CH_3SCH_2SCH_3$, the n_S orbital ionization potentials are separated by 2000 cm^{-1}, which splitting is reflected in the transitions to 4p at 51 700 and 53 700 cm^{-1} (*vert., 6.41 and 6.66 eV*) [O9]. The transitions of tetrahydrothiopyran (S-ring) have been sorted into *n*s, *n*p and *n*d Rydberg series showing a sharp vibronic structure up to the onset of ionization at 65 800 cm^{-1} (*adiabat., 8.16 eV*) [P32]. In the mixed chromophoric systems methyl vinyl sulfide and 4,4-dimethyl-2,3-dihydrothiopyran, the n_S orbital is mixed with the π MO of the C–C double bond; transitions from this mixed orbital to 4s, 4p and 3d Rydberg orbitals and to the π^* valence orbital are at lower frequencies than in the corresponding oxo compounds [P32]. The absorption coefficient of the metallic polymer SN_x has been measured out to 215 000 cm^{-1} (*27 eV*) [B55], and the reflectivity spectrum of orthorhombic S_8 confirms peaks reported earlier [S6]. The spectra of other sulfur-containing compounds are discussed in Section XII.C (thiones), Section XII.D (dithiones) and Section XV.A (thioamides).

Sulfur Hexafluoride

Though studied repeatedly in the last ten years, understanding of the electronic excitations in sulfur hexafluoride remains in a shambles. Our earlier work on this has contributed strongly to the confusion, since it contains a number of typographical errors, inaccurate data and flawed thinking [I, 293]. In particular, the assumption was made that valence $\mathscr{A}$ bands will not appear in the spectrum of a highly fluorinated species such as SF_6; because this assumption is incorrect, the entire analysis is in error. Others have since offered their own interpretations, but I feel these too are flawed in one way or another.

Let us make a fresh start, newly defining the data and our approach to its interpretation. In this, we begin with a discussion of the inner-shell spectrum, and then apply the lessons learned there to the outer-shell spectrum. In looking at the interpretations of the inner-shell spectra of heavily fluorinated systems such as CF_4 (Section IV.D), BF_3 (Section V.C), SiF_4 (Section IX.B) and XeF_6 (Section XXI.A) one sees the clear presence of broad, intense, often low-lying inner-shell valence excitations to σ^* with inner-shell (ϕ_i^{-1}, σ^*) term values which can be as large as 80 000 cm^{-1} (XeF_6), Table I.C-V. Not only are such excitations expected for SF_6 on general grounds, but the relatively low S–F bond energy (78.6

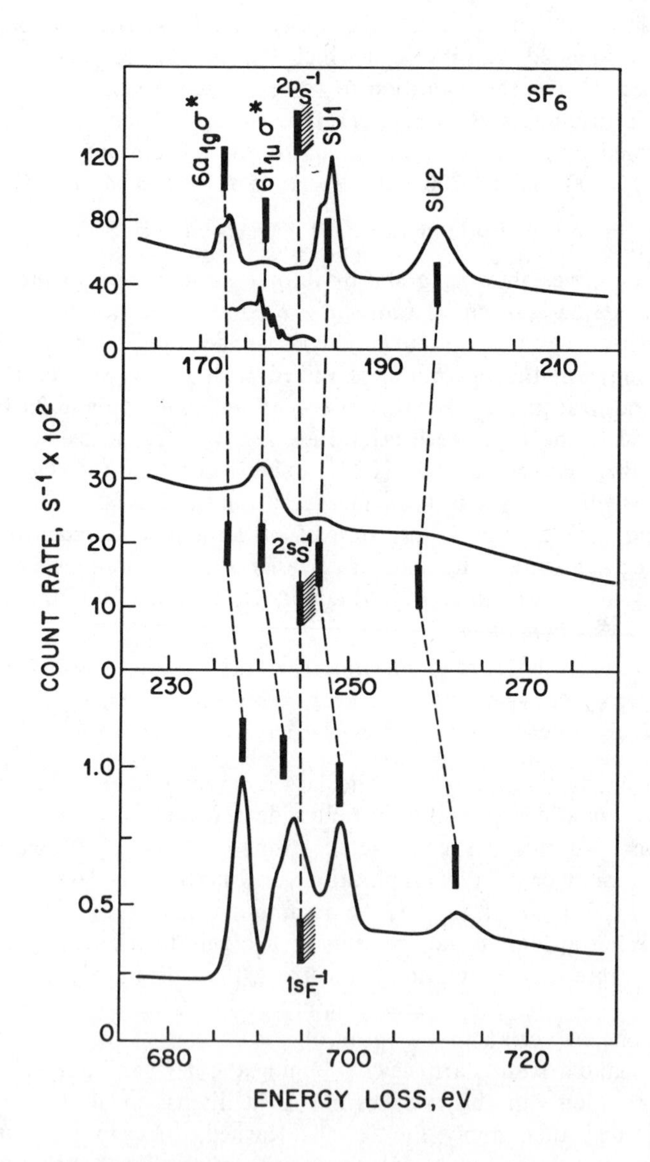

Fig. VIII.B-1. The inner-shell spectra of SF_6 as determined by electron energy loss at 2500 eV impact energy [H60], and drawn with the inner-shell ionization potentials aligned. The inset in the $2p_S$ spectrum is magnified to show the Rydberg fine structure resting upon the $2p_S \rightarrow 6t_{1u}\sigma^*$ transition. Supposed shake-up transitions are labelled SU1 and SU2.

kcal/mole) [S11] also indicates that they will fall at relatively low frequencies (Section I.C). Note as well that there are two virtual σ^* MO's in SF_6, $6a_{1g}\sigma^*$ and $6t_{1u}\sigma^*$, and if we follow the situation in H_2S (*vide supra*), the lower Rydberg orbitals of SF_6 will be 4s, 4p and 3d, with the last of these split into t_{2g} and e_g components. In octahedral symmetry, the inner-shell $2s_S \rightarrow 6t_{1u}\sigma^*$ and $2p_S \rightarrow 6a_{1g}\sigma^*$ valence transitions are dipole allowed, and each of the transitions from $1s_F$ (a_{1g}, e_g, t_{1u}) to $6t_{1u}\sigma^*$ and $6a_{1g}\sigma^*$ have allowed components. The pattern of inner-shell transitions to σ^* in SF_6, Fig. VIII.B-1, is readily interpreted along these lines. Consideration of not only the spectral widths, but the term values of the lowest-frequency bands in the inner-shell spectra [53 200 cm^{-1} ($1s_F$); 58 900 cm^{-1} ($2s_S$); 67 800 cm^{-1} ($2p_S$)] leaves no doubt that these valence excitations are $\mathscr{A}(\phi_i^{-1}, \sigma^*)$ bands. Their valence nature is further confirmed by the persistence of these bands in the spectra of neat SF_6 films [I, B33], and by the presence of related bands in the inner-shell spectrum of the isoelectronic PF_6^- ion in the solid state (Section XXI.A). The transitions from $1s_S$ to $6a_{1g}\sigma^*$ (forbidden) and to $6t_{1u}\sigma^*$ (allowed) also are seen at their expected positions in the spectrum of SF_6 [B12, L10], with a hint of the $1s_S \rightarrow 4p$ Rydberg transition atop the intense $1s_S \rightarrow 6t_{1u}\sigma^*$ band.

Careful work by Hitchcock and Brion [H60] reveals other weaker features in the SF_6 inner-shell spectrum corresponding to either low-intensity Rydberg bands or forbidden valence transitions. Thus the weak transitions at 175–180 eV, inset Fig. VIII.B-1, have term values of 25 000 and 19 000 cm^{-1}, respectively, and so are assigned as transitions from $2p_S$ to 4s and 4p.

Most intriguing is the presence in each of the inner-shell spectra of two discrete bands beyond the respective inner-shell ionization potentials, marked SU1 and SU2 in Fig. VIII.B-1. The bands in question have a nearly constant separation of 102 000 ± 4000 cm^{-1} (*12.6 ± 0.5 eV*) independent of the originating MO. Similar resonances beyond the outer-shell ionization potentials of SF_6 are reported by Gustafsson [G45]. Hitchcock and Brion propose that these transitions in the inner-shell spectra are valence transitions, the terminating MO's being the t_{2g} and e_g components of the 3d orbitals in an octahedral field. Also see [V16] and [I, D6] for similar assignments. Indeed, these bands are observed as well in the spectrum of solid SF_6 [I, B33], validating their valence character. However, because 4s and 4p are clearly a part of the Rydberg manifold and usually below 3d, it is something of a surprise to see 3d placed in the valence manifold. This view is supported by the theoretical work of Hay [H26], who calculated the outer-shell spectrum of SF_6 and reports that transitions to 3d are Rydberg, with quite ordinary term values in the

10 000−11 000-cm^{-1} range.† Thus we are inclined to reject assignments which terminate at 3d for the bands in question and propose instead that the SU1 and SU2 bands are due to shake-up processes terminating at valence MO's (Section II.C). Hitchcock and Brion [H60] have discounted the possibility that these bands are due to EXAFS structure. Shake-up transitions would be allowed from $a_{1g}\sigma$ MO's to $6a_{1g}\sigma^*$ and from $t_{1u}\sigma$ MO's to $6t_{1u}\sigma^*$. On the other hand, the observed intensities of these bands seem anomalously high for shake-up transitions, especially in the $2p_S$ spectrum. Understanding of these transitions would profit greatly from calculations aimed at determining whether the 3d orbitals of SF_6 are valence or Rydberg. Just this problem arises again in trying to interpret the TNI spectrum of SF_6 (*vide infra*).

Returning now to the outer-shell spectrum of SF_6, Fig. VIII.B-2 depicts the absorption spectrum as presented by Sasanuma *et al.* [S19], together with the ionization potential data of Gustafsson [G45]. Gustafsson presents a critical study of the orbital ordering in the outer shell of SF_6, which ordering we adopt here. In addition to the absorption peaks appearing in the curve of Fig. VIII.B-2, transitions are reported at positions **P** and **Q** by Lee *et al.* [L14], while Sasanuma *et al.* and Lee *et al.* report weak vibronic undulations at positions **G**′, **H**′ and **I**′. Many of the peaks in Fig. VIII.B-2 also appear in the photoionization spectrum of SF_6 [S20]. The photoabsorption cross sections of SF_6 have been discussed critically by Berkowitz [B34] and by Hitchcock and van der Wiel [H62].

Working with Fig. VIII.B-2, one finds first that it is not possible to force all the excitations into the Rydberg mold using term values and even if it were possible, several theoretically forbidden excitations are then found to have impossibly large observed oscillator strengths. However, with inner-shell (ϕ_i^{-1}, σ^*) term values of *ca.* 55 000 cm^{-1} and larger, it is clear that such transitions terminating at $6a_{1g}\sigma^*$ and $6t_{1u}\sigma^*$ in the outer-shell spectrum still will be low-lying with respect to the Rydberg excitations. Specifically, taking due account of the antishielding effects on valence (Section I.C) and Rydberg excitations (Section I.B), one expects term values of approximately 45 000 and 20 000 cm^{-1} for outer-shell excitations to $6a_{1g}\sigma^*$ and $6t_{1u}\sigma^*$, respectively, whereas for Rydberg excitations, the (ϕ_i, 4s), (ϕ_i, 4p) and (ϕ_i, 3d) term values will be approximately 25 000, 20 000 and 12 000 cm^{-1}. Indeed, in Hay's calculation of the outer-shell spectrum [H26] there are two low-lying transitions predicted to terminate

† Note however that the basis set did not include valence 3d orbitals.

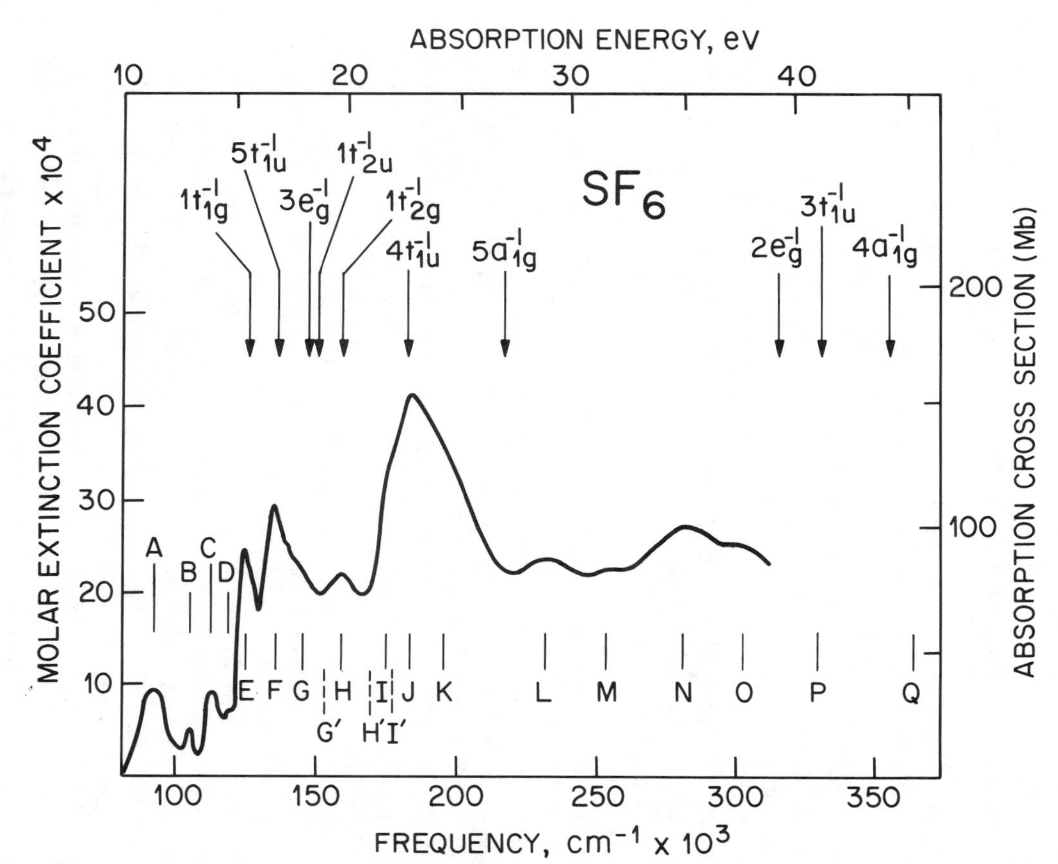

Fig. VIII.B-2. Quantitative absorption spectrum of SF_6, full curve [S19], and its vertical ionization potentials, arrows [G45].

at $6a_{1g}\sigma^*$ with term values of 34 000–47 000 cm^{-1}. As can be seen in Table VIII.B-I, this approach leads to immediate success, for all of the more intense features in the spectrum are seen to have the expected term values for allowed valence excitations to either $6a_{1g}\sigma^*$ or $6t_{1u}\sigma^*$. Since the cluster of uppermost filled MO's involve the 2pπ orbitals on the fluorine atoms, the transitions in question are $\mathscr{A}$(2pπ, σ^*(S–F)) bands, as already identified in chlorides, bromides and iodides, Section I.C. In addition to the optical bands shown in Fig. VIII.B-2, a weak energy loss is noted at 79 000 cm^{-1} (*vert., 9.8 eV*) [T17] which has a term value of 49 000 cm^{-1} with respect to the $(1t_{1g})^{-1}$ ionization potential, and so is assigned as $1t_{1g} \rightarrow 6a_{1g}\sigma^*$, vibronically allowed [H26]. It is clear as well from the spectrum of a solid film of SF_6 [**I**, B33] that bands **C**, **F** and **J** persist in the condensed phase, in accord with their valence upper-state assignments. Our assignments of bands **A** and **F** based on term-value arguments is identical to that of Hay based upon *ab initio* calculations. As listed in Table VIII.B-I, the remaining bands of SF_6 are readily assigned as Rydberg excitations on the basis of their term values and intensities.

A few small problems remain with the assignments summarized in Table VIII.B-I. *i*) Bands **L** and **M** have no obvious assignments, and possibly are related to the shake-up bands of negative term value found throughout the SF_6 spectrum, (*vide supra*). *ii*) The $(\phi_i, 6t_{1u}\sigma^*)$ term value is 33 000 cm^{-1} rather than 20 000 cm^{-1}. *iii*) In the spectrum of solid SF_6 [**I**, B33] there is a small but distinct peak at **H** which is difficult to assign as a valence transition. *iv*) The $4a_{1g} \rightarrow 6t_{1u}\sigma^*$ valence transition is expected at 320 000 cm^{-1} (*39.7 eV*), and though allowed, is not reported as yet. These problems are relatively minor, and suggest that though we are now much closer to the truth in regard the SF_6 spectrum, there is still room for more work on this problem.

The number and frequencies of TNI resonances are of direct importance to the questions of the number and energies of virtual valence orbitals in molecules, Section II.B. Experiments on SF_6 have shown several such TNI resonances [K12], but their interpretation is not clear as yet. The sharp resonance at 95 740 cm^{-1} (*vert., 11.87 eV*) in SF_6 has a frequency just slightly below that of the lowest observed Rydberg excitation, $1t_{1g} \rightarrow 4p$ (band **B** in Fig. VIII.B-2), and thereby is assigned the Feshbach configuration $^2(1t_{1g}, 4p^2)$. The three resonances below this, at 240, 20 600 and 56 900 cm^{-1} (*vert., 0.030, 2.56 and 7.05 eV*) must be valence TNI resonances by their frequencies. We note from Table VIII.B-I that the two σ^* MO's of SF_6 ($6a_{1g}$ and $6t_{1u}$) are separated by *ca.* 15 000 cm^{-1} in the neutral-molecule spectrum, and this split tempts us to assign the TNI resonances at 240 and 20 600 cm^{-1} as terminating at

TABLE VIII.B-I

ASSIGNMENTS AND TERM VALUES IN THE SPECTRUM OF SULFUR HEXAFLUORIDE

Band[a]	Assignment	a, f[b]	Term Value, cm^{-1}
-[c]	$1t_{1g} \rightarrow 6a_{1g}\sigma^*$	f	49 000
A	$5t_{1u} \rightarrow 6a_{1g}\sigma^*$	a	46 000
	$1t_{1g} \rightarrow 6t_{1u}\sigma^*$	a	33 000
B	$1t_{1g} \rightarrow 4pt_{1u}$	a	21 800
	$5t_{1u} \rightarrow 4sa_{1g}$	a	32 300
C	$5t_{1u} \rightarrow 4pt_{1u}$	f	23 800
	$3e_g \rightarrow 6t_{1u}\sigma^*$	a	31 000
	$1t_{1g} \rightarrow 3de_g, 3dt_{2g}$	f	12 200
D	$1t_{2u} \rightarrow 4sa_{1g}$	f	33 100
E	$1t_{2g} \rightarrow 4sa_{1g}$	f	33 500
	$3e_g \rightarrow 4pt_{1u}$	a	23 400
	$5t_{1u} \rightarrow 3de_g, 3dt_{2g}$	a	12 200
F	$1t_{2g} \rightarrow 4pt_{1u}$	f	25 000
	$3e_g \rightarrow 3de_g, 3dt_{2g}$	f	12 200
	$4t_{1u} \rightarrow 6a_{1g}\sigma^*$	a	48 000
G	$1t_{2g} \rightarrow 3de_g, 3dt_{2g}$	f	12 400
G'	$4t_{1u} \rightarrow 4sa_{1g}$	a	29 600–31 200
H	$4t_{1u} \rightarrow 4pt_{1u}$	f	24 000
H'	$4t_{1u} \rightarrow 5sa_{1g}$	a	12 400
	$4t_{1u} \rightarrow 4de_g, 4dt_{2g}$	a	12 400
I	$5a_{1g} \rightarrow 6a_{1g}\sigma^*$(?)	f	43 000
I'	$4t_{1u} \rightarrow 6sa_{1g}$	a	6000
	$4t_{1u} \rightarrow 4de_g, 3dt_{2g}$	a	6000
J	$5a_{1g} \rightarrow 6t_{1u}\sigma^*$	a	33 000
K	$5a_{1g} \rightarrow 4pt_{1u}$	a	21 800
L		-	
M		-	
N	$2e_g \rightarrow 4sa_{1g}$	f	32 300

TABLE VIII.B-I (continued)

Band[a]	Assignment	a, f[b]	Term Value, cm^{-1}
	$3t_{1u} \rightarrow 6a_{1g}\sigma^*$	a	48 000
	$2e_g \rightarrow 6t_{1u}\sigma^*$	a	32 000
O	$3t_{1u} \rightarrow 4pt_{1u}$	f	26 600
	$2e_g \rightarrow 3de_g, 3dt_{2g}$	f	12 200
P	$4a_{1g} \rightarrow 4pt_{1u}$	a	25 800

[a] See Fig. VIII.B-2.

[b] Allowed, forbidden.

[c] Weak band observed at 79 000 cm^{-1} (*vert., 9.8 eV*), but not shown in Fig. VIII.B-2.

these σ^* MO's. Left still unassigned is the resonance at 56 900 cm^{-1}, along with those at 218 000 and 400 000 cm^{-1} (*vert., 27 and 50 eV*). Tentatively, the latter two resonances can be assigned as core-excited valence TNI resonances involving the σ^* MO's in unspecified ways, while the first remains an enigma.

Using an X-α calculation, Dehmer *et al.* [D10] arrive at a totally different interpretation of the TNI resonances in SF_6; they do not assign the band at 240 cm^{-1} as terminating at a particular orbital, but then go on to assign the $6a_{1g}\sigma^*$ and $6t_{1u}\sigma^*$ resonances to the peaks observed at 20 600 and 56 900 cm^{-1}, while the resonances at 95 740 and 218 000 cm^{-1} are assigned as terminating at the t_{2g} and e_g components of an otherwise unoccupied nd orbital set. Because we maintain that the vacant nd orbitals of SF_6 are Rydberg in character and because electrons cannot be bound to such orbitals in TNI processes (Section II.B), we see that this approach also has its difficulties.

A study of the branching ratios and angular distributions of photo-ejected electrons in SF_6 led to the inferred presence of a shape resonance at 190 000 cm^{-1} (*vert., 23 eV*) [D11]. It is clear from Fig. VIII.B-2 and Table VIII.B-I that the resonance in question involves transition **J**, $5a_{1g} \rightarrow 6t_{1u}\sigma^*$.

The absorption spectrum of TeF_4 is discussed in Section XXI.A.

VIII.C. Addendum

M. Stuke, Ultrasensitive fingerprint detection of organometallic compounds by laser multiphoton ionization spectrometry, to appear in *Appl. Phys. Lett.*

CHAPTER IX

Silanes, Germanes and Stannanes

IX.A. Parent Compounds and Alkyl Derivatives

Two errors in our earlier discussion of the spectra of silanes [**I**, 296] have unnecessarily complicated the general picture of absorption in this class of substances. First, we seriously considered the weak absorption band reported for silane at 64 100 cm^{-1} (*vert., 7.95 eV*) as a $\sigma \rightarrow \sigma^*$ valence transition on the basis of its large term value (35 100 cm^{-1}), thereby assigning silane to Regime Z in Fig. I.D-3. More recent absorption [R28] and electron-impact work on silane [D27] show that this band is not present, and must have been due to an impurity absorption in the earlier work.[†] Consequently, as regards the relative Rydberg/valence ordering, silane is in Regime X or Y, but not in Z.[§] Second, we neglected

[†] Sandorfy [S16] wisely has questioned the reality of this band of silane. The small peak at 67 000 cm^{-1} (*8.3 eV*) in the earlier spectrum of hexamethyl disilane also is missing in more recent work [B18].

[§] More recent spectral studies (*vide infra*) show that silane is in Regime Y, Fig. I.D-3.

to account properly for multiplet splitting in the silicon atom when the claim was made that the term values for 4p and 3d orbitals were 20 000 cm^{-1} in each case. In fact, a value of 13 000 cm^{-1} is more nearly correct for the 3d outer-shell term values in silanes.

With the above two points cleared up, it is now possible to present a coherent set of spectral assignments in silane and its alkyl derivatives. These are listed in Table IX.A-I, along with comparable data on the corresponding alkanes. As in the alkanes, silane and its alkyl derivatives show a progression of outer-shell transitions to $(n + 1)$s, $(n + 1)$p and nd Rydberg orbitals. Moreover, the term values in the silanes closely approximate those of the corresponding alkanes. Because these transitions are associated with the lowest ionization potential, they originate with the σ(Si–H) or σ(Si–C) orbitals surrounding the silicon atom in alkyl silanes rather than the σ(C–H) or σ(C–C) MO's of the alkyl groups. We note the expected decrease of the (σ, 4s) term value upon alkylation of silane, the core splitting of the (σ, 4p) configurations in derivatives of low symmetry, and the propensity for $\Delta l = +1$ (p $\rightarrow$ d) excitation over $\Delta l = -1$ (p $\rightarrow$ s) excitation. As in alkanes, in no case is there term-value evidence for low-lying transitions to valence states in the outer-shell spectra of the silanes, and paralleling the situation in methane, the lowest singlet Rydberg state of silane ($2t_2\sigma$, $4sa_1$) dissociates into $SiH_2 + H_2$, both in valence states [G28].

As in methane (Section III.A), the excitations from the highest occupied a_1 valence orbital of silane ($3sa_1$) into the lower np Rydberg orbitals are vibronically structured, Fig. III.A-1. Dillon [D27], using electron impact at nonzero scattering angle, finds $3sa_1 \rightarrow 4$p and $3sa_1 \rightarrow 5$p excitations centered at 129 300 and 136 500 cm^{-1} (*vert., 16.03 and 16.93 eV*), each displaying 1700-cm^{-1} quanta of ν_1'. The transitions to 6p and higher np members are not discernible, and apparently are rapidly autoionized or dissociated. A Franck-Condon envelope identical to those of the $3sa_1 \rightarrow 4$p and $3sa_1 \rightarrow 5$p transitions is seen in the $(3sa_1)^{-1}$ band of the silane photoelectron spectrum [I, P30]. The $3sa_1 \rightarrow 4$s transition of silane appears in the electron-impact spectrum as a broad, featureless band (119 800 cm^{-1} *vert., 14.85 eV*) with a term value of 26 800 cm^{-1}. The electron-impact spectrum of silane also shows a peak (102 000 cm^{-1}, *12.65 eV*) which is beyond the first ionization potential and precedes the Rydberg excitations to the second. If this is a bound one-electron excitation and not just a maximum in the photoionization cross section, it must be valence since its term value with respect to the $(3sa_1)^{-1}$ ionization potential is 44 500 cm^{-1}. Because the (2p, σ^*) term value of silane is only 35 000 cm^{-1} (*vide infra*), it seems unlikely to us that the band at 102 000 cm^{-1} could be ($3sa_1$, σ^*). A related transition is observed in the outer-

TABLE IX.A-I

COMPARISON OF RYDBERG TERM VALUES IN ALKANES, ALKYLATED SILANES AND GERMANE

Compound	Ionization Potential, cm^{-1} (*eV*)	Term Value, cm^{-1} $(n+1)$s	$(n+1)$p	*n*d	References
CH_4	109 800 (*13.61*)	31 600	-	13 300	[I, 105]
	(*288.23*)[a]	29 600	20 700	-	
SiH_4	99 200 (*12.29*)	26 700	21 200	12 200	[D27]
	146 600 (*18.17*)	26 800	17 260	-	
	(*107.09*)[b]	26 600	17 700	12 800	[F33]
GeH_4	97 600 (*12.10*)	28 200	-	13 500	[D27]
	148 000 (*18.4*)	25 400	-	14 100	
CH_3CH_3	97 500 (*12.09*)	29 500	21 700	-	[I, 105]
	(*290.6*)[a]	28 900	21 800	-	
CH_3SiH_3	93 600 (*11.6*)	30 100	22 800	14 600	[R28]
		-	19 800	12 100	
$(CH_3)_2CH_2$	91 900 (*11.39*)	27 900	20 500	14 100	[I, 105]
$(CH_3)_2SiH_2$	90 300 (*11.2*)	27 800	21 300	11 300	[R28]
		-	17 600	-	
$(CH_3)_3CH$	89 800 (*11.1*)	25 700	20 900	15 800	[I, 105]
$(CH_3)_3SiH$	87 100 (*10.8*)	26 100	21 100	-	[R28]
		-	16 300	-	
$(CH_3)_4C$	87 900 (*10.9*)	23 900	19 900	13 900	[I, 105]
$(CH_3)_4Si$	83 900 (*10.4*)	24 700	22 900	12 200	[R28]
	85 600 (*10.62*)	26 500	-	13 700	[S59]
	111 300 (*13.8*)	25 700	-	-	[S59]
	(289.78)[a]	28 400	19 900	-	[S59]
$(C_2H_5)_2CH_2$	88 200 (*10.9*)	24 200	18 700	13 200	[I, 105]
$(C_2H_5)_2SiH_2$	83 100 (*10.3*)[c]	23 100	15 600	9100	[R28]
	87 100 (*10.8*)	27 100	19 600	13 100	

[a] Transitions originate at $1s_C$.

[b] Transitions originate at $2p_{Si}$.

[c] The term values appear more normal if computed with respect to the second ionization limit at 87 100 cm^{-1} (*10.8 eV*).

shell spectrum of germane as well, and is tentatively assigned as $t_2\sigma \rightarrow a_1\sigma^*$ (*vide infra.*).

Peaks in the electron-impact dissociative attachment curves for silane strongly suggest negative-ion resonances at 57 000 and 64 000 cm^{-1} (*vert., 7.0 and 8.0 eV*) [E2], and are reminiscent of the valence TNI resonances observed in methane at 52 700, 59 400 and 65 700 cm^{-1} (*vert., 6.53, 7.36 and 8.14 eV*) [B56], Section III.A. In the latter case, the peaks were assigned as the Jahn-Teller-split components of a valence TNI resonance having the odd electron in $t_2\sigma^*$; a parallel assignment would seem to hold as well for these bands of silane. The Feshbach resonance into the $^2(2t_2, 4s^2)$ configuration of silane is expected at about 69 000 cm^{-1} with a Jahn-Teller splitting of approximately 3600 cm^{-1}. A valence TNI resonance in silane at 17 000 cm^{-1} (*vert., 2.1 eV*) [G17] is calculated to involve the $t_2\sigma^*$ MO, said to be below $a_1\sigma^*$ in this molecule [T15]. In line with the destabilizing effect of methyl groups, the resonance to $^2(N, \sigma^*)$ is raised to 31 000 cm^{-1} (*vert., 3.9 eV*) in tetramethyl silane [G17].

Owing to the $^2P_{3/2} - {}^2P_{1/2}$ spin-orbit splitting of the $2p^5$ excited-state core configuration (4800 cm^{-1}) in silane, every inner-shell orbital transition from 2p is split into two distinct peaks. With due consideration of this, and using the term-value concept, the awkward lump of absorption beginning at 823 000 cm^{-1} (*102 eV*) in the $2p_{Si}$ inner-shell spectrum of silane (Fig. IX.A-1) can be separated into four components [F33, R31], two having 35 000-cm^{-1} term values with respect to the spin-orbit split ionization limits, and two having 26 600-cm^{-1} term values with respect to the same limits. Because the Rydberg excitations in the outer-shell spectrum of silane terminating at 4s have term values of 26 700 cm^{-1} (Table IX.A-I), it is reasonable to assign the inner-shell transition at 837 200 cm^{-1} (*vert., 103.8 eV*) [F33] having a term value of 26 600 cm^{-1} as $2p \rightarrow 4s$ with a $^2P_{3/2}$ core, whereas that having a term value of 35 000 cm^{-1} is assigned as an $\mathscr{A}(2p, \sigma^*)$ band with the same core configuration. The X-ray absorption spectrum of solid silane, Fig. IX.A-1, [F33, S60] in fact shows that the first transition persists, but the second does not, in accord with their respective valence and Rydberg characters. It is likely that the $2p \rightarrow 4s$ Rydberg transition has shifted to higher frequency and broadened in the solid, thereby contributing to the apparent width of the valence excitation. On the other hand, Schwarz and his coworkers [F33, S27] have calculated that the assignments proposed above are inverted, for they assign the first band as $2p \rightarrow 4s$ mixed with its valence-conjugate $(2p, a_1\sigma^*)$, and the second as $2p \rightarrow 3t_2\sigma^*$.

Following these features in the silane inner-shell spectrum, there are a number of relatively sharp bands which have been assigned as terminating at 4p and 3d Rydberg orbitals on the basis of their term values [R31];

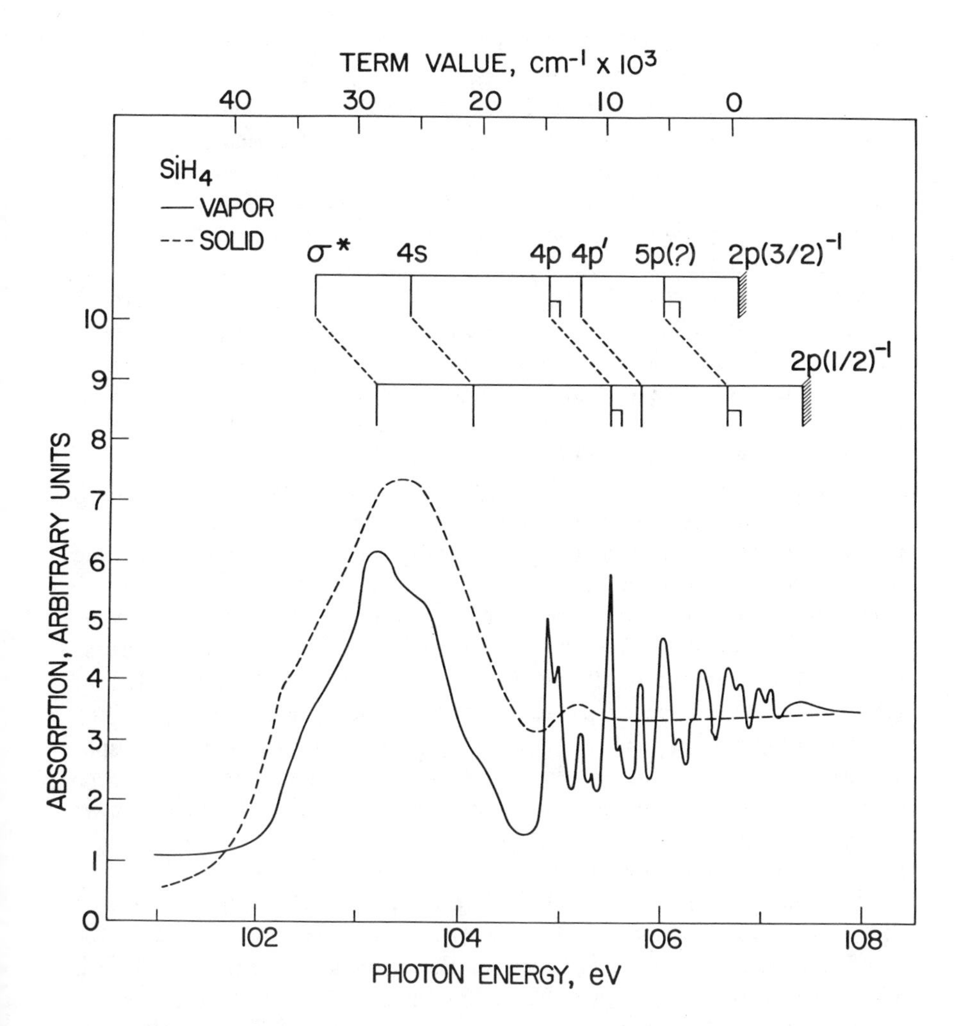

Fig. IX.A-1. Inner-shell absorption spectra of silane in the vapor phase (—) and as a polycrystalline film (- -) [F33, S60].

indeed these bands are obliterated in the spectrum of solid silane, Fig. IX.A-1. Again, the calculations of Schwarz *et al.* are at odds with the assignments based on term values, for these workers prefer not to assign any observed transition as terminating at 4p, while assigning the transition to 5s at *lower* frequency than that to 3d. Our assignments for the $2p_{Si}$ spectrum of silane closely parallel those for phosphine, Fig. VI.B-1. In the $2s_{Si}$ absorption region (150−160 eV) in silane, peaks are observed at 153.6 and 155.8 eV; the second is much weaker than the first [F33]. As in the

$2p_{Si}$ spectrum, the first of the two bands persists in the solid ($2s_{Si} \rightarrow \sigma^*$) while the second does not ($2s_{Si} \rightarrow 4s$). Assuming a (2s, 4s) term value equal to that of ($2p_{Si}$, 4s) leads to a $(2s_{Si})^{-1}$ ionization potential of 159.08 eV in silane; a value of 159.94 eV has been calculated [A5].

What is most enthralling here is that in the outer-shell spectrum of silane there is no experimental evidence whatsoever for a low-lying transition to a σ^* valence orbital, whereas in the inner-shell spectra the interpretation based upon term values places the excitation to σ^* as lowest of all! In both spectra, the placement of Rydberg excitations appears to be quite normal. Thus, it must be concluded that σ^*(Si—H) orbitals in the silanes may or may not be relevant to the observed spectra depending upon whether the originating orbital is an inner-shell or an outer-shell one. This statement probably carries over to other second-row hydrides as well, thereby placing them in Regime Y in Fig. I.D-3.

There is an oddity in the silane Rydberg term values, for in almost all molecular systems, the antishielding that results from the comparison of the outer-shell and inner-shell excited configurations raises the term value of the latter by several thousand cm^{-1}, Section I.B. However, in silane the exhaltation does not materialize. This behavior parallels that in methane and ethane, Table IX.A-I, where the term values actually decrease somewhat for the inner-shell excitations. One also notes in the Table that the sizeable difference of *n*s term values between methane and silane rapidly disappears as the central atoms are alkylated and the term values approach the alkyl limit. As discussed in [I, 51] this convergent behavior of the term values can be understood as due to the Rydberg orbital becoming progressively more alkyl-like (and therefore less central-atom-like) with increasing alkylation.

With the clear presence of low-lying σ^* MO's in the $2p_{Si}$ spectra of SiH_4 and SiF_4 (Section IX.B), it is odd indeed that clear $\mathscr{A}(2p_{Si}, \sigma^*)$ transitions seem not to be present in the spectrum of $Si(CH_3)_4$ [S59]. Two very weak features just above the noise have term values of *ca.* 32 000 cm^{-1} with respect to the spin-orbit components of the $(2p_{Si})^{-1}$ ionization potential and could be the $\mathscr{A}(2p_{Si}, \sigma^*)$ bands of $Si(CH_3)_4$, if indeed they are real. Excepting the transition to σ^*, the $2p_{Si}$ spectrum is otherwise normal, consisting of several Rydberg excitations of unremarkable term values. As expected, the outer-shell spectrum of $Si(CH_3)_4$ [I, 302] shows no sign of a transition to σ^* [S59].

The electron-impact energy loss spectrum of germane is of immediate interest because if $\sigma \rightarrow \sigma^*$ outer-shell valence excitations are to be found in the Group IV hydrides, clearly they will be found in molecules at the heavy end of the series. As with methane and silane, the germane spectrum is better resolved at the higher energy losses, Fig. III.A-1. A number of peaks in germane are resolvable below the first ionization potential, and as shown in Table IX.A-I, they have term values which lead to ($3t_2$, ϕ_j^{+1}) Rydberg-state assignments. Transitions with similar term values are found as well for Rydberg excitations originating from the uppermost a_1 orbital of germane. In both the $4a_1 \rightarrow \phi_j^{+1}$ energy losses and the $(4a_1)^{-1}$ photoelectron band [I, P30], long progressions of the ν_1' stretching vibration are excited. After accounting for the Rydberg excitations originating with the $3t_2$ and $4a_1$ MO's, there still remains an unassigned broad feature at 100 400 cm^{-1} (*vert., 12.45 eV*) which lies above the $(3t_2)^{-1}$ ionization limit and 48 000 cm^{-1} (*5.95 eV*) below that of the $4a_1$ MO. Tentatively, this is assigned as an allowed $t_2\sigma \rightarrow a_1\sigma^*$ transition of the outer-shell valence spectrum. A related feature is reported as well in the spectrum of silane, but not methane (*vide supra*). Though not reported yet, the inner-shell spectra of germane and stannane are certain to show low-lying $\mathscr{A}(\phi_i^{-1}, \sigma^*)$ valence excitations.

The spectra of $(SiH_3)_2S$ and $(GeH_3)_2S$ are discussed in Section VIII.B.

IX.B. Halides

As with silane, comparison of the $2p_{Si}$ spectra of SiF_4 in the vapor [I, 301] and solid phases, Fig. IX.B-1, is most useful in assessing the Rydberg/valence character of the various transitions [F34, S60]. We have seen already that going from CH_4 to SiH_4 introduces low-lying σ^* MO's in the inner-shell spectrum (Section IX.A), as does going from CH_4 to CF_4 (Section IV.D). Consequently, we expect transitions to one or more very low-lying σ^* MO's in the inner-shell spectrum of SiF_4. The first band of the $2p_{Si}$ inner-shell spectrum of SiF_4 has a 45 500 cm^{-1} term value and is almost unscathed in the spectrum of the solid, Fig. IX.B-1. These facts and ancillary calculations all point to this as being an $\mathscr{A}(2p_{Si}, a_1\sigma^*)$ band [F34, R31]. This transition of SiF_4 is closely related to the inner-shell valence transition of SiH_4 (*vide supra*), and comparison of their term values (45 500 *vs.* 35 000 cm^{-1}) illustrates the large exhaltation of inner-shell valence term values induced by fluorination (Table I.C-IV). In the $1s_F$ spectrum of SiF_4, the corresponding $\mathscr{A}(1s_F, a_1\sigma^*)$ band has a term value of 43 500 cm^{-1}. The broad inner-shell valence transition in the $2p_{Si}$ spectrum of SiF_4 is followed by a set of six sharp bands, the first of which has a term value (28 500 cm^{-1}) appropriate to a $2p_{Si} \rightarrow 4s$ Rydberg assignment. However, the spectrum of solid SiF_4 shows the first three features to be only modestly broadened and shifted to higher frequency, in every way resembling the behavior of the preceding inner-shell $\mathscr{A}(2p_{Si}, a_1\sigma^*)$ band. In contrast, the second three features of the vapor-

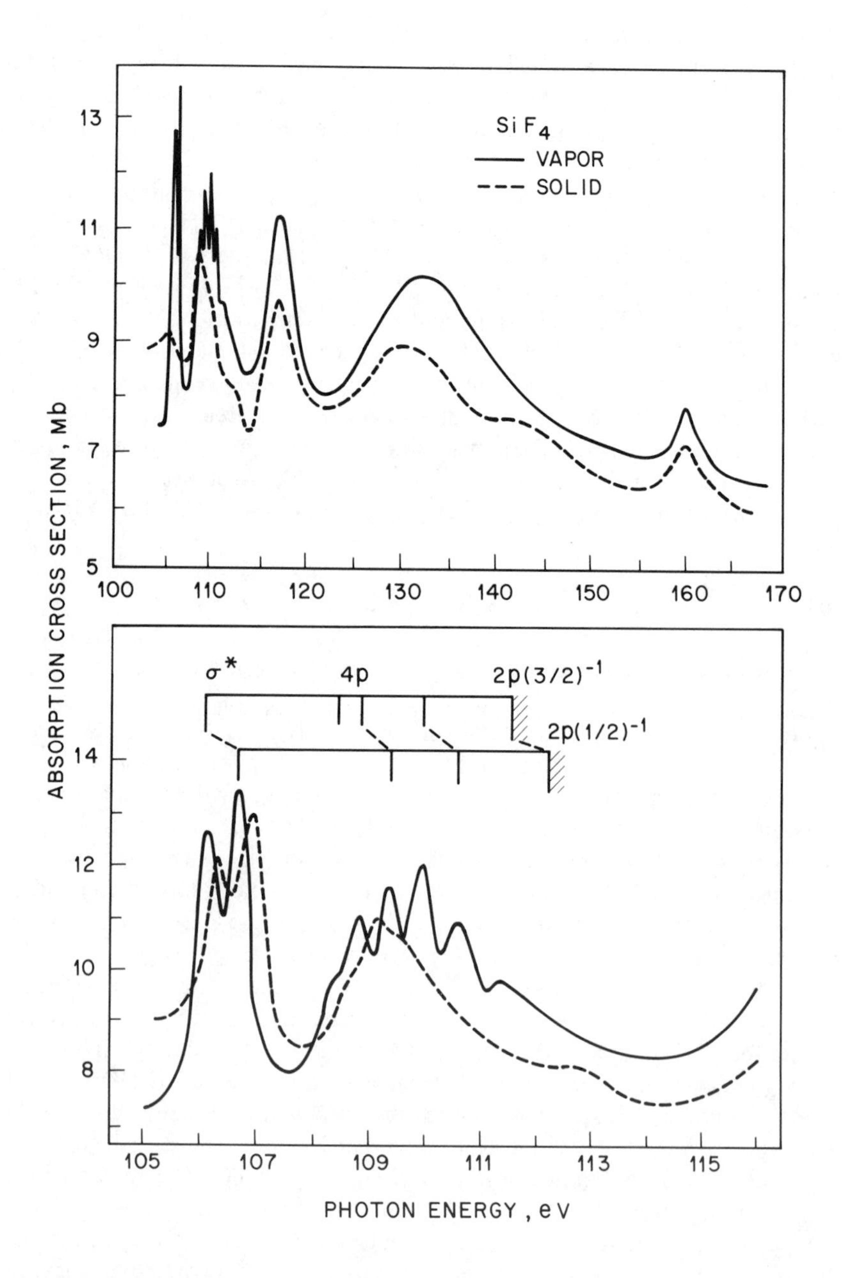

Fig. IX.B-1. Inner-shell absorption spectra of SiF_4 in the vapor phase (—) and as a polycrystalline film (- -) in the $(2p_{Si})^{-1}$ and $(2s_{Si})^{-1}$ regions, *upper panel*, and in detail for the $(2p_{Si})^{-1}$ region, *lower panel* [F34].

phase spectrum are totally missing in the spectrum of the solid, as appropriate for Rydberg excitations. Thus we assign the first three features as $2p_{Si} \rightarrow t_2\sigma^*$ with structure due to both $^2P_{3/2} - {}^2P_{1/2}$ spin-orbit splitting and to multiplets arising from the $(2p_{Si}, t_2\sigma^*)$ configuration, and the second three features to one or more Rydberg excitations. Friedrich *et al.* [F34] also assign the first three bands in question as $2p_{Si} \rightarrow t_2\sigma^*$, and then go on to assign the second cluster of three bands at 882 400 cm^{-1} (*109.4 eV*) as terminating at $t_2\sigma^*$ and at 4s, with an observed term value of 18 900 cm^{-1} for the first peak. Because a term value of 18 900 cm^{-1} is far too low for a transition terminating at 4s, but is commonly found for transitions terminating at 4p, we prefer a $2p_{Si} \rightarrow 4p$ assignment for the bands between 110 and 111 eV. Apparently the allowed transition from $2p_{Si}$ to 4s does not appear as a distinct feature in the vapor spectrum.

Several broad bands appear beyond the $(2p)^{-1}$ ionization potential of SiF_4, Fig. IX.B-1. Those in the 115–150 eV region have been assigned as EXAFS-type interferences [F34, S60] involving partial *d* waves, while that at *ca.* 160 eV in both vapor and solid would seem to be an $\mathscr{A}(2s_{Si}, \sigma^*)$ band, related to that at 153.6 eV in silane (Section IX.A). Similar bands are seen in the X-ray spectra of several other Group-IV halides [P18, P19, P20]. In parallel spectral studies of SiF_4 vapor and solid Na_2SiF_6, the claim is made for a one-to-one correspondence of the bands in their $1s_F$ spectra [D54], however this must be illusory for the anionic species SiF_6^{-2} will not display the Rydberg excitations which otherwise appear in the spectrum of SiF_4 (Section I.B).

In view of the close similarities of alkane and silane spectra in the vacuum ultraviolet, one expects that the outer-shell spectra of alkyl halides and silyl halides also will resemble one another closely, and this is so. In the series $SiF_x(CH_3)_{4-x}$ [R28], the outer-shell spectra look much like those of the alkyl silanes, but shifted to higher frequencies. Going from $x = 0$ to $x = 2$ shifts the lowest transition to 4s upward by *ca.* 4000 cm^{-1}, whereas the shift is 10 000 cm^{-1} on going from $x = 2$ to $x = 3$. This behavior parallels that observed in the fluoromethanes [R28, **I**, 179]. In accord with the higher penetration at fluorine, the outer-shell (σ, 4s) term value of $(CH_3)_3SiF$ (26 500 cm^{-1}) increases to 32 200 cm^{-1} in CH_3SiF_3. As in the fluoromethanes (Section IV.D), there are no σ^*(Si–F) valence orbitals lying below the Rydberg orbitals in the outer-shell spectra of the fluorosilanes, however, it is likely that this ordering will be reversed in the inner-shell spectra.

In compounds containing chlorine bonded to silicon [**I**, 311], the chlorine lone-pair orbitals are stabilized through interaction with the $3d\pi$ orbitals of silicon, thereby shifting the Rydberg spectra to higher frequencies compared to the corresponding alkyl chlorides. However, the valence

$\mathscr{A}(3p_{Cl}, \sigma^*)$ band is shifted upward even more strongly, for it cannot be identified as a peak in the spectra of any of the chlorosilanes [C10, R30]. The spectrum of $SiHCl_3$ does have a long tail in the 60 000–70 000 cm^{-1} (*7.4–8.7 eV*) region which may correspond to $\mathscr{A}(3p, \sigma^*)$ absorption; the lowest identifiable peak in this molecule comes at 71 000 cm^{-1} (*8.80 eV*, $3p \rightarrow 4s$) with a term value of 25 800 cm^{-1}. As discussed in Section I.C, the relatively high frequency of the $\mathscr{A}(3p, \sigma^*)$ band in the chlorosilanes would imply a relatively high Si–Cl bond strength; indeed, the Si–Cl bond strength varies between 126 and 108 kcal/mole in the series from H_3SiCl to Cl_3SiCl [B28], making it even stronger than the C–F bond. Thus, as in the fluoromethanes (Section IV.D), one expects the outer-shell $\mathscr{A}(3p_{Cl}, \sigma^*)$ bands of the chlorosilanes to lie beneath the more prominent Rydberg excitations, but to assume a more obvious role in the inner-shell spectra due both to the larger antishielding effect for valence excitations (Section I.C), and to their higher relative intensities in such spectra.

Once we admit to placing the $\mathscr{A}(3p, \sigma^*)$ bands of the chlorosilanes in the region of the $3p \rightarrow \phi_j^{+1}$ Rydberg conjugates, the possibility also becomes real that the $(3p, \sigma^*)$ valence configuration is totally dissolved in the conjugate Rydberg sea, in which case $\mathscr{A}(3p, \sigma^*)$ bands as such would not exist in any spectral region. Were this the case, then $SiCl_4$ could be looked upon as the methane of the second row. Further, if the $(3p, a_1\sigma^*)$ level of $SiCl_4$ is actually above the $(3p, \phi_j^{+1})$ Rydberg manifold, then the band will appear as a shape resonance and $SiCl_4$ could be looked upon as the BF_3 (Section V.C) of the second row!

Outer-shell Rydberg spectra of compounds in the $Si(CH_3)_xCl_yH_z$ series are broad but readily assigned to the lower Rydberg transitions arising from Cl lone-pair excitation. The spectrum of silicon tetrachloride is exceptional in regard the high values of its peak frequencies [C12]. The lowest excitation ($4t_2 \rightarrow 4s$) in $SiCl_4$, Fig. IX.B-2, is peaked at 72 100 cm^{-1} (*8.94 eV*) with a term value of 31 550 cm^{-1}. Though Causley and Russell [C12] have assigned this band as $4t_2 \rightarrow \sigma^*$, comparison of its term value with the entries in Table I.C-III, and the absence of the $\mathscr{A}$ band in all other silyl chlorides argue against this. Note as well, Fig. IX.B-2, that the Rydberg spectrum of $SiCl_4$ correlates nicely with those of CCl_4, $GeCl_4$ and $SnCl_4$ if one assigns the 72 100-cm^{-1} band of $SiCl_4$ as $3p \rightarrow 4s$, and that the latter three members of this group have low-lying $\mathscr{A}(np, \sigma^*)$ bands which are quite clear whereas that of $SiCl_4$ clearly is missing. Both $SiHCl_3$ and $SiCl_4$ show intense $3p \rightarrow 4p$ bands at 80 000–81 000 cm^{-1} (*vert., 9.92–10.0 eV*).

The $2p_{Si}$ X-ray absorption spectrum of $SiCl_4$ [**I**, 304] begins with a band having a term value of 51 000 cm^{-1} which clearly identifies it as $2p_{Si} \rightarrow a_1\sigma^*$. This term value is very close to that of 50 800 cm^{-1}

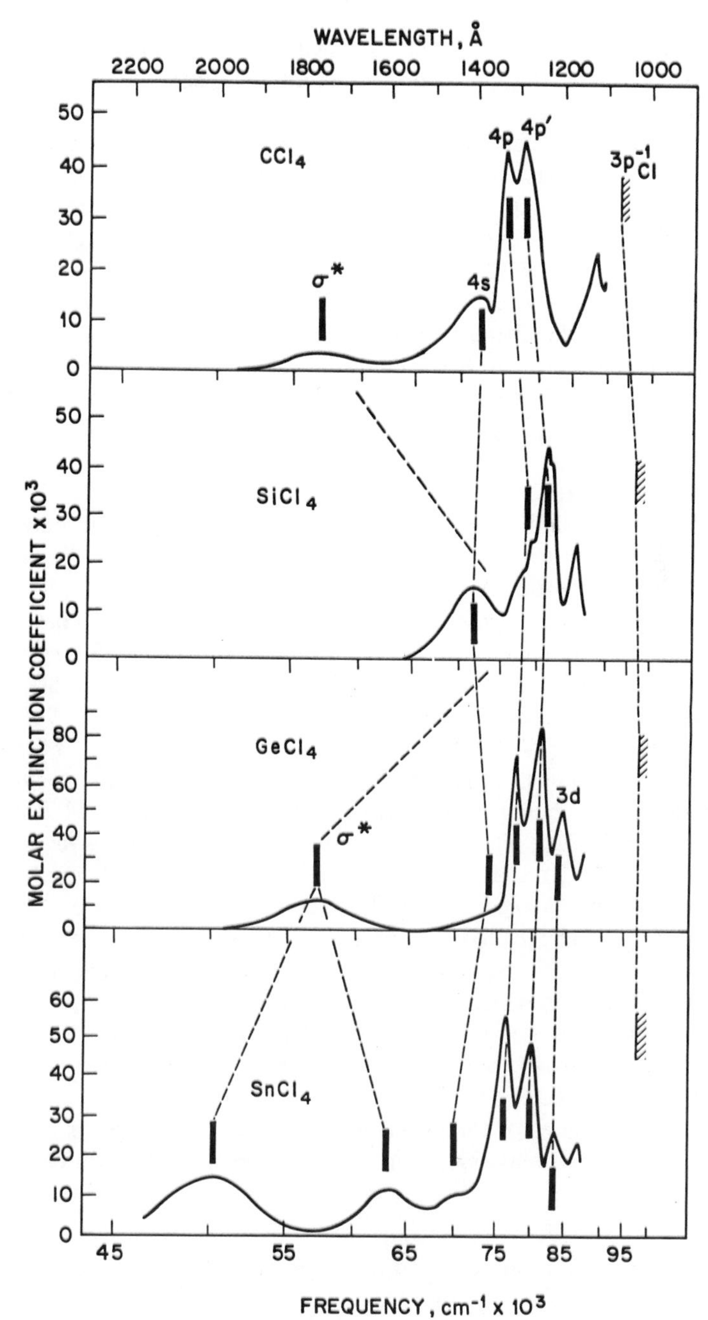

Fig. IX.B-2. Optical absorption spectra of the Group-IV tetrachlorides in the vapor phase [C12].

measured for the $1s_C \rightarrow a_1\sigma^*$ band of CCl_4 (Table I.C-II). Though it is at first sight surprising to find the $\mathscr{A}(2p_{Si}, \sigma^*)$ band of $SiCl_4$ with a 51 000-cm^{-1} term value when the $\mathscr{A}(3p_{Cl}, \sigma^*)$ band has a term value of less than 31 000 cm^{-1}, one must not forget the large exhaltation of valence term values that comes from antishielding of the inner shell (Section I.C). The next two inner-shell bands of $SiCl_4$ have term values of 37 500 and 27 800 cm^{-1} and it is difficult to decide which is the Rydberg transition to 4s, and which is the $\mathscr{A}(2p, t_1\sigma^*)$ valence band. The splitting of the σ^* manifold in the inner-shell spectrum of CCl_4 (7600 cm^{-1}, Section IV.A) was confirmed by the spacing between its σ^*(C—C) TNI resonances, and this assignment problem in $SiCl_4$ could be solved in the same way if TNI data were available for it. In any case, the splitting between $a_1\sigma^*$ and $t_2\sigma^*$ MO's in $SiCl_4$ is clearly much larger than in CCl_4, which possibly relates to the stronger bonds in the former. The σ^* splitting in the inner-shell spectrum of SiF_4 is 18 000 cm^{-1} (*vide supra*).

On going from $SiCl_4$ to $GeCl_4$ and $SnCl_4$, Fig. IX.B-2, the outer-shell Rydberg transitions to 4s and 4p† fall at nearly the same frequencies, in accord with first ionization potentials which are pegged at 97 270 ± 480 cm^{-1} (*advert., 12.06 ± 0.06 eV*) [C12] in this series of compounds. Strange then that in contrast to $SiCl_4$, but like CCl_4, the $\mathscr{A}(3p, \sigma^*)$ bands in $GeCl_4$ and $SnCl_4$ are at quite low frequencies: $GeCl_4$ (56 800 cm^{-1} *vert., 7.04 eV*, 44 800-cm^{-1} term value), $SnCl_4$ (50 250 and 62 900 cm^{-1} *vert., 6.23 and 7.80 eV*, 49 900- and 50 030-cm^{-1} term values). Moreover, in these compounds, the $\mathscr{A}$-band molar extinction coefficients ($\epsilon > 10\ 000$) are much larger than ordinarily encountered for this transition ($\epsilon = 1000$). The larger values of the outer-shell $\mathscr{A}$-band term values in CCl_4, $GeCl_4$ and $SnCl_4$ compared with that in $SiCl_4$ imply lower M—Cl bond strengths according to the argument presented in Section I.C. Indeed, in CCl_4 and $GeCl_4$ in which the $\mathscr{A}(3p, \sigma^*)$ bands have the same frequencies, the M—Cl bond strengths are 16 kcal/mole smaller than in $SiCl_4$, and in $SnCl_4$ in which the $\mathscr{A}(3p, \sigma^*)$ frequency is lowest, the M—Cl bond strength is 20 kcal/mole smaller than that in $SiCl_4$ [S11]. The inner-shell spectrum of $SnCl_4$ may well show (4p, σ^*) term values of 70 000 cm^{-1} or more.

In the tetrabromides of the Group-IV elements, the picture is much the same as in the tetrachlorides, with one exception. Absorption in the

† It is assumed that the transitions originate with the 3p orbitals of chlorine and that the lowest *n*s and *n*p orbitals therefore have $n = 4$.

Rydberg region (62 000–83 000 cm^{-1}, *7.7–10.3 eV*) is highly structured and complex, as anticipated from the complexities of the photoelectron spectra [C28]. Generally, the 4p → 5s transitions are weak to vanishing, whereas those to 5p and 4d are quite intense in these compounds. Longward of the Rydberg bands, several $\mathscr{A}(4p, \sigma^*)$ bands of large oscillator strength are observed with term values between 36 200 cm^{-1} ($SiBr_4$) and 52 700 cm^{-1} ($SnBr_4$). The ready identification of $\mathscr{A}(4p, \sigma^*)$ bands in $SiBr_4$ is in marked contrast to the situation in $SiCl_4$ where $\mathscr{A}(3p, \sigma^*)$ bands cannot be found at all! The difference would seem to lie with the Si–X bond strengths, for that in $SiBr_4$ is 14 kcal/mole less than that of $SiCl_4$ [S11], and as mentioned above, low bond strengths in halides promote low-lying $\mathscr{A}$ bands. However, it is surprising that the $\mathscr{A}(4p, \sigma^*)$-band term value of $SiBr_4$ is not larger than that of SiH_3Br, Section I.C. Possibly there is another $\mathscr{A}(4p, \sigma^*)$ band at 47 600 cm^{-1} (*vert., 5.90 eV*) in $SiBr_4$. Interpretation is straightforward in bromosilane [C11] where an $\mathscr{A}(4p, \sigma^*)$ band with a term value of 36 700 cm^{-1} appears at 52 400 cm^{-1} (*vert., 6.49 eV*), followed by Rydberg transitions to 5s, 5p and 4d, all with perfectly normal term values. Unlike the situation in CH_3Br, the $^2P_{3/2} - {}^2P_{1/2}$ spin-orbit intervals are not resolved in the spectrum of SiH_3Br.

IX.C. Addendum

T. Heinis, K. Börlin and M. Jungen, The threshold photoelectron spectrum of silane, to be published.

M. N. Piancastelli, P. R. Keller, J. W. Taylor, F. A. Grimm, T. A. Carlson, M. O. Krause and D. Lichtenberger, Trend of shape resonance-induced features in the angular distribution parameter as a function of photon energy for carbon, silicon and germanium tetrachlorides, *J. Electron Spectrosc. Related Phenomena* **34**, 205 (1984).

M. A. Dillon, R.-G. Wang, Z.-W. Wang and D. Spence, Electron impact spectroscopy of silane and germane, to appear in *J. Chem. Phys.*

CHAPTER X

Olefins

X.A. Ethylene

Mulliken [M82, M83], has reviewed the excited-state assignments in ethylene [**II**, 2, 324], focusing on the core splittings of the (π, 3p) and (π, 3d) states; his work is updated in Table X.A-I. The symmetries listed in the Table follow those earlier recommended by Merer and Mulliken [**II**, M26]. For purposes of the present discussion, the following set of axes are defined for the ethylene molecule: x is in the out-of-plane direction, z is along the C–C axis, and y is short axis, in-plane.

The vibronic structure of the N → V(π, π^*) band of ethylene remains something of a puzzle. Foo and Innes [F21] performed a high-resolution study of the N → V absorption in the deuterated ethylenes, and in all isotopomers except ethylene-d_4 they found that only the ν_4' torsion is active in the Franck-Condon envelope, whereas in the -d_4 compound some activity of ν_2' (C=C stretch) also was detected. From this they conclude that the C–C distance in the V state has expanded to only 1.41 Å, as was predicted theoretically [B78]. The torsional vibrations observed by Foo and Innes [F21] in the N → V transition were fit to a quadratic polynomial in the vibrational quantum number v and the origin then

TABLE X.A-I

EXCITED STATES OF ETHYLENE-h_4 BELOW THE FIRST IONIZATION POTENTIAL[a]

Frequency, cm^{-1} (*eV*)	Upper-State Configuration	Upper-state Symmetry	References
14 400 (*vert., 1.78*)	$^2(N, \pi^*)$[b]	$^2B_{2g}$	[C21, V11, W4]
35 200 (*vert., 4.36*)	$^3(\pi, \pi^*)$	$^3B_{1u}$	[F12, V11]
53 720 (*advert., 6.660*)	$^2(\pi, 3s^2)$[c]	$^2B_{3u}$	[D3, V11]
56 300 (*advert., 6.98*)	$^3(\pi, 3s)$	$^3B_{3u}$	[W26]
57 338 (*advert., 7.1088*)	$^1(\pi, 3s)$[d]	$^1B_{3u}$	[F21, M35, V11]
60 500 (*vert., 7.50*)	$^2(\pi, 3px^2)$[c]	^2A_g	[W4]
61 300 (*vert., 7.60*)	$^1(\pi, \pi^*)$[e]	$^1B_{1u}$	[II,2]
62 790 (*advert., 7.785*)	$^3(\pi, 3py)$	$^3B_{1g}$	[W26]
62 905 (*advert., 7.7990*)	$^1(\pi, 3py)$	$^1B_{1g}$	[D3, G10, M53, M83, V24, W26]
65 735 (*advert., 8.1499*)	$^3(\pi, 3px)$	^3A_g	[W26]
66 875 (*advert., 8.2913*)	$^1(\pi, 3px)$	^1A_g	[G10, J9, M35, M53, M83, W26]
69 080 (*advert., 8.565*)	$^3(\pi, 3d\sigma)$	$^3B_{3u}$	[W26]
69 531 (*advert., 8.6205*)	$^1(\pi, 3d\sigma)$[f]	$^1B_{3u}$	[M35, M53, W26]
71 813 (*advert., 8.9035*)	$^1(\pi, 3d\delta)$[g]	$^1B_{3u}$	[M35, S58]
75 250 (*advert., 9.3296*)	$^1(\pi, 3dxz)$	$^1B_{1u}$	[M53, V11]

[a] For excitations in ethylene above the first ionization potential, see Table X.A-II.

[b] A valence TNI resonance.

[c] A Feshbach resonance.

[d] Series to $n = 6$ observed.

[e] Origin at *ca.* 48 300 cm^{-1} (*5.99 eV*).

[f] n = 4 also observed.

[g] Series to $n = 6$ observed.

defined as the frequency when $v = 0$. This extrapolation places the $\pi \rightarrow \pi^*$ origin of ethylene-h_4 at 43 770 cm^{-1} (*5.427 eV*). On the other hand, *ab initio* calculations [P23] argue specifically against the result of such an extrapolation, instead placing the (0,0) band at 48 400 cm^{-1} (*6.000 eV*). McDiarmid has made a determined experimental effort to identify the N → V origin band using long paths of liquid ethylene [M31], and though some very weak absorption bands ($10^{-2} > \epsilon > 10^{-3}$) were observed in the 44 000–48 000-cm^{-1} (*5.45–5.95-eV*) region, they do not have a recognizable pattern and are low-frequency shifted from the vapor-phase spectrum by an unknown amount. Thus, there is no firm experimental evidence for an N → V origin below 48 000 cm^{-1} (*5.95 eV*) in vapor-phase ethylene. The N → V spectrum of crystalline ethylene-h_4 [D6] exhibits a long torsional progression (600–700 cm^{-1}) with the lowest reported feature appearing at 52 030 cm^{-1} (*6.451 eV*). It is highly unlikely that this is the electronic origin, for such an assignment would require a large shift to high frequency on going from the vapor to the solid, whereas for a valence upper state the opposite is expected (See Section X.C, for example). The presence of the torsional progression indicates that the V state of ethylene is twisted in the crystal as well as in the vapor phase. As expected, the $\pi \rightarrow$ 3s Rydberg transition does not appear in the neat-crystal spectrum, thereby revealing the N → V maximum unencumbered at 56 800 cm^{-1} (*7.04 eV*).

Interestingly, it has been calculated that the V state of ethylene is susceptible to "sudden polarization," meaning that the equilibrium geometry involves the strong pyramidalization of *one* methylene group by 65° and the consequent development of a large dipole moment (4.73 D) [T18]. There is no experimental data bearing directly on this interesting point, however the N → V resonance Raman results of Ziegler and Hudson [Z5] indicate that ethylene in the V state is only twisted about the somewhat lengthened C–C bond, and so would seem to argue against it.

Theoretical work continues on the electronic structure of the V state of ethylene. It is intriguing that though calculations [B75] show the $^1(\pi, \pi^*)$ wavefunction of ethylene is 45–50 % $^1(\pi, 3dxz)$, has a π^* $<x^2>$ value midway between those of the pure valence π^* and 3dxz Rydberg orbitals and has a $J_{\pi^*\pi^*}$ self-energy (Section I.D) far below the valence range but just at the upper limit of the Rydberg range, still the broadening behavior of the $\pi \rightarrow \pi^*$ band in a condensed phase is indistinguishable from that of a "pure" valence transition. All workers now agree that though there is an important admixture of the $^1(\pi, 3dxz)$ Rydberg configuration into $^1(\pi, \pi^*)$ in the V state, the resultant still functions phenomenologically as a valence state [B75, B78, B80, B81, M52, P23, P24]. In contrast to the situation in say water, where the $(1b_1, \sigma^*)$ valence level is totally dissolved in the conjugate $(1b_1, ns)$ Rydberg sea, in ethylene the mixing of (π, π^*) and its $(\pi, ndxz)$ conjugates is less that total due in part to energy mismatch and also to the fact that the nd orbitals are only very weakly penetrating and so have a poor overlap with π^*. Nonetheless, mixing with the higher conjugate Rydberg configurations lowers the calculated $\pi \rightarrow \pi^*$ frequency by *ca.* 6000 cm^{-1} and reduces the N → V oscillator strength by up to

50 %. As measured by the operator $\sum_i x_i^2$, the V state of ethylene has a calculated cross section of 20 a_0^2 (later increased to 29.5 a_0^2 [B79]), whereas the N and T states are calculated to amount to 12 a_0^2 and the 3dxz Rydberg orbital has a calculated cross section of 70 a_0^2. Consideration of $J_{\pi^*\pi^*}$, the Coulomb self-energy integral for the π^* orbital of the $^1(\pi, \pi^*)$ configuration also reveals a value intermediate between that calculated for π^* in $^3(\pi, \pi^*)$ and for 3dxz in $^1(\pi$, 3d$xz)$ [P24]. *Ab initio* calculation of the Franck-Condon factors in the N → V band is qualitatively correct, but quantitatively inadequate [M40].

Mulliken [M79] postulates that for N → V excitations in general in unsaturated systems, there will be a strong mixing of (π, π^*) and (σ, σ^*) configurations, and McMurchie and Davidson clearly state this is the case in their ethylene V-state calculations [M52]. The primary effect of mixing (π, π^*) with (σ, σ^*) excitations is to reduce the polarity of the (π, π^*) configuration and thereby reduce the diffuseness of the π^* MO. In contrast, Buenker *et al.* [B78, B81] state equally clearly and emphatically that (σ, σ^*) configurations are not important in describing the $^1(\pi, \pi^*)$ state of ethylene! These differences have been traced to differences in the basis sets [B80].

Ab initio theory [P23] also predicts that as the V state is twisted, the (π, π^*) and Rydberg $(\pi$, 3p$y)$ configurations are increasingly mixed [**II**, 12]. According to this calculation, the observed N → V profile of ethylene is really two overlapping transitions, $\pi \rightarrow \pi^*$ and $\pi \rightarrow$ 3py, which are strongly mixed. While we do agree that the transitions in question are overlapped, still there is no sign of an interaction between them, at least on that part of the torsional surface sampled by the Franck-Condon envelope of the $\pi \rightarrow$ 3py transition (*vide infra*). Mulliken also concludes that there is no spectroscopic evidence supporting the proposed $(\pi, \pi^*)/(\pi$, 3p$y)$ mixing [M82, M83].

Vibronic structure of the transition to the $^3(\pi, \pi^*)$ T state of ethylene is significantly different from that to $^1(\pi, \pi^*)$. Using large-angle electron scattering at low residual energy, Wilden and Comer [W24] uncover structure in the N → T transition consisting of very long progressions of ν_2', the C=C stretch, along with odd and even numbers of quanta of ν_4', the torsion, Fig. X.A-1. The relative importance of the ν_2' and ν_4' vibrations in the N → T transition is just opposite to that in the N → V transition! Note that since electron-impact energy losses at large scattering angles and low impact energy will not obey the optical selection rules, both odd and even numbers of ν_4' quanta can (and do) appear in the spectrum. The ν_2' Franck-Condon factors and frequencies imply a C–C bond length of 1.54 Å in the T state, whereas it is only 1.41 Å in the V state. The minimum in the torsional coordinate in the T state is presumably at 90°

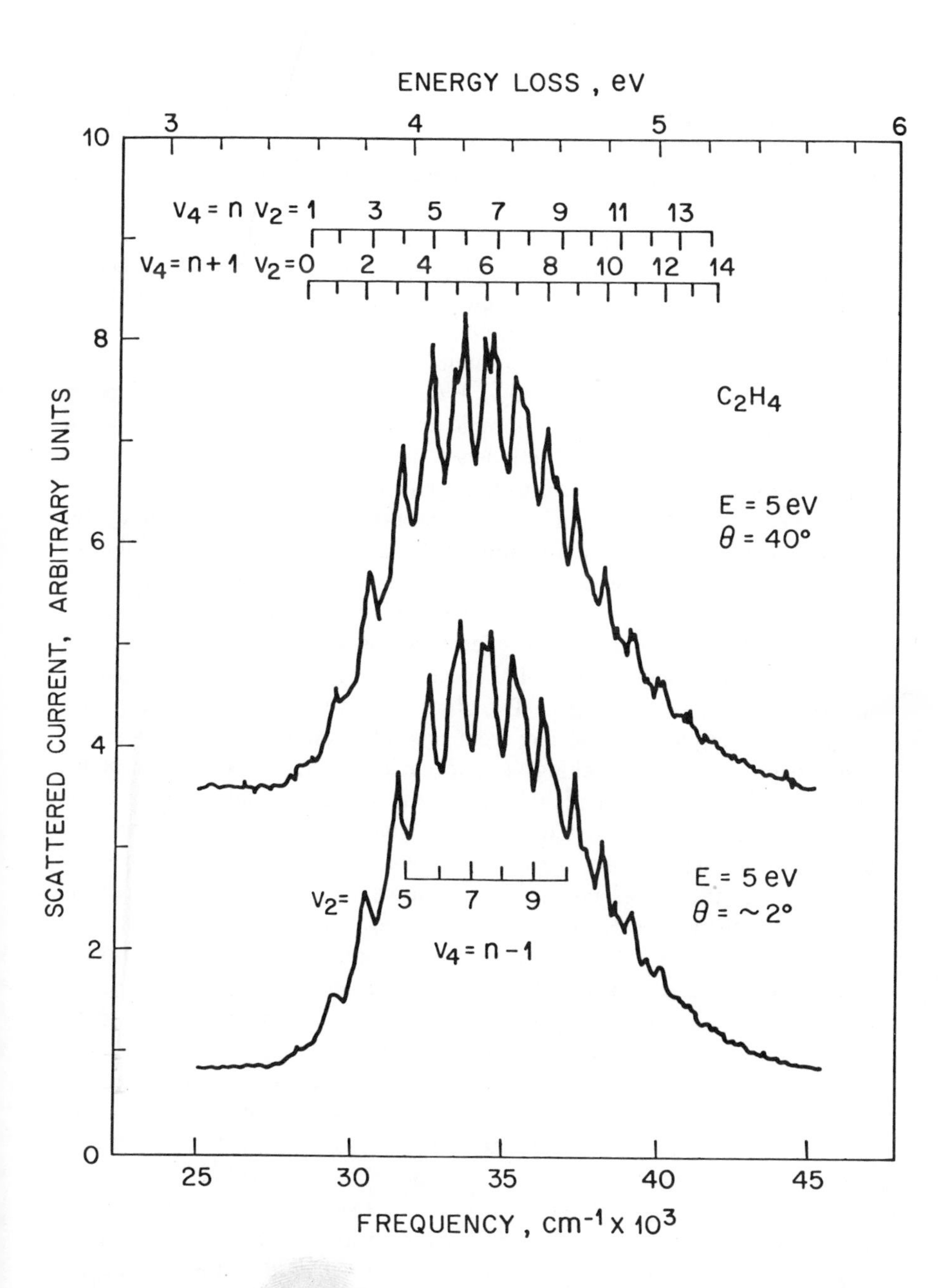

Fig. X.A-1. The N → T transition of ethylene as observed by inelastic electron scattering, including a vibronic analysis based upon progressions of ν_2' (C=C stretch) and ν_4' (H_2C–CH_2 torsion) [W24].

dihedral angle [**II**, 4], in which case the Franck-Condon factors for the ν_4' motion at large C–C distances must be nonzero over a very small range of dihedral angle. These vibronic and geometric differences in the transition to V and T may be due to the mixing of the (π, π^*) level with its $(\pi, 3dxz)$ Rydberg conjugate in the singlet manifold, but not in the triplet manifold.

The conventional assignment of the vibronic structure of the $\pi \rightarrow 3s$ profile of ethylene as involving a progression of ν_2' to each member of which there are attached multiple quanta of the torsion ($2\nu_4'$ and $4\nu_4'$) has been challenged by Watson and Nycum [W9]. These workers instead assign the "torsional" structure to intensity borrowed from the underlying $\pi \rightarrow \pi^*$ excitation through ν_8' and $\nu_4' + \nu_{10}'$ vibrations. In response, Findley *et al.* [F8] have calculated the vibronic envelopes of the $\pi \rightarrow 3s$ transitions in several of the deuterated ethylenes under the assumption that it is multiple quanta of ν_4' that are excited, and find excellent agreement with experiment. The Watson and Nycum proposal is refuted as well by the experimental results of Foo and Innes [F21], which show that the vibronic bands in question have out-of-plane polarization, whereas the vibronic-borrowing mechanism predicts long-axis in-plane polarization.

Though the $\pi \rightarrow 3s$ transition is forbidden as a two-photon absorption, it appears rather strongly in the MPI spectra of ethylene-h_4 and -d_4 resonant at the two-photon level [G10]. The absorption is vibronically induced by a single quantum of the ν_4 torsion; from this false origin there follows a long progression of C=C stretching (ν_2'), to each member of which are added the $2\nu_4'$ and $4\nu_4'$ twisting doublets, just as in the one-photon spectrum. Though the two-photon-forbidden $\pi \rightarrow \pi^*$ band could appear in the MPI spectrum through vibronic mixing with the allowed $\pi \rightarrow 3p$ resonances, the $\pi \rightarrow \pi^*$ excitation does not appear, presumably because the molecule in the V state dissociates before it can be ionized, Section II.A.

Foo and Innes [F21] have studied the one-photon $\pi \rightarrow 3s$ excitation at high resolution in ethylene and each of the deuterated isotopomers. Using band-contour analysis, they confirmed the $^1B_{3u}$ symmetry of the upper state and also derived an $H_2C{-}CH_2$ azimuthal angle of 37° for this state, with a lifetime of *ca.* 10^{-13} sec. The same set of deuterated ethylenes was studied by McDiarmid [M35], who delineated *n*s Rydberg series having $n = 3-6$ and best-fit quantum defects of 0.85. The $\pi \rightarrow 3s$ electronic origin shows an isotope shift of 72 cm^{-1} per D atom added [F21]. An odd feature is seen in the MCD spectrum of ethylene in the $\pi \rightarrow 3s$ region [B64], where the recognized $\pi \rightarrow 3s$ vibronic absorption bands appear as negative *B*-terms in the MCD, while three positive *B*-terms also appear among them. Though these positive *B*-terms have been taken to be

$\pi \rightarrow$ 3p excitations by some, they also appear in the vibronically-induced $\pi \rightarrow$ 3s two-photon spectrum, where they are readily assigned as $v\nu_2' + 3\nu_4'$, Fig. X.A-2 [G10]. Thus the negative B-terms appear in the MCD spectrum as a vibronically-induced part of the $\pi \rightarrow$ 3s transition rather than as a distinct $\pi \rightarrow$ 3p electronic transition.

The Rydberg excitations to $^3(\pi$, 3s) in ethylene-h_4 and -d_4 have been studied by Wilden and Comer [W26] using high-resolution electron impact energy-loss spectroscopy. The singlet-triplet nature of the excitation with origin at 56 300 cm^{-1} (*6.98 eV*) in ethylene-h_4 is confirmed by the behavior of its intensity with changing electron-impact energy and scattering angle. There appears to be a noticeable difference in the vibrational frequencies in the $^3(\pi$, 3s) and $^1(\pi$, 3s) states. Though transitions from the ground state involve ν_2' and $v\nu_4'$ vibrations in both, the ν_2' value of 1370 cm^{-1} in the singlet state becomes 1450 cm^{-1} in the triplet, whereas $2\nu_4'$ is 468 cm^{-1} in the singlet state and 400 cm^{-1} in the triplet. The singlet-triplet splitting of the (π, 3s) origins is 1040 cm^{-1} in ethylene-h_4, while a value of 1050 cm^{-1} has been calculated [F10].

A Feshbach resonance to the $^2(\pi$, 3s^2) configuration of the ethylene negative ion is observed to have its origin at 53 700 cm^{-1} (*6.66 eV*) [B83, V11, V24, **II**, AD163, AD38]. Such negative-ion resonances involving Rydberg excitation of the core usually fall about 4000 cm^{-1} (*0.5 eV*) below the corresponding neutral-molecule excitation (Section II.B), as observed here. The valence TNI resonance to the π^* MO of ethylene is observed at 13 900 cm^{-1} (*1.73 eV*) [J15], Fig. II.B-1; though resonances to the σ^* MO's have not been positively identified as yet, that at 46 000 cm^{-1} (*vert., 5.7 eV*) in ethylene is a good candidate [V11]. A tentative Feshbach resonance is observed in ethylene by Dance and Walker [**II**, AD38] at 62 000 cm^{-1} (*vert., 7.7 eV*), implying a parent Rydberg excitation at 66 000 cm^{-1} (*8.2 eV*); quite possibly this parent is (π, 3p), *vide infra*. Note however, that Walker *et al.* [W4] later assign this feature as a valence TNI resonance involving the lowest σ^* MO. Though this band appears homogeneous, it is clear from its extreme width (it stretches from 30 000 to 90 000 cm^{-1}) that it either is composed of several overlapping features, or has an extremely short lifetime.

Multiphoton ionization spectroscopy has proved to be the key to unlocking the puzzle of the excitations to (π, 3p) in ethylene. Though forbidden in one photon, excitations to (π, 3p) are allowed in two-photon absorption, and have been observed as such in the MPI spectra of ethylene-h_4 and -d_4 [G10]. The MPI spectrum of ethylene-h_4 reveals two-photon origins at 62 905 and 66 875 cm^{-1} (*7.7990 and 8.2912 eV*), each of which is followed by a progression in $m\nu_2' + v\nu_4'$ (v = 0, 2, 4). In

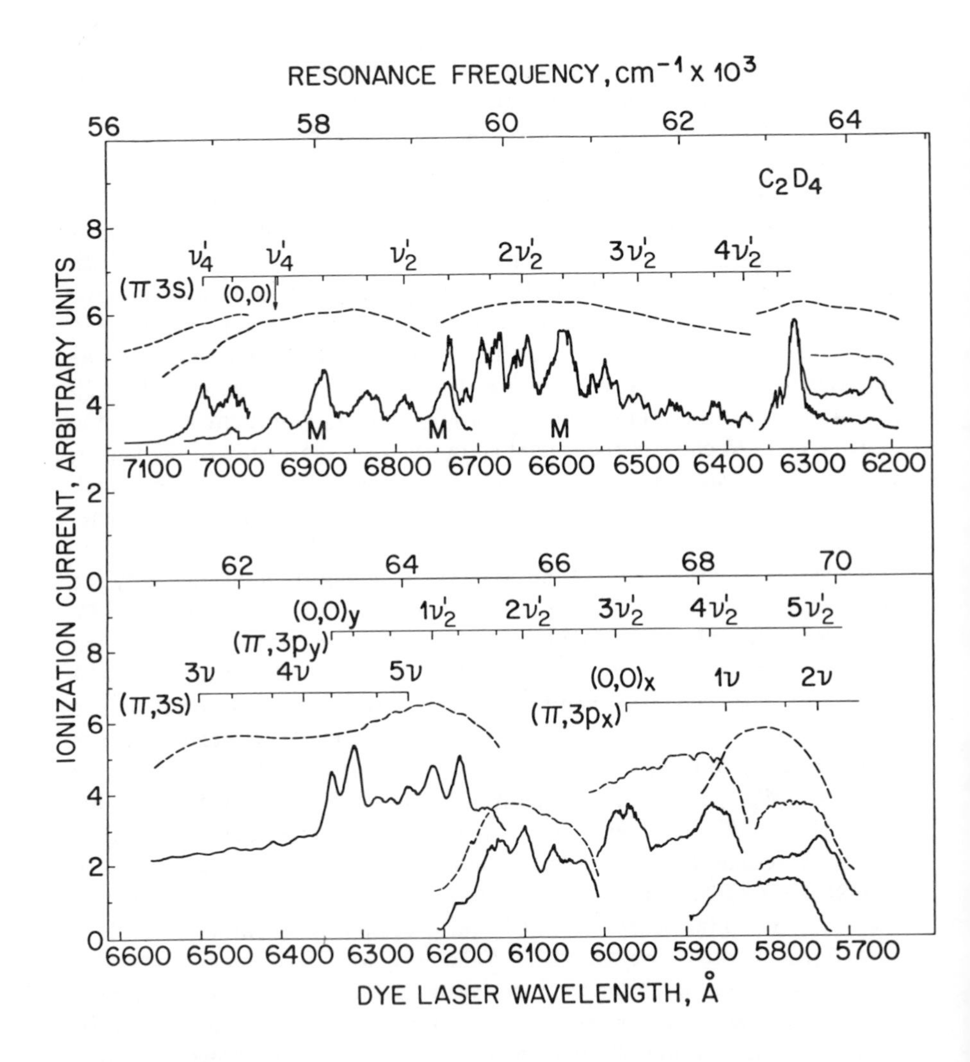

Fig. X.A-2. MPI spectra of ethylene-d_4 two-photon resonant in the $\pi \rightarrow 3s$ region [G10]. The dye-gain curves are shown as dashed lines. The MPI bands closest to the positive MCD bands observed by Brith-Lindner and Allen [B64] in ethylene-h_4 are marked **M**.

ethylene-d_4, the corresponding origin frequencies are 63 115 and 67 030 cm^{-1} (*7.8251 and 8.3105 eV*), Fig. X.A-2. Transitions very near these frequencies were reported prior to the MPI work. Thus threshold and near-threshold electron impact spectra show peaks at *ca.* 63 000 and 66 500 cm^{-1} (*7.81 and 8.24 eV*) [D3, J9, V11, W26], and McDiarmid

[M35] reports an origin in the one-photon spectrum at 66 591 cm^{-1} (*8.2560 eV*) which may be vibronically induced. The electron impact energy losses at 63 000 and 66 500 cm^{-1} have angular dependences characteristic of electric quadrupole-allowed transitions, as appropriate for $\pi \rightarrow$ 3p excitation [J9, V24, W26].

McMurchie and Davidson [M53] have calculated the core splitting of the $^1(\pi, 3p)$ configuration in ethylene; $(\pi, 3py)$ is predicted to fall at 63 960 cm^{-1} (*7.930 eV*), $(\pi, 3pz)$ at 64 610 cm^{-1} (*8.010 eV*) and $(\pi, 3px)$ at 67 430 cm^{-1} (*8.360 eV*). Later work by Nakatsuji [N12] is in almost perfect agreement with these figures. The two $\pi \rightarrow$ 3p MPI bands observed to date are thus assigned as $\pi \rightarrow 3py$ (62 905 cm^{-1} *adiabat., 7.7990 eV*) and $\pi \rightarrow 3px$ (66 875 cm^{-1} *adiabat., 8.2912 eV*). The corresponding excitations to $^3(\pi, 3p)$ states are reported by Wilden and Comer [W26] to have origins at 62 790 and 65 740 cm^{-1} (*7.785 and 8.150 eV*) in ethylene-h_4. The singlet-triplet split of only 115 cm^{-1} in the $(\pi, 3py)$ configuration offers no support for the suggestion that the $^1(\pi, \pi^*)$ and $^1(\pi, 3py)$ configurations are significantly mixed close to the ground-state geometry [P23]. Similarly, if $^1(\pi, \pi^*)$ and $^1(\pi, 3py)$ were mixed significantly, the ν_4' torsional frequency in the $^1(\pi, 3py)$ Rydberg state would differ from that in unmixed Rydberg states such as $(\pi, 3s)$ and $(\pi, 3px)$. However, there is no difference in the experimental ν_4' frequencies to which one can point in these Rydberg states. These arguments, of course, are dependent on the validity of the $(\pi, 3py)$ assignment of the 62 905-cm^{-1} upper state.

According to the arguments given above, there is little likelihood that the unusual MCD bands found in the $(\pi, 3s)$ region by Brith-Lindner and Allen [B64] are part of the $(\pi, 3p)$ manifold, in spite of arguments to the contrary [M82, M83]. The questionable quadrupole transition reported by Ross and Lassettre [**II**, R27] at *ca.* 60 000 cm^{-1} (*vert., 7.4 eV*) in ethylene now seems even more to be an artifact. Johnson and coworkers [J8] explain this spurious band as due to the finite spectral bandwidth of the earlier experiment, and Wilden and Comer [W26] also question its reality after failing to find any angular-scattering evidence for a quadrupole excitation in the vicinity of 60 000 cm^{-1}. Theoretical calculations [M53] also argue against placing any $\pi \rightarrow$ 3p excitation in ethylene below 63 000 cm^{-1} (*7.81 eV*).

Specification of the $\pi \rightarrow$ 3d Rydberg states of ethylene is fragmentary but encouraging. McDiarmid [M35] has assigned the absorption spectra of the deuteroethylenes using the constancy of the isotope shift for electronic origins, and thereby assembled two short Rydberg series having $\delta = 0.30$ and 0.08; she assigns these to $\pi \rightarrow nd\sigma$ and $\pi \rightarrow nd\delta$, respectively. Members of the second series are said to show both even and odd quanta of ν_4', contrary to the D_{2h} optical selection rules. Additionally, ν_3' appears in these bands, but not in the $\pi \rightarrow 3s$ band, whereas ν_2' is prominent in $\pi \rightarrow 3s$, but not in $\pi \rightarrow 3d$. The $n = 3$ members of the $nd\sigma$ and $nd\delta$ series are observed optically at 69 531 and 71 813 cm^{-1} (*adiabat., 8.6205 and 8.9035 eV*), while Wilden and Comer [W26] report dipole-allowed electron energy losses at 69 400 and 71 780 cm^{-1} (*8.600 and 8.899 eV*) which most likely correspond to the $\pi \rightarrow 3d$ optical bands of McDiarmid. The energy-loss spectra also reveal a triplet state having an origin at 69 080 cm^{-1} (*8.565 eV*), which complements the $\pi \rightarrow 3d$ singlet at 69 531 cm^{-1} (*8.6205 eV*).

The *ab initio* calculations on the (π, 3d) manifold of ethylene are most helpful and interesting [F10, M53]. The (π, 3d) manifold is split into five components, with transitions allowed to four of these from the ground state. The lowest frequency transition is to 3dσ (3dz^2), and it is predicted to be far weaker than the transitions to the other components. Based on the frequency of its leading member [70 980 cm^{-1} (*8.800 eV*) predicted; 69 531 cm^{-1} (*8.6205 eV*) observed], the series having $\delta = 0.30$ is assigned as $\pi \rightarrow nd\sigma$ [M53], in agreement with McDiarmid's recommendation. According to the calculations, the first member of the $\pi \rightarrow nd\delta$ ($ndx^2 - y^2$) series is predicted to come at 72 510 cm^{-1} (*8.9899 eV*), corresponding to the band observed at 71 813 cm^{-1} (*advert., 8.9035 eV*). The MCD spectrum of ethylene in the $\pi \rightarrow 3d$ region [S58] reveals a pseudo-A term at 71 800 cm^{-1} (*8.90 eV*) which can come about only through the agency of two closely-spaced B-terms of opposite sign. Presumably, these are two components of (π, 3d) separated by approximately 200 cm^{-1}.

The excitation to 3dπ (3dxz) lies highest in the $\pi \rightarrow 3d$ manifold of ethylene due to its mixing with the (π, π^*) valence conjugate configuration at lower frequency. This injection of valence character into the Rydberg upper state results in a predicted singlet-triplet split of 2580 cm^{-1} for (π, 3dxz) [N12], whereas the other members of the (π, 3d) manifold have splits calculated to be only *ca.* 110 cm^{-1}. According to the *ab initio* calculations [N12], the $\pi \rightarrow 3dxz$ transition has an oscillator strength of only 0.085 after mixing with the $\pi \rightarrow \pi^*$ promotion. Buenker and Peyerimhoff [B78] explain that the oscillator strength lost by the $\pi \rightarrow \pi^*$ promotion owing to its assumption of partial Rydberg character is spread evenly among all members of the $\pi \rightarrow ndxz$ series, so that $\pi \rightarrow \pi^*$ remains relatively the most intense. Experimental observation of the $\pi \rightarrow 3dxz$ transition is uncertain, however the energy loss reported by van Veen [V11] at 75 250 cm^{-1} (*vert, 9.329 eV*) correlates nicely with the predicted value of 75 090 cm^{-1} (*9.310 eV*).

Higher Rydberg excitations in ethylene as determined by electron impact [J9] are assigned in Table X.A-II based on their symmetry and term values. To these must be added the Rydberg series of six members observed as autoionizing peaks in the photoionization mass spectrum of ethylene [W35]. This series converges on the $(1b_{3u}\sigma)^{-1}$ ionization potential (128 000 cm^{-1} *vert, 15.87 eV*); with $\delta = 1.02$ and a term value of 28 230 cm^{-1} for the $n = 3$ member, the autoionizing series clearly involves dipole-allowed transitions from $1b_{3u}\sigma$ to ns orbitals. The absorption spectrum of ethylene at very high frequencies, 155 000–555 000 cm^{-1} (*19.2–68.8 eV*) [L13] shows no structure, but only a smooth decrease in absorption cross section. The absorption spectrum reported for ethylene in this region is very nearly superposable upon that of ethane [L13]. Because the $(2sa_1)^{-1}$ ionization potential of ethylene comes at 189 000 cm^{-1} (*vert., 23.4 eV*) [I, B59] some low-frequency structure in absorption is possible, *e.g.*, 2s → 3p, but none is reported. As in methane (Section III.A), such bands in ethylene undoubtedly will appear in the low-energy, nonzero-angle electron scattering spectrum. The absorption spectrum of the dipositive ethylene cation is presented in [B30], and theoretical calculations on the spectral properties of twisted ethylene are discussed in Section X.B.

Inner-shell excitations in ethylene and the olefins have been observed as satellites in photoelectron spectra and as absorption bands and energy losses on both sides of the inner-shell ionization potential. On the low-frequency side of the $(1s_C)^{-1}$ ionization potential of ethylene, Fig. X.A-3, the banded absorption begins with an intense feature having a term value of 48 900 cm^{-1} [E1, H56, T19]. Because this is far too large for the $(1s_C, 3s)$ Rydberg excited state, the band must be assigned as a valence $1s_C \rightarrow \pi^*$ transition. Theory confirms this assignment, and also assigns the vibrational structure observed in this band to C–H stretching in the upper state [B15]. The polarization of this transition in an oriented sample of polybutadiene is out-of-plane, as expected for a $1s_C \rightarrow \pi^*$ assignment [S37]. It is most interesting to see that the Coulomb self-energy for the π^* MO in the $^1(1s_C, \pi^*)$ configuration is computed to be 93 500 cm^{-1} (*11.6 eV*), Table I.D-I, for this is very close to that for the π orbital of the ground state (94 400 cm^{-1}, *11.7 eV*) and the π^* orbital in the $^3(\pi, \pi^*)$ T state (88 700 cm^{-1}, *11.0 eV*). Since the self-energy for the π^* orbital in the $^1(\pi, \pi^*)$ state is only 32 500 cm^{-1} (*4.03 eV*) owing to its strong Rydberg admixture, clearly the π^* orbital of ethylene does not take on Rydberg character when excited from $1s_C$. In accord with the discussion in Section (II.B), methylation of ethylene destabilizes the π^* MO such that in the butenes, the $(1s_C, \pi^*)$ term values are reduced to 46 000 cm^{-1} [H66].

TABLE X.A-II

ASSIGNMENT OF THE HIGHER RYDBERG EXCITATIONS IN ETHYLENE AND THE METHYL ETHYLENES[a]

Compound	Ionization Potential, cm^{-1} (*eV*)	Energy Loss, cm^{-1} (*eV*)	Term Value, cm^{-1}	Terminating Orbital
Ethylene	84 750 (*10.51*)	57 340 (*7.109*)	27 410	3s
		62 905 (*7.7990*)	21 845	3p
		66 875 (*8.2913*)	17 875	3p′
		73 011 (*9.0520*)	11 740	3d
	102 400 (*12.70*)	79 100 (*9.81*)	23 300	3p
		83 100 (*10.3*)	19 300	3p′
	118 600 (*14.70*)	90 300 (*11.2*)	28 300	3s
		96 000 (*11.9*)	22 600	3p
Propylene	80 000 (*9.92*)	53 100 (*6.58*)	26 900	3s
		61 200 (*7.59*)	18 800	3p′
	99 290 (*12.31*)	75 820 (*9.40*)	23 470	3p
	106 700 (*13.23*)	81 500 (*10.1*)	25 200	3s
		87 100 (*10.8*)	19 600	3p′
	116 800 (*14.48*)	94 400 (*11.7*)	22 400	3p
	128 200 (*15.89*)	101 600 (*12.6*)	26 600	3s
		105 700 (*13.1*)	22 500	3p
		114 500 (*14.2*)	13 700	3d
Isobutene	75 900 (*9.41*)	49 800 (*6.17*)	26 100	3s
		56 300 (*7.00*)	19 600	3p′
		57 200 (*7.10*)	18 700	3p′
	95 260 (*11.81*)	69 400 (*8.60*)	25 900	3s
	104 000 (*12.89*)	79 800 (*9.9*)	24 200	
		85 500 (*10.6*)	18 500	3p′
	111 300 (*13.80*)	89 500 (*11.1*)	21 800	3p
		95 200 (*11.8*)		
	123 400 (*15.30*)	104 900 (*13.0*)	18 500	3p′
	126 600 (*15.70*)	113 000 (*14.0*)	13 600	3d
trans-Butene-2	75 100 (*9.31*)	49 420 (*6.13*)	25 680	3s
		52 100 (*6.46*)	23 000	3p
		59 830 (*7.42*)	15 270	
	96 470 (*11.96*)	74 200 (*9.2*)	22 270	3p
		79 050 (*9.8*)	17 420	3p′
	106 500 (*13.20*)	87 900 (*10.9*)	18 600	3p′
	117 000 (*14.50*)	99 200 (*12.3*)	17 800	3p′
cis-Butene-2	74 900 (*9.29*)	48 300 (*6.00*)	26 600	3s
		51 140 (*6.34*)	23 760	3p
		59 800 (*7.41*)	15 100	
	94 000 (*11.6*)	73 400 (*9.1*)	20 600	3p′

TABLE X.A-II (continued)

Compound	Ionization Potential, cm^{-1} (*eV*)	Energy Loss, cm^{-1} (*eV*)	Term Value, cm^{-1}	Terminating Orbital
	102 900 (*12.76*)	79 000 (*9.8*)	23 900	3p
	110 500 (*13.70*)	87 100 (*10.8*)	23 400	3p
		91 200 (*11.3*)	19 300	3p′
	116 100 (*12.9*)	104 000 (*12.9*)	12 100	3d
	122 600 (*12.9*)	104 000 (*12.9*)	18 600	3p′
Trimethyl ethylene	71 200 (*8.8*)	48 660 (*6.03*)	22 540	3s
		52 000 (*6.45*)	19 200	3p′
		59 200 (*7.34*)	12 000	3d
	91 140 (*11.3*)	71 800 (*8.9*)	19 300	3p′
		77 400 (*9.6*)	13 700	3d
	100 000 (*12.4*)	77 400 (*9.6*)	22 600	3p
	104 000 (*12.9*)	85 500 (*10.6*)	18 500	3p′
		89 500 (*11.1*)	14 500	3d
	126 600 (*15.7*)	102 400 (*12.7*)	24 200	3s
Tetramethyl ethylene	67 900 (*8.42*)	45 850 (*5.68*)	22 050	3s
		48 640 (*6.03*)	19 260	3p′
		52 200 (*6.47*)	15 700	3p″
	88 080 (*10.9*)	69 400 (*8.6*)	18 700	3p′
	102 000 (*12.6*)	79 000 (*9.8*)	23 000	3s
		83 100 (*10.3*)	18 900	3p′
		87 900 (*10.9*)	14 100	3d

[a] Reference [J9].

The remainder of the bands in ethylene up to the $(1s_C)^{-1}$ ionization potential are readily assigned to Rydberg excitations terminating at 3s, 3p, 4p, *etc.*, with term values resembling those observed for π excitation, but slightly larger due to the exhaltation of Rydberg term values caused by inner-shell antishielding (Section I.B). In the $(1s_C, 3p)$ inner-shell configuration, the $3p\sigma$ and $3p\pi$ orbitals are observed to be split by 4100 cm^{-1} ($3p\sigma$ below $3p\pi$), as compared with 3970 cm^{-1} in the $(\pi, 3p)$ outer-shell configuration [G10]. Also in line with the similarity of transitions originating with $1s_C$ and π in ethylene, the $1s_C \rightarrow \pi^*$ and $\pi \rightarrow \pi^*$ valence transitions each are about ten times more intense than any Rydberg excitation originating at $1s_C$ or π, and considerably broader as well.

Going beyond the $(1s_C)^{-1}$ ionization potential of ethylene, a satellite band is observed at 67 700 cm^{-1} (*vert., 8.39 eV*) higher frequency in the XPS spectrum [B10, C5]. In propylene and butene-1, this band moves down to 58 900 cm^{-1} (*vert., 7.30 eV*) whereas in the saturated molecule

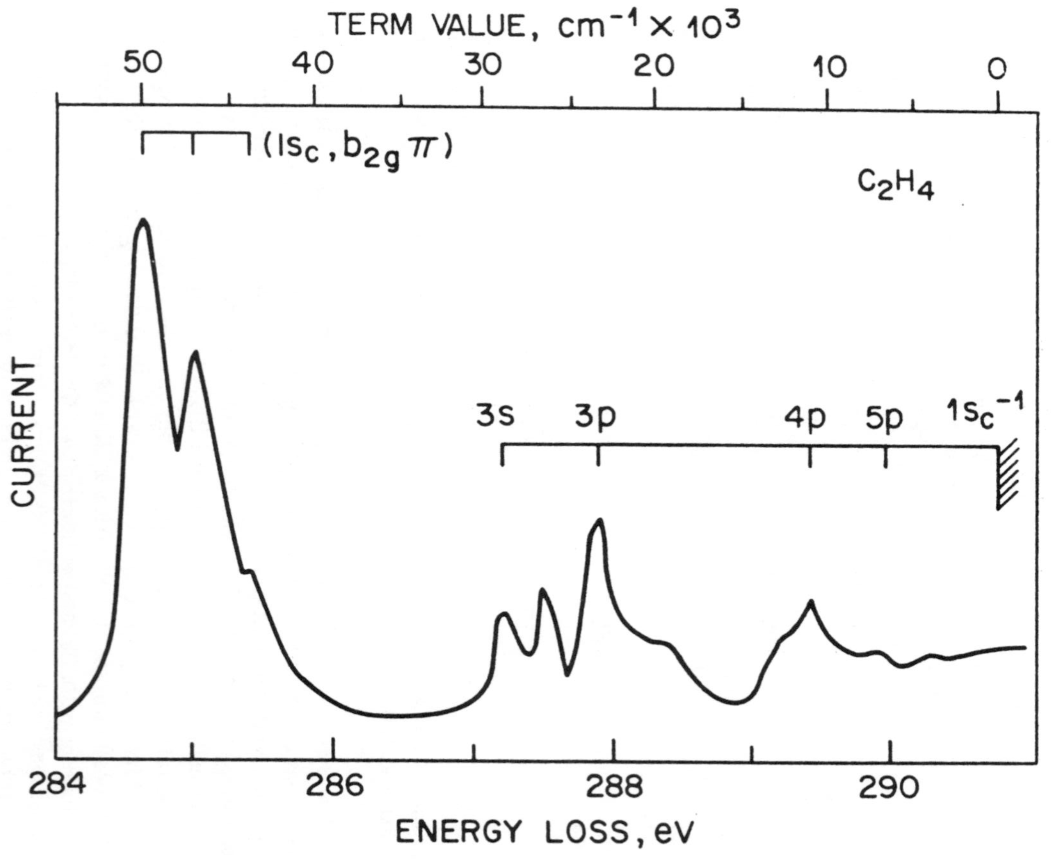

Fig.X.A-3. Electron energy-loss spectrum of ethylene in the $(1s_C)^{-1}$ region at an impact energy of 1500 eV [T19].

propane, it does not appear at all. Hence, the excitation is assigned as a $\pi \rightarrow \pi^*$ shake-up satellite in the $(1s_C)^{-1}$ positive ion (Section II.C). This excitation is calculated to require 49 500 cm^{-1} (*6.14 eV*) in the $(1s_C)^{-1}$ ethylene ion [B15]. In the $(1s_C)^{-1}$ cations, the (π, π^*) configuration is now so far below the conjugate (π, nd_{xz}) Rydberg configurations that there is no mixing whatsoever. In each of the above-mentioned molecules, saturated or not, there are broad peaks which are 145 000 and 218 000 cm^{-1} (*18.0 and 27.0 eV*) beyond the $(1s_C)^{-1}$ ionization potential which probably are due to maxima in the photoionization cross sections. Looking beyond the $(1s_C)^{-1}$ ionization potential of ethylene in the electron energy-loss spectrum, Hitchcock and Brion [H56] report two bands separated from the ionization limit by 16 000 and 37 100 cm^{-1} (*1.98 and 4.60 eV*). Barth *et al.* [B15] feel that the first of these is likely to be a double excitation ($1s_C \rightarrow \pi^*$; $\pi \rightarrow \pi^*$), but there is also a possibility that such bands result from σ(C–H) $\rightarrow \pi^*$ transitions in the cation, where the originating MO is shared between C^{+1} and H while the terminating orbital is totally on C^{+1}.

X.B. Alkyl Olefins

Using large scattering angles and low impact energies, the electron-impact spectra of the methyl ethylenes reveal transitions to $^3(\pi, \pi^*)$ states in the 33 100–34 800 cm^{-1} (*vert., 4.10–4.31 eV*) interval [F12, J9, K43]. With methylation, the $^3(\pi, \pi^*)$ levels shift far less than do the $^1(\pi, \pi^*)$ and $^1(\pi, 3s)$ levels or the $(\pi)^{-1}$ ionization potentials, as is also the case in the fluoroethylenes (Section X.C). Careful search has failed to uncover any higher triplets in the methyl ethylenes. As expected group theoretically, the excitation to $^1(\pi, \pi^*)$ in the vinyl group of oriented polybutadiene is polarized along the C=C axis, though not as strongly as one expects [H19]. The low polarization observed here may be due to the tipping of the transition moment away from the C=C direction by the pendant alkyl group [**II**, P4].

Johnson *et al.* [J9] have studied the electron-impact energy-loss spectra of the methyl ethylenes in detail out to 120 000 cm^{-1} (*14.9 eV*), with emphasis on tracing the Rydberg transitions from one molecule to the next. Thus they have followed the $\pi \rightarrow 3p_x$ band of ethylene (66 500 cm^{-1} *vert., 8.24 eV*) through the methyl ethylenes, down to 50 900 cm^{-1} (*vert., 6.31 eV*) in tetramethyl ethylene. These assignments expand upon those given in [**II**, 24]. Broad features beyond the first ionization potentials of the methyl ethylenes were assigned as valence $\sigma \rightarrow \pi^*$

transitions since many of them track the higher $(\sigma)^{-1}$ ionization potentials. However, because such behavior is much more reasonable for Rydberg than valence excitations, we prefer Rydberg assignments here. As seen in Table X.A-II, the term values of the broad features fit nicely for $n = 3$ Rydberg excitations, as do the relative intensities. This interpretation also parallels that used so successfully in assigning the spectra of the fluoroethylenes beyond the first ionization potentials (Section X.C). Based on their term values, many of the autoionizing peaks in the photoionization efficiency curves of propylene [K36] and cyclopropene [P16] also can be assigned as Rydberg transitions to $n = 3$ levels leading to higher ionization potentials. An autoionizing Rydberg series with $\delta = 1.15$ is reported in *trans*-butene-2 which is said to converge upon the ionization potential at 142 200 cm^{-1} (*vert., 17.63 eV*) [W35]. The Rydberg states of all the methylated ethylenes originating at the π MO likely are twisted about the C=C bond, for that is the case in the $(\pi)^{-1}$ radical cations [T14].

In earlier work [**II**, 45], it was observed that in highly strained cyclic olefins, the σ MO manifold is raised with respect to that of the π MO's, and in consequence of this, $\sigma \rightarrow \pi^*$ becomes a relatively low-frequency promotion. Thus in cyclopropene and in tricyclo[3.3.0.0^{2,6}]oct-3-ene a $\sigma \rightarrow \pi^*$ valence band of low oscillator strength appears below the $\pi \rightarrow \pi^*$ band, in the vicinity of 40 000–50 000 cm^{-1} (*4.96–6.20 eV*). Another such $\sigma \rightarrow \pi^*$ band possibly appears in the strained olefin benzvalene. A smooth band peaks at 46 500-cm^{-1} (*5.76 eV*) with an oscillator strength of only 0.065 in this olefin, and upon this there rests a sharply structured feature, Fig. X.B-1 [G39, H16, H17]. Absorption in the 38 000–53 000-cm^{-1} (*4.7–6.6-eV*) region of benzvalene was tested using both high-pressure helium gas and an argon matrix and it was found that the response of the structured band was that of a Rydberg excitation whereas the underlying smooth absorption band responded as would a valence excitation [H16]. With an ionization potential of 68 900 cm^{-1} [*8.54 eV vert.*; $(4b_1\pi)^{-1}$, the pi MO] [H15], the term value of the structured band of benzvalene amounts to *ca.* 22 400 cm^{-1}, strongly suggesting an electronically-allowed $4b_1\pi \rightarrow 3sa_1$ assignment. The vibronic pattern however, is complex, and the band in question also may contain components of the $4b_1\pi \rightarrow 3p$ excitation. Assignment of the 46 500-cm^{-1} band as $\pi \rightarrow \pi^*$ would be acceptable based on its frequency, however its oscillator strength of only 0.065 is suspiciously low. Since the $(10a_1\sigma)^{-1}$ ionization potential is only 8400 cm^{-1} below that of $(4b_1\pi)^{-1}$, it is also possible that the 46 500-cm^{-1} band corresponds to $10a_1\sigma \rightarrow \pi^*$, a symmetry-forbidden valence excitation. This would seem to be an excellent opportunity for theory to solve an intriguing problem.

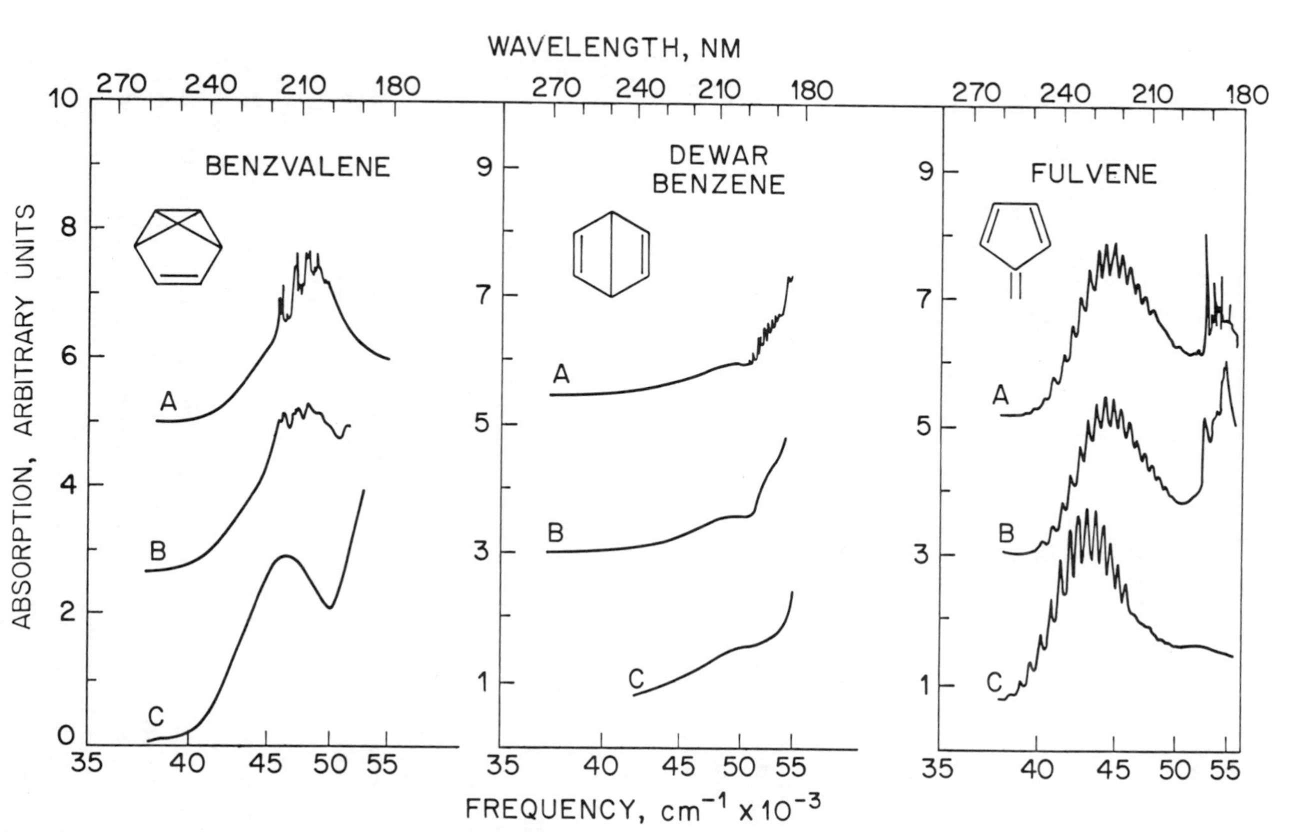

Fig. X.B-1. Absorption spectra of three isomers of benzene [H16]. Curves labeled **A** are the vapor spectra, whereas those labeled **B** and **C** are determined in argon matrices and under 1000 psi He pressure, respectively.

Hirayama and Lipsky [H53] have succeeded in observing the very weak fluorescence of many alkylated ethylenes when excited at 54 083 cm^{-1} (1849 Å, *6.705 eV*). On going from propylene (vapor) to 2-methyl pentene-2 (neat liquid), the fluorescence maximum moves from 44 400 to 40 700 cm^{-1} (*5.50 to 5.05 eV*), with quantum yields of the order of 10^{-6}. No emission was observed from ethylene. The fluorescence frequency shifts in the alkyl olefins follow the $\pi \rightarrow 3s$ frequencies more regularly than they do the $\pi \rightarrow \pi^*$ frequencies, and an oscillator strength of 0.02 is estimated for the fluorescent transition. Consequently, it appears that the fluorescent transition in question is $3s \rightarrow \pi$ rather than $\pi^* \rightarrow \pi$. On the other hand, while the $\pi \rightarrow 3s$ absorption is strongly perturbed by solution in nonpolar solvents, the apparent $3s \rightarrow \pi$ luminescence shows no solvent effects! This can be rationalized if, after the absorption into the (π, 3s) level, the solvent expands to accommodate the Rydberg bubble state so that the $3s \rightarrow \pi$ transition then takes place essentially in vacuum, Section II.D. A similar argument has been used to explain fluorescence experiments involving the $\pi \leftrightarrow 3s$ transitions of *tetrakis*-dimethylaminoethylene in solution [N9] and of NO in rare-gas matrices [G24, G25].

In a planar alkyl olefin, the $\pi \rightarrow \pi^*$ excitation fails to show a circular dichroism (CD) band because the transition has an electric moment (z, long axis) but no magnetic moment, whereas a nonzero rotatory strength in CD demands both. The observation of the $\pi \rightarrow \pi^*$ band in the CD spectra of chiral olefins arises through the mixing of the (π, π^*) configuration with another (or others) having magnetic-dipole transition moment components parallel to the electric-dipole transition moment of the $\pi \rightarrow \pi^*$ band. An anomalously large g-factor ($g = \Delta\epsilon/\epsilon$) of approximately 0.01 is symptomatic of a CD transition which is magnetic-dipole allowed, but electric-dipole forbidden in the achiral chromophore.

In the simplest case in which an electric-dipole transition and a magnetic-dipole transition mix to make both active in the CD spectrum, one speaks of a "CD couplet." The bands in a CD couplet will have equal and opposite rotatory strengths. There has been considerable discussion of the involvement of the $\pi \rightarrow \pi^*$ bands of chiral olefins in such CD couplets, and the nature of the other states coupled to them. The difficulty with this rests in the fact that there are not only several states in ethylene (Rydberg and valence) from which to choose the coupling partner, but that the chiral molecules studied to date always contain large alkyl groups which themselves have CD bands in the 50 000–70 000-cm^{-1} (*6.2–8.7-eV*) region, Chapter III. Thus it is not surprising that there is no consistent pattern of CD couplets among the chiral alkyl olefins.

An excellent study of the absorption and CD spectra of chiral olefins has been performed by Drake and Mason [D45]. Quite apart from the information on the olefins, the findings of Drake and Mason are of general interest in regard the external perturbations of Rydberg states. In this work, they first detail the spectral effects of placing a chiral olefin into a solvent, and then the effects of lowering the temperature of the solution. As expected, the $\pi \rightarrow \pi^*$ valence band shifts to lower frequencies upon solution, whereas the $\pi \rightarrow 3s$ Rydberg band broadens and shifts in the opposite direction. When the solution is cooled, the Rydberg excitation continues to shift to higher frequency, and these solvent and temperature shifts may be so large that the Rydberg/valence ordering of the vapor phase actually may be reversed in solution at very low temperature. It is also found that whereas a Rydberg excitation cannot be discerned in the absorption spectrum of a molecule in solution, a broad CD band, shifted to higher frequency and corresponding to the Rydberg excitation, often can be observed, Section II.D.

Spectra of the symmetric olefin bicyclohexylidene 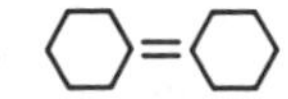 have proved perplexing [**II**, 47]. To recapitulate the earlier work [**II**, S42], bicyclohexylidene in the vapor phase shows intense, structured bands centered at 48 000 and 55 000 cm^{-1} (*5.95 and 6.82 eV*) [**II**, 47]. In the crystal, reflection bands appear at 48 000 and 51 000 cm^{-1} (*5.95 and 6.32 eV*), each with perfect long-axis polarization; it was assumed by Snyder and Clark [**II**, S42] and later workers as well that these two crystal transitions correspond to the two vapor-phase transitions. It has been suggested that the lowest band in the crystal spectrum of bicyclohexylidene is the $\pi \rightarrow 3s$ Rydberg transition vibronically coupled to the N $\rightarrow$ V band [W9], however, the simple fact of its appearance in a crystal argues instead for a valence upper state. In the chiral derivatives of bicyclohexylidene in the vapor and solution phases, Drake and Mason find apparent CD couplets with components centered at 48 000 and *ca.* 55 000 cm^{-1} (*5.95 and ca. 6.8 eV*) [D45]. Noting that such a couplet is impossible for two intense transitions with parallel electric-dipole polarizations, they argue that one of these is really a magnetic-dipole transition intensified by mixing with the electric-dipole allowed transition via the chiral perturbation. This argument fails, however, because the achiral parent molecule itself, bicyclohexylidene, shows both transitions to be far too intense for either to warrant a magnetic-dipole assignment. It is clear from the absorption and CD spectra of derivatives such as cyclohexylidene borane and cyclohexylidene fenchane in the vapor and solution phases [D45] that the "48 000-cm^{-1} transition" in fact consists of two components, centered at *ca.* 48 000 and 51 000 cm^{-1} (*5.95 and 6.32 eV*).

The composite nature of the low-lying excitations in bicyclohexylidene is clearly revealed in the electron energy loss and photoelectron spectral studies of Allan *et al.* [A12], Fig. X.B-2. The $(\pi)^{-1}$ ionization of bicyclohexylidene falls at 65 630 cm^{-1} (*vert., 8.137 eV*), with successive quanta of 1340 cm^{-1} (C=C stretch) peaking at $v' = 2$. As shown in the Figure, just this pattern is found repeatedly in the electron impact spectra recorded at fixed residual energy, and term values are readily determined along with the Rydberg symmetry type. Because of the heavy alkylation of the double bond in bicyclohexylidene, the transitions to 3s and 3p are separated by only 1500 cm^{-1} and four origins to $n = 3$ Rydberg states fall in the 43 000−51 000 cm^{-1} (*5.3−6.3 eV*) region.

According to the above analysis, all the resolvable peaks in the vapor spectrum of bicyclohexylidene are assigned to Rydberg excitations originating at the uppermost π MO, as are the CD peaks of the derivatives studied by Drake and Mason [D45]. It is odd that there is no distinct $\pi \rightarrow \pi^*$ band found here either in absorption, electron impact or CD. Even stranger are the highly structured, long-axis-polarized transitions of the crystal spectrum in the 41 000−54 000 cm^{-1} (*5.1−6.1 eV*) region [**II**, S42]. According to our analysis, these bands of the crystal spectrum do not correlate with any of the structured features of the vapor spectrum, and so remain among the best unsolved puzzles of the vacuum ultraviolet.

Theoretical work on the CD spectrum of *trans*-cyclooctene [**II**, 40] has been carried forward using twisted butene-2 as a model. Bouman and Hansen [B58] performed such model calculations in a valence basis using the random-phase approximation and found that the (π, π^*) and $(\sigma(C{-}H), \pi^*)$ configurations form a CD couplet with a separation of 12 000−16 000 cm^{-1}. These correlate nicely with the CD bands observed in *trans*-cyclooctene at 50 700 and 64 000 cm^{-1} (*vert., 6.28 and 7.93 eV*) having equal and opposite rotatory strengths. In contrast, there are no $\pi \rightarrow \sigma^*$ valence excitations calculated to have significant rotatory strength. Left unexplained by this valence calculation on *trans*-cyclooctene are the weaker CD bands at 47 200 and 56 200 cm^{-1} (*vert., 5.85 and 6.97 eV*). The CI calculation of Liskow and Segal [L28] on *trans*-cyclooctene confirms the valence CD couplet assignment quoted above, and also assigns the 47 200-cm^{-1} and 56 200-cm^{-1} features as $\pi \rightarrow 3s$ and $\pi \rightarrow 3d$, respectively. These Rydberg assignments are in agreement with those given earlier [**II**, 41] based upon term-value arguments. Another Rydberg transition at 62 000 cm^{-1} (*vert., 7.7 eV*) possibly terminates at 4s. Though the band at 50 700 cm^{-1} has a term value appropriate for a $\pi \rightarrow 3p$ assignment, it is calculated that $\pi \rightarrow 3p$ is magnetically too weak

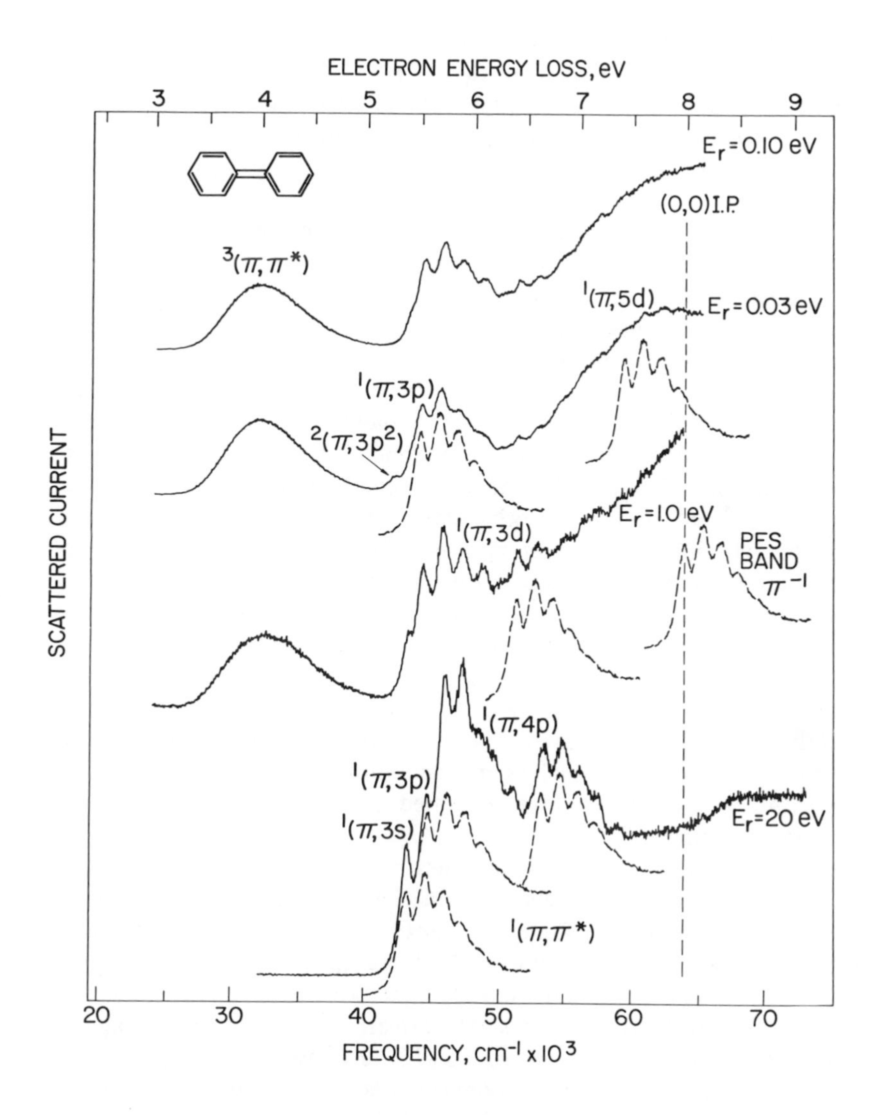

Fig. X.B-2. Electron impact energy loss spectra of bicyclohexylidene at constant residual energy, E_r. The dashed curves are the $(\pi)^{-1}$ photoelectron band envelopes, positioned so as to coincide with the Rydberg excitations originating at the π MO [A12].

to function effectively in a couplet with $\pi \rightarrow \pi^*$, and moreover, the absorption coefficient observed for the band at 50 700 cm^{-1} is far too large to support such a Rydberg assignment. Liskow and Segal also find $\pi \rightarrow \sigma^*$ transitions to be weakly rotating, however they predict that the oscillator strength of the CD-active $\sigma \rightarrow \pi^*$ excitation ($^1A_g \rightarrow {}^1B_{1g}$) in twisted ethylene is 25 % larger than that of $\pi \rightarrow \pi^*$, whereas Bouman and Hansen predict that it is 90 % smaller. The latter seems more likely since the lowest frequency $\sigma \rightarrow \pi^*$ electric-dipole transition is forbidden in planar ethylene and gains its oscillator strength by mixing with $\pi \rightarrow \pi^*$ on twisting.

As in *trans*-cyclooctene, five bands are observed in the CD spectrum of 3-methyl cyclopentene [L21], here alternating in sign. In this system, weak Rydberg transitions are observed from π to 3s (48 800 cm^{-1} *vert.*, *6.05 eV*; absorption and CD), to 3d (56 500 cm^{-1} *vert.*, *7.00 eV*; absorption) and to 4s (60 000 cm^{-1} *vert.*, *7.44 eV*; absorption). The $\pi \rightarrow \pi^*$ excitation (54 000 cm^{-1} *vert.*, *6.69 eV*) appears to form a CD couplet with the CD band at 62 500 cm^{-1} (*vert.*, *7.75 eV*) and so would seem to parallel the situation in *trans*-cyclooctene. If this analogy is correct, then the 62 500-cm^{-1} band of 3-methyl cyclopentene can be assigned as $\sigma \rightarrow \pi^*$. In their work, Levi *et al.* [L21] calculate that the state at 62 500 cm^{-1} which mixs with (π, π^*) to form the CD couplet is (π, 3p), but the observed frequency is much too high for this; furthermore, Liskow and Segal calculate that the $\pi \rightarrow$ 3p transition (in *trans*-cyclooctene) is magnetically too weak to function effectively in a couplet with $\pi \rightarrow \pi^*$. Levi *et al.* also note that the $\pi \rightarrow$ 3s transition displays an anomalously large *g*-factor ($g = 4 \times 10^{-3}$) for a transition which is inherently electric-dipole allowed, and argue that the band is overlapped by the $\sigma \rightarrow$ 3s transition, which is magnetic-dipole allowed. Since the π and σ MO's of cyclopentene are separated by 14 000 cm^{-1} [K18], this explanation is unlikely.

The simple picture given above becomes more complex once we look at the CD spectra of some open-chain chiral olefins. In particular, we focus on *S*,3-methyl-1-pentene, *S*,4-methyl-1-hexene and *S*,5-methyl-1-heptene [G42], Fig. X.B-3. In these compounds, there appear oppositely rotating $\pi \rightarrow$ 3s (*ca.* 53 000 cm^{-1} *vert.*, *6.6 eV*) and $\pi \rightarrow \pi^*$ (*ca.* 56 000 cm^{-1} *vert.*, *6.9 eV*) transitions, and the state normally coupled to (π, π^*) again is present at *ca.* 61 000 cm^{-1} (*vert.*, *7.6 eV*). However, in the open-chain 1-alkenes with their higher ionization potentials, the 61 000-cm^{-1} bands are $\pi \rightarrow$ 3d, while the bands at 66 000 cm^{-1} are $\sigma \rightarrow \pi^*$. The assignments of the $\pi \rightarrow$ 3s and $\pi \rightarrow$ 3d transitions in the methyl-1-alkenes are aided by the fact that in heavily alkylated systems, the 3s$-$3d separation is approximately 9000 cm^{-1} regardless of the ionization potential. The methyl-1-alkene CD spectra differ from those of *trans*-cyclooctene and 3-methyl cyclopentene, for in the methyl-1-alkenes, the $\sigma \rightarrow \pi^*$ band at 66 000 cm^{-1} (*vert.*, *8.2 eV*) is involved in a CD couplet with $\pi \rightarrow$ 3d

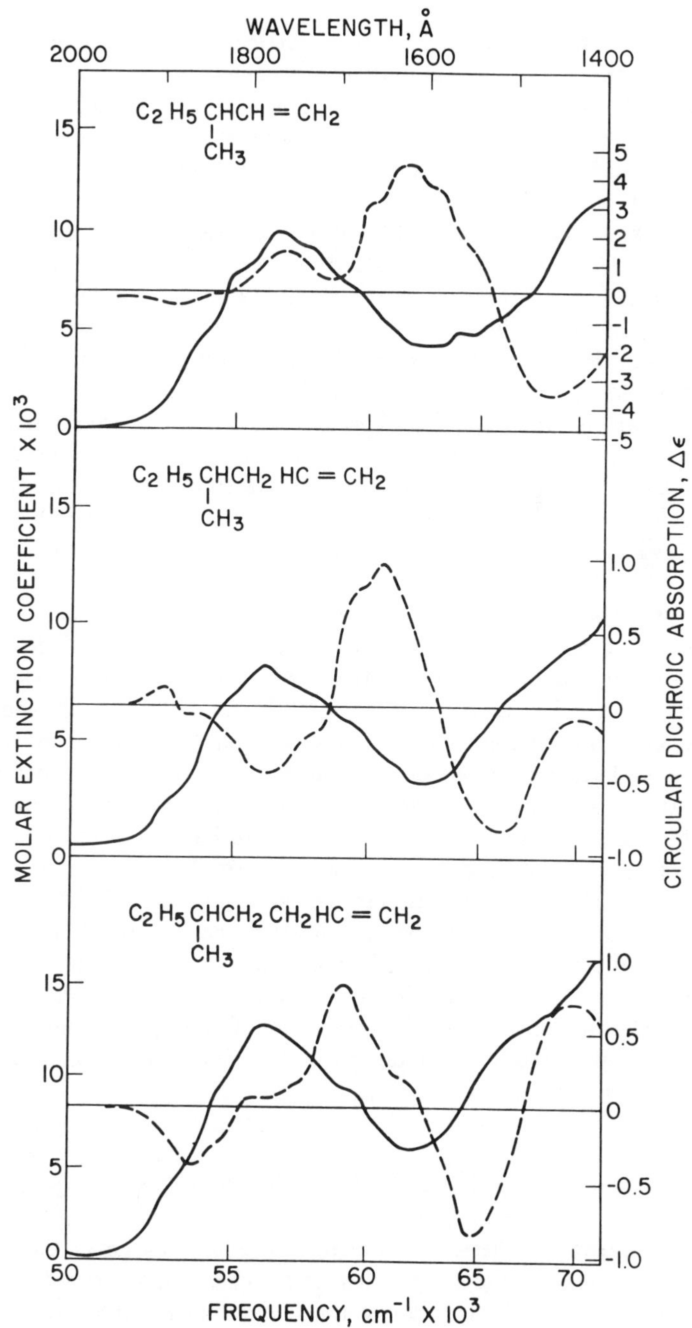

Fig. X.B-3. Absorption (*full line*) and CD spectra (*dashed line*) of the methyl-1-alkenes in the vapor phase [G42].

rather than with $\pi \rightarrow \pi^*$. This may related to the Rydberg/valence conjugate nature of the (π, 3d) and (π, π^*) configurations.

Though its double bond is situated within a six-membered ring, the CD spectrum of α-pinene [S55] fits into the scheme described above for the open-chain alkenes, with the bands at 55 000 and 64 000 cm^{-1} (*vert., 6.8 and 7.9 eV*) functioning as a CD couplet [**II**, 40]. As noted above, these two bands are assigned as $\pi \rightarrow$ 3d and $\sigma \rightarrow \pi^*$, respectively. In β-pinene, [S55] the CD couplet is formed instead by the (π, π^*) (50 000 cm^{-1} *vert., 6.1 eV*) and (π, 3d) (56 000 cm^{-1} *vert., 6.4 eV*) states [**II**, 40]. The observed signs of the first two features in the CD spectra of α- and β-pinene imply that the first is electric-dipole polarized x and the second z, as expected for $\pi \rightarrow$ 3s and $\pi \rightarrow \pi^*$ transitions.

Appropriate deuteration of norbornene [**II**, 31] yields chiral materials showing CD bands near 50 000 cm^{-1} (*vert., 6.2 eV*) [P8]. In all cases, the $\pi \rightarrow$ 3s band does not appear in the CD spectrum, however a strongly-rotating $\pi \rightarrow \pi^*$ excitation is observed. This latter is thought to be half of a CD couplet formed with a higher lying, magnetic-dipole allowed transition, presumably $\sigma \rightarrow \pi^*$.

The problems are manifold in trying to find a common, unifying principle in olefin CD spectra. First, we see that CD couplets (and triplets) often are formed, but between different pairs of transitions in different olefins. Next, though theory seems to explain the various CD spectra, there are notable contradictions between the detailed predictions. Moreover, all the chiral olefins bear bulky alkyl groups as either chiral perturbations or geometric constraints. The assumption has been made that all the CD bands up to *ca.* 70 000 cm^{-1} (*8.7 eV*) in such systems are associated with the olefinic double bond, whereas it is known that alkanes themselves display CD bands far below 70 000 cm^{-1} [A14, D45], Chapter III, and hence could be complicating the picture. Perhaps most puzzling is the bizarre tangle of bands at low frequencies in crystalline bicyclohexylidene and their absence in the vapor spectrum. Such a complicated situation in this otherwise uncomplicated olefin suggests that the spectral interpretations so far given for other alkyl olefins may be grossly oversimplified.

X.C. Haloethylenes

Electron-impact studies by Kuppermann and coworkers [C32, C33, K43] have added much new spectroscopic information on the fluoroethylenes [**II**, 50]. In each of the fluoroethylenes, the excitation to

$^3(\pi, \pi^*)$ is found at 33 700−37 700 cm^{-1} (*vert., 4.18−4.67 eV*). A second triplet state was claimed at 51 600 cm^{-1} (*vert., 6.40 eV*) in fluoroethylene, and the first inclination is to assign it to $^3(\pi, 3s)$. However, such an assignment results in a singlet-triplet split of 4700 cm^{-1}, whereas 1000−2000 cm^{-1} is the more normal range for Rydberg configurations [**I**, 22]. Indeed, using the trapped-electron technique, Verhaart and Brongersma [V24] report the $^3(\pi, 3s)$ origin to be at 55 200 cm^{-1} (*6.85 eV*) in fluoroethylene. Appropriately, this origin is only 1500 cm^{-1} below that of $^1(\pi, 3s)$. Inasmuch as the $\mathscr{A}(\pi, \sigma^*)$ bands of the fluoroethylenes fall in the 55 000−57 000 cm^{-1} (*6.8−7.1 eV*) region (*vide infra*), it is reasonable to assume that the triplet at 51 600 cm^{-1} in fluoroethylene has the (π, σ^*) configuration. A Feshbach resonance to the $^2(\pi, 3s^2)$ negative-ion state has its origin at 52 100 cm^{-1} (*6.46 eV*) in fluoroethylene and displays several quanta of 1530 cm^{-1} (C=C stretch); it is 4700 cm^{-1} below its $^1(\pi, 3s)$ parent state, as in ethylene.

The pattern of levels observed in the trapped-electron spectrum of 1,1-difluoroethylene [V24] is identical to those in ethylene and fluoroethylene:- the $^2(\pi, 3s^2)$ Feshbach resonance with an origin at 50 200 cm^{-1} (*6.22 eV*) is 3100 cm^{-1} below the $^1(\pi, 3s)$ parent state and the transition to the $^3(\pi, 3s)$ origin, 53 200 cm^{-1} (*6.60 eV*), is 900 cm^{-1} below that to $^1(\pi, 3s)$. Additionally, a broad resonance at 67 800 cm^{-1} (*vert., 8.4 eV*) in 1,1-difluoroethylene can be assigned as $\pi \rightarrow 3p$ based on an 18 500-cm^{-1} term value. The $^3(\pi, 3s)$ origin in chlorotrifluoroethylene is observed at 47 300 cm^{-1} (*5.86 eV*), while the $^1(\pi, 3s)$ adiabatic excitation frequency is 48 700 cm^{-1} (*6.04 eV*) [V24].

Many Rydberg excitations are observed below the lowest $(\pi)^{-1}$ ionization potentials in the energy-loss spectra of the fluoroethylenes [C32, C33, K43], and have been assigned as in previous optical studies using term-value arguments [**I**, 50]. Many energy-loss peaks also were observed beyond the $(\pi)^{-1}$ ionization potentials, and were readily assigned as Rydberg excitations going to the lowest $(\sigma)^{-1}$ ionization potentials, again owing to their term values. These (π, ϕ_j^{+1}) and (σ, ϕ_j^{+1}) term values show an interesting trend; in general, one sees that the $(\pi, 3s)$ and $(\sigma, 3s)$ term values are very nearly equal when the number of fluorine atoms is low, but that the $(\pi, 3s)$ term value becomes larger than that of $(\sigma, 3s)$ by up to 4000 cm^{-1} with increasing fluorination. This is counter to the conclusion one draws by noting that σ in general is a precursor in the core for 3s [**I**, 13] whereas π is not. In contrast, the $(\sigma, 3p)$ term values are 1000−4000 cm^{-1} larger than their corresponding $(\pi, 3p)$ counterparts regardless of the level of fluorination. (The 3p component being measured here is unspecified.)

Coggiola *et al.* [C33] also report that the $\pi \rightarrow \pi^*$ band of trifluoroethylene (61 400 cm^{-1} *vert., 7.61 eV*) does not shift on going to trifluorochloroethylene (62 800 cm^{-1} *vert., 7.78 eV*), thus illustrating that the 10 200-cm^{-1} shift of the $\pi \rightarrow \pi^*$ absorption to high frequency on going from trifluoroethylene to tetrafluoroethylene [**II**, 52] is due to electronic rather than steric effects. According to Salahub [S5], the $\pi \rightarrow \pi^*$ frequency shift can be understood to be the result of the much higher frequencies of the (σ, σ^*) configurations in C_2F_4 compared to those in any fluoroethylene having one or more C—H bonds. Another possible factor here is that the C=C and C—F distances of C_2F_4 are 0.027 and 0.029 Å shorter than the C=C distance in C_2H_4 and the C—F distance in CH_2=CHF, respectively. As pointed out by Chiu *et al.* [C21], this shrinkage has the effect of raising the π^* level of C_2F_4 by 10 200 cm^{-1} in the ground-state geometry, compared to ethylene.†

McDiarmid has made a detailed study of the absorption peaks in *cis*- and *trans*-difluoroethylene-h_2 and -d_2 [M33]. In the *cis* systems, transitions to the n = 4 and 5 members of an ns series are identified as having $\delta = 0.89$, while the bands at 52 400 cm^{-1} (*advert., 6.49 eV*) are discounted as the n = 3 members of the series because they correspond to $\delta = 1.09$. We maintain that this transition is $\pi \rightarrow 3s$, the difference in δ's being due to the effect of the conjugate $\pi \rightarrow \sigma^*$ valence excitations which significantly perturb only the lowest (n = 3) member of the (π, ns) series. The Rydberg character of the 52 400-cm^{-1} band is confirmed by its behavior in condensed phases [D7]. In the *trans* compounds, McDiarmid attributed the deviation of the lower ns members of the (π, ns) series to mixing with the (σ, 3s) configuration (said to be at 80 390 cm^{-1} *advert., 9.967 eV*), however, this is not possible for symmetry reasons.

The matrix and crystal spectra of the fluoroethylenes are most interesting and relevant. Dauber and Brith [D7] report that in the crystalline phases of the fluoroethylenes all sharp-line Rydberg absorptions disappear, whereas the N → V transitions shift to lower frequencies but show no vibrational structure. The low-frequency spectra of *cis*-difluoroethylene and trifluoroethylene in the vapor and solid phases show

† The electron transmission spectra of the fluoroethylenes show that the π^* MO is destabilized in a regular way on going from ethylene (π^* TNI resonance at 14 000 cm^{-1}, *1.73 eV*) to tetrafluoroethylene (π^* TNI resonance at 24 200 cm^{-1}, *3.00 eV*) [C21]. Calculations show, however, that though this is indeed the trend for the vertical excitation frequencies, the adiabatic frequencies in fact *decrease* with fluorination [P1]!

that the band at 52 600 cm^{-1} (*vert., 6.52 eV*) disappears in the condensed phase as is usual for Rydberg excitations, but that a valence excitation persists at 55 200 cm^{-1} (*vert., 6.84 eV*). This behavior confirms the earlier assignment of the bands at 55 000−57 000 cm^{-1} in these compounds as $\pi \rightarrow \sigma^*$ valence excitations [**II**, 55]. In the present work, such excitations are called $\mathscr{A}(\pi, \sigma^*)$ bands (Section I.C). The term values of the $\mathscr{A}(\pi, \sigma^*)$ excitations in *cis*- and *trans*-difluoroethylene and in trifluoroethylene with respect to their π-orbital ionization potentials, Table X.C-I, are remarkably close to those deduced for $\mathscr{A}(\pi, \sigma^*)$ band excitations in various chloro, bromo and iodoethylenes. Only the term value for the $\mathscr{A}(\pi, \sigma^*)$ transition of tetrafluoroethylene is outside this pattern. The corresponding $\mathscr{A}(2p_F, \sigma^*)$ bands in the fluoroethylenes are expected at approximately 107 000 cm^{-1} (*vert., 13.5 eV*) on the basis of $(2p_F)^{-1}$ ionization potentials of 137 000 cm^{-1} (*vert., 17 eV*) and 30 000 cm^{-1} term values.

Peaks observed at 50 000 and 92 700 cm^{-1} (*vert., 6.2 and 11.5 eV*) in the dissociative-electron attachment spectrum of tetrafluoroethylene [I1] have Feshbach decrements of *ca.* 4000 cm^{-1} with respect to their parent Rydberg configurations, and are assigned therefore as having $^2(b_{3u}\pi, 3s^2)$ and $^2(6a_g\sigma, 3s^2)$ negative-ion upper states. By default, attachment peaks at 24 000, 32 000 and 40 000 cm^{-1} (*vert., 3, 4 and 5 eV*) must involve valence TNI configurations with the incident electron captured in π^* and σ^* MO's. Related Feshbach and TNI valence resonances are observed as well in the various chlorofluoroethylenes and in tetrachloroethylene [I1, K9]. A dissociative attachment peak in the latter at 62 000 cm^{-1} (*vert., 7.7 eV*) is a Feshbach resonance having the $^2(3p, 4s^2)$ configuration as judged from its decrement (Section II.B).

The $(1s_C, \pi^*)$ term value in ethylene is observed to be 48 900 cm^{-1} (Section X.A) and alkylation as in the butenes destablizes π^* so that the term values drop to 46 000 cm^{-1} (Section X.B). On the other hand, fluorination to form perfluorobutene-2 has the opposite effect, for the $(1s_C, \pi^*)$ term value in this case is 54 800 cm^{-1} [H66]. However, as in the inner-shell spectrum of CF_4 (Section IV.D), the inner-shell spectrum of perfluorobutene-2 is complicated by low-lying $\mathscr{A}(1s_C, \sigma^*)$ bands which cluster in the region 40 000 cm^{-1} or so below the $(1s_C)^{-1}$ ionization limit of the CF_3 group.

The photographic work of Walsh and coworkers on the spectra of the chloroethylenes [**II**, 56] now is complemented by quantitative photoelectric recording of all such spectra in the 38 500−71 000 cm^{-1} (*4.8−8.8 eV*) region [B35]. In several of these spectra, Fig. X.C-1, there is an unexpected band on the low-frequency wing of the $\pi \rightarrow \pi^*$ excitation,

TABLE X.C-I

$\mathscr{A}$-BAND TERM VALUES AND ASSIGNMENTS IN THE HALOETHYLENES

Molecule	Absorption Frequency, cm^{-1}	Ionization Potential, cm^{-1}	Term Value, cm^{-1}	Assignment	Reference
cis-FHC=CHF	57 000	84 100	27 100	$\mathscr{A}(\pi, \sigma^*)$	[D7, **II**, 55]
trans-FHC=CHF	55 000	83 700	28 700	$\mathscr{A}(\pi, \sigma^*)$	[**II**, 55]
$F_2C{=}CHF$	57 000	84 900	27 900	$\mathscr{A}(\pi, \sigma^*)$	[D7, **II**, 55]
$F_2C{=}CF_2$	62 000	84 900	22 900	$\mathscr{A}(\pi, \sigma^*)$	[**II**, 55]
trans-ClHC=CHCl	48 500	77 800	29 300	$\mathscr{A}(\pi, \sigma^*)$	[B35]
$Cl_2C{=}CCl_2$	43 500	76 700	33 200	$\mathscr{A}(\pi, \sigma^*)$	[B35]
	61 900	91 700	29 800	$\mathscr{A}(3p\pi, \sigma^*)$	
$H_2C{=}CHBr$	52 080	79 600	27 500	$\mathscr{A}(\pi, \sigma^*)$	[F3, S22]
$H_2C{=}CBr_2$	48 700	78 900	30 200	$\mathscr{A}(\pi, \sigma^*)$	[F3, S22]
	57 100	86 600	29 500	$\mathscr{A}(4p\pi, \sigma^*)$	
cis-HBrC=CHBr	47 600	77 700	30 200	$\mathscr{A}(\pi, \sigma^*)$	[F3, S22]
	58 800	86 600	29 500	$\mathscr{A}(4p\pi, \sigma^*)$	
trans-HBrC=CHBr	47 300	77 000	29 700	$\mathscr{A}(\pi, \sigma^*)$	[F3, S22]
	60 600	89 100	28 500	$\mathscr{A}(4p\pi, \sigma^*)$	
$H_2C{=}CHI$	46 200	75 200	29 000	$\mathscr{A}(\pi, \sigma^*)$	[**II**, 67]
$F_2C{=}CFCl$	50 000	79 400	29 400	$\mathscr{A}(\pi, \sigma^*)$	[**II**, S25]
$F_2C{=}CFBr$	50 800	81 600	30 800	$\mathscr{A}(\pi, \sigma^*)$	[S23]
$F_2C{=}CFI$	48 500	77 120	28 600	$\mathscr{A}(\pi, \sigma^*)$	[S23]
	61 300	90 300	29 000	$\mathscr{A}(5p, \sigma^*)$[a]	
$H_2C{=}CHCH_2Cl$	52 100	82 270	30 200	$\mathscr{A}(\pi, \sigma^*)$	[W36]
	63 100	90 900	27 800	$\mathscr{A}(3p, \sigma^*)$	
$H_2C{=}CHCH_2Br$	51 800	80 700	28 900	$\mathscr{A}(\pi, \sigma^*)$	[W36]
	57 100	86 700	29 600	$\mathscr{A}(4p, \sigma^*)$	
$H_2C{=}CHCH_2I$	45 000	74 600	29 600	$\mathscr{A}(5p, \sigma^*)$	[W36]
	48 100	78 100	30 000	$\mathscr{A}(5p, \sigma^*)$	
	53 400	82 800	29 400	$\mathscr{A}(\pi, \sigma^*)$	

[a] This broad feature is assigned as 5p → 6s in [S23].

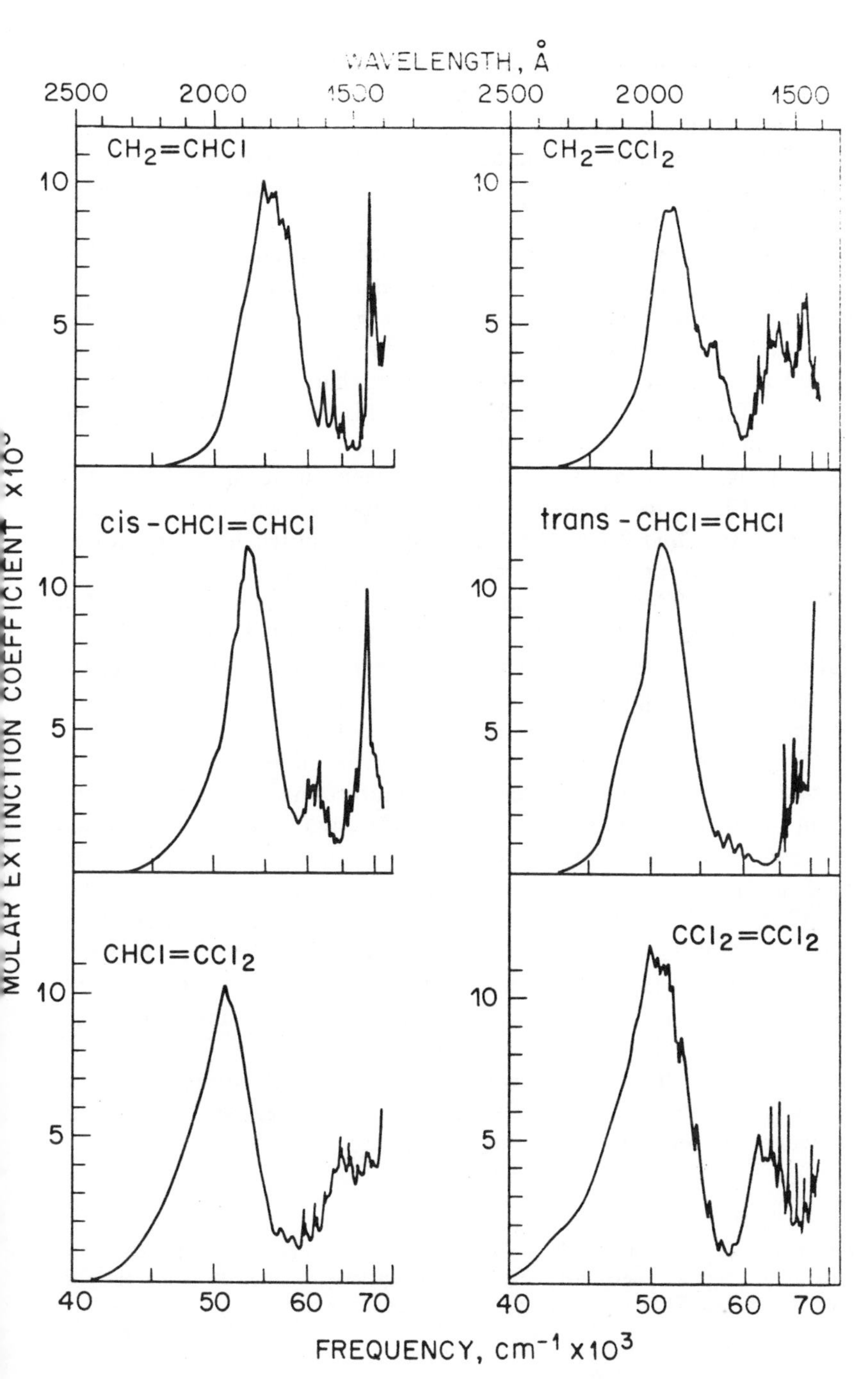

Fig. X.C-1. Absorption spectra of the chloroethylenes in the vapor phase [B35].

most easily seen at 48 500 cm^{-1} (*vert., 6.01 eV*) in *trans*-dichloroethylene and at 43 500 cm^{-1} (*vert., 5.39 eV*) in tetrachloroethylene. In the latter, the band at 43 500 cm^{-1} is more evident in the Kr matrix and neat-crystal spectra [D7], where it shows vibrational quanta of 1550 cm^{-1}. This is then followed by a progression of 1100–1300 cm^{-1} which characterizes the N → V(π, π^*) excitation.

By its frequency and by its behavior in condensed phases, the 43 500-cm^{-1} band of tetrachloroethylene must be valence, involving either $3p\pi_{Cl}$ or π(C=C) as originating MO's, and either π^*(C=C) or σ^*(C–Cl) as terminating MO's. Another likely $\mathscr{A}$ band appears in tetrachloroethylene at 61 900 cm^{-1} (*advert., 7.67 eV*) as a sharp but unstructured feature, Fig. X.C-1. Dauber and Brith [D7] observe this band clearly in the neat crystal and in a Kr matrix where it is shifted to the low-frequency side of the vapor-phase band, as is the $\pi \to \pi^*$ valence band at 47 600 cm^{-1} (*vert., 5.90 eV*). By contrast, the $\pi \to$ 3s excitation† between 50 000 and 57 000 cm^{-1} (*6.2 and 7.1 eV*) is high-frequency shifted in a Kr matrix, and does not appear at all in the spectrum of the neat crystal.

Note that the π(C=C) MO ($4b_{3u}$) is above the uppermost $3p\pi_{Cl}$ MO ($5b_{3g}$) by 15 000 cm^{-1} in tetrachloroethylene [K18] and that the separation between the $\mathscr{A}$ bands is 18 400 cm^{-1}. This near-equality of level separations strongly suggests that the $\mathscr{A}$-band transitions originate at π and $3p_{Cl}$ orbitals and terminate at a common level, either π^*(C=C) or σ^*(C–Cl). Since the $\pi \to \pi^*$ assignment is preempted by the intense band at 47 600 cm^{-1}, we conclude that the features at 43 500 and 61 900 cm^{-1} in tetrachloroethylene are $\mathscr{A}(\pi, \sigma^*)$ and $\mathscr{A}(3p, \sigma^*)$ bands, respectively. The 48 500-cm^{-1} $\mathscr{A}$ band of *trans*-dichloroethylene is assigned a (π, σ^*) upper state. Though it is quite unexpected, the $\mathscr{A}(np, \sigma^*)$ and $\mathscr{A}(\pi, \sigma^*)$ term values of the haloethylenes are observed to be quite constant (28 500 ± 1500 cm^{-1}) irrespective of the halogen's atomic number or the extent of halogenation, Table X.C-I. These term values are not enhanced by fluorination, but this is not so surprising since the $\mathscr{A}$-band term values for the fluoroethylenes themselves are no larger than those for the other haloethylenes. There is no explanation as yet for such behavior. $\mathscr{A}(\pi, \sigma^*)$ and $\mathscr{A}(np, \sigma^*)$ bands related to those of the haloethylenes are found as well in the halobenzenes (Section XIX.C).

† In the haloethylenes, we consider the lowest Rydberg excitation originating at π(C=C) to terminate at 3s, whereas those originating with the np lone-pair orbitals of the halogen terminate at $(n + 1)$s. The difference is largely one of perspective.

The MPI spectrum of tetrachloroethylene two-photon resonant at the (π, 3s) level has been analyzed vibrationally [H32]. Unlike the situation in ethylene, there is no activity of the torsional vibration, and it is concluded that the (π, 3s) state is planar in this molecule. Similarly, since the $\pi \rightarrow \pi^*$ band at 47 600 cm^{-1} (*5.90 eV*) displays a long C–C stretching progression, but no torsional modes, it too has been assigned a planar structure [D7]. Dissociative attachment spectra of the chloroethylenes [K9] show several low-lying excitations to σ^*(C–Cl) valence TNI levels as well as higher-frequency bands in chloroethylene and tetrachloroethylene which can be interpreted as Feshbach resonances using the 4000 cm^{-1} decrement as a guide (Section II.B).

The perfluoro effect has been investigated by comparing the electronic spectra of bromoethylene and iodoethylene [**II**, 66] with their perfluorinated counterparts trifluorobromoethylene and trifluoroiodoethylene [S23]. Analysis of the Rydberg term values appropriate to the $(\pi_1)^{-1}$ and $(\pi_2)^{-1}$ ionization potentials [**II**, 67] is revealing. In both of the nonfluorinated compounds, the (π_1, 3s) and (π_2, 3s) term values are 24 500 ± 500 cm^{-1}, whereas upon perfluorination, they shift upward to 29 250 ± 250 cm^{-1}. This is in keeping with the general rules of the perfluoro effect, stating that the binding energy of σ orbitals (3s) increase significantly upon perfluorination, whereas the binding energies of π orbitals remain almost unshifted upon perfluorination [**I**, B67]. Within the 3p manifolds in the nonfluorinated compounds, two core-split components are observed, with term values of 17 400 ± 600 and 20 100 ± 500 cm^{-1}. In the perfluoro compounds, one 3p component remains at 17 300 ± 500 cm^{-1} while the second increases to 21 700 ± 400 cm^{-1}. Applying the arguments of the perfluoro effect to these 3p term values (ionization potentials), one concludes that the (π, 3pσ) term-value shift on fluorination will be significantly larger than that for (π, 3pπ). Thus the excitations with 17 000 cm^{-1} term values terminate at npπ, whereas those with 20 000–22 000 cm^{-1} term values terminate at npσ. Of course, one does expect that the npπ Rydberg orbitals will have less penetration than npσ Rydberg orbitals (Section I.B), and hence will have the smaller term values.

In analyzing the spectra of trifluoroiodoethylene and trifluorobromoethylene, Schander and Russell [S23] assign a relatively weak band at *ca.* 48 000–50 000 cm^{-1} (*vert., 5.9–6.2 eV*) in each as $\pi \rightarrow \pi^*$. In trifluoroethylene and trifluorochloroethylene [**II**, 65] the olefinic $\pi \rightarrow \pi^*$ transitions are clearly seen at 63 000 cm^{-1} (*vert., 7.81 eV*) in each [C33]. Because a 13 000-cm^{-1} shift on replacing the chlorine atom with either bromine or iodine seems unlikely, we propose that the 50 000-cm^{-1} bands in these compounds instead are excitations to $\mathscr{A}(\pi, \sigma^*)$ states, in line with our assignments in the fluoro- and chloroethylenes, and in line with their term values of *ca.* 30 000 cm^{-1}, Table X.C-I. Though the $\mathscr{A}(\pi, \sigma^*)$ bands come at the same frequency regardless of whether σ^* is σ^*(C–F), σ^*(C–Br) or σ^*(C–I) in the mixed haloethylenes, this is not so for the various $\mathscr{A}(n\text{p}, \sigma^*)$ bands. The lowest frequency $\mathscr{A}(n\text{p}, \sigma^*)$ bands are expected at 66 000 cm^{-1} (*vert., 8.2 eV*) in trifluorobromoethylene and at 60 000 cm^{-1} (*vert., 7.4 eV*) in the iodo compound based on 30 000-cm^{-1} term values and their reported $(n\text{p}\pi)^{-1}$ ionization potentials [S23]. The latter suggests a 5p → σ^* assignment for the band at 61 300 cm^{-1} (*vert., 7.60 eV*) otherwise assigned by Schander and Russell as 5p → 6s. In trifluorobromoethylene, the olefinic $\pi \rightarrow \pi^*$ transition is most likely still at approximately 61 000 cm^{-1} (*vert., 7.56 eV*), beneath the 4p → 5p Rydberg excitation, whereas in trifluoroiodoethylene, it is at 58 000 cm^{-1} (*vert., 7.2 eV*), beneath the 5p → 6p Rydberg excitation. In C_2X_4 systems, there will be five $\pi \rightarrow \pi^*$ excitations if one includes $n\text{p}\pi \rightarrow \pi^*$.

The ultraviolet spectra of bromoethylene and the various dibromoethylenes have been analyzed [F3, S22, W32]. Because the bromine lone-pair and the C=C pi-bond ionization potentials are so close together, Rydberg series terminating at s, p and d Rydberg orbitals and leading to the first through third or fourth ionization potentials are observed in close proximity. In *cis*-dibromoethylene, Felps *et al.* [F3] assign bands at *ca.* 50 000 and *ca.* 71 500 cm^{-1} (*vert., 6.2 and 8.86 eV*) as $\pi \rightarrow \pi^*$ excitations, choosing the former as an $\mathscr{A}(4\text{p}\pi, \pi^*)$ band and the latter as N → V. Schander and Russell [S22] again place $\pi \rightarrow \pi^*$ in the bromoethylenes at *ca.* 50 000 cm^{-1} (*vert., 6.2 eV*), assuming a V upper-state description rather than $\mathscr{A}(n\text{p}\pi, \pi^*)$.

Let us instead approach the $\mathscr{A}$-type bands of the bromoethylenes in the manner parallel to that used above for the chloroethylenes, which is to say, we expect $\mathscr{A}(\pi, \sigma^*)$ bands to have term values of *ca.* 29 000 cm^{-1} with respect to the $\pi(\text{C=C})^{-1}$ ionization potentials, while $\mathscr{A}(4\text{p}, \sigma^*)$ bands will have the same term values with respect to the $(4\text{p}\pi)^{-1}$ ionization potentials. Using the data in [S22], Table X.C-I, it is seen that this approach works quite well for the four bromoethylenes which have been studied if the $\mathscr{A}$ bands in the vicinity of 48 000 cm^{-1} (*6 eV*) are assigned as $\mathscr{A}(\pi, \sigma^*)$ and that at *ca.* 58 000 cm^{-1} (*7.2 eV*) is assigned as $\mathscr{A}(4\text{p}\pi, \sigma^*)$. Note, however, that the absorption intensities near 50 000 cm^{-1} are quite high (ϵ approximately 8000) in these compounds, and so the N → V excitations tentatively must be placed in this region along with the $\mathscr{A}(\pi, \sigma^*)$ bands.

It is most surprising to find that the constant term values of *ca.* 29 000 cm^{-1} for the $\mathscr{A}(np, \sigma^*)$ and $\mathscr{A}(\pi, \sigma^*)$ bands of the haloethylenes discussed above also apply to the spectra of the allyl halides $H_2C{=}CHCH_2X$, Table X.C-I. At first thought, one would expect a composite spectrum in these systems, say those of propylene and CH_3X, with little or no interaction. However, as seen in the Table, not only are $\mathscr{A}(np, \sigma^*)$ bands assignable with 29 000 cm^{-1} term values, but so are $\mathscr{A}(\pi, \sigma^*)$ bands as well with the same term values! Once these $\mathscr{A}$ bands are assigned in the allyl halides, the remaining bands are easily seen to be the lowest Rydberg excitations and $\pi \rightarrow \pi^*$, based on their term values and intensities.

X.D. Addendum

I. D. Petsalakis, G. Theodorakopoulos, C. A. Nicolaides, R. J. Buenker and S. D. Peyerimhoff, Nonorthonormal CI for molecular excited states. I. The sudden polarization effect in 90° twisted ethylene, *J. Chem. Phys.* **81**, 3161 (1984).

S. Kadifachi, Trends in temporary negative ion states in linear and cyclic alkenes, *Chem. Phys. Lett.* **108**, 233 (1984).

P. A. Snyder, P. N. Schatz and E. M. Rowe, Assignment of $\pi \rightarrow$ 3d Rydberg transitions in ethylene around 9 eV using MCD and synchrotron radiation, *Chem. Phys. Lett.* **110**, 508 (1984).

C. F. Koerting, K. N. Walzl and A. Kuppermann, An electron-impact investigation of the singlet $\rightarrow$ triplet transitions in the chloro-substituted ethylenes, *Chem. Phys. Lett.* **109**, 140 (1984).

CHAPTER XI

Azo, Imine and Oxime Compounds

XI.A. Azo Compounds

The optical spectrum of *trans*-diimide HN=NH [**II**, 68] has been re-examined under conditions which give much cleaner spectra than previously obtained. The sharp band with origin at 57 926.5 cm^{-1} (*7.18180 eV*) was analyzed rotationally [N20] and assigned as $n_+ \rightarrow 3p$ ($^1A_g \rightarrow {}^1B_u$) [F36, N20], where n_+ is the upper component resulting from the through-bond interaction between the two n_N lone pairs and 3p is either in-plane $b_u\sigma$ or in-plane $b_u\pi$; the lower component of the lone-pair splitting is written as n_-. The $n_+ \rightarrow 3p$ transition excites progressions of symmetric H—N=N bending and N=N stretching in an upper state which is somewhat predissociated. According to a very large calculation [V12], $3p\sigma$ is the lowest of the 3p components in diimide, with the other two 3800 and 5500 cm^{-1} higher. The same calculation places the $\pi \rightarrow \pi^*$ band of diimide more than 15 000 cm^{-1} above the $(n_+)^{-1}$ ionization potential but below the $(\pi)^{-1}$ ionization potential.

A second band in the diimide optical spectrum has vibronic structure closely similar to that of the $n_+ \rightarrow 3p$ transition; with an origin at 67 894 cm^{-1} (*8.4176 eV*), it has a term value of 9420 cm^{-1}. This band is assigned as $n_+ \rightarrow 4p$ rather than $n_+ \rightarrow 3d$ because its term value is too small for a transition terminating at 3d, and also because $n_+ \rightarrow 3d$ is parity forbidden. The valence transition $n_- \rightarrow \pi^*$ in diimide is predicted to fall between $n_+ \rightarrow 3p$ and $n_+ \rightarrow 4p$ [V12], but there is no sign of it or of the $n_+ \rightarrow 3s$ forbidden transition expected at *ca.* 50 000 cm^{-1} (*6.2 eV*).

The situation in methyl diazene ($CH_3N{=}NH$) is especially interesting due to its lower symmetry; in the C_s point group of this molecule, the $n_+ \rightarrow 3s$ transition is electronically allowed. The 58 000-cm^{-1} band of diimide ($n_+ \rightarrow 3p$, 19 400-cm^{-1} term value) now appears in methyl diazene at 58 800 cm^{-1} (*vert., 7.29 eV*, 18 400-cm^{-1} term value) [F36, V25]. Though no Rydberg absorption at lower frequency has been assigned in diimide, in methyl diazene there is a strong feature peaked at 48 100 cm^{-1} (*5.96 eV*) which is most likely $n_+ \rightarrow 3s$. The band is decorated with a long progression of ν_{10}' (CNN deformation) and has a term value of 29 100 cm^{-1}. The $n_N \rightarrow 3s$ Rydberg transition in the related molecule dimethyl amine has a term value of 28 800 cm^{-1} [**I**, 210].

Both trapped-electron and electron-impact techniques have been applied to azomethane [**II**, 69] with interesting results [M72, M74]. The two intense optical transitions of azomethane labelled I and II ($n_+ \rightarrow 3s$ and $n_+ \rightarrow 3p$) are observed as broad peaks in electron impact at 54 100 and 64 000 cm^{-1} (*vert., 6.71 and 7.90*), respectively. In addition to these, an intense transition having a singlet upper state is found at 76 600 cm^{-1} (*9.50 eV*). Since it has a term value of 18 900 cm^{-1} with respect to the $(\pi)^{-1}$ ionization potential at 95 500 cm^{-1} (*vert., 11.8 eV*), it is assigned as $\pi \rightarrow 3p$. Fine structure in the photoionization efficiency curve of azomethane in the 72 000–88 000-cm^{-1} (*8.9–10.9-eV*) region [F23] in part may involve the $\pi \rightarrow 3p$ excitation.

In earlier work [**II**, 69] it was surmised that the symmetry-forbidden $n_+ \rightarrow 3s$ transition of azomethane would fall close to 44 000 cm^{-1} (*5.4 eV*). A band of low relative intensity having a singlet upper state is found in the electron-impact spectrum at 48 500 cm^{-1} (*vert., 6.01 eV*) [M72, M74]. However, if this is the $n_+ \rightarrow 3s$ excitation, its term value of 23 900 cm^{-1} is surprisingly low, for one expects the shift of the (n_+, 3s) term value on going from methyl diazene (29 100 cm^{-1}) to azomethane to resemble closely that on going from acetaldehyde (27 508 cm^{-1}) to acetone (27 150 cm^{-1}), or from propylene (26 900 cm^{-1}) to *trans*-butene-2 (25 680 cm^{-1}). Because of this term-value discrepancy, we lean more toward a valence $n_- \rightarrow \pi^*$ assignment for the 48 500-cm^{-1} band. If the $^1(n_+, 3s)$ state is sufficiently stable, it readily should be seen as a two-photon resonance in the MPI spectrum. The excitations to $^3(n_+, \pi^*)$ and probably $^3(\pi, \pi^*)$ are observed at 22 200 and 39 000 cm^{-1} (*vert., 2.75 and 4.84 eV*) in azomethane [M72, M74, V12].

The trapped-electron spectrum of azomethane has a broad peak approximately 1500 cm^{-1} below the transition to $^1(n_+, 3p)$ which could be the triplet complement of this Rydberg band. Another prominent band in the trapped-electron spectrum at 48 500 cm^{-1} (*vert., 6.01 eV*) is at the proper frequency for assignment to the $^2(n_+, 3s^2)$ Feshbach resonance, however Mosher *et al.* [M72, M74] consider it to be a neutral-molecule transition instead.

Bands I and II [**II**, 69] in azo-*t*-butane come at 49 000 and 59 000 cm^{-1} (*vert., 6.08 and 7.31 eV*), whereas when the $(CH_3)_3C-$ groups are replaced by $(CH_3)_3Si-$ groups, the ionization potential drops by 9000 cm^{-1} (*1.1 eV*) and bands I and II shift downward to 40 000 and 51 000 cm^{-1} (*vert., 4.96 and 6.3 eV*) while maintaining their relative spacing [B50]. Note that the absorption spectrum of the silyl derivative was measured in pentane solution where Rydberg absorptions generally are not expected to survive.

The spectral effects of fluorinating the azo chromophore have been calculated by Vasudevan and Kammer [V13]. Fluorination of diimide to form *trans*-difluorodiazine (FN=NF) [**II**, 73] acts to *lower* the predicted $\pi \rightarrow \pi^*$ excitation to 68 100 cm^{-1} (*8.44 eV*), and Vasudevan and Kammer have assigned this to the peak observed at 66 000 cm^{-1} (*vert., 8.18 eV*) and earlier assigned as $n_+ \rightarrow \sigma^*$ in FN_2F [**II**, 73]. This lowering of the $\pi \rightarrow \pi^*$ frequency in the azo chromophore by *ca.* 25 000 cm^{-1} is quite unexpected in view of the zero shift of the $\pi \rightarrow \pi^*$ band observed on going from ethylene to *trans*-difluoroethylene and the strongly positive shift of the $\pi \rightarrow \pi^*$ frequency observed on going from ethylene to tetrafluoroethylene [**II**, 52]. The CI calculation shows that the (π, π^*) level in difluorodiazine is heavily mixed with $(n_+, 3s)$ and (n_+, σ^*) configurations, as often happens with (π, π^*) states in molecules bearing lone-pair electrons. In contrast, on going from the cyclic azo compound diazirine (N_2CH_2) to difluorodiazirine (N_2CF_2) [**II**, 73], the $\pi \rightarrow \pi^*$ vertical excitation (strongly mixed with $n_+ \rightarrow \sigma^*$) is predicted to move to higher frequency by 10 000 cm^{-1} [V13].

XI.B. Imines and Oximes

The imine group $R_2C{=}NR$ is closely related electronically to the olefin, keto and azo groups, but with an optical spectrum unlike any of them [**II**, 74]. New optical and photoelectron spectroscopic data recently has become available on three alkylated imines: $C_2H_5CH{=}NC_2H_5$, $(CH_3)_2CHCH{=}NC_2H_5$ and $(CF_3)_2C{=}NH$ [S17, V29]. These molecules display valence $n_N \rightarrow \pi^*$ excitations peaking in the 41 000–43 000-cm^{-1}

(*5.1–5.3-eV*) region followed by intense $\pi \rightarrow \pi^*$ excitations at *ca.* 58 000 cm^{-1} (*vert., 7.2 eV*) in the alkylated imines and at 69 000 cm^{-1} (*vert., 8.6 eV*) in the fluorinated system. The high intensity of the latter band ($\epsilon > 10\,000$) and its persistence in solution [N19] leaves little doubt that it involves a (π, π^*) valence upper state. If the carbon atom of the C=N group bears two alkyl groups, then the $\pi \rightarrow \pi^*$ transition in the vapor is observed at 55 200–58 100 cm^{-1} (*vert., 6.84–7.20 eV*), and at *ca.* 55 000 cm^{-1} (*vert., 6.8 eV*) in *n*-heptane solution [N19].

With $(n_N)^{-1}$ ionization potentials of *ca.* 75 000 cm^{-1} (*vert., 9.3 eV*) in the alkylated imines, the $n_N \rightarrow 3s$ and $n_N \rightarrow 3p$ transitions of these systems fall within the low-frequency and high-frequency wings of the intense $\pi \rightarrow \pi^*$ excitation and hence are not seen. Possible excitations to 3d appear at *ca.* 64 000 cm^{-1} (*vert., 7.9 eV*). Note however, that in $(CF_3)_2C{=}NH$ [ionization potential 95 100 cm^{-1} (*vert., 11.8 eV*)], the $n_N \rightarrow 3s$ and $n_N \rightarrow 3p$ excitation regions are relatively uncluttered, yet no bands corresponding to these excitations can be seen. By contrast, the corresponding $n_O \rightarrow 3s$ band of the isoelectronic molecule hexafluoroacetone is readily observed [**II**, 80]. With $(\pi)^{-1}$ ionization potentials of *ca.* 83 500 cm^{-1} (*vert., 10.4 eV*), the alkyl imines show prominent $\pi \rightarrow 3p$ excitations at *ca.* 64 000 cm^{-1} (*vert., 7.9 eV*), paralleling the situation in the azo chromophore. Spectra of conjugated polyimines are discussed in Section XVII.E.

Addition of the –OH or –OR groups to an aldehyde results in the corresponding carboxylic acid or ester and a characteristic shift of spectral levels. Since the imine group is isoelectronic with the keto group, the same sorts of shifts are expected when an –OH or –OR group is added to the nitrogen atom of an imine to form an oxime. [Oximes and amides form an isomeric pair.] In particular, based on the shifts observed on going from ketones to acids/amides (Chapter XV), it is expected in oximes that, *i*) the $(n_N)^{-1}$ and $(\pi)^{-1}$ ionization potentials will be much closer than in imines and may even invert their order, *ii*) as in amides and acids, the singlet manifold of oximes will be ordered as W, R_1, V_1, R_2, Q, *iii*) the $n_N \rightarrow \pi^*$ bands will be shifted strongly toward high frequencies in oximes compared with imines, and *iv*) the $\pi \rightarrow \pi^*$ bands will be shifted strongly toward low frequencies in oximes compared with imines. These expectations are realized qualitatively.

In the oximes formaldoxime ($H_2C{=}NOH$) and acetaldoxime ($H_3CHC{=}NOH$) investigated by Dargelos and Sandorfy [D4], the ionization potentials are inverted, with $(\pi)^{-1}$ *above* $(n_N)^{-1}$ by *ca.* 5500 cm^{-1}. The $n_N \rightarrow \pi^*$ transitions (W) in these oximes are observed at *ca.* 48 000 cm^{-1} (*vert., 6.0 eV*), Fig. XI.B-1, compared to 41 000 cm^{-1} (*vert., 5.1 eV*) in imines. Moreover, the $\pi \rightarrow \pi^*$ ($N \rightarrow V_1$) excitation is shifted from 58 000 cm^{-1} (*vert., 7.2 eV*) in imines to *ca.* 55 000 cm^{-1} (*vert., 6.8 eV*) in oximes. These shifts are in the expected directions, but are significantly smaller than in ketone/acid pairs. Possibly, this is owing to nonplanarity in the oximes.

Where are the R_1 ($\pi \rightarrow 3s$) and R_2 ($\pi \rightarrow 3p$) bands in the oximes? In formaldoxime, a band appears at 61 000 cm^{-1} (*advert., 7.56 eV*), Fig. XI.B-1, displaying the same 1250-cm^{-1} vibrations seen in the first photoelectron band [D4]. With respect to the $(\pi)^{-1}$ ionization potential at 85 660 cm^{-1} (*advert., 10.62 eV*), it has a term value of 24 600 cm^{-1}. This is a rather awkward value, since the (π, 3s) term value in this molecule should be close to that in formamide (29 200 cm^{-1}), or if the transition is to (π, 3p), then closer to 20 000 cm^{-1}. Because the similarly structured band in acetaldoxime has a term value of 20 170 cm^{-1}, we are inclined to go with a $\pi \rightarrow 3p$ assignment for this band. In the case of formaldoxime, the 61 000-cm^{-1} band has the requisite term value for assignment as the second excitation to 3s ($n_N \rightarrow 3s$, 29 000-cm^{-1} term value) and a parallel assignment is possible as well for the 61 000-cm^{-1} band of acetaldoxime. Structureless transitions at 67 900 and 70 400 cm^{-1} (*vert., 8.42 and 8.73 eV*) in formaldoxime have term values of 21 870 and 19 370 cm^{-1} with respect to the $(n_N)^{-1}$ ionization potential at 89 700 cm^{-1} (*vert., 11.13 eV*), and so are assigned as components of the $n_N \rightarrow 3p$ complex. A $\pi \rightarrow 3d$ Rydberg band is observed at 68 200 cm^{-1} (*vert., 8.46 eV*) in acetaldoxime.

Oximes in which the OH group is replaced by OR should maintain the spectral characteristics of the parent oximes. In $(CH_3)_3CCH{=}NOCH_3$ [D56, S17], Fig. XI.B-1, the W band is buried beneath the intense $\pi \rightarrow \pi^*$ band centered at 51 000 cm^{-1} (*vert., 6.3 eV*). Again, no Rydberg transition to 3s is seen, but $\pi \rightarrow 3p$ bands are observed at 56 200 (*vert., 6.97 eV*, 18 000 cm^{-1} term value) and 57 500 cm^{-1} (*vert., 7.13 eV*, 16 700 cm^{-1} term value). Two further broad bands at 63 300 and 69 000 cm^{-1} (*vert., 7.85 and 8.55 eV*) have term values of 18 200 and 12 500 cm^{-1}, respectively, with respect to the $(n_N)^{-1}$ ionization potential at 81 500 cm^{-1} (*10.1 eV*), and so are assigned as $n_N \rightarrow 3p$ and $n_N \rightarrow 3d$. The spectra of conjugated oximes are discussed in Section XVII.E.

We conclude that absorption in the chromophoric systems $R_2C{=}NR$ and $R_2C{=}NOR$ parallel those in ketones and amides, respectively, except that in imines the $\pi \rightarrow \pi^*$ band is below the lowest ionization potential, but above it in ketones, and in oximes and imines the lower transitions to 3s are not observed, whereas in amides they are readily identified.

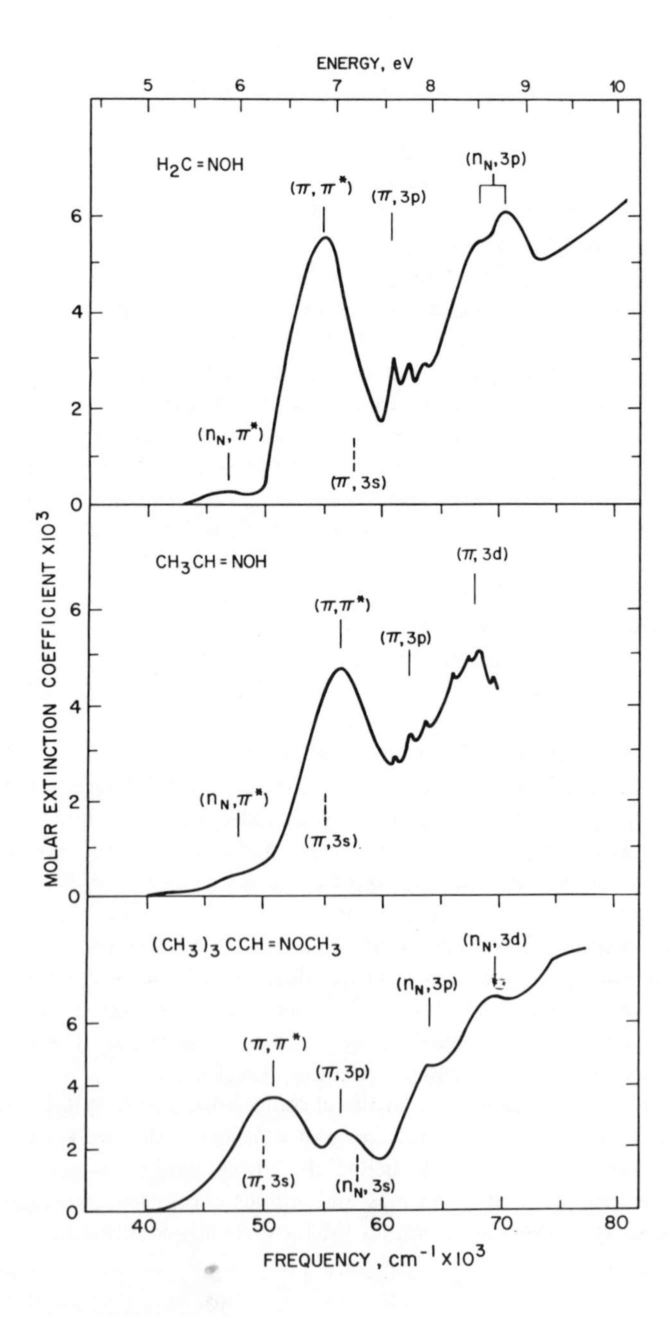

Fig. XI.B-1. Absorption spectra of assorted oximes in the vapor phase [D4, D56]. The assignments indicated by dashed lines are for bands anticipated but not yet observed.

CHAPTER XII

Aldehydes and Ketones

XII.A. Formaldehyde

Spectroscopic work on the formaldehyde molecule [**II**, 82, 327] moves forward on both the experimental and theoretical fronts. The long history of the spectroscopy of this molecule to 1975 is presented in a detailed review by two prominent workers in the field, D. C. Moule and A. D. Walsh [M75] and is updated to 1977 in [L20].

For formaldehyde, it has been calculated [H14] that $<r^2>$ for the 3s orbital in the $^1(2b_2, 3s)$ configuration is 44 a_o^2, whereas it is 62 a_o^2 for 3p in $^1(2b_2, 3p)$ and 115 a_o^2 for 3d in $^1(2b_2, 3d)$. (The $2b_2$ MO of formaldehyde is the uppermost lone-pair orbital on the oxygen atom). For comparison, the value of $<r^2>$ for a hydrogen atom in $n = 3$ is 126 a_o^2. Clearly, the decreasing penetration of the Rydberg orbital with increasing l increases its size. The large size of the molecular Rydberg orbital in general makes it highly susceptible to perturbation by solvent molecules. In a nice example of this, Messing *et al.* [M54, M55] have shown that for formaldehyde as a solute in liquid argon, the $2b_2 \rightarrow 3s$ Rydberg transition is vibronically sharp at a density of 0.17 gm/cm^3, whereas at a density of 0.94 each of the vibronic components has broadened so strongly to the high-frequency side that no structure is visible, Fig. II.D-2. Further increase of the density to 1.4 shifts the entire band to yet higher frequencies. Electric-field studies on the $2b_2 \rightarrow 3s$ origin in formaldehyde vapor (57 310 cm^{-1}, *7.105 eV*) [C13, C14] show the transition to be polarized perpendicular to the C–O axis.

In a C_{2v} molecule such as formaldehyde, the transitions from $2b_2$ to 3p orbitals are allowed to $3pyb_2$ (y is in-plane, perpendicular to the C–O line) and to $3pza_1$ orbitals (z is along the C–O line), whereas the transition to $3pxb_1$ (x is out-of-plane) is electric-dipole forbidden. Lessard and Moule [L19] first identified the forbidden $2b_2 \rightarrow 3px$ transition in formaldehyde as having two false origins induced by the nontotally symmetric vibrations ν_5' and ν_6'. The bands are polarized out-of-plane as deduced from the rotational analyses. The same band system also has been observed in a high-resolution electron impact experiment by Taylor *et al.* [T7], who state that the variation of intensities of these transitions with impact variables labels them as electric-quadrupole in nature. Thus they place the $2b_2 \rightarrow 3px$ origin at 67 540 cm^{-1} (*8.374 eV*, Q_1 in Fig. XII.A-1), and assign the second band of Lessard and Moule (Q_2 in Fig. XII.A-1) as a quantum of the totally symmetric ν_3' vibration rather than as a second origin. The confirmation of the frequency of the $2b_2 \rightarrow 3px$ origin in formaldehyde most easily could be achieved by observing its two-photon resonance in an MPI experiment, for it will be strongly allowed under these circumstances, and the vibronic analysis should be straightforward. Just as observed for the $2b_2 \rightarrow 3pxb_1$ transition, the $2b_2 \rightarrow 3db_1$ band also is forbidden in the optical spectrum, but is detected as an electric-quadrupole transition in the electron-scattering spectrum, Table XII.A-I [T7].

In a high-resolution optical study of formaldehyde beyond 77 000 cm^{-1} (*9.5 eV*), Drury-Lessard and Moule [D47] report many sharp-line features which lie above the Rydberg transitions terminating at $n = 3$. Originating at $2b_2$, these were assigned to the following series: ns ($n \leqslant 5$), np ($n \leqslant 11$), nd ($n \leqslant 7$) and nf ($n \leqslant 12$). The nf series, with a quantum defect of less than 0.10, is rarely seen in polyatomic molecules. (However, see butadiene, Section XVII.A.) The $2b_2 \rightarrow n$f transitions of formaldehyde would be forbidden by the $\Delta l = \pm 1$ rule if $2b_2$ were a pure 2p orbital, but due to the antibonding interaction of $2b_2$ with the C–H bonds, it has partial "3d" character and d → f satisfies the selection rule on Δl [D47].

Harding and Goddard [H14] and Nakatsuji *et al.* [N10] calculate some rather useful aspects of the formaldehyde spectrum. Thus, in the $^1(2b_2, 3p)$ manifold the core splitting amounts to 1900 cm^{-1} and the transition to 3py is most intense. The splitting pattern is just that

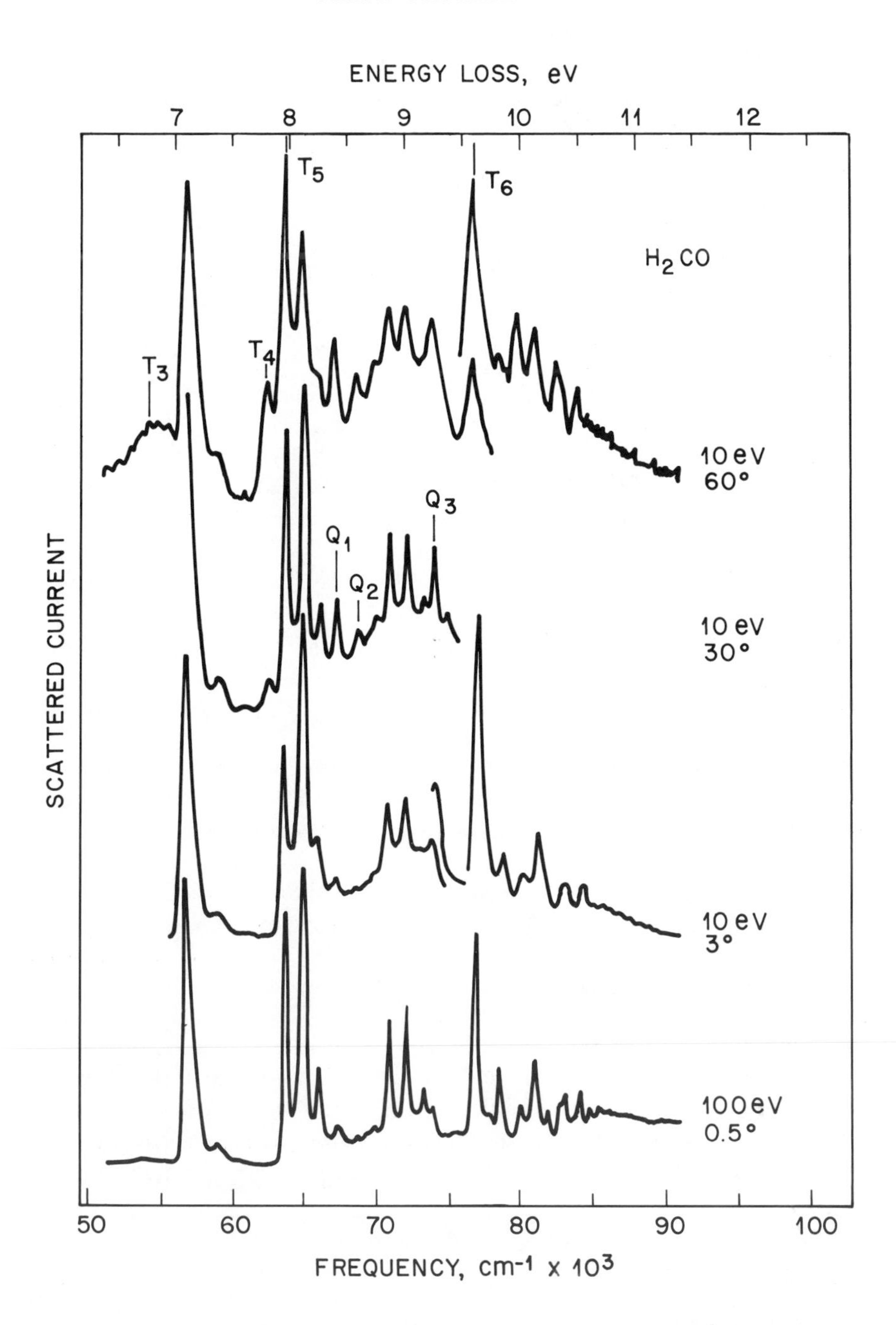

Fig. XII.A-1. Energy loss spectrum of formaldehyde taken under various combinations of impact energy and scattering angle [T7].

TABLE XII.A-I

TERM VALUES IN THE RYDBERG AND VALENCE SINGLET SPECTRA OF SELECTED ALDEHYDES AND KETONES

Ionization Potential, cm^{-1} $(\phi_i)^{-1}$	Term Values, cm^{-1} $(\phi_i, 3s)$	$(\phi_i, 3p)$	$(\phi_i, 3d)$	(ϕ_i, π^*)	$(\phi_i, \phi_j)^a$	References
		FORMALDEHYDE				
87 787 $(2b_2)^{-1}$	30 477	23 520 22 127 20 250	16 187 13 410	55 000	− 36 000	[C14, L19, L20, M75, T7]
116 100 $(1b_1)^{-1}$	30 600	21 700	12 900	28 800[b]	-	[C38, H14, L4]
2.3751×10^6 $(1s_C)^{-1}$	34 800	26 300 22 800 19 100	14 400	68 200	− 51 900	[H64]
4.3510×10^6 $(1s_O)^{-1}$	32 000	26 700	14 500	69 700	− 36 800	[H64]
		ACETALDEHYDE				
82 504 $(n_O)^{-1}$	27 508	22 334 20 000	14 474	48 000	-	[C14, E5, H30, T1]
2.3486×10^6 $(1s_C)^{-1}$ [c]	-	23 300 17 600	13 600	-	− 30 800	[H64]
2.3699×10^6 $(1s_C)^{-1}$ [d]	29 900	23 600 21 200	12 900	60 700	− 57 900	[H64]
4.3428×10^6 $(1s_O)^{-1}$	26 800	20 900	14 100	58 800	− 28 800 − 49 000	[H64]
		ACETONE				
78 420 $(n_O)^{-1}$	27 150	18 310	13 170	42 700	-	[C14, D35, H47, S73, T2]
112 000 $(\pi)^{-1}$	27 500	-	13 000	-	-	[II, 87]
143 000 $(\sigma)^{-1}$	26 200	-	-	-	-	[II, 87]
2.3478×10^6 $(1s_C)^{-1}$ [c]	-	21 900 16 800	-	-	− 46 100	[H64, W21]
2.3690×10^6 $(1s_C)^{-1}$ [d]	25 700	18 600	14 600	55 800	− 64 400	[H64, W21]
4.3386×10^6 $(1s_O)^{-1}$	-	17 800	-	53 300	− 59 600	[H64, W21]

[a] Term values for apparent shape resonances (Section II.C).

[b] Calculated value [C38].

[c] Transitions originating at 1s on the carbon atom of the methyl group.

[d] Transitions originating at 1s on the carbon atom of the carbonyl group.

anticipated from the spatial anisotropy of the molecule, *i.e.*, $3pza_1$ is lowest and $3pxb_1$ is highest [Y1, H14]. The 3d manifold is split by 1400 cm^{-1} and all Rydberg singlet-triplet splits are calculated to be less than 800 cm^{-1}. Continua observed at 85 500, 94 400 and 103 200 cm^{-1} (*vert., 10.6, 11.7 and 12.8 eV*) in formaldehyde are calculated to have (π, 3s), (π, 3p) and (π, 3d) upper states; the corresponding term values computed with respect to the $(1b_1\pi)^{-1}$ ionization potential at 116 000 cm^{-1} (*vert., 14.39 eV*) strongly support these assignments. Miller [M61] has calculated the dependence of the generalized oscillator strength upon momentum transferred on electron impact for the lower Rydberg states of formaldehyde as well as the optical oscillator strengths for Rydberg excitations originating at $2b_2$ and $1b_1$ MO's. These calculated oscillator strengths are in generally good agreement with experiment.

The valence TNI resonance (Section II.B) of formaldehyde involving the π^* MO comes at 6900 cm^{-1} (*vert., 0.86 eV*), and as with most π^* resonances, the frequency increases with methylation [J16, V10]. The Feshbach resonance of formaldehyde to the $^2(2b_2, 3s^2)$ configuration has not been reported as yet, but is expected to have an intense origin at 53 000 cm^{-1} (*6.57 eV*). Nakatsuji *et al.* [N10] have calculated several low-lying TNI resonances to fall in the 0–9800-cm^{-1} (*0–1.2-eV*) region of formaldehyde, however they assign the terminating MO's as Rydberg orbitals and this is contrary to our point of view (Section II.B). A possible resonance at 68 000 cm^{-1} (*vert., 8.4 eV*) in formaldehyde [V10] is close to that calculated for attachment of an electron into the σ^*(C–H) MO [N10].

The most pressing problems in the spectrum of formaldehyde are those of the transition frequencies to the $^1(\pi, \pi^*)$ and $^1(n_O', \pi^*)$ states.† Less pressing but no less interesting are the questions as to where the valence transitions to the various σ^* MO's might fall. *Ab initio* calculations are now very sophisticated and the most recent of these agree on the nature of the $^1(\pi, \pi^*)$ state of formaldehyde, though not on its frequency. The most impressive calculation is that of Colle *et al.* [C38], who account for 95 % of the correlation energy in the relevant states. They predict that the $\pi \rightarrow \pi^*$ excitation has a frequency of 87 270 cm^{-1} (*vert., 10.82 eV*) and that the π^* orbital composition is 85 % valence/15 % Rydberg in this

† n_O' is the second nonbonding MO ($5a_1$) of the keto group, having its unique axis along the C–O line.

state. An essentially identical result is obtained by Nakatsuji *et al.* [N10]. The $<r^2>$ value for the π^* orbital in the $^1(\pi, \pi^*)$ state is calculated to be 23.2 a_0^2, whereas those of the valence π and the Rydberg 3d orbitals are 3.1 and 115 a_0^2, respectively. In contrast, Langhoff and coworkers [L3, L4] calculate that the $\pi \rightarrow \pi^*$ excitation of formaldehyde correlates with the finely-structured energy loss at 106 000 cm^{-1} (*13.1 eV*), which we earlier assigned as $5a_1 \rightarrow 3p$ [**II**, 80]. Because the $^1(\pi, \pi^*)$ state is calculated to fall above the $(2b_2)^{-1}$ ionization potential and is autoionized by its continuum, the transition from the ground state to $^1(\pi, \pi^*)$ should be very broad [H14, L6]. Adding to the experimental difficulty in finding the transition to $^1(\pi, \pi^*)$ is the theoretical prediction that it is unbound along the C—O coordinate [M75].

One would like more corroboration for the assignment of the $n_O' \rightarrow \pi^*$ valence transition in formaldehyde to a set of bands in the 69 400–73 000-cm^{-1} (*8.60–9.05-eV*) region [M75]. A valence assignment is suggested for these bands [M55] because they broaden and shift to lower frequency under external perturbation rather than to higher frequency as most Rydberg excitations do. Though support for an $n_O' \rightarrow \pi^*$ assignment of the bands at 69 400–73 000 cm^{-1} comes from the theoretical calculations of Langhoff *et al.* [L3, L4] and of Nakatsuji *et al.* [N10], both groups calculate an oscillator strength of only 0.0048 or less for this transition. On the other hand, the term values suggest $2b_2 \rightarrow 3d$ assignments for the bands observed in this region.

Yet another open question in the formaldehyde spectrum is the location of the $2b_2 \rightarrow a_1\sigma^*$ valence excitation conjugate to the $2b_2 \rightarrow 3s$ Rydberg band, if indeed there is a state with enough $(2b_2, a_1\sigma^*)$ character to be called by that name. In our earlier work [**II**, 81], it was surmised that the valence conjugate is near-degenerate with $(2b_2, 3s)$ in ketones and aldehydes and affects it in a way not experienced by the other Rydberg states. In this regard, it is interesting to note the comment of Moule *et al.* [L20, M75] that the vibrational frequencies in the $^1(2b_2, 3s)$ state are very different from those in all other Rydberg states; a large but selective mixing of $(2b_2, \sigma^*)$ with $(2b_2, 3s)$ solely would manifest itself as aberrant vibrational frequencies in the $(2b_2, 3s)$ state. Though there still may be a germ of truth in the above argument, the initial premise that $(2b_2, 3s)$ and its conjugate configuration are near-degenerate is refuted by the theoretical work of Langhoff *et al.* [L3, L4], who compute that all the valence transitions to $5a_1\sigma^*$(C—O) appear as shape resonances in the 97 000–129 000-cm^{-1} (*12–16-eV*) region of the formaldehyde spectrum.

Calculations on formaldehyde [C38, H14, Y1] show that the two lowest singlet $\rightarrow$ triplet promotions are $n_O \rightarrow \pi^*$ and $\pi \rightarrow \pi^*$, Fig. XII.A-2, and

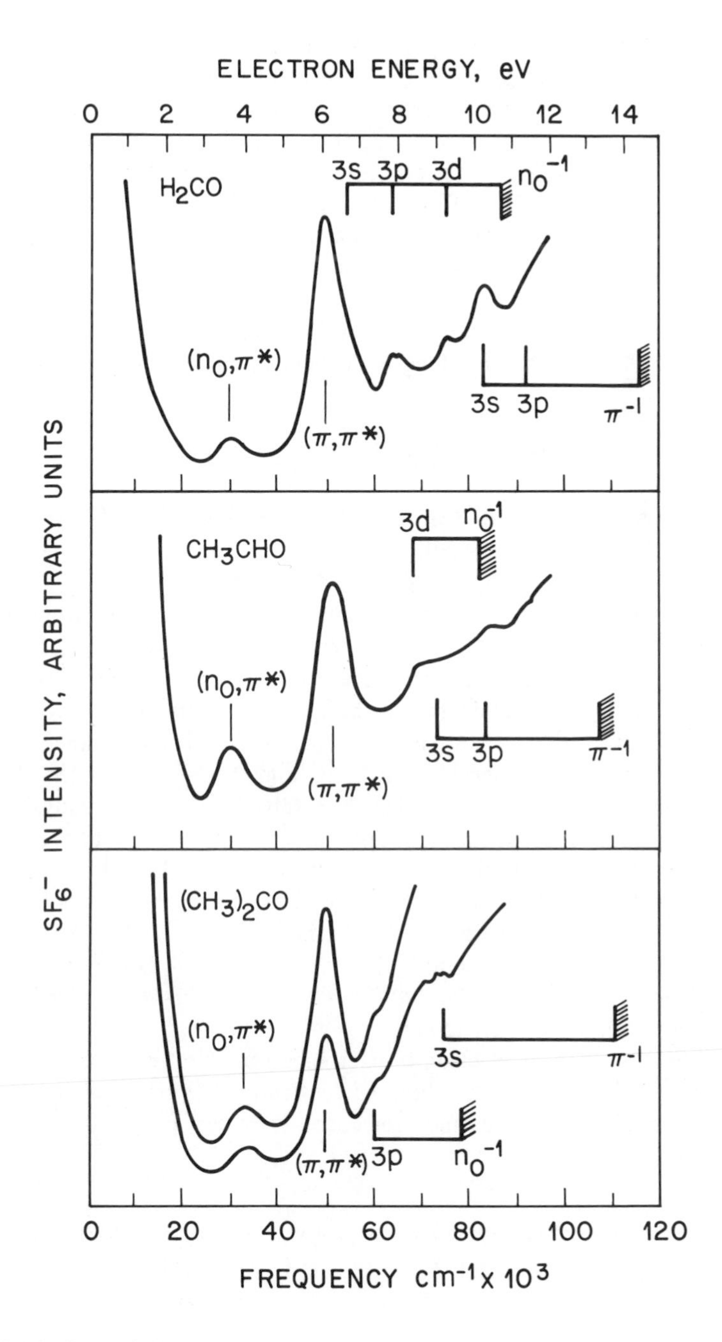

Fig. XII.A-2. Trapped-electron spectra of selected aldehydes and acetone interpreted in terms of excitations to triplet valence and Rydberg states [S69].

this will hold for most aldehydes and ketones. A number of the triplets observed in the trapped-electron spectrum of formaldehyde [S69], Fig. XII.A-2, have been observed at considerably higher resolution in electron impact by Taylor *et al.* [T7], Fig. XII.A-1. These workers are the first to resolve a long progression in the C—O stretch (855 cm^{-1}) in the transition to $^3(\pi, \pi^*)$. This and other singlet → triplet excitation frequencies for formaldehyde are listed in Table XII.A-II, where it is seen that the next six triplet states have been assigned Rydberg configurations originating at $2b_2n_O$ or $1b_1\pi$ MO's.

The inner-shell spectrum of formaldehyde and its term values are discussed in Section XII.B, along with those of acetaldehyde and acetone.

XII.B. Alkyl and Halogen Derivatives

The proposals of Miller [M60] and of Kim *et al.* [K17] that a plot of generalized oscillator strength versus momentum transfer on electron impact will show minima for electric-dipole allowed Rydberg transitions has been tested and confirmed for the $n_O \rightarrow 3s$, 3p and 3d transitions of acetaldehyde [A22].† The $n_O \rightarrow 3s$ transitions of acetaldehyde [**II**, 75] and its deuterated analogs recently were studied as two-photon resonances using the multiphoton ionization technique [E5, H30], Section II.A. This two-photon excitation is very intense due to the near-resonance with the (n_O, π^*) state at the one-photon level, and closely resembles the one-photon spectrum, but is much better resolved. Vibronic analysis shows that the $^1(n_O, 3s)$ state has a planar *s-cis* configuration and a methyl-group torsional barrier of approximately 760 cm^{-1} [G30]. As expected group theoretically, the $n_O \rightarrow 3s$ transition of acetaldehyde is polarized in-plane [C14]. The spectra of the higher aldehydes, Fig. XII.B-1, are closely related to that of acetaldehyde, except that the larger systems show broad continua in the 70 000–80 000-cm^{-1} (*8.8–10-eV*) region owing to transitions within the alkyl groups.

As the resolution of electron-impact spectrometers increases, the results have come to rival those obtained optically. Optical-quality electron impact spectra are reported for acetaldehyde, propionaldehyde and

† The $2b_2$ lone-pair orbital of formaldehyde is written as n_O in the higher aldehydes and ketones.

TABLE XII.A-II.

FREQUENCIES OF TRIPLET STATES IN SELECTED ALDEHYDES AND KETONES, cm^{-1} (*vert., eV*)

	F_2CO[a]		H_2CO[b]		CH_3CHO[c]		$(CH_3)_2CO$[c]	
$^3(n_O, \pi^*)$	48 400	(*6.0*)	28 250	(*3.502*)	30 250	(*3.75*)	33 470	(*4.15*)
$^3(\pi, \pi^*)$	48 400	(*6.0*)	47 260	(*5.859*)	51 600	(*6.40*)	50 400	(*6.25*)
$^3(n_O, 3s)$	66 900	(*8.3*)	55 060	(*6.827*)[d]	-		50 400	(*6.25*)
$^3(n_O, 3p)$	-		62 830	(*7.790*)[e]	-		60 500	(*7.50*)
			64 160	(*7.955*)[f]				
$^3(n_O, 3d)$	-		-		68 560	(*8.50*)	-	
$^3(n_O, 4p)$	-		77 340	(*9.589*)[g]	-		-	
$^3(\pi, 3s)$	83 100	(*10.3*)	82 700	(*10.2*)	-		75 000	(*9.30*)
$^3(\pi, 3p)$	99 200	(*12.3*)	91 100	(*11.3*)	83 900	(*10.4*)	-	
$^3(\pi, 3d)$	105 700	(*13.1*)	-		-		-	

[a] References [D40, V14].

[b] References [C24, S69, T7].

[c] Reference [S69] and Fig. XII.A-2.

[d] Band T_3 in Fig. XII.A-1.

[e] Band T_4 in Fig. XII.A-1.

[f] Band T_5 in Fig. XII.A-1.

[g] Band T_6 in Fig. XII.A-1.

isobutyraldehyde [**II**, 86], Fig. XII.B-1, and for various ketones [D35, S73, T1, T2]. In the $n_O \rightarrow 3p$ transitions of acetone, two components are allowed and one is forbidden. Only one component is observed[†] [D35] and

[†] Actually, Tam and Brion [T2] report a second $(n_O, 3p)$ component 1000 cm^{-1} above the first, and this is within the calculated 3p core splitting [H47]; however Doering and McDiarmid [D35] assign it instead to a vibronic component of the lowest excitation to $(n_O, 3p)$.

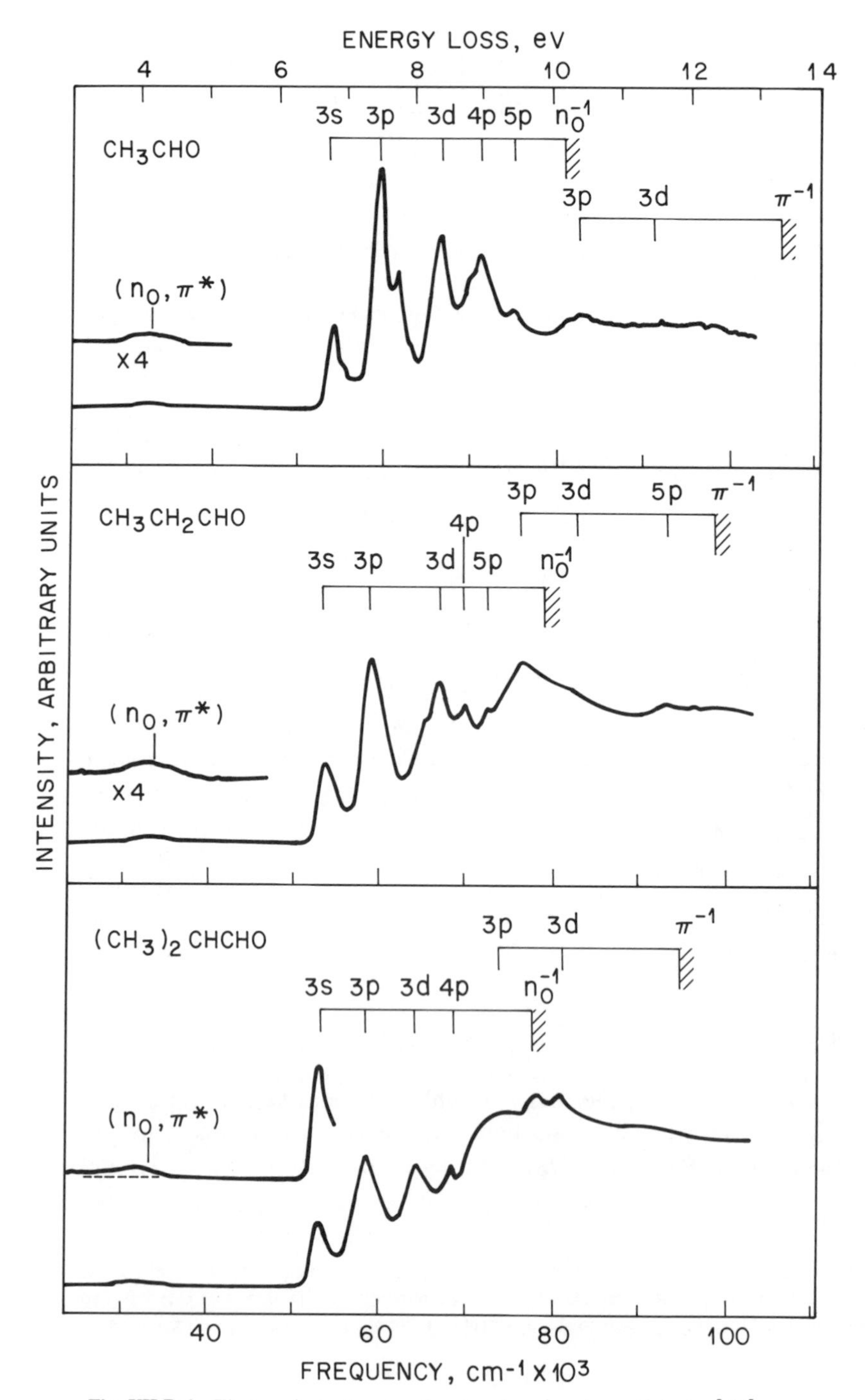

Fig. XII.B-1. Electron impact energy loss spectra of various aldehydes [T1].

surprisingly, it is said to be the forbidden component to 3p*x*! This conclusion is based on careful analyses of the vibronic structures in acetone-h_6 and -d_6, which are said to show both true and false origins in the electron impact spectra. If this is so, then the appearance of the "true origin" of the forbidden excitation likely is due to an electric-quadrupole mechanism as in formaldehyde. Note however, that in formaldehyde [L19, T7], the $n_O \rightarrow 3px$ transition in electron impact appears as an electric-quadrupole-allowed origin accompanied by quanta of totally symmetric vibrations, rather than accompanied by a false origin promoted by a nontotally symmetric vibration. It is especially interesting to see that according to calculation [H47], the out-of-plane hydrogen atoms of acetone act to invert the 3p core-splitting pattern with respect to that in formaldehyde (Section I.B). The $\pi \rightarrow \pi^*$ excitation of acetone is calculated to fall in the 80 000–90 000-cm^{-1} (*9.9–11.1-eV*) region [H47], and as in formaldehyde, is beyond the $(n_O)^{-1}$ ionization limit and would be autoionized. Using the elegant electric-field techniques on the Rydberg absorption in acetone, Scott and Russell [S31] have demonstrated that the polarization of the 51 270-cm^{-1} (*6.353-eV*) band is perpendicular to the C–O line, as expected for an $n_O \rightarrow 3s$ promotion. *Ab initio* calculations on acetone [H47] confirm the Rydberg assignments presented in [**II**, 80].

In the lower symmetry of methyl ethyl ketone [**II**, 90], the component of $n_O \rightarrow 3p$ which is forbidden in acetone becomes formally allowed and the transitions from n_O to all three of the 3p levels are observed within a span of 1700 cm^{-1} in this molecule [D35, T2]. In the mixed-chromophoric system $CH_3OCH_2COCH_3$ [O9], both of the oxygen lone-pair orbitals are nearly degenerate, and the only distinct feature in the absorption spectrum is an overlapping of excitations to 3s in the 50 000–54 000-cm^{-1} (*6.2–6.6-eV*) region. The spectrum of the thiol-ketone analog $CH_3SCH_2COCH_3$ shows the two lowest transitions to *n*s separated by 7000 cm^{-1}.

In [O1] there are posted $n_O \rightarrow 3s$ and $n_O \rightarrow 3p$ oscillator strengths for a large number of ketones. These values [0.008–0.04 for transitions to $(n_O, 3s)$; 0.005–0.16 for transitions to $(n_O, 3p)$] are within the limits of the Rydberg intensity rule quoted previously [**I**, 29]. In looking at a number of highly alkylated ketones, Tam and Brion [T2] propose that the oscillator strengths for $n_O \rightarrow 3p$ transitions are larger than those for $n_O \rightarrow 3s$ if the alkyl group is branched at the α-carbon atom, but smaller if there is no branching at this position. The optical data of Table IV.C-II in reference [**II**, 83] argue against this.

Electric-field experiments [C15] on the series of cyclic ketones from cyclobutanone to cycloheptanone [**II**, 92] show that the $n_O \rightarrow 3s$

excitations in all of them are polarized perpendicular to the C—O line, as predicted group theoretically. The larger ketones in the series have larger changes of dipole moment in the (n_O, 3s) upper states owing to the extension of the 3s orbitals resulting from their orthogonality to the core orbitals [I, 19]. Vibronic analysis of the $n_O \rightarrow 3s$ transition in cyclobutanone shows prominent sequences of ν_{20} ring puckering and a planar C_4 ring in the upper state [D48].

The spectra of the halogenated ethylenes, Section X.C, exhibit two interesting types of valence excitation:- *i*) $\mathscr{A}(n\text{p}, \sigma^*)$ bands in which an electron is excited from the halogen $n\text{p}_X$ orbital into σ^*(C—X), and *ii*) $\mathscr{A}(\pi, \sigma^*)$ bands in which an electron is excited from the π(C=C) MO into σ^*(C—X), Section I.C. Generally, the $\mathscr{A}(\pi, \sigma^*)$ bands are found at lower frequencies than the $\mathscr{A}(n\text{p}, \sigma^*)$ bands. Irrespective of the halogen X and the degree of halogenation, the $\mathscr{A}(\pi, \sigma^*)$ and $\mathscr{A}(n\text{p}, \sigma^*)$ term values are very constant at 29 500 cm^{-1} (Table X.C-I). Let us now search for $\mathscr{A}$ bands in the halogenated derivatives of the keto chromophore. In the spectrum of phosgene (Cl_2CO), a broad band is centered at 65 000 cm^{-1} (*8.1 eV*) having a peak molar extinction coefficient of approximately 20 000. Because its term value is too large for a $\pi \rightarrow 3s$ assignment, it was assigned as $\pi \rightarrow \pi^*$ by default [II, 99]. Note however that the $(\pi)^{-1}$ ionization potential of 95 400 cm^{-1} (*vert., 11.83 eV*) in Cl_2CO leads to a 30 400 cm^{-1} term value for the 65 000-cm^{-1} band, suggesting an alternate assignment as an $\mathscr{A}(\pi, \sigma^*)$ band. The $\pi \rightarrow \sigma^*$ and $\pi \rightarrow \pi^*$ transitions possibly can be distinguished by their polarizations, *i.e.*, out-of-plane and in-plane, respectively. According to [II, 100], the only other band like this in ketone spectra is that at 67 570 cm^{-1} (*vert., 8.377 eV*) in acetyl chloride. With a $(\pi)^{-1}$ ionization potential of 96 800 cm^{-1} (*vert., 12.00 eV*) [C18], this band of acetyl chloride has a 29 200-cm^{-1} term value, again suggesting that it is an $\mathscr{A}(\pi, \sigma^*)$ excitation. The $\mathscr{A}(3\text{p}, \sigma^*)$ band of Cl_2CO is expected at 72 000 cm^{-1}, where it might rest beneath the more intense $n_O \rightarrow 3p$ bands. The $\mathscr{A}(\pi, \sigma^*)$ band of F_2CO is expected at 88 000 cm^{-1} (*vert., 11 eV*), in a spectral region in which this compound has yet to be investigated.

Reasonable assignments of high-lying molecular triplet states to Rydberg upper-state configurations often are accomplished by first estimating the frequencies of the relevant singlet Rydberg excitations [I, 65] using term-value arguments, and then applying a singlet-triplet splitting of *ca.* 300—4000 cm^{-1} depending upon the Rydberg penetration (Section I.B). As with the singlet manifold, this approach leaves very few triplet excitations unassigned. Trapped-electron spectra of formaldehyde, acetaldehyde and acetone [S69, V10] reveal many high-lying triplets, the lowest of which also are observed by nonthreshold electron impact [C24].

The various higher singlet → triplet excitations of acetaldehyde and acetone [S69] are readily assigned to Rydberg configurations, Table XII.A-II, using the term value and singlet-triplet splitting arguments quoted above, whereas the first two are valence transitions to $^3(n_O, \pi^*)$ and $^3(\pi, \pi^*)$ [N10]. The triplet assignments in the higher aldehydes are derived directly from those of formaldehyde, Table XII.A-II.

Using the trapped-electron technique to detect excited-state triplets, Doiron and McMahon [D40] assign a triplet at 48 400 cm^{-1} (*vert., 6.00 eV*) in carbonyl difluoride as $^3(\pi, \pi^*)$, Fig. XII.B-2; assignment to $^3(n_O, \pi^*)$ corresponding to $^1(n_O, \pi^*)$ at 50 000 cm^{-1} (*vert., 6.2 eV*) [**II**, 96] is less likely. A weak shoulder in this spectrum at 66 900 cm^{-1} (*vert., 8.29 eV*) could well have the $^3(n_O, 3s)$ upper-state configuration, with the transition to $^1(n_O, 3s)$ coming appropriately at 67 900 cm^{-1} (*vert., 8.42 eV*). While this assignment gives a singlet-triplet split of the expected size, it also results in an $^1(n_O, 3s)$ term value of 41 800 cm^{-1}, which is unexpectedly large, even for a fluorinated system. Three further triplets are reported in F_2CO at 83 100, 99 200 and 106 000 cm^{-1} (*vert., 10.3, 12.3 and 13.1 eV*). Inasmuch as they have term values of 34 700, 18 600 and 11 800 cm^{-1} with respect to the $(\pi)^{-1}$ ionization potential at 117 800 cm^{-1} (*vert., 14.60 eV*), they are assigned $^3(\pi, 3s)$, $^3(\pi, 3p)$ and $^3(\pi, 3d)$ configurations. (See Table XII.A-II for a comparison with formaldehyde).

Trapped-electron spectra of 1,1,1-trifluoroacetone and of hexafluoroacetone [**II**, 80] reveal a number of high-lying triplet states [D40]. What are apparently the transitions to $^3(\pi, \pi^*)$ are observed at 50 800 cm^{-1} (*vert., 6.30 eV*) in trifluoroacetone and at 49 200 cm^{-1} (*vert., 6.10 eV*) in hexafluoroacetone. These triplets are followed by another at 64 500 cm^{-1} (*vert., 8.00 eV*) in the trifluoro derivative and at 70 200 cm^{-1} (*vert., 8.70 eV*) in the hexafluoro derivative. With respect to the first ionization potentials [Y3], these triplets have term values of only 24 200 and 27 300 cm^{-1}, respectively, which are somewhat low for $^3(n_O, 3s)$ excitations in fluorinated systems, and certainly too high for $^3(n_O, 3p)$.† The band at 95 200 cm^{-1} (*vert., 11.80 eV*) in trifluoroacetone does have the appropriate term value (16 800 cm^{-1}) with respect to the $(\pi)^{-1}$ ionization potential at 112 000 cm^{-1} (*vert., 13.89 eV*) to be assigned as

† Actually, since the singlet-triplet split of Rydberg configurations scales with the magnitude of the penetration of the Rydberg orbital, unusually large triplet term values may occur in heavily fluorinated systems where the Rydberg penetration is large.

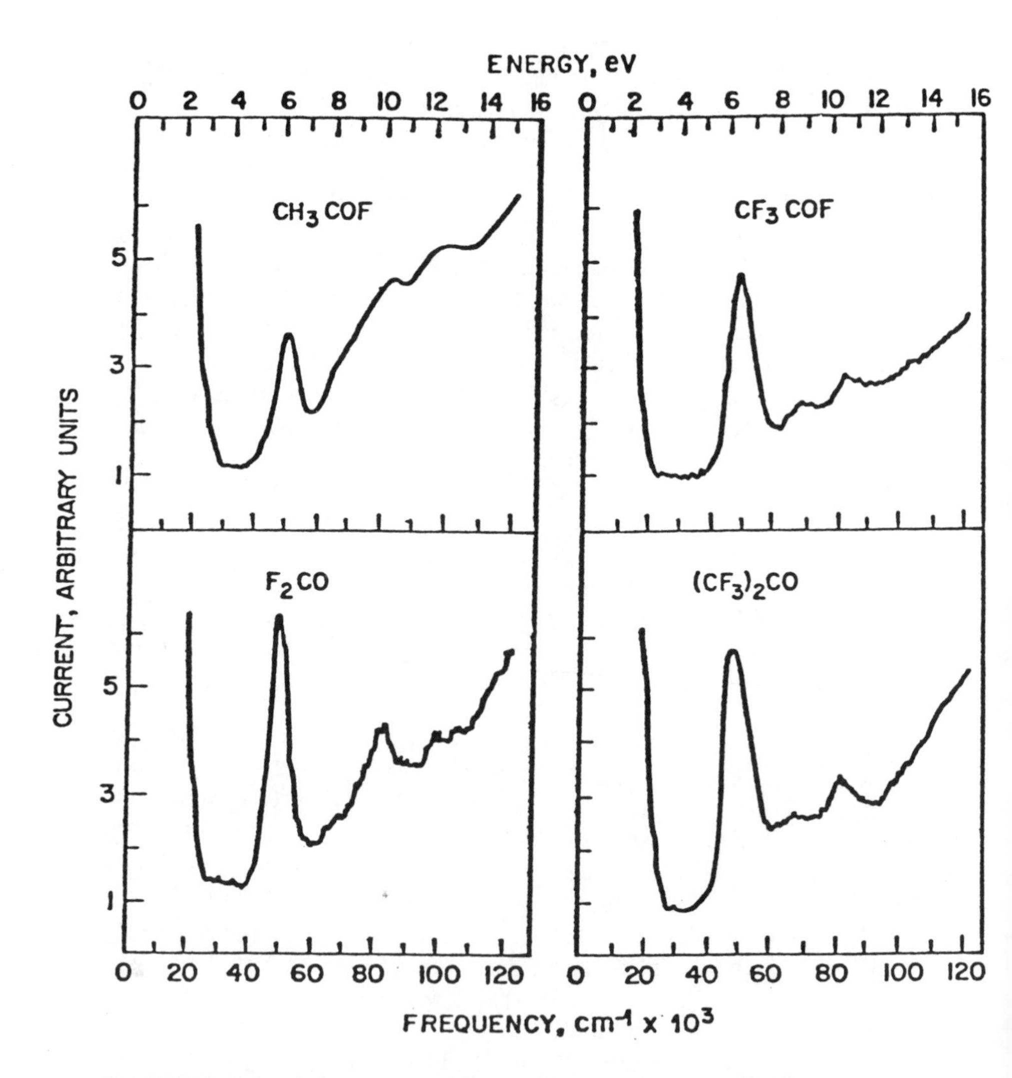

Fig. XII.B-2. Trapped-electron spectra of various fluorinated carbonyls [D40].

$^3(\pi, 3p)$, while that at 104 000 cm^{-1} (*vert.*, *12.9 eV*, term value of 12 000 cm^{-1}) would seem to be the transition from a methyl-group orbital (116 000 cm^{-1}, *14.4 eV* ionization potential) to 3d. A band at 83 900 cm^{-1} (*vert.*, *10.4 eV*) in the trapped-electron spectrum of hexafluoroacetone is assigned as $^3(n_O, 3d)$ based on its 13 600-cm^{-1} term value.

The $1s_C$ and $1s_O$ inner-shell spectra of formaldehyde commence with intense bands, assigned as $1s \rightarrow \pi^*$ inner-shell valence excitations [H64]. One sees from Table XII.A-I that the (ϕ_i^{-1}, π^*) inner-shell term values are very nearly equal when the hole is in $1s_C$ and $1s_O$ for formaldehyde, but is about 13 900 cm^{-1} smaller when the hole is in the n_O outer-shell orbital. Since the π^* MO is localized largely on the C–O fragment, excitations to it from the $1s_C$ orbitals on the methyl groups of acetaldehyde and acetone have no detectable intensity. In both acetaldehyde and acetone, the term values of the $(1s_C, \pi^*)$ and $(1s_O, \pi^*)$ excited states are again equal, and about 11 800 cm^{-1} smaller than those of the (n_O, π^*) valence states. The $\pi \rightarrow \pi^*$ shake-up frequencies in the $(1s_C)^{-1}$ and $(1s_O)^{-1}$ XPS of formaldehyde are interpretable in terms of the AO compositions of the π and π^* MO's [B16].

Consideration of the experimental inner-shell term values shows that the π^* orbital is more tightly bound (*ca.* 13 000 cm^{-1}) when the complementary open shell is either $1s_C$ or $1s_O$, than when it is n_O. This follows naturally from Slater's rules for the effective nuclear charge [E15], wherein a 1s electron is far more effective in shielding an outer electron from the core than is one in 2p (n_O). This exhaltation of the (ϕ_i, π^*) term values in the inner-shell spectra of aldehydes and ketones is similar to that found for inner-shell excitations to σ^* in alkyl halides (Section I.C and Table I.C-II). Note also that as formaldehyde is alkylated, the (ϕ_i, π^*) term value decreases in a regular way, showing that the alkyl groups destabilize the π^* MO. On going from formaldehyde to acetone, the π MO is destabilized by 15 300 cm^{-1} according to the photoelectron spectra [K18]; the inner-shell spectra terminating at π^* show that the π^* destabilization is 12 300 cm^{-1} in this series, while the π^* orbital electron-affinity difference amounts to 5200 cm^{-1} [J16, V10]. The quantitative differences between these figures result from the fact that all the $1s \rightarrow \pi^*$ excitations in the inner-shell spectra involve compact π^* MO's, much like the compact π MO's involved in the photoelectron spectra, whereas the π^* MO's involved in the $^1(\pi, \pi^*)$ and π^* TNI configurations are much more expanded. Indeed, one sees in Table XII.A-I that the (π, π^*) term value of formaldehyde is calculated to be only about half as large as those observed for the other (ϕ_i, π^*) valence levels, presumably owing to the substantial mixing of the (π, π^*) configuration with its (π, nd) Rydberg conjugates. These differences of π^* character in the various (ϕ_i, π^*) configurations should be reflected clearly in the values computed for the $J_{\pi^*\pi^*}$ self-energy integrals (Section I.D).

Another possible valence excitation in aldehydes and ketones is $1s \rightarrow \sigma^*$ where σ^* spans the whole molecule and so is accessible from all heavy atoms. The σ^* MO lies so high in these molecules that excitations to it appear only as extremely broad shape resonances 40 000–80 000 cm^{-1} (*5–10 eV*) beyond the ionization potential (Section II.C). There may be more than one of these for each originating MO. In the aldehydes and ketones, Hitchcock and Brion [H64] have assigned the inner-shell shape resonances as terminating at what is the $6a_1\sigma^*$(C–O) MO in formaldehyde, in parallel with the theoretical work of Langhoff *et al.* [L4] on the outer-shell spectrum of formaldehyde. Looking at Table XII.A-I, one sees that 1s → σ^* is most energetic when the 1s orbital is on the keto C atom [50 000–65 000 cm^{-1} (*6.2–8.0 eV*) beyond the ionization potential], and appreciably less so when the originating orbitals are either $1s_C$ of the alkyl groups or $1s_O$. This suggests that the σ^* MO has a relatively low electron density on the keto C atom, and this fits far better for the $3b_2\sigma^*$(C–H) MO than for $6a_1\sigma^*$(C–O) [I, S37].

We return now to the Rydberg excitations in the outer- and inner-shell spectra, remembering that Rydberg term values will be dependent upon the location of the hole orbital (Section I.B). Beginning with formaldehyde, we note that the $1s_C$ → 3s excitation is noticeably broader than the other Rydberg bands, suggesting that there is another excitation overlapping this band. In both the $1s_C$ and $1s_O$ spectra there are 1s → 3p excitations with very large term values of over 26 000 cm^{-1}, whereas 23 000 cm^{-1} is the expected maximum value. The core splitting of the 3p manifold by $(1s)^{-1}$ holes seems to be larger than normally found in the valence shell. As in formaldehyde, the keto ($1s_C$, 3s) term value in acetaldehyde is rather high, as are the corresponding ($1s_C$, 3p) term values. With these inner-shell exhaltations aside, the remainder of the Rydberg transitions in the core spectra are readily assigned based on their regular term values, Table XII.A-I. The inner-shell exhaltation effect (Section I.B) is clearly working on the Rydberg term values of formaldehyde and acetaldehyde, increasing the inner-shell (ϕ_i^{-1}, 3s) and (ϕ_i^{-1}, 3p) term values by up to 4000 cm^{-1}, while in acetone, the effect is no longer active. As explained in Section I.C, the Rydberg exhaltation is much smaller than that for valence excitations.

XII.C. Thio Derivatives

Comparison of the spectra of formaldehyde and its sulfur analog thioformaldehyde reveals the close parallels predicted by *ab initio* theory [B72, B86]. Moule and coworkers [D49, D50, J19] have studied the thioformaldehyde spectrum up to 56 000 cm^{-1} (*6.94 eV*), observing sharp features at 47 100, 53 200 and 55 000 cm^{-1} (*advert., 5.84, 6.60 and 6.82 eV*). With respect to the $(3b_2)^{-1}$ sulfur lone-pair ionization potential at 75 600 cm^{-1} (*9.37 eV*), these features have term values of 28 500, 22 400 and 20 600 cm^{-1}, which compare closely with the corresponding quantities in formaldehyde: 30 480 ($2b_2$, 3s), 23 520 ($2b_2$, 3py), and 22 130 cm^{-1} ($2b_2$, 3pz). These figures match much more closely than do the corresponding ones for the CH_3OH/CH_3SH pair [**I**, 259; **I**, 284]. Analysis of the $3b_2 \rightarrow 4s$ rotational structure in thioformaldehyde shows it to be a perpendicular transition, and the Stark effect on this band [G21] reveals a − 3.85 D change of dipole moment upon excitation from the ground state. All of the Rydberg excitations in thioformaldehyde are extremely vertical, though according to the calculations of Burton *et al.* [B86], the Rydberg excitation observed at 55 000 cm^{-1} ($3b_2 \rightarrow 4p$) is vibronically induced, *i.e.*, the terminating orbital is 4pxb_1.

Most interestingly, the excitation to $^1(\pi, \pi^*)$ so long sought without success in formaldehyde is readily apparent in thioformaldehyde. A highly structured band, with an origin at 45 170 cm^{-1} (*5.600 eV*) and peaking at 50 000 cm^{-1} (*6.2 eV*), is assigned the $^1(\pi, \pi^*)$ upper-state valence configuration [B86, D50]. Rotational analysis identifies the transition as $^1A_1 \rightarrow {}^1A_1$, while vibrational analysis shows that it consists of a long progression of 476 cm^{-1} (ν_3', C–S stretch, 1063 cm^{-1} in the ground state). Appropriately, this strongly allowed valence excitation has an oscillator strength (f = 0.38 [S71]) approximately 10 times larger than those of the intense Rydberg excitations in its vicinity. The corresponding excitation to $^3(\pi, \pi^*)$ has not been observed for thioformaldehyde but theory places it at 26 500 cm^{-1} (*3.28 eV*) [B86]. It is noteworthy that though the $^3(3b_2, \pi^*)$ and $^1(3b_2, \pi^*)$ configurations of thioformaldehyde are 11 000 ± 1000 cm^{-1} below those of formaldehyde, just as are the various corresponding Rydberg transitions and the lone-pair ionization potential, this difference is much larger for the valence $\pi \rightarrow \pi^*$ promotions. As expected, the $^1(\pi, \pi^*)$ configuration of thioformaldehyde is strongly mixed with its Rydberg conjugate, $^1(\pi, 3d)$ [B86].

An *ab initio* calculation of the electronic excitations in thioacetone [B71] reveals a spectrum much like that of thioformaldehyde, with the predicted excitation to $^1(\pi, \pi^*)$ at 58 000 cm^{-1} (*7.19 eV*) preceded by excitations to (n_S, 4s) and (n_S, 4p) Rydberg states. The experimental spectrum of thioacetone [J21] is not too different from this: a broad, structureless $\pi \rightarrow \pi^*$ band is centered at 45 500 cm^{-1} (*5.64 eV*) resting upon which are the sharply structured $n_S \rightarrow 4s$ and $n_S \rightarrow 4py$, 4pz excitations at 44 300, 51 600 and 52 600 cm^{-1} (*advert., 5.49, 6.40 and 6.52 eV*), respectively. The sharp Rydberg band origins anchor short progressions of C–S stretching and C–C–C bending vibrations. The assignment of the $n_S \rightarrow 4p$ components of thioacetone follows from the calculation [B71], which predicts an ordering the reverse of that in thioformaldehyde. This inversion among the 4p levels is due to the out-of-plane hydrogen atoms of the methyl groups (Section I.B), and is seen as well in formaldehyde/acetone. In thioacetaldehyde [J21], the $\pi \rightarrow \pi^*$ continuum is centered at 45 500 cm^{-1} (*5.9 eV*). In contrast to the situation in thioacetone, the $n_S \rightarrow 4s$ transition of thioacetaldehyde is quite nonvertical, with an origin at 45 200 cm^{-1} (*5.61 eV*) and an extended progression of C–S stretching. In the C_S symmetry of thioacetaldehyde, the (π, π^*) and (n_S, 4s) configurations can mix so as to perturb the latter. Two sharp transitions to components of 4p appear at 51 500 and 52 000 cm^{-1} (*advert., 6.39 and 6.54 eV*). Spectra of the thioacetone-like thiones $(CH_3)_3CCSCH_3$ and $(CH_3)_2CHCSCH_3$ [B73] reveal the broad $\pi \rightarrow \pi^*$ excitation peaked at 45 400 cm^{-1} (*5.64 eV*) with $f = 0.15$, while transitions from n_S to 4s, 4py and 4pz are much sharper and occur at 44 700, 50 100 and 51 800 cm^{-1} (*vert., 5.54, 6.21 and 6.42 eV*).

The $\pi \rightarrow \pi^*$ excitations in several halogenated derivatives of thioformaldehyde behave in an interesting manner. Thus, on going from thioformaldehyde to thiocarbonyl difluoride (F_2CS), the $\pi \rightarrow \pi^*$ origin moves to lower frequency (39 868 cm^{-1}, *4.9429 eV*) [L18], whereas the effect of fluorination generally is to shift excitations to higher frequency (see, for example, [**II**, 51]). The assignment of this band of thiocarbonyl difluoride as $\pi \rightarrow \pi^*$ is strongly supported by its oscillator strength of 0.24. On the other hand, $\pi \rightarrow \pi^*$ bands with origins at 35 277 and 34 500 cm^{-1} (*4.3737 and 4.28 eV*) for ClFCS [C31] and for Cl_2CS [J20], respectively, have oscillator strengths of only 0.087 and 0.059 [S71]. Such low oscillator strengths are puzzling in view of the value of 0.38 reported for thioformaldehyde itself. $\mathscr{A}$-band assignments at such low frequencies are unrealistic.

A recent calculation of the Rydberg and valence excitations in carbonyl difluoride (F_2CO) [V14] confirms the earlier assignments [**II**, 96] of the bands at 50 000 and 67 000 cm^{-1} (*6.2 and 8.3 eV*) as having $^1(n_O, \pi^*)$ and $^1(n_O, 3s)$ upper states, respectively, where n_O is the generic label for what is the $2b_2$ lone-pair orbital in formaldehyde. However, these workers assign the 59 000-cm^{-1} (*7.3-eV*) optical band as terminating at $^3(\pi, \pi^*)$ in spite of its observed intensity. The corresponding excitation to $^1(\pi, \pi^*)$ is assigned to the continuum centered at 78 000 cm^{-1} (*9.7 eV*). As in thiocarbonyl difluoride, the $\pi \rightarrow \pi^*$ excitation in carbonyl difluoride is calculated to be 5000–10 000 cm^{-1} lower in frequency than in the corresponding dihydro compound. There are no predicted excitations to valence σ^* orbitals below 100 000 cm^{-1} (*12.4 eV*) in carbonyl fluoride, however an $\mathscr{A}(\pi, \sigma^*)$ band is expected at 88 000 cm^{-1} (*vert., 11 eV*) on term-value grounds.

XII.D. Conjugated Aldehydes and Ketones; Diones and Vinyl Ketones

Because there is more than one $\pi \rightarrow \pi^*$ excitation possible in conjugated aldehydes and ketones [**II**, 103], we will use the $N \rightarrow V_n$ notation in this Section. The electron-impact spectrum of acrolein has been extended beyond the earlier optical studies by two groups [F31, T3]. Though the experimental spectra of these workers are in close agreement, their assignments are not. Fridh *et al.* [F31] assign the intense feature in acrolein at 51 200 cm^{-1} (*vert., 6.35 eV*) as $N \rightarrow V_1$ whereas Tam and Brion [T3] assign it as $n_O \rightarrow 3s$, Fig. XII.D-1. Its term value, 30 300 cm^{-1}, is too large for a Rydberg transition,† and its high intensity and behavior in solution [**II**, 103] are appropriate only for a $\pi \rightarrow \pi^*$ transition. In acrolein and methyl vinyl ketone, Tam and Brion argue against $N \rightarrow V_1$ assignments for the bands at 50 000–51 000 cm^{-1} (*6.2–6.3 eV*) using one-electron valence term-value arguments involving $^1(n_O, \pi^*)$ and $^1(\pi, \pi^*)$ configurations. However, such arguments are invalid due to the very significant difference in the nature of the π^* orbitals in these two singlet configurations. On the other hand, the assignment of the 57 200-cm^{-1} band as $n_O \rightarrow 3s$ as suggested in [**II**, 103] and by Fridh *et al.* is somewhat odd, for its term value is only 24 300 cm^{-1}. The spectrum of methyl vinyl ketone, Fig. XII.D-1, displays more regular term values of 26 600 cm^{-1} for Rydberg excitations terminating at 3s. A *n*s series in acrolein ($\delta = 1.12$) converging on the $(\pi_2)^{-1}$ ionization potential (88 160 cm^{-1} *advert., 10.93 eV*) also has been delineated [F31], and several broad continua beyond the first ionization potential are reported to be centered at 87 000, 99 000 and 111 000 cm^{-1} (*10.8, 12.3 and 13.8 eV*). The first of these broad bands has been assigned by Fridh *et al.* as valence $N \rightarrow V_4$ whereas Tam and Brion assign all of them as Rydberg excitations. We assign the first two of these to $\pi_1 \rightarrow 3p$ and $\pi_1 \rightarrow 4p$ transitions, where π_1 is the orbital corresponding to the third ionization potential of the molecule.

Trapped-electron spectra of glyoxal and biacetyl [V23] reveal many of their higher-lying triplet states. The $N \rightarrow V_1$ singlet band of glyoxal at 60 000 cm^{-1} (*vert., 7.44 eV*) [**II**, 104] has its corresponding triplet at 42 000 cm^{-1} (*5.2 eV*); a triplet also occurs at this frequency in biacetyl

† The $(n_O, 3s)$ term value in acrolein will be close to those in other ketones containing the same number of C and O atoms. Thus, it will lie near those of acetone (27 150 cm^{-1}), propionaldehyde (26 190 cm^{-1}) and cyclopropenone (26 300 cm^{-1}) [**II**, 80; **II**, 328].

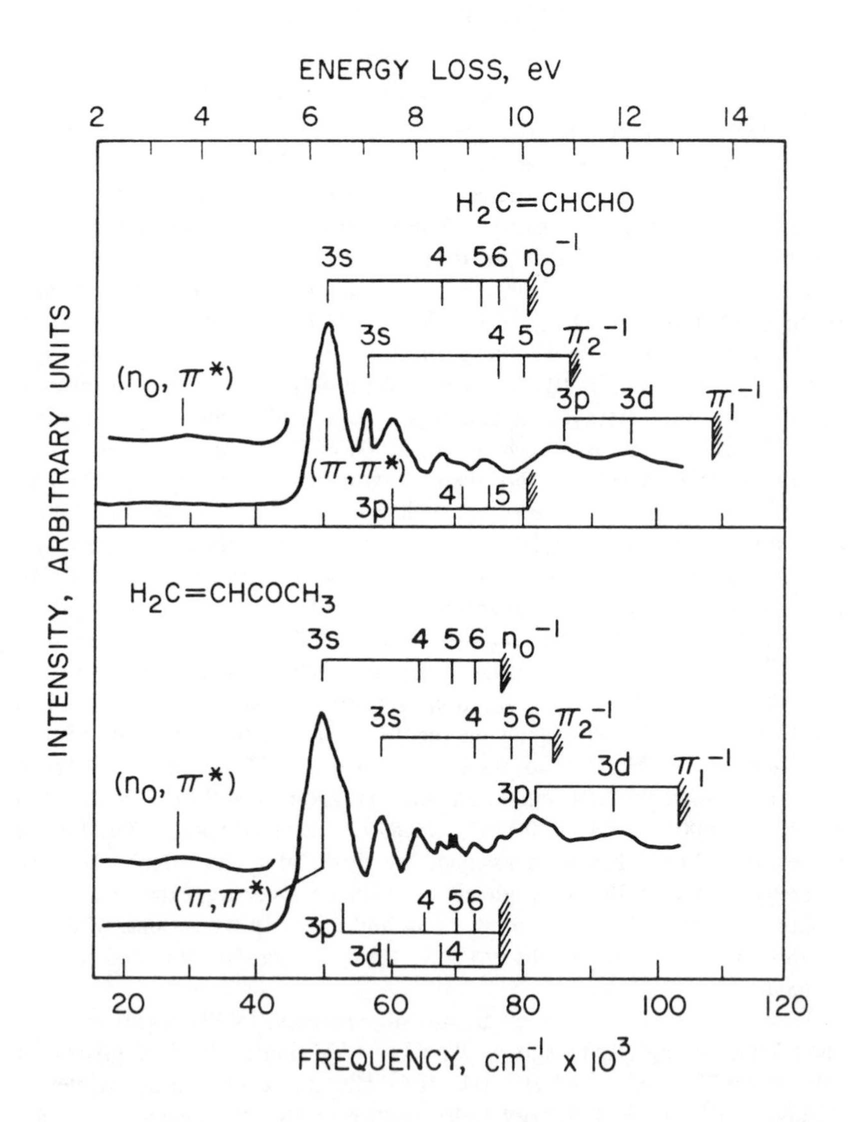

Fig. XII.D-1. Electron impact energy-loss spectra of acrolein (*upper*) and methyl vinyl ketone (*lower*) recorded at 100 eV impact energy and 2° scattering angle [T3].

[II, 105]. Triplet excitations at 50 000 cm^{-1} (*vert., 6.2 eV*) in glyoxal and biacetyl are too low for the orbitally forbidden $n_g \rightarrow 3s$ excitation and probably correspond to $n \rightarrow \pi^*$ valence excitations. A similar statement holds for the band at 53 200 cm^{-1} (*vert., 6.6 eV*) in glyoxal, said to have a singlet upper state. Broad bands observed at 64 000 cm^{-1} (*vert., 7.9 eV*) in both glyoxal and biacetyl were assigned as $\pi \rightarrow \pi^*$. Verhaart and Brongersma [V23] go one step further and assign the 60 000-cm^{-1} band of glyoxal to the forbidden $n_O \rightarrow 3s$ promotion, but its apparently high intensity argues against this. Transitions of unknown spin multiplicity appear at 73 400 cm^{-1} (*9.1 eV*) in glyoxal and at 69 400 cm^{-1} (*8.6 eV*) in biacetyl. The most recent calculation on glyoxal places the excitation to $^1(\pi, \pi^*)$ at 78 900 cm^{-1} (*vert., 9.78 eV*) [H51]. Surprisingly, the CD spectrum of the dione *d*-camphor in the vacuum ultraviolet reveals even fewer features than are obvious in the absorption spectrum [G8].

At first sight, the ultraviolet absorption of tetramethyl cyclobutanedione-1,3 (TMCBD) O = ◇ = O [B62, V3] would seem to be in accord with expectations regarding Rydberg absorption. A smooth band at 49 380 cm^{-1} (*vert., 6.122 eV*), Fig. XII.D-2, has a term value of 21 600 cm^{-1} with respect to the ionization at 71 000 cm^{-1} (*advert., 8.80 eV*) and so is a good candidate for assignment as the $n_O \rightarrow 3s$ excitation at the alkyl limit. (In acetone, the corresponding excitation falls at 51 300 cm^{-1} (*6.36 eV*), with a term value of 27 050 cm^{-1} [D49]; this large reduction of the $(\phi_i^o, 3s)$ term value on dimerizing the molecule has been observed repeatedly in the past.) The $n_O \rightarrow 3s$ band of TMCBD has an extinction coefficient of only 200. This band of the dione is preceded by a smooth, weak band at 44 800 cm^{-1} (*vert., 5.55 eV*) which appears unperturbed in various solvents and in the crystal spectrum [V3], and thus must be a valence excitation. The transition at 53 750 cm^{-1} (*advert., 6.664 eV*) has a vibronic profile matching that of the first band in the photoelectron spectrum, and with a term value of 17 250 cm^{-1} it is clearly the $n = 3$ member of an np Rydberg series. The vibrational symmetry in this band is totally symmetric and the extinction coefficient at the origin is 2500. A band at 55 660 cm^{-1} (*advert., 6.900 eV*) with a term value of 15 340 cm^{-1} is claimed to be the $n = 3$ member of an nd series [B62], but it can more easily be assigned as a component of 3p based on its large term value and high intensity.

The two lowest Rydberg transitions identified in TMCBD (49 380 and 53 750 cm^{-1}) have every indication of being symmetry forbidden and symmetry allowed, respectively. This implies that the highest occupied molecular orbital in this molecule, an oxygen lone-pair MO, has g symmetry. Hückel calculations on this dione [C46, T16] predict two lone-pair MO's (b_{2u} and b_{3g}), but place the b_{2u} lone-pair orbital *above* the b_{3g} lone-pair orbital. This calculated ordering contradicts the conclusions drawn from the observed Rydberg spectrum, and we hold that

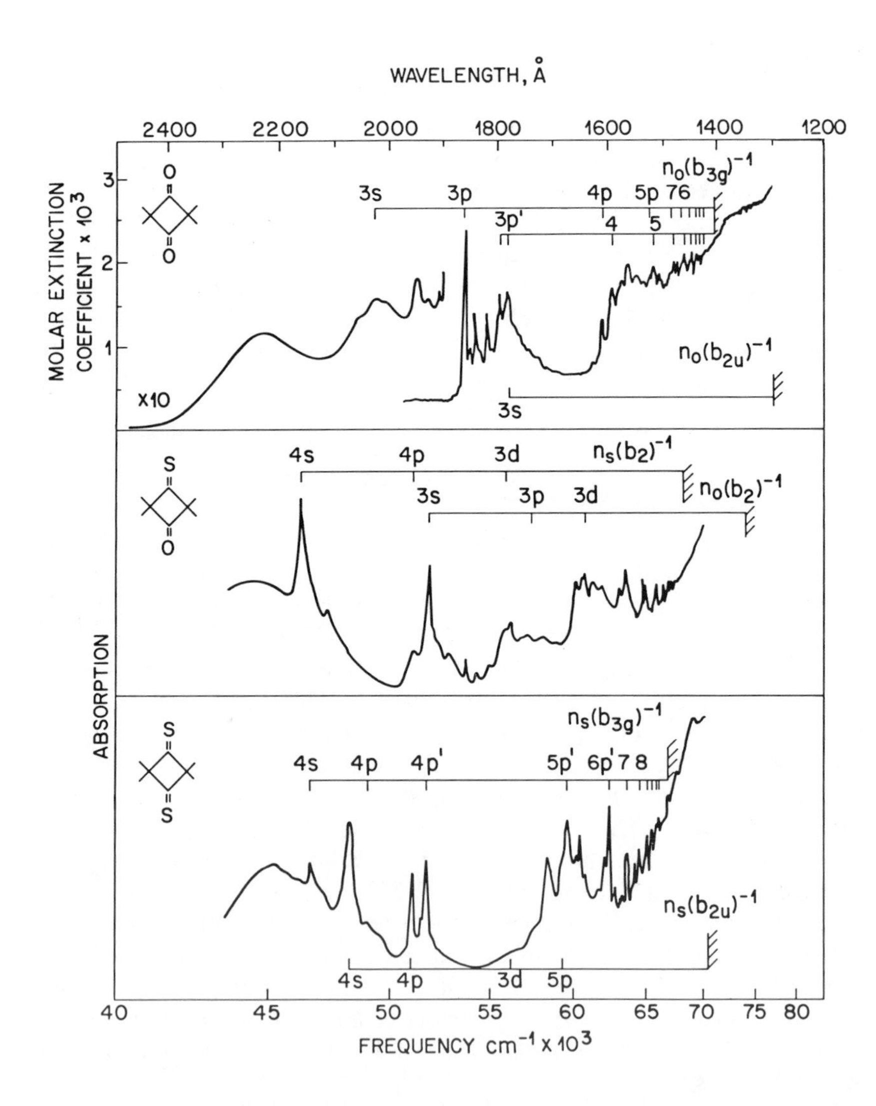

Fig. XII.D-2. Optical absorption spectra of tetramethyl cyclobutanedione-1,3 and its mono- and di-thio derivatives in the vapor phase [B62, T16, V3].

the semiempirical calculation must be in error on this point. A recent *ab initio* calculation on TMCBD by Wadt in fact places the lone-pair b_{3g} MO about 8000 cm^{-1} *above* the b_{2u} combination [W3], in agreement with the spectral intensities, and a similar calculation on cyclobutanedione-1,3 [B19] also places the b_{3g} level above b_{2u}. The only excitation originating at $n_O(b_{2u})$ which can be assigned is the broad, underlying $n_O \rightarrow 3s$ absorption in the vicinity of 56 000 cm^{-1} (*vert., 6.9 eV*).

In the dithione derivative of TMCBD, the two lone-pair orbitals are separated by only 3400 cm^{-1}, again with b_{3g} above b_{2u} according to an *ab initio* calculation [B19]. However, past interpretation of the vacuum-ultraviolet spectrum of the dithione [T16], Fig. XII.D-2, was based upon the assumption of an inverse ordering of the lone-pair MO's. Placing b_{3g} uppermost, we assign the weak features at *ca.* 47 000 cm^{-1} (*vert., 5.8 eV*) in the dithione as $b_{3g} \rightarrow 4s$ resting on top of unspecified valence absorption. The corresponding allowed transition from b_{2u} to 4s appears as a sharp feature centered at 48 207 cm^{-1} (*5.9768 eV*) with a term value of 22 540 cm^{-1}. The transition from b_{3g} to 4p is symmetry allowed, and is readily assigned to the intense features at 49 000−52 000 cm^{-1} (*6.1−6.4 eV*) having *ca.* 16 000 cm^{-1} term values with respect to the $(b_{3g})^{-1}$ ionization potential of 67 350 cm^{-1} (*vert., 8.35 eV*). All the features discussed above have been placed into Rydberg series having symmetry assignments [T16] different from those given here. A similar conflict exists for the interpretation of the spectrum of the mixed system tetramethyl-3-thiocyclobutane-1-one [T16], Fig. XII.D-2.

Spectra of the acyclic 1,3-diones do not bear close relationships to that of TMCBD, for they exist predominantly as enol structures in the vapor phase. The simplest of these, malonaldehyde, (written as $OHCCH_2CHO$ in its keto form) is thought to have a symmetrical hydrogen bridge in the enol form. Its spectrum shows an intense valence $\pi \rightarrow \pi^*$ ($N \rightarrow V_1$) band at 38 000 cm^{-1} (*vert., 4.7 eV*) and a second structured band at 52 700 cm^{-1} (*vert., 6.53 eV*) [S40] with an oscillator strength reported to be equal to or greater than 0.2. This latter band has a term value of 26 000 cm^{-1} with respect to the $(\pi)^{-1}$ ionization potential at 78 700 cm^{-1} (*9.76 eV*) [N26], and so is tentatively assigned as $\pi \rightarrow 3s$. The reported oscillator strength is totally out of line with that expected for a Rydberg excitation however, and one is tempted to invoke underlying valence intensity. However, alkylated derivatives of malonaldehyde show no valence absorption in this region. A (π, 3s) Rydberg term value of 26 000 cm^{-1} for malonaldehyde is quite in line with the value of 26 500 cm^{-1} observed for methyl acetate, a molecule also having three C and two O atoms.

Symmetric methylation of malonaldehyde yields acetyl acetone [N5, N6]. As in malonaldehyde, the $N \rightarrow V_1$ valence excitation of acetyl acetone again falls at 38 000 cm^{-1} (*4.7 eV*) and is intense ($f = 0.24$), Fig. XII.D-3. This is followed by a very weak $\pi \rightarrow 3s$ band at 47 170 cm^{-1} (*vert., 5.848 eV*, term value of 26 310 cm^{-1}) and a $\pi \rightarrow 3p$ transition at 53 300 cm^{-1} (*vert., 6.608 eV*, term value of 20 180 cm^{-1}). Another possible component of 3p appears weakly at 56 000 cm^{-1} (*vert., 6.94 eV*).

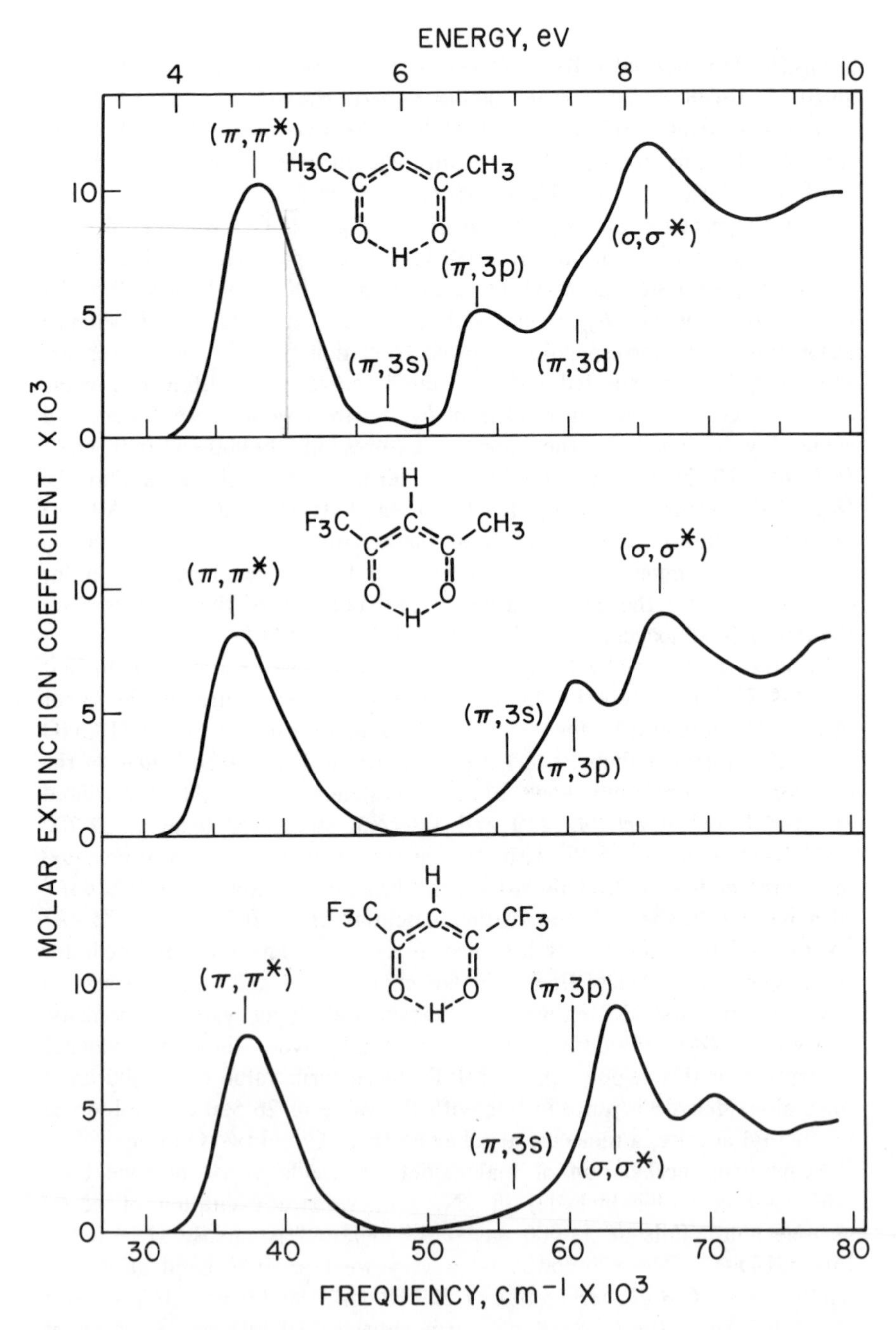

Fig. XII.D-3. Optical absorption spectra of acetyl acetone and its fluorinated derivatives in the vapor phase, all in the enol form [N5, N6].

None of the above Rydberg transitions are visible in the spectrum of acetyl acetone dissolved in perfluorohexane, as appropriate for big-orbit excitations. In the keto form of acetyl acetone, the transition to 3s moves down to 52 000 cm^{-1} (*vert., 6.45 eV*). In the enol form, a very intense (f = 0.5) valence excitation is found at 65 200 cm^{-1} (*vert., 8.08 eV*), which Nakanishi *et al.* [N6] assign as $\sigma \rightarrow \sigma^*$, largely on the basis of its intensity.

The $\sigma \rightarrow \sigma^*$ band of enolic acetyl acetone at 65 200 cm^{-1} is very interesting from two points of view. First, on the basis of semiempirical calculation, it is predicted to be stronger by a factor of 5 than any other transition in the molecule. Thus its assignment to the strongest feature observed in the spectrum is reasonable. This is one of the few $\sigma \rightarrow \sigma^*$ valence bands to be observed with any certainty in an organic molecule. Second, the upper state of this transition involves large amounts of oxygen → hydrogen charge transfer in the symmetrical O···H···O hydrogen bond.

In 1,1,1-trifluoroacetyl acetone, Fig. XII.D-3, the $(\pi)^{-1}$ ionization potential moves upward to 80 100 cm^{-1} (*9.93 eV*) [N26], the transition to (π, 3s) remains weak at *ca.* 56 000 cm^{-1} (*vert., 6.94 eV*), while that to (π, 3p) is intense and peaks at 60 600 cm^{-1} (*7.51 eV*) [N6]. An intense valence excitation falls at 66 700 cm^{-1} (*vert., 8.27 eV*), and probably is due to overlapping $\pi \rightarrow \pi^*$ and $\sigma \rightarrow \sigma^*$ promotions. In hexafluoroacetyl acetone, corresponding valence transitions appear with high intensity at 63 700 and 70 400 cm^{-1} (*vert., 7.90 and 8.73 eV*). If these assignments are correct, it is surprising to see that the $\sigma \rightarrow \sigma^*$ excitation of acetyl acetone remains at *ca.* 65 000 cm^{-1} as the molecule is fluorinated. With an ionization potential of 86 460 cm^{-1} (*10.72 eV*) [L31, N26] in hexafluoroacetyl acetone one sees that the transition to 3p in this molecule is buried beneath the strong valence excitation at 63 700 cm^{-1} and that the transition to 3s probably corresponds to the shoulder centered at *ca.* 57 000 cm^{-1} (*7.1 eV*). Note how fluorination increases the (π, 3s) term value on going from acetyl acetone (26 300 cm^{-1}) to hexafluoroacetyl acetone (29 500 cm^{-1}). The spectra of heavy-metal complexes of hexafluoroacetyl acetone are discussed in Section XXI.C.

As might be expected for an easily reduced molecule such as benzoquinone with four nondegenerate π^* MO's, the valence TNI spectrum is a rich one [A10], with five resonances between 5600 and 44 000 cm^{-1} (*0.7 and 5.45 eV*). Additionally, the electron-impact energy loss spectrum shows two sharp but weak bands at 56 000 and 60 500 cm^{-1} (*vert., 7.0 and 7.5 eV*) which qualify for $n_u \rightarrow 3s$ and $n_u \rightarrow 3p$ assignments on the basis of their term values with respect to the $(n_u)^{-1}$ lone-pair ionization potential at 81 500 cm^{-1} (*advert., 10.11 eV*). This region also contains a broad valence transition which persists in hydrocarbon solution [K41]. Comparison of the vapor-phase and solution spectra of phenanthrenequinone to 52 000 cm^{-1} (*6.45 eV*) and of anthraquinone to 59 000 cm^{-1} (*7.3 eV*) [K42] show no solvent effects characteristic of Rydberg excited states (Section II.D).

XII.E. Addendum

W. Eberhardt, T. K. Sham, R. Carr, S. Krummacher, M. Strongin, S. L. Weng and D. Wesner, Site-specific fragmentation of small molecules following soft X-ray excitation, *Phys. Rev. Lett.* **50**, 1038 (1983).

S. Paone, D. C. Moule, A. E. Bruno and R. P. Steer, Vibronic analyses of the Rydberg and lower intravalence electronic transitions in thioacetone, *J. Mol. Spectrosc.* **107**, 1 (1984).

P. R. Keller, J. W. Taylor, F. A. Grimm and T. A. Carlson, Angle-resolved photoelectron spectroscopy of formaldehyde and methanols, *Chem. Phys.* **90**, 147 1984.

R. Dressler and M. Allan, CH_3^- formation through predissociation of Feshbach resonances in acetaldehyde, to be published in *Chem. Phys. Lett.*

CHAPTER XIII

Acetylenic Compounds

XIII.A. Acetylene and Derivatives

Acetylene and Alkyl Derivatives

Symmetry dictates that the four Rydberg series terminating at ns, ndσ, ndπ and ndδ upper orbitals and converging upon the $^2\Pi_u$ ionic ground state of acetylene will be electric-dipole allowed. Excitations to npσ and npπ from the $1\pi_u$ MO on the other hand, are allowed instead as electric-quadrupole transitions. Refined experimental work on acetylene using C and H isotopic substitution, rotational analyses at high resolution, and low temperature has led to an identification and extension of the electric-dipole allowed Rydberg series [H42, H43]. As discussed earlier [**II**, 107], the $1\pi_u \rightarrow 3$s transition has its origin at 65 790 cm^{-1} (*8.157 eV*) and initiates a long Rydberg series ($3 \leqslant n \leqslant 12$) having $\delta = 0.974$. According to calculation [D14], the nd manifold is split with ndσ lowest, ndδ highest and ndπ intermediate in energy; the experimental work confirms this ordering. The transitions from π_u to 3dσ and 3dπ, at 74 500 and 74 754 cm^{-1} (*adiabat., 9.2366 and 9.2681 eV*), are spectrally overlapped, and at high n values, the two transitions cannot be resolved. The quantum defects for these series, 0.48 and 0.49, are anomalously high, making them look more like transitions to np than to nd. The transitions to ndδ are similarly peculiar, beginning at 80 458 cm^{-1} (*9.9753 eV*) and forming a series having $\delta = -0.10$. It is an unanswered question as to why the core splitting in the ($1\pi_u$, 3d) manifold of acetylene is so large (*ca.* 6000 cm^{-1}), eclipsing that of the ($1\pi_u$, 3p) configuration by almost a factor of four. Two singlet-triplet excitations observed by electron impact at 74 000 and 77 350 cm^{-1} (*advert., 9.17 and 9.59 eV*) [W27] probably terminate at the 3dσ/3dπ and 3dδ levels, respectively. The transition frequencies from $1\pi_u$ to ndσ in propyne also are anomalously high, with $\delta = 0.32$, while the transitions from $1\pi_u$ to npσ and npπ in this molecule have more normal δ's of 0.67 and 0.60, respectively [D14].

Early members of the transitions from $1\pi_u$ to npσ and npπ orbitals are observed as electric-quadrupole transitions in the electron impact spectrum of acetylene recorded at low impact energy and large scattering angle [W27], Fig. XIII.A-1. The calculated ordering is npσ below npπ [D14] as expected from the electrostatic model, Section I.B. Wilden *et al.* [W27] spot $1\pi_u \rightarrow 3$pσ at 72 700 cm^{-1} (*9.01 eV*) with the corresponding triplet 1000 cm^{-1} lower, while for $1\pi_u \rightarrow 3$pπ the triplet and singlet absorption frequencies are 73 100 and 74 300 cm^{-1} (*9.06 and 9.21 eV*), respectively. The ^{3}($1\pi_u$, 3pσ) configuration is the initial member of a short series having $\delta = 0.62$, a perfectly normal value for an np series. The absorption intensities of over a dozen Rydberg excitations in acetylene have been measured photoelectrically and integrated vibration-by-vibration to yield oscillator strengths [S85]; in a given Rydberg series, the f-values decrease smoothly with increasing n, and the largest is less than 0.08 per degree of spatial degeneracy [I, 30].

The MCD spectrum of acetylene in the vapor phase is most interesting [G6]. Because the $\pi \rightarrow \pi^*$ valence excitation $^1\Sigma_g^+ \rightarrow {}^1A_u$ in the vicinity of 50 000 cm^{-1} (*6.2 eV*) shows no MCD spectrum, the first MCD feature is the electronic origin at 54 050 cm^{-1} (*6.701 eV*), just as in the matrix absorption spectrum [II, 115]. This MCD valence excitation contains a complex mixture of bands of both signs, and no assignments are given. The $1\pi_u \rightarrow 3$s origin at 65 814 cm^{-1} ($^1\Sigma_g^+ \rightarrow {}^1\Pi_u$, *8.1597 eV*) shows the expected pure A-term MCD profile, as do all the vibronic components built upon it. There seems to be a weak continuum underlying the transition to 3s.

Jungen [II, AD97] has suggested that there is a dipole-forbidden $1\pi_u \rightarrow 3\sigma_u^*$ valence excitation in acetylene in the vicinity of 72 000 cm^{-1} (*9 eV*) and Wilden *et al.* [W27] do indeed find a quadrupole-allowed band at just the suggested frequency, Fig. XIII.A-1. The calculation of Machado *et al.* [M1] also supports this valence assignment.

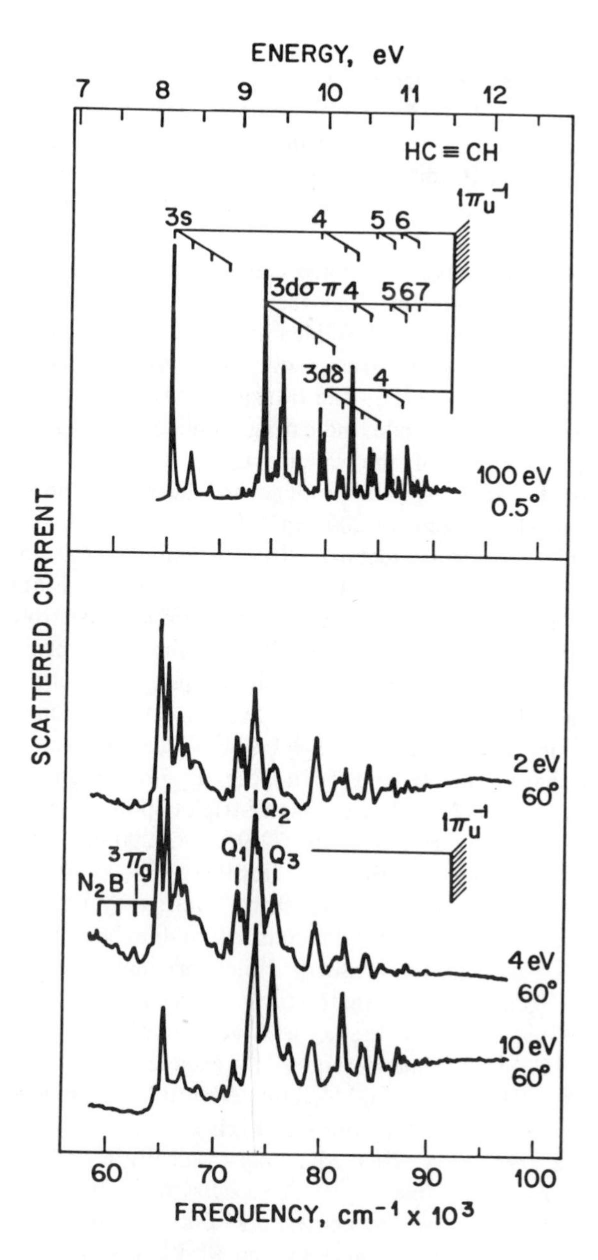

Fig. XIII.A-1. Electron impact energy loss spectra of acetylene. In the upper panel, the spectrum was determined using 100-eV electrons incident on the molecule and scattered inelastically at 0.5°. Spectra in the lower panel were run at fixed residual energy (as shown) after scattering through 60° [W27]. The bands marked Q_n are electric quadrupole-allowed excitations to 3p.

Okabe reproduces the acetylene spectrum at modest resolution in the 75 000–87 000 cm^{-1} (*9.3–11 eV*) region [O5], while the albedos of Jupiter and Saturn show the acetylene bands in the 52 000–63 000 cm^{-1} (*6.4–7.9 eV*) region [C30]. The intense valence transition to $^1\Sigma_u^+$ still has not been observed, but is predicted to fall at 81 000 cm^{-1} (*10 eV*) by Åsbrink *et al.* [A28].

More triplet states have been uncovered for acetylene [**II**, 112] than for any other polyatomic system other than water, largely due to the superb electron-impact spectroscopy of Wilden *et al.* [W27]. At low energies, two valence triplet upper states are uncovered at 41 800 and 49 200 cm^{-1} (*vert., 5.1 and 6.1 eV*) [W23]. The transitions display long progressions of the 700 cm^{-1} bending mode indicating nonlinear upper states; in the *trans*-bent point group of the valence upper states (C_{2h}), these levels transform as ^3B_u and ^3A_u. In propyne and butyne-1, the corresponding triplets fall at 41 900 and 47 200 cm^{-1} (*vert., 5.19 and 5.85 eV*) [F16]. The next higher triplet in acetylene (65 100 cm^{-1} *advert., 8.07 eV*) first appeared in the threshold electron spectrum [D2, V8] and was later confirmed by low-energy, large-angle electron impact spectroscopy [W27]. The triplet state at 65 100 cm^{-1} exhibits a vibrational progression of ν_2 (1780 cm^{-1}) each member of which is just 800 cm^{-1} below the corresponding vibronic band of the $^1(1\pi_u, 3s)$ state. Consequently, the 65 100-cm^{-1} level is assigned the $^3(1\pi_u, 3s)$ configuration. The higher triplets terminating at 4s and 5s appear at 79 900 and 84 900 cm^{-1} (*advert., 9.91 and 10.22 eV*) with singlet-triplet splits of only 160 cm^{-1} in each case. Four other triplets in the 72 000–83 000 cm^{-1} (*8.9–10.3 eV*) region are tentatively identified as Rydberg based on their vibronic intervals and nearness to singlet Rydberg levels; they probably terminate at components of 3d and 4d. In contrast to the bent geometry of the valence $\pi \rightarrow \pi^*$ triplets, the Rydberg triplets originating at $1\pi_u$ are linear. However, a Renner-Teller interaction between the $^1(1\pi_u, 3d\sigma)$ and $^1(1\pi_u, 4s)$ states involving the ν_4 and ν_5 bending vibrations has been documented [C37]. Geometries of many excited states of acetylene are computed by Demoulin [D15], in essential agreement with the experimental work of Herman and Colin [H41].

A transient absorption observed in electronically excited acetylene has been touted as a possible $T_1 \rightarrow T_n$ spectrum [L9]. If so, then the terminating triplets T_n would come at 91 000 and 100 000 cm^{-1} (*vert., 11.3 and 12.4 eV*), however there are no Rydberg singlets in this region and so the assignment of the T_n states as Rydberg triplets is hereby ruled out.

Temporary-negative-ion (TNI) resonances (Section II.B) in acetylene reported by van Veen and Plantenga [V8] and by Dance and Walker [D2] are in good agreement with one another. These workers claim that the binding of the incident electron by the $1\pi_g^*$ MO of ground-state acetylene corresponds to the valence TNI resonance at 13 700 cm^{-1} (*vert., 1.70 eV*). On the other hand, Jordan and Burrow [J16] report a resonance at 21 000 cm^{-1} (*vert., 2.6 eV*), so one is not sure which involves σ^* and which involves π^*. Because the 21 000-cm^{-1} resonance leads to a more symmetrical splitting pattern for the two π^* valence TNI resonances of diacetylene observed at 8000 and 45 000 cm^{-1} (Section XIII.B), we are inclined to assign it to π^* and the 13 700-cm^{-1} resonance to σ^*. In accord with these assignments, the 21 000-cm^{-1} band moves to higher frequencies in the alkylated derivatives, Table XIII.A-I, as π^* TNI resonances are wont to do (Section II.B). Excitation from the neutral-molecule ground state to the Feshbach-resonance configuration $^2(1\pi_u, 3s^2)$ of the negative ion occurs at 60 600 cm^{-1} (*vert., 7.52 eV*), just 4000 cm^{-1} below the neutral-molecule transition to $^1(1\pi_u, 3s)$. van Veen and Plantenga claim yet another Feshbach state at 71 900 cm^{-1} (*vert., 8.92 eV*) in acetylene, but the data of Dance and Walker [D2] does not confirm this. If real, the 71 900-cm^{-1} resonance of acetylene is in a perfect position for assignment to the Feshbach configuration $^2(1\pi_u, 3d^2)$, being just *ca.* 2700 cm^{-1} below the neutral molecule $^1(1\pi_u, 3d)$ states at *ca.* 74 600 cm^{-1} [W27].

Several of the negative-ion resonances quoted above for acetylene appear as well in the threshold electron spectra of the alkyl acetylenes, Table XIII.A-I [D2]. One sees that the alkyl groups destabilize the π_g^* anion ground state (as in the alkyl olefins, Section X.B) and that the $^2(1\pi_u, 3s^2)$ Feshbach states have term values which decrease somewhat with increasing alkylation in a more regular way than do the $^1(1\pi_u, 3s)$ neutral-molecule term values. From the term-value trends in the Table, one calculates that the unobserved $(1\pi_u, 3s^2)$ Feshbach resonance in dimethyl acetylene will come at 49 100 cm^{-1} (*6.09 eV*). The threshold [D2] and electron impact energy loss [F16, S76] spectra of propyne look much alike, except for the presence of triplets and TNI resonances in the former which are missing in the latter. These similarities hold as well for butyne-1 but not butyne-2. Many of the super-excited states seen as energy losses [F16] in the alkyl acetylenes can be assigned as Rydberg excitations to higher ionization potentials as judged by their term values.

Beyond the first ionization potential of acetylene, Hayaishi *et al.* [H28] find a few members of an autoionizing Rydberg series ($\delta = 0.6$, $3\sigma_g \rightarrow np\pi$) at *ca.* 115 000−125 000 cm^{-1} (*vert., 14−15 eV*), preceded by a broad and intense $3\sigma_g \rightarrow 3\sigma_u^*$ valence excitation centered at 107 500 cm^{-1} (*13.3 eV*) and followed by another valence excitation ($2\sigma_u \rightarrow 1\pi_g^*$)

TABLE XIII.A-I

TEMPORARY-NEGATIVE ION RESONANCES IN ACETYLENE AND ALKYL DERIVATIVES[a]

Molecule	$(\pi_u)^{-1}$ Ionization Potential, cm^{-1} (*eV*)	TNI Resonance, cm^{-1} (*eV*)	Parent State, cm^{-1} (*eV*)	Feshbach Decrement, cm^{-1}	Assignment[b]
$HC{\equiv}CH$	92 030 (*11.41*)	13 700 (*1.7*)	-	-	$^2(N, \sigma^*)$
		21 000 (*2.6*)	-	-	$^2(N, \pi^*)$
		61 700 (*7.65*)	66 000 (*8.18*)	4300	$^2(\pi, 3s^2)$
$HC{\equiv}CCH_3$	83 560 (*10.36*)	24 200 (*3.0*)	-	-	$^2(N, \pi^*)$
		54 000 (*6.7*)	58 100 (*7.2*)	4000	$^2(\pi, 3s^2)$
$HC{\equiv}CC_2H_5$	82 110 (*10.18*)	22 600 (*2.8*)	-	-	$^2(N, \pi^*)$
		54 000 (*6.7*)	55 600 (*6.9*)	1600	$^2(\pi, 3s^2)$
$H_3CC{\equiv}CCH_3$	77 100 (*9.56*)	29 000 (*3.6*)	-	-	$^2(N, \pi^*)$
		-	51 000 (*6.3*)	-	-

[a] Reference [D2].

[b] See Section II.B for definitions of TNI configurations.

centered at 123 500 cm^{-1} (*15.3 eV*), Fig. XIII.A-2. The vapor-phase excitation at 123 500 cm^{-1} is also present as an autoionizing resonance in the photoemission spectrum of solid acetylene [F19], thereby documenting its valence nature. These valence excitations of acetylene also have been observed in photoelectron experiments [H68, P17], and appear as peaks in calculations of the $(3\sigma_g)^{-1}$ and $(2\sigma_u)^{-1}$ photoionization cross sections [L5, L32, M1]. Notice however, that we refrain from calling them shape resonances because they fall below the ionization potentials of their originating MO's, Section II.C. Though the band at 107 500 cm^{-1} has been assigned the $(3\sigma_g, 3p\sigma)$ Rydberg upper-state configuration [P17], an np Rydberg assignment for the terminating level of this band is discordant with its term value (25 200 cm^{-1}) and its high intensity. On the other hand, Unwin *et al.* [U3] reject a Rydberg assignment for the feature at 115 000 cm^{-1} on energy grounds, while Langhoff *et al.* [L5] reject it for intensity reasons. If the valence assignments of these high-lying excitations of acetylene at 107 500 and 123 500 cm^{-1} are correct, they

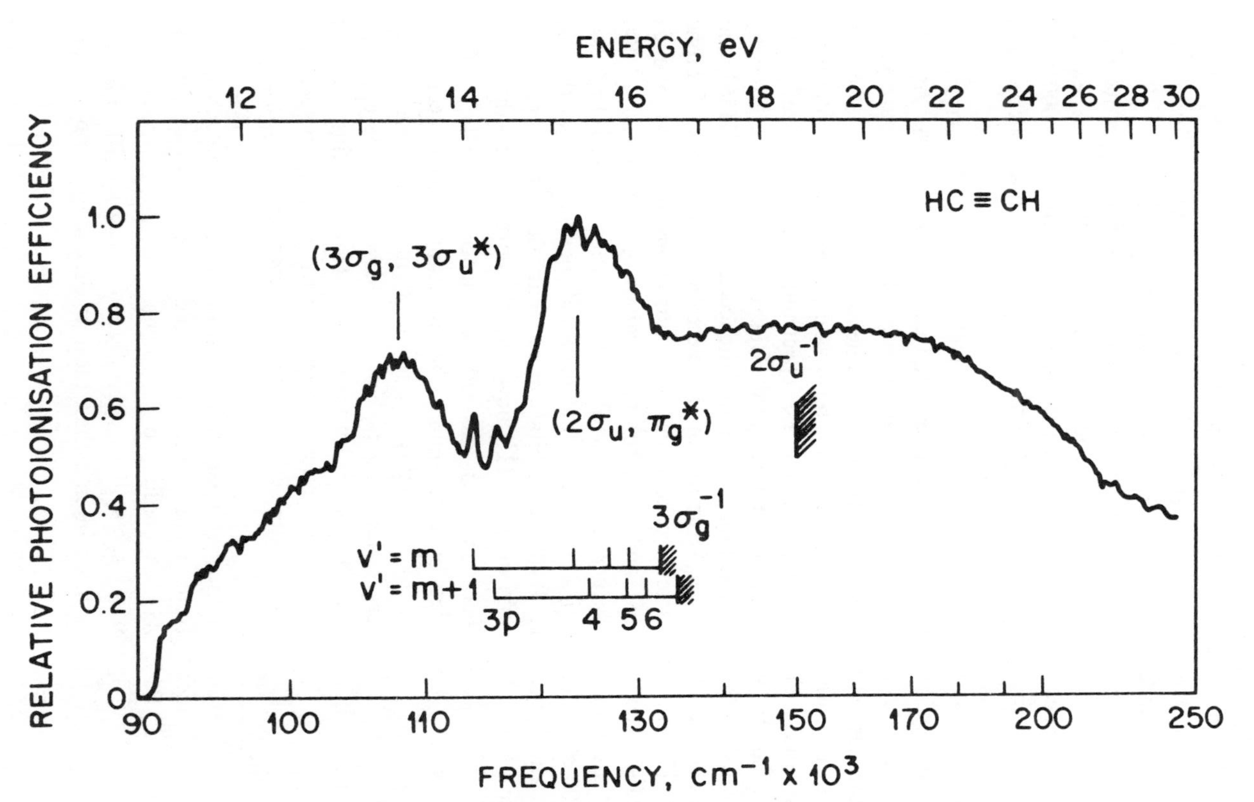

Fig. XIII.A-2. Valence excitations in the photoionization yield curve of acetylene [H28].

supply the first clues as to where $\sigma \rightarrow \sigma^*$ and $\sigma \rightarrow \pi^*$ excitations are to be found in small, unsaturated molecules. Also missing is the $^1\Sigma_g^+ \rightarrow {^1\Sigma_u^+}$ $(1\pi_u \rightarrow 1\pi_g^*)$ excitation in acetylene, which is predicted to appear as a shape resonance beyond the $(1\pi_u)^{-1}$ ionization limit [L32].

Fluorescence and absorption spectra of acetylene and many of its alkyl derivatives are reported [H9]. The emission from the lowest excited singlet valence state is broad and structureless, with peaks at 26 000 cm^{-1} (*3.2 eV*) in the derivatives and at 27 000 cm^{-1} (*3.3 eV*) in acetylene. The fluorescence efficiency is only 1 %.

As one might expect, the $1s_C$ inner-shell spectrum of acetylene is qualitatively much like those of ethylene and benzene [E1, H56]. As with these unsaturated systems, the inner-shell absorption of acetylene commences with an intense $1s_C \rightarrow 1\pi_g^*$ valence excitation, followed by weaker transitions to 3s (26 800 cm^{-1} term value) and to 3p (19 300 cm^{-1} term value), Fig. XIII.A-3. In the $1s_C$ spectrum observed optically, the transitions to 3s and 3p are not resolved [E1], however these are clearly separated in the electron impact work of Hitchcock and Brion [H56], and even show clear vibronic structure in the electron impact work of Tronc *et al.* [T19]. The latter workers also find transitions to 4s, 4p and 5s orbitals. The ($1s_C$, 3s) and ($1s_C$, 3p) term values of 26 800 and 19 300 cm^{-1} are somewhat larger than those from the valence region [($1\pi_u$, 3s), 26 000 cm^{-1}; ($1\pi_u$, 3p), 17 500 cm^{-1}] as expected considering that the $1s_C$ hole antishields the Rydberg orbital more than does the $1\pi_u$ hole, Table I.B-II. Machado *et al.* [M1], however, choose to assign the band here taken as Rydberg $1s_C \rightarrow 3s$ instead as valence $1s_C \rightarrow 4\sigma_g^*$. This question can be settled by determining the spectrum of solid acetylene, Section II.D. In view of the highly anomalous term values observed for the $1\pi_u \rightarrow nd$ Rydberg series (*vide supra*) it would be most interesting to measure such quantities for $1s_C \rightarrow nd$ transitions. However, this is not to be, for all of the Rydberg promotions observed in the inner-shell spectrum of acetylene [T19] are assignable to ns and np upper levels, with normal term values.

Ab initio studies of the $1s_C$ inner-shell spectrum of acetylene confirm the analysis deduced from term values and add several other bits of information [B15, M1]. According to the calculations, the ($1s_C$, 3pσ) and ($1s_C$, 3pπ) configurations are split by *ca.* 4000 cm^{-1}, and the transition to $1\pi_g^*$ is 50−100 times more intense than those to 3s and 3p, as observed.[†]

[†] Note however, that the calculation of [M1] predicts that the lowest $1s_C \rightarrow \sigma_g$ band terminates at the valence MO $4\sigma_g$, rather than at the 3s Rydberg MO. In another calculation [D14] the $4\sigma_g$ MO is described as a "borderline case" in regard Rydberg/valence character.

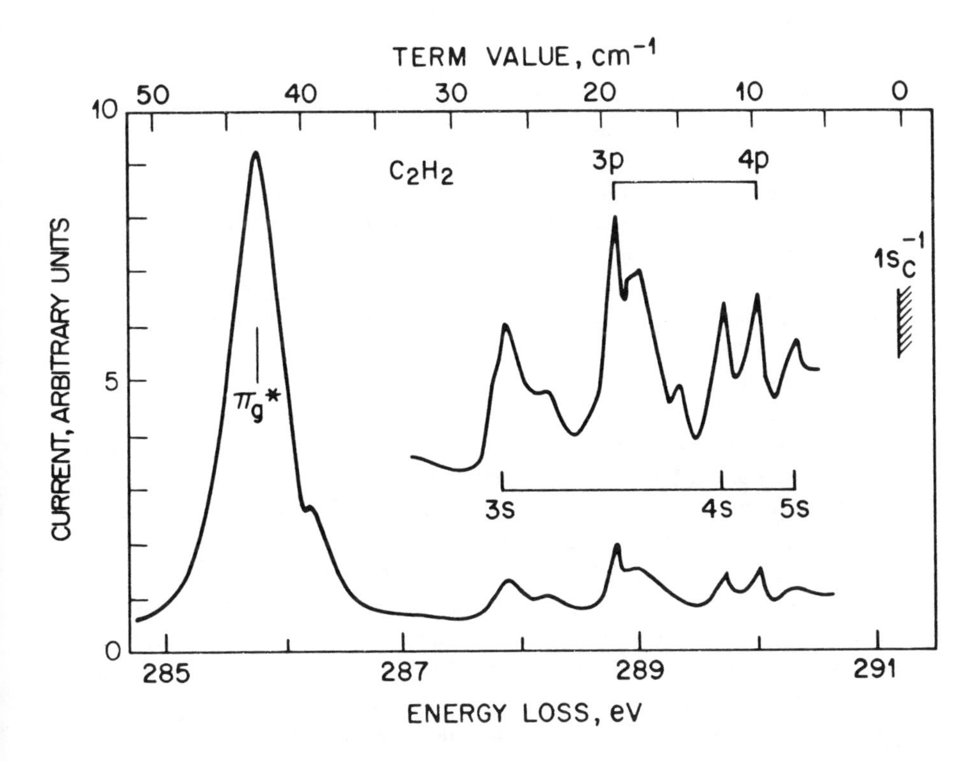

Fig. XIII.A-3. Inner-shell energy loss spectrum of acetylene at an impact energy of 1500 eV [T19].

These calculations are especially valuable in assigning the many shake-up bands found beyond the $(1s_C)^{-1}$ ionization limit. Thus two excitations 58 100 and 136 000 cm^{-1} (*vert., 7.2 and 16.9 eV*) beyond the limit are assigned as $1\pi_u \rightarrow 1\pi_g^*$ and $\sigma \rightarrow \sigma^*$ shake-up excitations in the $(1s_C)^{-1}$ positive ion, while the transitions from $1\pi_u$ to 3p and from $2\sigma_u$ to 3s have frequencies of 122 000 and 178 000 cm^{-1} (*15.1 and 22.1 eV*) [B15, H56]. It is calculated that the transition from $1s_C(\sigma_g)$ to an unspecified σ_g valence orbital will appear as a shape resonance some 160 000 cm^{-1} (*20 eV*) beyond threshold [L32]. The $\sigma \rightarrow \sigma^*$ transition in the $(1s_C)^{-1}$ ion at 136 000 cm^{-1} is the analog of the $3\sigma_g \rightarrow 3\sigma_u^*$ band of the neutral molecule at 107 500 cm^{-1} [H28], Fig. XIII.A-2. Note that symmetry-changing excitations such as $1\pi_u \rightarrow 1\pi_g^*$ formally are forbidden as shake-up processes in a symmetric molecule such as acetylene, but appear because the 1s hole is localized and thereby lowers the molecular symmetry to $C_{\infty v}$.

Haloacetylenes

Spectra of the haloacetylenes Cl–C≡C–H and Br–C≡C–H are especially interesting because they are a complex mixture of valence and Rydberg promotions studied at a time when $\mathscr{A}$-band and Rydberg term-value concepts were not appreciated. Consequently, to this time, work on these spectra has focused upon cataloging the absorption lines and continua and determining the origin frequencies using isotope shifts and vibronic analysis [E9, T10, T11]. Abbreviated Rydberg series have been proposed for both chloroacetylene and bromoacetylene [E9], however, the convergence limits do not agree with the results of photoelectron spectroscopy [H3].

As with the haloethylenes (Section X.C), both $\mathscr{A}(n\text{p}, \sigma^*)$ and $\mathscr{A}(\pi, \sigma^*)$ valence excitations are expected in the haloacetylenes. Note however, that this classification scheme assumes that the $n\text{p}\pi$ and π(C≡C) orbitals are not strongly mixed, so that peaks in the photoelectron spectra can be unambiguously assigned to $n\text{p}\pi$ and π(C≡C) orbitals. This is not so for all the haloacetylenes, HC≡CX, according to Haink *et al.* [H3]. For X = F and Cl, the highest filled orbital is π(C≡C) followed by $n\text{p}\pi$, whereas for X = I, $n\text{p}\pi$ is uppermost. The $\mathscr{A}$-band classification scheme will work for the above systems, with due care for the orbital ordering; however for X = Br, the mixing is so strong that $n\text{p}\pi$ and π(C≡C) orbitals cannot be defined as such, and one must forego the $n\text{p}\pi/\pi$(C≡C) distinction in this compound when discussing its $\mathscr{A}$ bands. Note also that the $\mathscr{A}(\pi, \sigma^*)$ band of HC≡C–X has the same symmetry as the $\mathscr{A}(n\text{p}, \sigma^*)$ band of CH_3–X and thus will have four spin-orbit components as in the methyl halides. However, in this case the splitting will be governed by spin-orbit interactions on the HC≡C– part of the molecule rather than at the halogen, and consequently should be quite small from the spin-orbit point of view. Principal quantum numbers appropriate to Rydberg excitations in the haloacetylenes are chosen as per the footnote on page 241.

The initial clue to assigning the chloroacetylene spectrum is the observation of Thomson and Warsop [T10] that the nature of the absorption bands change at 60 000 cm^{-1} (*7.5 eV*) from weak-and-broad to strong-and-narrow. Additionally, the vibronic band structures below 60 000 cm^{-1} imply strongly bent upper states (valence) whereas those above 60 000 cm^{-1} imply linear upper states (Rydberg). With respect to the $(\pi(\text{C}{\equiv}\text{C}))^{-1}$ ionization potential at 85 740 cm^{-1} (*advert., 10.63 eV*) [H3], the first linear → linear transition at 61 520 cm^{-1} (*advert., 7.6273 eV*) in chloroacetylene has a term value of 24 220 cm^{-1}, suggesting a (π, 3s) upper state. This term value at first sight is somewhat smaller than that one might expect for a transition terminating at 3s in such a small molecule, however, the corresponding term value in chloroethylene is only 23 300 cm^{-1} [**II**, 52] and is 25 270 cm^{-1} in the isoelectronic system cyanogen chloride [A31, M46]. Rydberg excitations to the second ionization potential in chloroacetylene will lie above 88 000 cm^{-1} (*11 eV*) and so will not overlap those converging upon the first ionization limit.

Searching for the $\pi \rightarrow$ 3p excitation in chloroacetylene, we note origins reported at 65 705 and 66 397 cm^{-1} (*advert., 8.1462 and 8.2320 eV*) with term values of 19 340 and 20 030 cm^{-1}. These transitions are forbidden in acetylene, but allowed in the lower symmetry of the haloacetylenes. In acetylene itself, it is calculated that the tight cluster of transitions to 3pπ lies 2600 cm^{-1} above the transition to 3pσ [D14], and this ordering should hold for chloroacetylene as well. A cluster of bands with origins at 71 893, 73 159 and 74 399 cm^{-1} (*advert., 8.9134, 9.0704 and 9.2241 eV*) in chloroacetylene would appear to be transitions to 3d. Note that these bands have unexceptional term values of 13 850, 12 580 and 11 340 cm^{-1}, whereas the three $1\pi_u \rightarrow$ 3d transitions of acetylene itself have anomalously high term values of 17 475 and 17 221 cm^{-1} and a more normal value of 11 517 cm^{-1} [H43]. A (π, ns) Rydberg series in chloroacetylene can be assembled from the origins at 61 520, 75 157, 79 534 and 81 672 cm^{-1}, (*7.6273, 9.3181, 9.8607 and 10.126 eV*), with a quantum defect of 0.80 characterizing the series. The transitions at 75 589 and 76 389 cm^{-1} (*9.3716 and 9.4708 eV*) are components of the transition to (π, 4p).

At frequencies between 40 000 and 60 000 cm^{-1} (*5.0 and 7.5 eV*) weak absorptions appear in chloroacetylene displaying regular vibrations, irregular vibrations and a continuum in successive order. These constitute the $\pi \rightarrow \pi^*$ and $\mathscr{A}(\pi, \sigma^*)$ valence excitations of chloroacetylene, with the latter having a 34 000 cm^{-1} term value with respect to the $(\pi(C{\equiv}C))^{-1}$ ionization potential. Another weak continuum appears at *ca.* 65 000 cm^{-1} (*vert., 8.1 eV*) [E9] and possibly is 3p$\pi \rightarrow \sigma^*$, in which case its term value is 48 600 cm^{-1}.

The situation in bromoacetylene [E9, T11] is much like that in chloroacetylene described above. Using term-value arguments, it is trivial to deduce that the sharp origin band at 57 556 cm^{-1} (*advert., 7.1359 eV*, 25 030 cm^{-1} term value) is the 4p $\rightarrow$ 5s promotion, while that at 64 252 cm^{-1} (*7.9660 eV*, 18 340 cm^{-1} term value) is the 4p $\rightarrow$ 5p promotion. Felps *et al.* [F6] have identified three of the four spin-orbit components of the bromoacetylene (4p, 5s) configuration. Several intense origins in the region of 70 000−72 000 cm^{-1} (*8.7−8.9 eV*) involve 4p $\rightarrow$ 4d excitations. We are unable to assemble a satisfactory 4p $\rightarrow$ ns series from the origins quoted in [E9]. Of course, as discussed above, the π and 4p MO's of bromoacetylene are very strongly mixed, and so the "4p" designation used here must be understood to reflect this.

In the valence region below 57 500 cm^{-1} (*7.13 eV*) in bromoacetylene, two broad continua are reported, centered at 47 600 and 56 200 cm^{-1} (*5.90 and 6.97 eV*). The lower of these two continua has a term value of 35 000 cm^{-1}, which is a very reasonable value for a $(4p\pi, \sigma^*)$ configuration in a bromide (Table I.C-III), whereas the term value of the 56 200-cm^{-1} continuum (26 400 cm^{-1}) would be anomalously low for an $\mathscr{A}$-band in a bromide, and is more likely due to a $\pi \rightarrow \pi^*$ excitation. The corresponding $\mathscr{A}(5p\pi, \sigma^*)$ term value in iodoacetylene [**II**, 116] is 38 540 cm^{-1}.

It is interesting to note that the pattern of absorption bands described above for X–C≡CH closely resembles that in the isoelectronic system X–C≡N [A31, F5, K22, K23, M46]. Note especially that in X–C≡N, the $\mathscr{A}(\pi, \sigma^*)$ bands have a constant term value of 45 000 cm^{-1} for X = Cl, Br and I, and that the equivalent $\mathscr{A}$-band term values in X–C≡CH are nearly constant at 34 000 ± 2000 cm^{-1}. The term value difference implies that the C–X bond strengths are higher in the cyanogen halides than in the corresponding haloacetylenes.

XIII.B. Polyacetylenes and Derivatives

Photolysis of acetylene in the vapor phase and in an argon matrix leads to the formation of diacetylene [**II**, 330] as demonstrated by the spectrum of the photoproduct [C19, O7]. The bands of diacetylene in the matrix have origins at 60 241 and 71 633 cm^{-1} (*adiabat., 7.4688 and 8.8812 eV*), which are said to correlate with the vapor-phase Rydberg origins at 60 767 and 73 638 cm^{-1} (*7.5340 and 9.1297 eV*), respectively. Though the vibronic patterns in the vapor phase and in the matrix support such assignments, it is very odd that the matrix perturbation produces shifts to *lower* frequencies. More normal shifts result if the matrix transitions instead are correlated with the Rydberg origins at 58 540 and 69 152 cm^{-1} (*7.2579 and 8.5736 eV*) in the vapor phase. Okabe [O7] and Kloster-Jensen *et al.* [K25] present spectra of diacetylene determined quantitatively in the region 55 500–83 300 cm^{-1} (*6.9–10.3 eV*). One notes in the spectrum of diacetylene that there is a monotonic decrease of the spectral half widths of the bands with increasing frequency, running from *ca.* 1800 cm^{-1} at 60 000 cm^{-1} (*7.4 eV*) to *ca.* 350 cm^{-1} at 83 000 cm^{-1} (*10.3 eV*) [K25, O7]. Valence TNI resonances involving the $e_u\pi^*$ and $e_g\pi^*$ MO's of diacetylene are observed at 8000 and 45 200 cm^{-1} (*vert., 1.0 and 5.60 eV*) [A11], while the corresponding resonances in the isoelectronic molecule cyanogen are reported at 4600 and 43 300 cm^{-1} (*vert., 0.58 and 5.37 eV*) [T20].

High quality electron-impact spectra of diacetylene, Fig. XIII.B-1, add considerably to our understanding of its excited states. Allan [A9] reports the vertical $\pi \rightarrow \pi^*$ transition to $^3\Delta_u$ at 33 900 cm^{-1} (*4.2 eV*), with a possible stepout to $^3\Sigma_u^+$ at 21 800 cm^{-1} (*vert., 2.7 eV*). The optically forbidden $\pi \rightarrow \pi^*$ transition to $^1\Delta_u$ is found to have an origin at 40 800 cm^{-1} (*5.06 eV*) in the electron impact spectrum, in agreement with the value deduced indirectly from the optical spectrum [H4]. Following these $1\pi_g \rightarrow 2\pi_u^*$ excitations, Allan identifies three new Rydberg series of undetermined spin multiplicity, in addition to those already known optically [**II**, AD170]. Note especially that one series has $\delta = 0.87$ with its $n = 3$ member at 57 800 cm^{-1} (*advert., 7.16 eV*). With a term value of 24 300 cm^{-1}, this is clearly $1\pi_g \rightarrow 3s$. Two Feshbach resonances in diacetylene at 54 100 and 56 500 cm^{-1} (*advert., 6.71 and 7.00 eV*, $\mathbf{r}_1$ and $\mathbf{r}_3$ in Fig. XIII.B-1) have decrements of 3600 and 4400 cm^{-1} with respect to the ($1\pi_g$, 3s) and ($1\pi_g$, 3p) parent Rydberg states and so are assigned as $^2(1\pi_g, 3s^2)$ and $^2(1\pi_g, 3p^2)$, respectively. These Feshbach decrements strongly suggest that the corresponding parent states are singlets rather than triplets.

The first two TNI resonances in dimethyl diacetylene [N21], at 12 700 and 41 500 cm^{-1} (*vert., 1.40 and 5.15 eV*), involve capture of the incident electron in the $2\pi_u^*$ and $2\pi_g^*$ MO's while the next resonance at 68 600 cm^{-1} (*vert., 8.50 eV*) is too close to the $(1\pi_g)^{-1}$ ionization potential of 71 900 cm^{-1} (*advert., 8.91 eV*) to be a Feshbach resonance involving the $1\pi_g$ MO. However, it fits nicely as $^2(1\pi_u, 3p^2)$ with a $(1\pi_u)^{-1}$ ionization potential of 92 400 cm^{-1} (*vert., 11.46 eV*). The corresponding transition to the ($1\pi_u$, 3p) parent state is expected at 72 600 cm^{-1} (*9.0 eV*) in dimethyl diacetylene.

Cyanoacetylene N≡C—C≡CH [**II**, 331] is isoelectronic with both cyanogen and diacetylene and occupies a spectrally intermediate position. The first clue to the interpretation of the cyanoacetylene spectrum is the report that the structured features in the 61 000—66 000-cm^{-1} (*7.6—8.2-eV*) region are unaffected by the application of 1500 psi He pressure and hence must be $\pi \rightarrow \pi^*$ promotions, whereas the sharp features beyond 66 000 cm^{-1} are broadened to higher frequencies as appropriate for Rydberg excitations [C43]. Among the sharp Rydberg bands is a broader valence transition peaked at 69 000 cm^{-1} (*8.6 eV*). In our earlier work [**II**, 331], we assigned the 69 000-cm^{-1} feature of cyanoacetylene as $\pi_2 \rightarrow 3s$, but it is now clear that this very intense feature ($\epsilon = 60\,000$) is a $\pi \rightarrow \pi^*$ band, and that the $\pi_2 \rightarrow 3s$ Rydberg transition falls at 66 550 cm^{-1} (*advert., 8.251 eV*) with a term value of 27 010 cm^{-1}. Connors *et*

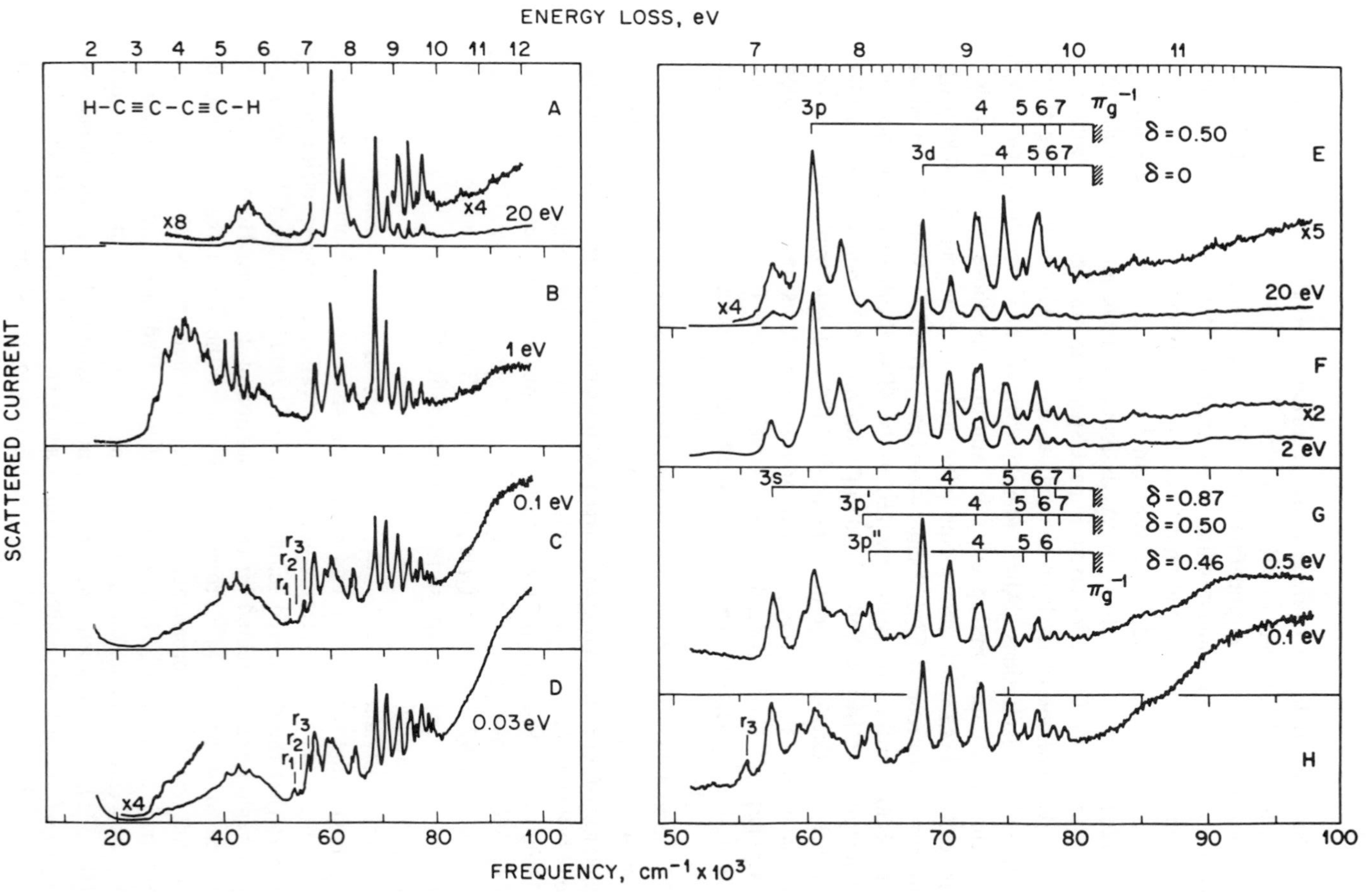

Fig. XIII.B-1. Energy loss spectra of diacetylene **(A–D)** recorded at fixed residual energy as shown. Spectra **E–H** are on an expanded scale [A9]. Bands marked $\mathbf{r_1}$, $\mathbf{r_2}$, and $\mathbf{r_3}$ are Feshbach resonances.

al. [C43] construct a Rydberg series having $\delta = 1.10$ for which the 66 550-cm^{-1} band is the $n = 3$ member. A second series has $\delta = 0.40$ ($\pi_2 \rightarrow n$p) with its $n = 3$ member at 77 400 cm^{-1} (*advert., 9.60 eV*); a second (π_2, 3p) component is found *ca.* 4000 cm^{-1} higher. The general (ϕ_i^o, 3p) splitting pattern for molecules of this type has 3pσ below 3pπ (Section I.B). The intense band at 69 000 cm^{-1} in cyanoacetylene is clearly the ${}^1\Sigma_g^+ \rightarrow {}^1\Sigma_u^+$ ($\pi \rightarrow \pi^*$) excitation, correlating with those at 78 000 cm^{-1} (*advert., 9.7 eV*) in cyanogen and 60 790 cm^{-1} (*advert., 7.537 eV*) in diacetylene [K25].

The spectral parallels drawn above between diacetylene, cyanoacetylene and cyanogen hold as well for the higher homologs, Table XIII.B-I. Thus one sees in these extended systems a few weak $\pi \rightarrow \pi^*$ excitations leading up to very intense ${}^1\Sigma_g^+ \rightarrow {}^1\Sigma_u^+$ bands ($f = 0.5-1$) which are at higher frequencies in the cyanoacetylenes than in the isoelectronic polyacetylenes. In accord with their lower ionization potentials, the Rydberg excitations in the polyacetylenes are found at much lower frequencies (when they can be found at all) than in the cyanoacetylenes, though with nearly identical term values.†

XIII.C. Addendum

C. Y. R. Wu and D. L. Judge, Photoabsorption cross section of acetylene in the EUV region, to be published.

† Note that the three Rydberg series proposed for dicyanodiacetylene [C43] converge upon an ionization limit which is 2500 cm^{-1} higher than the PES value [K26], and so must be reassigned as regards both n and δ.

TABLE XIII.B-I

EXCITATIONS IN THE POLYACETYLENES AND THEIR ISOELECTRONIC NITRILES

Compound	Ionization Potential, cm^{-1} (*eV*)	Frequency, $^1\Sigma_g^+ \rightarrow {}^1\Sigma_u^+$, cm^{-1}	Frequency (Term Value), cm^{-1} $\pi \rightarrow 3s$	$\pi \rightarrow 3p$	$\pi \rightarrow 3d$	References
$H(C{\equiv}C)_2H$	82 110 (*10.18*)	60 790	57 800 (24 300)	60 900 (21 200)	-	[A9, K25, O7]
$HC{\equiv}C{-}C{\equiv}N$	93 560 (*11.60*)	69 000	66 550 (27 010)	77 400 (16 160)	-	[C43, **II**, T21]
$N{\equiv}C{-}C{\equiv}N$	107 800 (*13.36*)	78 000	-	85 980 (21 820)	95 740 (12 060)	[C43]
$H(C{\equiv}C)_3H$	76 600 (*9.50*)	54 600	-	-	-	[H35, K25]
$N{\equiv}C{-}C{\equiv}C{-}C{\equiv}N$	95 260 (*11.81*)	62 150	69 000 (26 260)	75 200 (20 060)	81 000 (14 260)	[C43, **II**, T21]
$H(C{\equiv}C)_4H$	73 320 (*9.09*)	48 000	-	55 370 (17 750)	-	[H35, K25]
$N{\equiv}C{-}(C{\equiv}C)_2{-}C{\equiv}N$	87 755 (*10.88*)	52 000	61 300 (26 460)	69 300 (18 460)	74 550 (13 210)	[C43, K26]
$H(C{\equiv}C)_5H$	-	*ca.* 44 000[a]	-	-	-	[K25]

[a] Vapor-phase frequency estimated from an absorption maximum at 39 800 cm^{-1} (*advert., 4.93 eV*) in *n*-pentane solution.

CHAPTER XIV

Nitriles

Recent optical [N29, R19] and electron impact [F30, S77] studies on acetonitrile [**II**, 117] have extended our understanding of its spectrum, Fig. XIV-1. Nuth and Glicker [N29] have assigned several long Rydberg series converging on the two lowest ionization potentials [$(2e\pi)^{-1}$ and $(7a_1\sigma)^{-1}$, the latter being the lone pair orbital of the C≡N group]. In this work, a Rydberg series going to the $(7a_1)^{-1}$ ionization limit with $\delta = 0.97$ was assigned as $7a_1 \rightarrow np\sigma$, but $7a_1 \rightarrow ns$ is much more likely (see also [F30, R19, S77]). The $n = 3$ member of this series is observed at 78 364 cm^{-1} (*adiabat., 9.7156 eV*), just 1760 cm^{-1} above the dissociative attachment peak of the CN^- fragment ion formed at 76 600 cm^{-1} (*vert., 9.50 eV*) [S75]. The frequency difference suggests a $^2(7a_1, 3s^2)$ Feshbach configuration for the TNI resonance (Section II.B). Excitations going to the second and third ionization potentials of acetonitrile also appear as autoionization lines in the photoionization efficiency curves [R19], and are readily placed into ns ($\delta = 0.93$ and 1.06) and nd ($\delta = 0.2$) Rydberg series. Several of these bands also are observed in the energy-loss spectrum [R14].

Using electron impact, Stradling and Loudon [S77] observe a possible singlet-triplet excitation in acetonitrile at 49 000 cm^{-1} (*vert., 6.08 eV*) and another at 66 100 cm^{-1} (*vert., 8.20 eV*). Though these same bands also

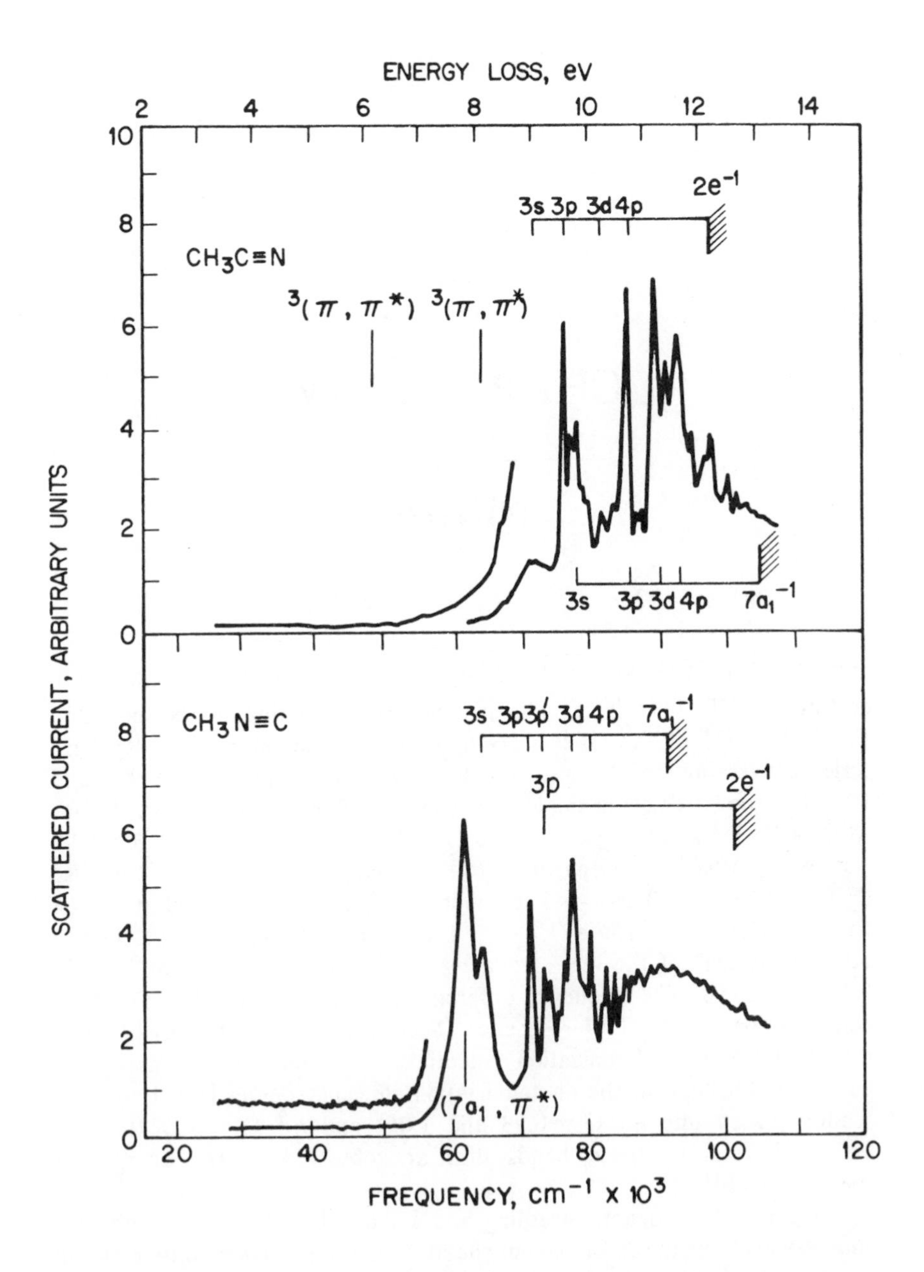

Fig. XIV-1. Electron impact energy loss spectra of acetonitrile (*upper*; 70 eV impact energy, $\theta = 0^\circ$) and methyl isonitrile (*lower*; 45 eV impact energy, $\theta = 3^\circ$) [F30]. Positions of the triplet levels taken from [R14, S77].

are observed in propionitrile and butyronitrile, Fridh [F30] was unable to find the 49 000-cm^{-1} band of acetonitrile in his duplication of the earlier experiment. However, Rianda *et al.* [R14] recently reported a triplet excited state in the 42 000–60 500 cm^{-1} (*5.2–7.5 eV*) region of acetonitrile. Note also that in polyacrylonitrile $(-CH_2CHCN-)_x$ a band centered at 49 000 cm^{-1} (*6.08 eV*) again is observed in the electron impact spectrum [R27], and at 51 000 cm^{-1} (*vert., 6.3 eV*) optically [H20]. If the bands at *ca.* 49 000 cm^{-1} are genuine and have triplet upper states, they must be valence as witnessed by their low frequencies; possibly they are the triplet complements of the singlet transitions at *ca.* 60 000 cm^{-1} (*vert., 7.4 eV*). In an oriented film of polyacrylonitrile [**II**, 118], the band at 63 000 cm^{-1} (*vert., 7.8 eV*; $f = 0.04$) is polarized perpendicular to the C≡N axis, suggesting a $^1\Sigma \rightarrow {}^1\Pi$ ($n_N \rightarrow \pi^*$) assignment for it [H20], and $^1\Sigma \rightarrow {}^3\Pi$ for that at 49 000 cm^{-1}. Yet another triplet state of acetonitrile [R14, S77] (65 300 cm^{-1} *vert., 8.2 eV*) is too far below the lowest singlet Rydberg excitation ($2e \rightarrow 3s$, 72 900 cm^{-1} *vert., 9.04 eV*) to merit consideration as its triplet complement, and so is assigned as $\pi \rightarrow \pi^*$.

As might be expected, the spectrum of methyl isonitrile ($CH_3N{\equiv}C$) [A30, F30] appears rather like that of acetonitrile, Fig. XIV-1. The key difference here is that the $7a_1\sigma$ MO is above the $2e\pi$ occupied MO in the isonitrile rather than below it as in acetonitrile. The isonitrile spectrum begins with an intense $7a_1 \rightarrow 3e\pi^*$ valence transition† at 63 100 cm^{-1} (*vert., 7.82 eV*) followed by $7a_1 \rightarrow 3s$ at 65 150 cm^{-1} (*vert., 8.077 eV*). The term value of the 65 150-cm^{-1} band of the isonitrile (25 530 cm^{-1}) matches closely the ($2e$, 3s) and ($7a_1$, 3s) term values in acetonitrile (26 460 and 27 540 cm^{-1}). The origin of the $2e \rightarrow 3s$ band of the isonitrile falls at 74 960 cm^{-1} (*advert., 9.294 eV*), showing a 25 600-cm^{-1} term value. Fridh calculates that the $\pi \rightarrow \pi^*$ bands of the isonitrile will occur above 80 000 cm^{-1} (*9.9 eV*). A rather intense feature in the optical spectrum of the isonitrile peaking at 54 850 cm^{-1} (*6.800 eV*) [A31] does not appear in Fridh's electron impact spectrum. If genuine, this is a valence excitation, quite possibly $n_C \rightarrow \pi^*$.

Trifluoroacetonitrile is a very transparent substance [A30, A31], Fig. XIV-2, and should be an excellent solvent for vacuum-ultraviolet spectroscopy. The material shows a weak valence transition centered at

† The observed transition may be too intense for the $7a_1 \rightarrow 3e\pi^*$ assignment, however.

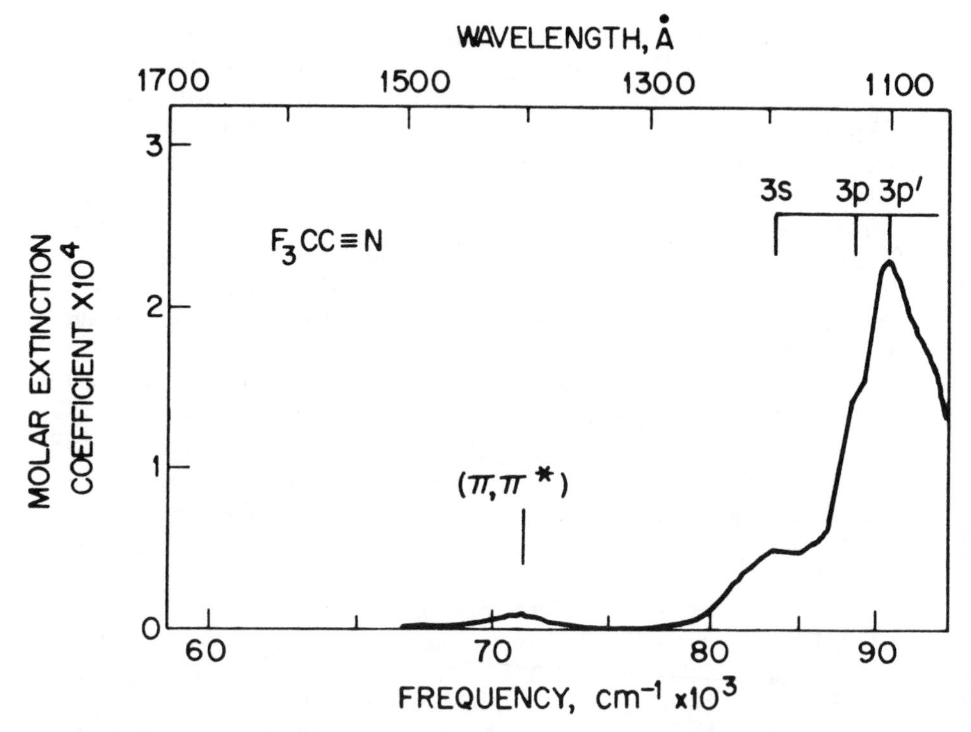

Fig. XIV-2. Absorption spectrum of trifluoroacetonitrile in the vapor phase [A31], illustrating its extreme transparency. The lowest ionization potential is at 111 300 cm^{-1} (*vert., 13.80 eV*).

71 000 cm^{-1} (*8.8 eV*) which is probably[†] $2e\pi \rightarrow \pi^*$. Since the $7a_1\sigma$ and $2e\pi$ MO's in this molecule are accidentally degenerate at 111 300 cm^{-1} (*vert., 13.80 eV*) [A30, A31], one looks for the lowest Rydberg excitations at quite high frequencies and expects overlapping transitions. The transitions to 3s come at 83 800 cm^{-1} (*vert., 10.3 eV*) with 28 000 cm^{-1} term values, and transitions to 3p occur at 88 900 cm^{-1} (*11.0 eV*, 22 400 cm^{-1} term value) and 90 900 cm^{-1} (*11.27 eV*, 20 400 cm^{-1} term value). Ashfold and Simons [A30] point out that the spectrum of trifluoroacetonitrile closely resembles that of hydrogen cyanide.

[†] We carry over the acetonitrile MO designations to describe the equivalent MO's in trifluoroacetonitrile.

Recent work on the optical [C43, N29] and electron impact spectra [F28, H63] of cyanogen [**II**, 120] are readily interpreted using valence calculations [B27] and Rydberg term values. According to calculations by Bell [B27] and by Fridh *et al.* [F28], the lowest excited configuration $1\pi_g^3 2\pi_u^1$ gives rise to the two triplet and two singlet states in the 33 000–50 000 cm^{-1} (*4.1–6.2 eV*) region which previously have been observed optically. The same configuration also yields a $^1\Sigma_u^+$ valence state assigned as the upper level in the intense transition centered at 78 000 cm^{-1} (*9.7 eV*, f = 2.0); another valence excitation ($5\sigma_g \rightarrow 2\pi_u$) peaks at 60 500 cm^{-1} (*7.50 eV*) [F28]. Connors *et al.* have studied cyanogen in the interval 56 000–86 000 cm^{-1} (*6.9–10.7 eV*) when pressurized with 1500 psi He, and report the region to be free of Rydberg absorption [C43]. This is reasonable, since by term-value arguments one knows that the lowest Rydberg excitations will reside in the region between 80 000 and 100 000 cm^{-1} (*10.2 and 12.4 eV*). Based on term-value considerations, the lower members of several Rydberg series in cyanogen can be assigned as in Table XIV-I. The $1\pi_g \rightarrow 3s$ transition will fall at *ca.* 80 000 cm^{-1} (*9.8 eV*), where it would be buried by the intense $\pi \rightarrow \pi^*$ valence transition to $^1\Sigma_u^+$. The electron impact spectrum of malononitrile, $N{\equiv}CCH_2C{\equiv}N$, resembles that of acetonitrile rather than cyanogen due to the lack of conjugation through the methylene group [R14].

The TNI resonance spectrum of cyanogen is understandable in large part. A resonance at 72 600 cm^{-1} (*9.00 eV*) [N21] is claimed to be too broad to be a Feshbach resonance, but it is to be noted that it has a 6400 cm^{-1} Feshbach decrement with respect to the ($1\pi_g$, 3s) Rydberg state expected at 79 000 cm^{-1} (Table XIV-I). Consequently, we prefer a $^2(1\pi_g, 3s^2)$ Feshbach assignment (Section II.B) for the 72 600-cm^{-1} band. Two other resonances observed at 4700 and 43 300 cm^{-1} (*vert., 0.58 and 5.37 eV*) are assignable as valence TNI states involving the $2\pi_u^*$ and $2\pi_g^*$ MO's of cyanogen, considering that the $e\pi^*$ level of HCN comes at 18 200 cm^{-1} (*vert., 2.26 eV*). For comparison, the two π^* valence TNI resonances in isoelectronic diacetylene are reported to fall at 8000 and 45 200 cm^{-1} (*vert., 1.0 and 5.6 eV*) [A11]. By default, a fourth resonance at 58 900 cm^{-1} (*vert., 7.3 eV*) [T20] in cyanogen either involves attachment of the incident electron to a σ^* MO or is due to a core-excited valence TNI resonance.

Excitations from $1\pi_g$ to $2\pi_u^*$ account for the valence excitations in cyanogen at frequencies below the $(1\pi_g)^{-1}$ ionization potential, whereas the transitions from $1\pi_g$ to the $2\pi_g^*$ and $5\sigma_u^*$ virtual MO's lie above the $(1\pi_g)^{-1}$ ionization. The $1\pi_g \rightarrow 5\sigma_u^*$ shape resonance in cyanogen is deduced to lie at approximately 140 000 cm^{-1} (*vert., 17 eV*) as revealed by the angular dependence of the photoelectron spectrum [K37].

TABLE XIV-I

RYDBERG EXCITATION FREQUENCIES AND TERM VALUES IN CYANOGEN

Transition	Frequency, cm^{-1} (eV)	Term Value, cm^{-1}	References
$(1\pi_g)^{-1}$ I.P.	107 800 (*13.36*)	-	
$1\pi_g \rightarrow 3p$	85 980 (*10.66*)	21 820	[C43, F28, H63, N29]
$1\pi_g \rightarrow 3d$	95 740 (*11.87*)	12 100	
$(4\sigma_u)^{-1}$ I.P.[a]	116 900 (*14.49*)	-	
$4\sigma_u \rightarrow 3s$	88 500 (*10.97*)	28 400	
$4\sigma_u \rightarrow 3p$	96 300 (*11.94*)	20 600	
$4\sigma_u \rightarrow 3d$	105 600 (13.09)	11 300	
$(5\sigma_g)^{-1}$ I.P.[a]	119 800 (*14.85*)	-	
$5\sigma_g \rightarrow 3s$	91 300 (*11.32*)	28 500	
$5\sigma_g \rightarrow 3p$	100 740 (*12.49*)	19 100	
$5\sigma_g \rightarrow 3d$	107 000 (*13.30*) 107 500 (*13.33*)	12 800 12 300	
$(1\pi_u)^{-1}$ I.P.	124 800 (*15.47*)	-	
$1\pi_u \rightarrow 3s$	96 830 (*12.00*)	28 000	
$1\pi_u \rightarrow 3p$	103 400 (*12.82*) 105 600 (*13.09*)	21 400 19 200	
$1\pi_u \rightarrow 3d$	112 900 (*14.00*)	11 900	
$(1s_C)^{-1}$ I.P.	2.3754×10^6 (*294.50*)	-	[H63]
$1s_C \rightarrow 3s$	2.3431×10^6 (*290.50*)	32 200	
$1s_C \rightarrow 3p$	2.3528×10^6 (*291.70*)	22 600	
$1s_C \rightarrow 3d$	2.3495×10^6 (*291.30*)	11 300	
$(1s_N)^{-1}$ I.P.	3.2860×10^6 (*407.44*)	-	
$1s_N \rightarrow 3s$	3.2537×10^6 (*403.40*)	32 200	
$1s_N \rightarrow 3p$	3.2642×10^6 (*404.70*)	21 800	
$1s_N \rightarrow 3d$	3.2715×10^6 (*405.60*)	14 500	

[a] According to the theoretical work of Bell [B27], the $5\sigma_g$ orbital lies slightly above the $4\sigma_u$ orbital, however, we find that the observed Rydberg relative intensities are much more understandable if the ordering is reversed as in the Table.

Pi-electron molecules can show four types of excitation in the 1s inner-shell spectrum: *i)* 1s → π^* valence excitation, intense and below the ionization limit, *ii)* 1s → ϕ_j^{+1} Rydberg excitation, weak but sharp and below the ionization limit, *iii)* 1s → σ^* shape resonance, very broad and beyond the ionization limit, and *iv)* shake-up excitation involving $(1s)^{-1}$ ionization and simultaneous bound excitation of a second electron. Possibly, cyanogen shows all these [H63]. At *ca.* 67 000 cm^{-1} (*8.3 eV*) below the $(1s_C)^{-1}$ and $(1s_N)^{-1}$ ionization limits of cyanogen there are observed intense bands assigned as 1s → $2\pi_u^*$. These bands are followed by inner-shell Rydberg excitations, as shown in Table XIV-I. Beyond the $(1s_C)^{-1}$ ionization potential, three broad features extend another 150 000 cm^{-1} (*18.6 eV*). A broad feature of this sort also is observed in the valence region, centered at 137 000 cm^{-1} (*17.0 eV*) [F28], and as mentioned above, is assigned as a shape resonance. As regards the other broad features, it is difficult to know whether they are shake-up excitations, photoionization cross-section maxima, or shape resonances.

The tabulation of Table XIV-I is interesting from several points of view. First, one sees that all the Rydberg excitations have term values of either 30 000 ± 2000 cm^{-1} (ϕ_i → 3s), 20 000 ± 2000 cm^{-1} (ϕ_i → 3p) or 12 000 ± 1000 cm^{-1} (ϕ_i → 3d), and that approximately the same term values are observed in the outer-shell and inner-shell spectra. Actually, with regard to the latter point, the (ϕ_i^{-1}, 3s) inner-shell term values are systematically larger than their outer-shell counterparts by approximately 4000 cm^{-1}. This exhaltation of the (ϕ_i^{-1}, 3s) inner-shell term value is due to the antishielding effect that results from the creation of an inner-shell hole, Section I.B. A slight exhaltation of the inner-shell (ϕ_i^{-1}, 3p) term value also is apparent in the data of Table XIV-I.

Spectra of various cyanoacetylenes are discussed in Section XIII.B, where they are compared with those of the polyacetylenes.

CHAPTER XV

Amides, Acids, Esters and Acyl Halides

XV.A. Amides

It has been proposed [II, 122] that the amide-group spectrum is characterized by a progression of five singlet excited states:- ^{1}W (n_O, π_3^*), $^1R_1$ (n_O, 3s) / (π_2, 3s), $^1V_1$ (π_2, π_3^*), $^1R_2$ (n_O, 3p) / (π_2, 3p), and ^{1}Q (π_1, π_3^*) descending into the vacuum ultraviolet.† The $^1(\pi_1, \pi_3^*)$ assignment of the band labelled ^{1}Q remains doubtful (*vide infra*). Depending upon the patterns of alkyl-group substitutions in amides, pairs of neighboring bands may be badly overlapped or even inverted in order, nonetheless, work to the present has confirmed in large part the general ordering stated above. In the triplet manifold of the amide group, the $^3(n_O, \pi^*)$ configuration remains lowest, while the $^3(\pi_2, \pi_3^*)$ configuration

† The configurations of the ^{1}R upper states are written as (n_O, ϕ_j^{+1}) / (π_2, ϕ_j^{+1}) because the n_O and π_2 MO's in most amides are near-degenerate, and the $n_O \rightarrow \phi_j^{+1}$ and $\pi_2 \rightarrow \phi_j^{+1}$ Rydberg transitions cannot be resolved.

drops onto or below $^3R_1$ since the triplet Rydbergs are only a 1000 cm^{-1} or so below their singlet complements, whereas the singlet-triplet splits in the valence (π, π^*) manifold are more than an order of magnitude larger; the frequency of the ^3Q band is again uncertain.

Using the trapped-electron technique, various singlet → triplet excitations in formamide and dimethyl formamide have been observed [S69], Fig. XV.A-1. The lowest triplet in formamide falls at 42 700 cm^{-1} (*vert., 5.30 eV*), and is assigned a $^3(n_O, \pi_3^*)$ upper state; the corresponding singlet excitation is observed at 45 600 cm^{-1} (*vert., 5.65 eV*). The second and third triplets occur at 53 200 cm^{-1} (*vert., 6.60 eV*) [$^3(n_O$, 3s) and $^3(\pi_2$, 3s) overlapped] and at 80 700 cm^{-1} (*vert., 10.0 eV*). In broad spectra such as these, Rydberg excitations to (n_O, 3s) and (π_2, 3s) will not be resolved, for the $(n_O)^{-1}$ and $(\pi_2)^{-1}$ ionization potentials in formamide [**II**, 123] are separated by only 1600 cm^{-1}. In any event, the peak of the 53 200-cm^{-1} triplet excitation is only 800 cm^{-1} below that of the corresponding (n_O, 3s) / (π_2, 3s) singlets, illustrating the very small singlet-triplet split exhibited by Rydberg excitations [**I**, 22]. It is likely that the excitation to $^3(\pi_2, \pi_3^*)$ in formamide also falls in the vicinity of 50 000 cm^{-1} (*6.2 eV*), for numerous theoretical calculations place it between 40 000 and 50 000 cm^{-1} (*5.0 and 6.2 eV*), more likely at the higher end [H12, L26, O10, S72]. The peaks labelled π^* in the Figure correspond to valence TNI resonances into the vacant π^* MO's [L26].

Assignment of the triplet state at 80 700 cm^{-1} (*vert., 10.00 eV*) in formamide is uncertain. Since the $^3(n_O$, 3p) – $^3(\pi_2$, 3p) complex will come at *ca.* 63 000 cm^{-1}, it can be eliminated, as can the triplet excitations to 4s. On the other hand, considering a $^3(\pi_1$, 3s) Rydberg term value of about 32 000 cm^{-1} with respect to the $(\pi_1)^{-1}$ ionization at 114 400 cm^{-1} (*vert., 14.18 eV*) in formamide, leads one to expect the π_1 → 3s triplet excitation approximately 82 400 cm^{-1} above the ground state. It is also possible that the excitation to $^3(\pi_1, \pi_3^*)$ comes in the 80 000-cm^{-1} region, but $^3(\pi_1$, 3s) seems a safer assignment.

In dimethyl formamide [**II**, 132], the alkylation of the amino group inverts the ordering of the π_2 and n_O orbitals, placing π_2 higher by 4000 cm^{-1}. The trapped-electron spectrum of this material shows three bands, Fig. XV.A-1, again assumed to have triplet upper states [S69]. The first triplet, at 40 300 cm^{-1} (*vert., 5.00 eV*), has the $^3(n_O, \pi_3^*)$ configuration. The second triplet, at 48 400 cm^{-1} (*vert., 6.00 eV*), is 2000 cm^{-1} below the corresponding $^1R_1$ band, while the third, at 56 500 cm^{-1} (*vert., 7.00 eV*), is 2200 cm^{-1} below the $^1R_2$ band. Thus the second and third trapped-electron bands of dimethyl formamide have $^3(\pi_2$, 3s) and $^3(\pi_2$, 3p) configurations, respectively. Once again, there is no clear transition to $^3(\pi_2, \pi_3^*)$.

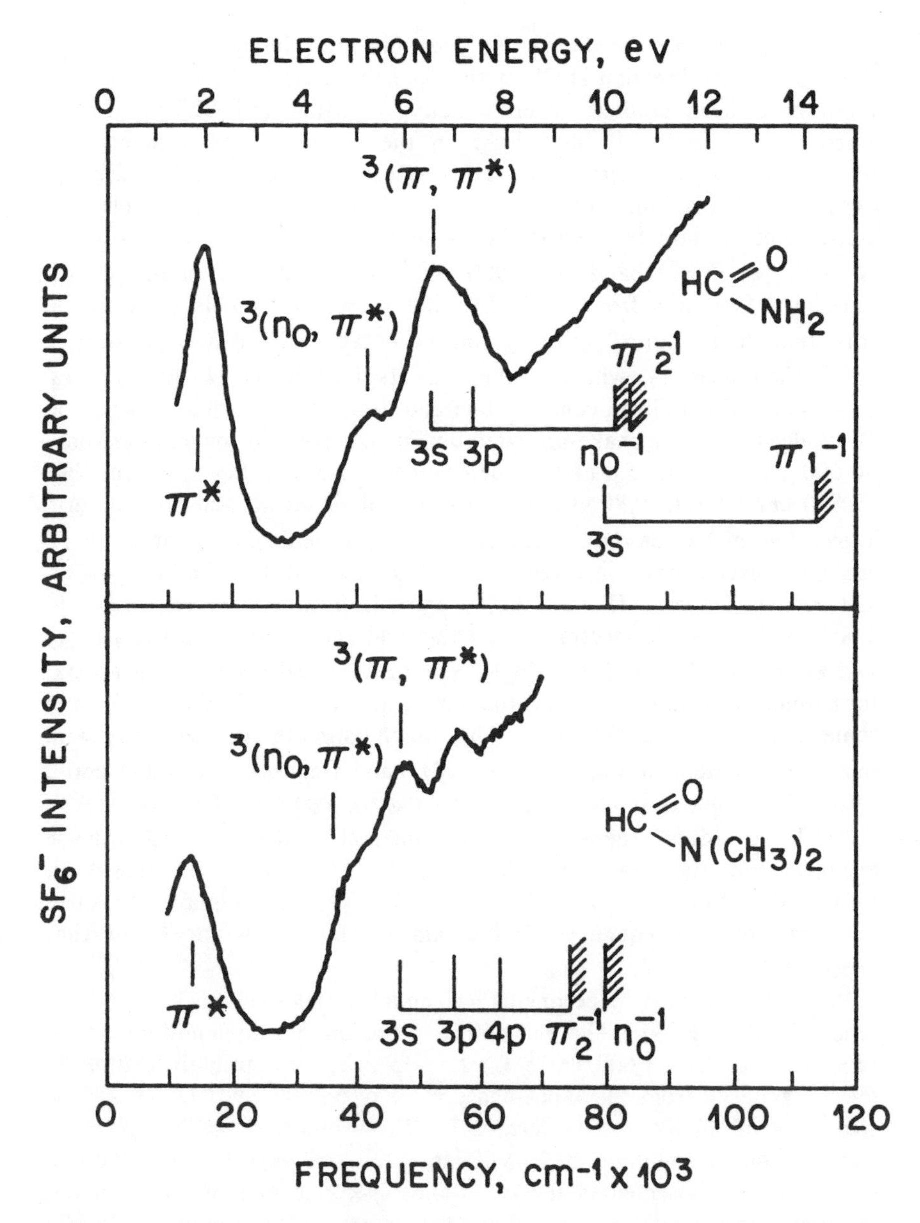

Fig. XV.A-1. Trapped-electron spectra of formamide and dimethyl formamide [S69], interpreted in terms of excited triplet states and π^* valence TNI resonances.

Basch [B17] has predicted the locations and intensities of valence shake-up bands (Section II.C) in the K-shell spectra of formamide, and these have been studied experimentally by Mills and Shirley [M62]. These studies indirectly shed light on the $N \rightarrow V_2$ band frequency in formamide, which otherwise is rather uncertain. For this, we use the experimental shake-up spectra of the 1s-hole ions and the pi-electron densities of formamide given in the Table in [**II**, 132]. We begin with the $N \rightarrow V_1$ ($\pi_2 \rightarrow \pi_3^*$) band at 58 900 cm^{-1} (*vert., 7.30 eV*) in the neutral molecule. One sees from the Table that when the 1s hole is on C the transition from π_2 to π_3^* will be stabilized because π_3^* has a larger density on C than does π_2, whereas when the 1s hole is on N the $\pi_2 \rightarrow \pi_3^*$ transition is destabilized considerably, and destabilized somewhat when on O. Indeed, the $^1V_1$ shake-up excitation in the positive ion requires only 40 000 cm^{-1} (*vert., 5.0 eV*) as determined from the $1s_C$ spectrum, but 79 800 cm^{-1} (*vert., 9.89 eV*) when the 1s hole is on N, and 59 700 cm^{-1} (*vert., 7.40 eV*) when on O. Similar trends are predicted for the $\pi_1 \rightarrow \pi_3^*$ shake-up excitations, observed at 77 400 cm^{-1} (*vert., 9.60 eV*; $1s_C^{-1}$), 121 000 cm^{-1} (*vert., 15.0 eV*; $1s_N^{-1}$), and 112 000 cm^{-1} (*vert., 13.88 eV*, $1s_O^{-1}$) in the K-shell spectra. From this and the electron densities in π_1 and π_3^*, one predicts that the singlet $\pi_1 \rightarrow \pi_3^*$ vertical excitation in neutral formamide will fall above 77 400 cm^{-1}, far below 121 000 cm^{-1} and somewhat below 112 000 cm^{-1}. This rough estimate accords nicely with our earlier statement that the $^1(\pi_1, \pi_3^*)$ band frequency (105 000 cm^{-1}, *13.0 eV*) is approximately twice that of the $^1(\pi_2, \pi_3^*)$ band (58 000 cm^{-1}, *7.2 eV*). To date, there is no experimental evidence for a valence excitation in the 100 000–120 000 cm^{-1} (*12.4–14.9 eV*) region in formamide. Moreover, the identity of the ^1Q band remains an open question, for its frequency (73 000 cm^{-1}, *9.05 eV*) is far below that expected for $\pi_1 \rightarrow \pi_3^*$.

Vacuum-ultraviolet experiments on liquid formamide also have been reported [B48, H39]. The reflectance spectrum of the liquid shows an $N \rightarrow V_1$ peak at 55 600 cm^{-1} (*vert., 6.89 eV*), appropriately shifted to lower frequency from the vapor-phase value of 58 900 cm^{-1} (*7.30 eV*). A frequency of 54 500 cm^{-1} (*vert., 6.76 eV*) is observed for the $N \rightarrow V_1$ band of solid formamide at 24.5 K [**II**, 129]. As expected, no Rydberg excitations are observed in the reflectance spectrum of liquid formamide, however, another peak in the ϵ_2 spectrum is seen in the liquid at 129 000 cm^{-1} (*vert., 16.0 eV*) and the corresponding energy loss appears centered at 220 000 cm^{-1} (*27 eV*). This latter is thought to be a volume plasmon; the corresponding surface plasmon of the liquid was later assigned to the peak in the 120 000–150 000 cm^{-1} (*15–19 eV*) region [B48]. Integration of ϵ_2 up to 198 000 cm^{-1} (*24.5 eV*) in liquid formamide corresponds to 12 electrons participating in electronic excitations. Since this figure equals the number of electrons in the molecule outside the 1s cores, all outer-shell excitations therefore fall below 198 000 cm^{-1} in this molecule.

Substitution of the keto oxygen of formamide by sulfur produces thioformamide. With respect to the first ionization potential (n_S) of thioformamide at 70 170 cm^{-1} (*vert., 8.700 eV*), the two absorption bands in this molecule at 43 020 and 50 000 cm^{-1} (*5.334 and 6.199 eV*) [T24] have term values of 27 150 and 20 170 cm^{-1}, respectively, suggesting $^1(n_S, 4s)$ and $^1(n_S, 4p)$ Rydberg assignments. Less likely is a $^1(\pi_2, 4s)$ assignment for the 50 000-cm^{-1} band, for this results in a term value (25 000 cm^{-1}) which falls below that expected of 4s in a compound of such small size. It is also possible that the $\pi_2 \rightarrow \pi_3^*$ excitation falls at 50 000 cm^{-1} in thioformamide.

On going from formaldehyde to thioformaldehyde, there is a remarkably constant low-frequency shift (11 000 ± 1000 cm^{-1}) of all transitions originating with the lone-pair orbitals (Section XII.C). Though data is much more sparse for the formamide/thioformamide pair, again a constant low-frequency shift (12 500 ± 1500 cm^{-1}) is apparent. This constant decrement can be applied to the observed $^3(n_O, \pi^*)$ and $^1(n_O, \pi^*)$ frequencies of formamide to predict corresponding transition frequencies of 30 200 and 33 100 cm^{-1} in thioformamide. The $^1(n_S, \pi^*)$ state of thioformamide is observed at 37 000 cm^{-1} [T24].

Several calculations on the amide group [N27, O10, S72] demonstrate that the π_3^* orbitals in the singlet and triplet (π_2, π_3^*) configurations are quite different. In the singlet manifold, there is heavy mixing of (π_2, π_3^*) with $(n_O, 3s)$ and $(n_O, 3p\sigma)$ configurations resulting in considerable big-orbit character in the singlet state which is absent in the triplet. This configuration mixing is affected by alkylation [N28]. The large difference in the π^* MO's of complementary singlet and triplet (π, π^*) configurations appears to be quite general for pi-electron chromophores if the upper singlet state is ionic. (However, see butadiene, Section XVII.A.) The Rydberg/valence mixing in formamide is an unusual one since the levels involved are nonconjugate, though of the same state symmetry (Section I.D). Nonetheless, it is clear from the large differences in the $(n_O/\pi_2)^{-1}$ PES and $n_O/\pi_2 \rightarrow 3s$ Rydberg absorption band shapes [**II**, B52] that the Rydberg configuration is strongly mixed with the valence excitation of like symmetry (Section I.D).

An extensive *ab initio* calculation on the spectrum of urea [E6] predicts an $n_O \rightarrow \pi_4^*$ transition at 59 300 cm^{-1} (*7.35 eV*), followed by the $\pi_3 \rightarrow \pi_4^*$ and $\pi_2 \rightarrow \pi_4^*$ allowed transitions at 72 400 and 87 800 cm^{-1} (*8.98 and 10.9 eV*). On this basis, the one observed peak in urea at 58 400 cm^{-1}

(*vert., 7.24 eV*) [**II**, 140] has been assigned as $n_O \rightarrow \pi_4^*$. However, since the calculation did not use diffuse π orbitals in the basis set, the $\pi \rightarrow \pi^*$ excitations as calculated are undoubtedly too high by 10 000–15 000 cm^{-1} (*1–2 eV*) [S72]. Because of this, a $\pi_3 \rightarrow \pi_4^*$ assignment is still quite viable for the 58 400-cm^{-1} band; a more secure assignment awaits a measurement of the oscillator strength of this transition.

In a polypeptide [**II**, 140], the $N \rightarrow V_1$ excitations of the individual amide groups are strongly coupled and split, resulting in two "$N \rightarrow V_1$" peaks in the 50 000–65 000 cm^{-1} (*6.2–8.0 eV*) region. As expected from such an excitonic splitting, the second of these has an intensity which is conformation dependent [M51]. Photoelectron spectra of polypeptide films reveal electron binding energies significantly lower than those of the corresponding vapor-phase amides. Thus, in films of poly(glycine), poly(L-alanine) and poly(L-valine) the thresholds for electron emission are 59 000, 59 000 and 58 000 cm^{-1} (*7.3, 7.3 and 7.2 eV*), whereas in the vapor phase, the ionization thresholds for formamide, N-methyl formamide and N,N-dimethyl formamide are 81 500, 78 000 and 73 000 cm^{-1} (*10.1, 9.7 and 9.1 eV*) [I10, I11]. The figures quoted for the polypeptide films probably apply to both $(n_O)^{-1}$ and $(\pi_2)^{-1}$ ionizations in each case. The ionization threshold for each of the polypeptides falls on the high-frequency tail of the $N \rightarrow V_1$ amide band centered at *ca.* 52 600 cm^{-1} (*6.5 eV*).

The amide group $R_2N{-}C\overset{O}{\underset{R}{}}$ is isoelectronic with the suitably alkylated oxime $R_2C{=}N{-}O{-}R$, and their vacuum-ultraviolet spectra are similar, Section XI.B.

XV.B. Acids and Esters

Our previous experience was that amide spectra [**II**, 122] and acid/ester/acyl halide spectra [**II**, 146] stood in one-to-one correspondence. This remains a valid point of departure for discussion of the spectra of acids and esters.

Several experimental reports on the spectrum of formic acid improve upon the older data [**II**, 152]. Fridh [F29] presents the electron impact spectrum shown in Fig. XV.B-1, along with a semiempirical calculation which predicts a ^1W, $^1R_1$, $^1V_1$, $^1R_2$, $^1V_2$ ordering of the bands. Among the structured features, Fridh constructs an ns Rydberg series with $\delta = 1.14$ and three np series with quantum defects of 0.94, 0.66 and 0.61, all originating at n_O. A δ value of 0.94 is surprising large for an np series, and is based upon an $n = 3$ member at 67 400 cm^{-1} (*vert., 8.36 eV*)

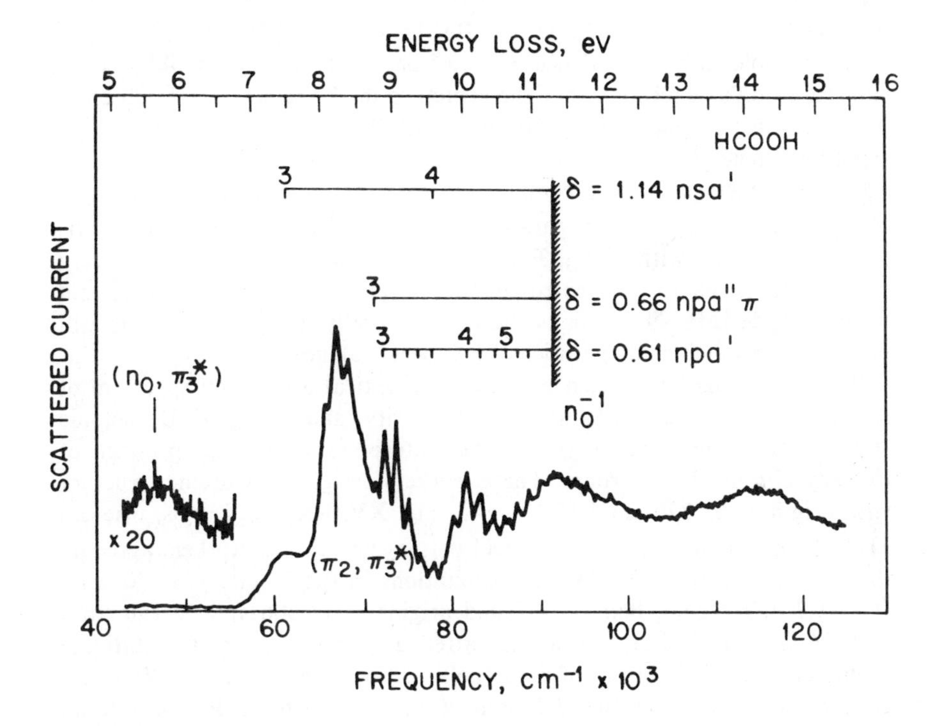

Fig. XV.B-1. Electron impact energy-loss spectrum of formic acid, with Rydberg transitions assigned according to Fridh [F29] in large part.

which we earlier had assigned as the vibronic structure of the $N \rightarrow V_1$ band. Fridh also observes a broad continuum at 115 000 cm^{-1} (*vert., 14.3 eV*) in the vapor spectrum (see also [A22]) which perhaps is related to the band at 129 000 cm^{-1} (*16.0 eV*) in liquid formamide [H39], or possibly might be the $\pi_1 \rightarrow \pi_3^*$ excitation expected in this frequency regime in formamide (Section XV.A). In any event, the band is too intense by far to support a Rydberg assignment, even though its term value suggests an (8*a*′, 3s) upper-state configuration.

Similar, but less-resolved electron-impact spectra of formic acid are reported by Ari and Hasted [A22], who studied the generalized oscillator strengths of various peaks as a function of momentum transfer. They invoked the assumption that Rydberg excitations will show local minima in the generalized oscillator strength as a function of the momentum transfer, whereas valence transitions will not. Proper minima were observed for excitation at 61 900 cm^{-1} (*7.68 eV*, $n_O \rightarrow 3s$), 72 200 cm^{-1} (*8.95 eV*,

$n_O \rightarrow 3p$) and 81 790 cm^{-1} (*10.14 eV*, $n_O \rightarrow 4p$). However, the same minimum was observed as well for the peak at 67 400 cm^{-1} (*8.36 eV*), which we hold to be the valence $\pi_2 \rightarrow \pi_3^*$ band! This result would seem to support Fridh's suggestion that the structure in the region of 67 000 cm^{-1} has a Rydberg upper state. However, all these data have the "wrong" slope for dipole-allowed excitations as the momentum transfer goes to zero, and so the picture is more complicated than first thought. Furthermore, the valence $^1(\pi_2, \pi_3^*)$ configuration in formic acid is expected to be strongly mixed with $^1(n_O, 3s)$, $^1(n_O, 3p\sigma)$ and $(\pi_2, 3p\pi)$ Rydberg configurations, as for the corresponding states in formamide [N27] and formaldehyde [M61]. If this is the case generally, then such minima will not be worth much as diagnostics of Rydberg/valence character.

Bell *et al.* [B26] have reinvestigated the optical absorption spectrum of formic acid at high resolution, high purity and using H/D isotopic substitution. The result is yet another interpretation of the spectrum of this key chromophoric group. These workers assign the vibronic structure with origin at 65 370 cm^{-1} (*8.105 eV*), Fig. XV.B-1, as $n_O \rightarrow 3s$, whereas Fridh took it as $n_O \rightarrow 3p$ and we feel it is $\pi_2 \rightarrow \pi_3^*$. All workers agree on the presence of $n_O \rightarrow 3p$ Rydberg excitations in the 71 000–73 000-cm^{-1} (*8.8–9.1 eV*) region; Bell *et al.* found origins at 71 306 and 72 196 cm^{-1} (*8.8406 and 8.9510 eV*) with the latter displaying a type C rotational profile characteristic of out-of-plane polarization. Taking the 67 400-cm^{-1} band as terminating at $(n_O, 3s)$, Bell *et al.* constructed a Rydberg series with $\delta = 0.85$, in addition to a series with $\delta = 0.60$ (np) and a series having $\delta = 0.14$ (nd).

Upon reconsideration of the total picture in formic acid, we are led to affirm the assignments first given in [**II**, 146]. If for no other reason than its oscillator strength ($f = 0.37$), it is clear that the 67 400-cm^{-1} band of formic acid is the valence $\pi_2 \rightarrow \pi_3^*$ excitation, and not a Rydberg transition. Fridh's interpretation of this band as a component of $n_O \rightarrow 3p$ leads to an improbably high term value of 25 400 cm^{-1} and a 3p core splitting of 6900 cm^{-1}, whereas values of 20 000 and 2500 cm^{-1} are more normal. The assignment by Bell *et al.* of the 67 400-cm^{-1} band as $n_O \rightarrow 3s$ results in a term value which is unreasonably small for a molecule the size of formic acid, and offers no explanation for the transition observed at 60 000 cm^{-1}. Finally, the 67 400 cm^{-1} band of formic acid is clearly related to the intense band at 59 000 cm^{-1} (*7.37 eV*) in formamide, which has been shown to be a valence excitation by its behavior in neat crystal and matrix spectra [**II**, 129]. Calculations on formic acid confirm that the bands at 70 000–73 000 cm^{-1} (*8.7–9.1 eV*) are $n_O \rightarrow 3p$, that $n_O \rightarrow 3s$ comes below $\pi_2 \rightarrow \pi_3^*$ and that no $n_O \rightarrow \sigma^*$ valence excitations appear below 130 000 cm^{-1} (*16.1 eV*) [D17].

The $n_O \rightarrow 3p$ band of acetic acid [**II**, 152] is found in the region between 65 800 and 71 300 cm^{-1} (*8.16 and 8.84 eV*) with several quanta of C=O stretch attached, while the transition from n_O to 3d occupies the region between 72 600 and 74 800 cm^{-1} (*9.00 and 9.28 eV*) [B26]. Many other sharp features were recorded but not assigned at the high-frequency end of this spectrum.

Multilayer films of stearic acid show a broad $\pi_2 \rightarrow \pi_3^*$ absorption at 58 100 cm^{-1} (*vert., 7.20 eV*) followed by the alkyl-chain absorption beginning at *ca.* 64 000 cm^{-1} (*7.9 eV*) and rising to a maximum at 69 400 cm^{-1} (*8.60 eV*) [N1, N3]. This latter closely tracks the absorption in polyethylene and lengthy alkanes such as $C_{36}H_{74}$ [**I**, 135 and Section III.D]. Using oriented multilayers of barium stearate, Nagahira *et al.* [N3] have measured polarized absorption and interpret the results to show that in the alkyl-group absorption region, the polarization is short-axis at the low-frequency end (64 000−72 000 cm^{-1}; *7.9−8.9 eV*) but tends to become more long-axis in the 72 000−80 000 cm^{-1} (*8.9−9.9 eV*) region. Unfortunately, this does not agree with the polarized absorption results on polyethylene itself, Fig. III.D-1. Spectra also are reported of other heavy-metal salts of fatty acid multilayers [N2].

The vacuum-ultraviolet bands of acrylic acid ($H_2C{=}CHCO_2H$) are broad and poorly resolved [M68]. The conjugation of the vinyl and carboxyl groups in this molecule results in a low-frequency shift of the carboxyl $n_O \rightarrow \pi_3^*$ band from 46 000 cm^{-1} (*vert., 5.70 eV*) to 41 100 cm^{-1} (*vert., 5.10 eV*) and a low-frequency shift of the vinyl $\pi \rightarrow \pi^*$ transition from 61 600 to 54 100 cm^{-1} (*vert., 7.64 to 6.71 eV*). The intense $\pi \rightarrow \pi^*$ band persists in the solid, as does a second valence excitation at 61 800 cm^{-1} (*vert., 7.66 eV*) which has been assigned as a carboxyl-to-vinyl intramolecular charge transfer [M68]. A Rydberg band at 60 200 cm^{-1} (*vert., 7.46 eV*; term value, 27 000 cm^{-1}; $n_O \rightarrow 3s$) in acrylic acid is very sensitive to pressure and phase. Addition of a second *cis* carboxyl group to acrylic acid to form maleic acid ($HO_2CCH{=}CHCO_2H$) further shifts the lowest $\pi \rightarrow \pi^*$ band to 51 700 cm^{-1} (*vert., 6.41 eV*), where its oscillator strength is 0.35 [M69].

Spectra of mixtures of triethyl amine and acetic acid vapors [N4] show an intense band ($f = 0.2$) at 60 600 cm^{-1} (*vert., 7.51 eV*) which is absent in the spectra of the constituents. The new band occurs at 58 800 cm^{-1} (*vert., 7.29 eV*) in heptane and acetonitrile solutions, and is due to an amine → carboxyl charge transfer in the hydrogen-bonded complex. Since the π_3^* valence MO is vacant in the carboxyl group, the amine → carboxyl charge transfer is a valence transition. Charge-transfer transitions of this sort no longer are relevant in the spectra of the solid amino acids, for the amino lone pair electrons are spectroscopically inactive in the zwitterionic forms $R{-}C\begin{smallmatrix}NH_3^+ \\ CO_2^-\end{smallmatrix}$. Spectra of thin films of the aliphatic α-amino acids [I2] display a clear carboxylate $\pi_2 \rightarrow \pi_3^*$ band at 56 000−63 000 cm^{-1} (*vert., 7.0−7.8 eV*) depending upon alkylation. A second band of some prominence in the 63 000−69 000 cm^{-1} (*7.8−8.5 eV*) region is possibly a charge-transfer band involving $-NH_3^+$ and $-CO_2^-$ groups [M5] or less likely an exciton component of the $\pi_2 \rightarrow \pi_3^*$ transition.

Polymethyl methacrylate (PMMA) [R22, **II**, 155] contains the ester chromophore and consequently its energy loss spectrum has an $n_O \rightarrow \pi_3^*$ transition at 46 300 cm^{-1} (*vert., 5.74 eV*) [compare to 47 000 cm^{-1} (*vert., 5.83 eV*) in acetic acid vapor] and an intense $\pi_2 \rightarrow \pi_3^*$ band at 60 300 cm^{-1} (*vert., 7.48 eV*) [compare to 58 100 cm^{-1} (*vert., 7.20 eV*) in stearic acid and 64 500 cm^{-1} (*8.00 eV*) in methyl formate vapor]. Several other energy-loss features observed in the 65 000−100 000 cm^{-1} (*8.1−12.4 eV*) region of PMMA are due to excitations in the alkyl portion of the polymer. The authors of this work have compared the spectra of vapor-phase methyl formate and solid PMMA, but because they failed to recognize that the prominent Rydberg bands of the vapor-phase ester will not appear in the PMMA spectrum, the comparison is suspect. As with other substances of this type (Section II.D), PMMA also shows a plasmon-like collective excitation in the vicinity of 130 000−160 000 cm^{-1} (*16−20 eV*).

XV.C. Acyl Halides

As with the amides, trapped-electron spectra of acetyl fluoride and trifluoroacetyl fluoride [D40], Fig. XII.B-2, display a series of triplet excitations. In CF_3COF, the first three bands are observed at 50 000, 71 000 and 83 000 cm^{-1} (*vert., 6.2, 8.8 and 10.3 eV*). Since the lowest of these is *above* the $^1(n_O, \pi_3^*)$ state, a $^3(n_O, \pi_3^*)$ assignment is highly unlikely, and a tentative $^3(\pi_2, \pi_3^*)$ assignment is offered instead for the lowest band. The $n_O \rightarrow 3s$ and $n_O \rightarrow 3p$ singlet bands in CF_3COF occur at 72 100 and 85 700 cm^{-1} (*8.94 and 10.6 eV*) (*vert.*), respectively [**II**, 146]. As in the amides described above (Section XV.A), these singlet Rydberg bands of CF_3COF are complements to the triplet bands observed in the trapped-electron spectrum at 71 000 and 83 000 cm^{-1}, with the singlet-triplet split amounting to only a few thousand cm^{-1} in each case.

In acetyl fluoride [D40], the transition to $^3(\pi_2, \pi_3^*)$ is assigned to the band at 49 000 cm^{-1} (*vert., 6.1 eV*), with further triplets peaking at 63 000, 82 300 and 97 600 cm^{-1} (*7.8, 10.2 and 12.1 eV*), Fig. XII.B-2.

The $^3(\pi_2, \pi_3^*)$ state is calculated to come at 49 600 cm^{-1} (*6.15 eV*) in acetyl fluoride [V15]. Though there is no published singlet spectrum of acetyl fluoride, we do have a photoelectron spectrum [C18], and can estimate the triplet Rydberg frequencies from this and the term-value concept. Thus, the 63 000-cm^{-1} triplet has a 32 200 cm^{-1} term value with respect to the ionization potential at 95 200 cm^{-1} (*11.8 eV*), and the two triplets at 82 300 and 97 600 cm^{-1} (*10.2 and 12.1 eV*) are respectively 33 000 and 17 700 cm^{-1} removed from the ionization potential at 115 300 cm^{-1} (*14.30 eV*). The bands with term values of 32 000−33 000 cm^{-1} terminate at 3s Rydberg orbitals, while that with a term value of 17 700 cm^{-1} terminates at 3p.

The broad, intense band at 67 570 cm^{-1} (*vert., 8.377 eV*) in acetyl chloride was earlier assigned as a possible $\pi_2 \rightarrow \pi_3^*$ transition [**II**, 100], however, it should be noted that its term value with respect to the $(\pi_2)^{-1}$ ionization potential (29 200 cm^{-1}) strongly suggests a $\pi_2 \rightarrow \sigma^*$(C−Cl) $\mathscr{A}(\pi, \sigma^*)$-band transition, Table I.C-III.

CHAPTER XVI

Oxides of Nitrogen

In the nitrate ion, the doubly degenerate transition $e\pi_{2,3} \rightarrow a_2''\pi_4^*$ is strongly allowed at 50 000 cm^{-1} (*vert., 6.2 eV*) [**II**, 164], while in ethyl nitrate the bands appear to remain degenerate at 52 600 cm^{-1} (*vert., 6.52 eV*) despite the lower symmetry. The corresponding transition in nitric acid is a composite one, with one intense component centered at 54 000 cm^{-1} (*6.70 eV*) and a second much weaker one coming at 63 300 cm^{-1} (*vert., 7.85 eV*) [S86]. According to MO calculations [H18], the lower-frequency $\pi \rightarrow \pi^*$ component should be the more intense and corresponds to a local transition within the $-NO_2$ group, whereas the higher-frequency component is the $RO \rightarrow NO_2$ charge-transfer band discussed by Kaya *et al.* [**II**, K10]. Yet another intense valence band ($\pi_1 \rightarrow \pi_4^*$) appears in HNO_3 at 73 300 cm^{-1} (*vert., 9.08 eV*), and is related to the intense feature at 81 000 cm^{-1} (*vert., 10 eV*) in the nitrate ion [**II**, 274]. These broad, featureless valence transitions of HNO_3 are followed by several weaker but structured entities which are undoubtedly Rydberg in nature. Indeed, the first photoelectron band of HNO_3 is highly structured vibrationally, and is said to consist of three overlapping ionizations having a center of gravity at 100 800 cm^{-1} (*vert., 12.50 eV*) [F35]. With respect to this narrow triplet of ionization peaks, the first two Rydberg bands have term values of 23 200 and 19 000 cm^{-1}, identifying them as terminating at 3p. The lowest transition to 3s is buried beneath the $\pi_1 \rightarrow \pi_4^*$ excitation, however, a prominent OH fluorescence-excitation peak is found at 72 200 cm^{-1} (*8.95 eV*) and this has the expected term value (28 600 cm^{-1}) for assignment to 3s.

Nitrous acid, O=N–OH, is isoelectronic with formic acid (Section XV.B) and their spectra should be related. Nitrous acid shows the $n_O \rightarrow \pi^*$ (^1W) band at 28 000 cm^{-1} (*vert., 3.47 eV*) and the $N \rightarrow V_1$ ($\pi_2 \rightarrow \pi_3^*$) band at 46 000 cm^{-1} (*vert., 5.76 eV*) [L7]. Because the ionization potential is so high (88 000 cm^{-1} *vert., 10.9 eV*) the Rydberg spectrum of nitrous acid will begin about 10 000 cm^{-1} beyond the center of the $N \rightarrow V_1$ absorption, in a region as yet uninvestigated.

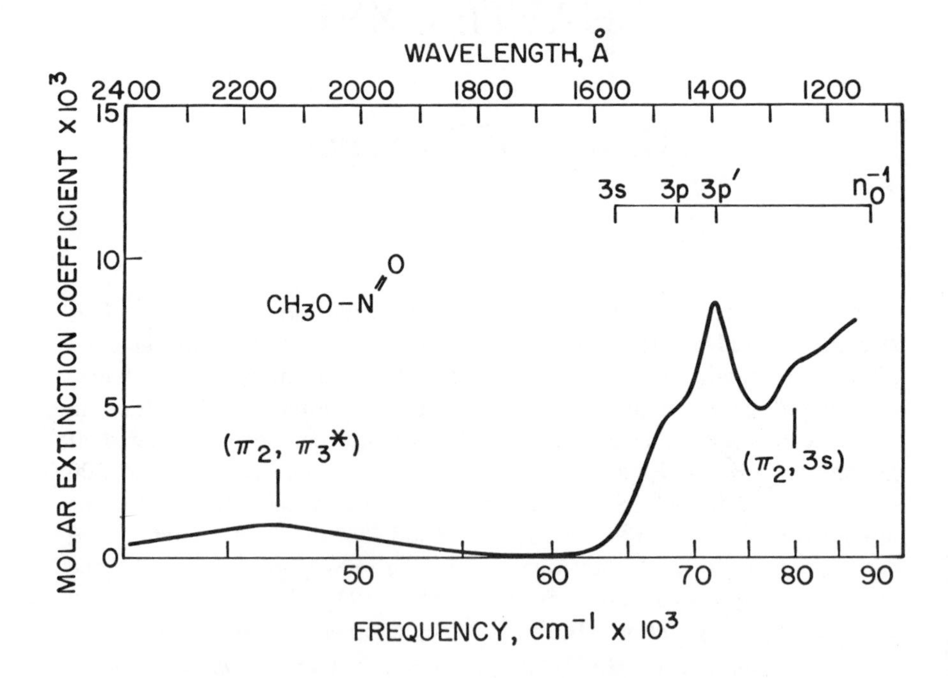

Fig. XVI-1. Absorption spectrum of methyl nitrite in the vapor phase [L1].

Spectral data on the methyl ester of nitrous acid (methyl nitrite, CH_3ONO) [**II**, 162] is available, Fig. XVI-1. Lahmani *et al.* [L1] report the $\pi_2 \rightarrow \pi_3^*$ excitation in this system at 46 700 cm^{-1} (*vert., 5.79 eV*), in agreement with its position in nitrous acid. Bands in methyl nitrite peaking at 68 000 and 71 900 cm^{-1} (*8.49 and 8.91 eV*) have term values of 20 200 and 16 800 cm^{-1} with respect to the ionization potential of the

n_O orbital (88 700 cm^{-1} *vert., 11.0 eV*) [B33]. Such term values are far more characteristic of $n_O \rightarrow 3p$ transitions than $n_O \rightarrow 3s$, and it must be concluded that $n_O \rightarrow 3s$ is weak and hiding in the low-frequency tail of the $n_O \rightarrow 3p$ absorption near 65 000 cm^{-1} (*8.1 eV*), as suggested in the Figure. Another Rydberg transition at 80 000 cm^{-1} (*vert., 9.92 eV*) in methyl nitrite has the proper term value (24 400 cm^{-1}) with respect to the second ionization potential at 94 370 cm^{-1} (*vert., 11.74 eV*) for classification as $\pi_2 \rightarrow 3s$.

CHAPTER XVII

Dienes and Higher Polyenes

XVII.A. Butadiene

The two lowest triplet excited states of butadiene (T_1 and T_2) are observed in electron impact at 25 800 and 39 500 cm^{-1} (*vert., 3.20 and 4.90 eV*) [M71], and are calculated to terminate at $^3B_u(\pi, \pi^*)$ and $^3A_g(\pi, \pi^*)$ valence states [B77]. In the methylated derivatives of butadiene, these two triplet excitations appear unshifted [M66] provided the chromophore remains planar; if skewed as in 2,4-dimethyl-1,3-pentadiene, the T_1 frequency increases whereas that of T_2 decreases, so as to converge upon T_1 of butene-2 (33 800 cm^{-1} *vert., 4.19 eV*). The lowest Rydberg triplet in butadiene, $^3(b_g, 3s)$, is expected at *ca.* 49 000 cm^{-1} (*6.1 eV*), where it is covered by the intense N → $V_1(\pi_2, \pi_3^*)$ band.

Further experimental and theoretical work has been performed on the N → V_1 band of butadiene [**II**, 166, 355] in an effort to extract more structural information. Using a UV-IR double resonance technique [R34], the N → V_1 electronic origin was shown to lie at 46 300 cm^{-1} (*5.74 eV*), with only ν_4'' (symmetric C=C stretch) excited in the hot-band region. The absence of ν_{12}'' (CH_2 torsion) suggests that the V_1 state is not twisted, as is suggested as well by preresonance Raman-scattering experiments

[H40]. The most recent calculation of the butadiene N → V_1 Franck-Condon envelope [D28] concludes that V_1 may be nonplanar but if so, the terminal bond twist angle is less than 30°. The N → V_1 absorption profile of butadiene does not sharpen upon cooling in a jet expansion [V2], paralleling the lack of response shown on going from the vapor to the neat-crystal spectrum at 24 K [**II**, 169]. It is deduced from experiments such as these that the V_1 state of butadiene is short-lived leading to a homogeneously broadened N → V_1 band. A short lifetime for the V_1 state is consistent with the fact that the N → V_1 transition does not appear as a three-photon resonance in the MPI spectrum of butadiene even though it is symmetry-allowed as such [R46]. The detailed shape of the N → V_1 envelope in electron impact varies with electron energy in a minor but reproducible way in butadiene and its methyl derivatives [D30, M34], and is said to be due to a varying Rydberg/valence orbital composition across the band, rather than to hidden transitions.

The spectrum of solid butadiene [**II**, 169] shows a weak valence feature at 57 100 cm^{-1} (*vert., 7.08 eV*) which was assigned earlier as N → V_2 [**II**, 168]. What is apparently the same feature now has been identified by Doering and McDiarmid [D32, D34] in the vapor-phase electron impact spectrum as a symmetry-forbidden band of appreciable width centered at 60 000 cm^{-1} (*7.4 eV*). This band lies beneath a cluster of sharp Rydberg features, and is evident in the electron impact spectrum only when run under non-optical conditions, *i.e.*, at low impact energy and nonzero scattering angle. These workers also assign it as N → V_2 ($^1A_g \rightarrow {}^1A_g$), or less likely as $b_2\pi_2 \rightarrow$ 3d. Based on *ab initio* calculation, it has been suggested [B59] that the 60 000-cm^{-1} band of butadiene might be assigned as $a_g\sigma \rightarrow a_u\pi_3^*$ ($^1A_g \rightarrow {}^1A_u$), intensified by vibronic borrowing, however, such a transition is formally allowed, whereas the electron-impact data argues for a symmetry-forbidden excitation at this frequency.

The earlier confusion surrounding the 50 000-cm^{-1} (*6.2-eV*) feature in the butadiene spectrum finally has been sorted out. We first identified it as having a Rydberg upper state involving excitation from $b_g\pi_2$ on the basis of the condensed-phase and perfluoro effects, but carelessly called it ($b_g\pi_2$, 3p) in one place [**II**, 173] and ($b_g\pi_2$, 3s) in another [**II**, 174]. A transition to the first of these is allowed, whereas it is forbidden to the second. McDiarmid [M28] has studied the band in question using H/D isotopic substitution and concludes that the transition is forbidden electronically, but is made allowed by the excitation of one quantum of ν_{12}', the a_u CH_2 torsion. In butadiene-h_6, the origin frequency is deduced to be 50 144 cm^{-1} (*6.2169 eV*). In this work, McDiarmid first assigned the forbidden band to the N → V_2 ($^1A_g \rightarrow {}^1A_g$) valence promotion [M28], but later adopted a $b_g\pi_2 \rightarrow$ 3s Rydberg assignment [M29]. The 23 100-

cm^{-1} term value observed for this transition supports the $^1(b_g\pi_2, 3s)$ assignment of the upper state. Doering's [D30] tentative assignment of a singlet → triplet energy loss at 49 000 cm^{-1} (*vert., 6.1 eV*) is compatible with a transition to $^3(b_g\pi_2, 3s)$.

Being one-photon forbidden, the 50 144-cm^{-1} band of butadiene will be two-photon allowed in the MPI spectrum. Indeed, an intense origin is found in the MPI spectrum of butadiene-h_6 at 50 100 cm^{-1} (*6.21 eV*), followed by a parade of totally symmetric vibrations [J11, R46, V1]. This vibrational envelope closely resembles those in the one-photon allowed Rydberg excitations from $b_g\pi_2$ to np. As appropriate for a Rydberg terminating state, the 50 100-cm^{-1} transition broadens to high frequency upon the application of high pressure, and is totally obliterated in the thermal-lensing spectrum of the neat liquid [V1]. Additionally, three-photon resonances in the MPI spectrum are observed terminating at 3s and 4s upper orbitals, owing to an allowing quantum of an a_u vibration in each case [R46]. MPI work at higher frequencies has uncovered the higher members of the $b_g\pi_2 \rightarrow n$s series ($n = 4-12$) characterized by $\delta = 0.91$ [M3].

The assignment of the 50 100-cm^{-1} band of butadiene as $b_g\pi_2 \rightarrow 3s$ is amply confirmed by *ab initio* theory [B77, N16], which places the $^1(b_g\pi_2, 3s)$ state at 50 600 cm^{-1} (*6.27 eV*). The apparent success of such calculations, however, does not carry over to the $^1(b_g\pi_2, 3p)$ manifold. Experimentally, using a variety of spectroscopic techniques, $b_g\pi_2 \rightarrow n$p Rydberg series with quantum defects of 0.67 and 0.42 are observed to converge upon the $(b_g\pi_2)^{-1}$ ionization potential [F17, J11, M29, R46]. The $n = 3$ members of these series have origins at 53 697 and 57 009 cm^{-1} (*6.6574 and 7.0680 eV*) respectively, each followed by vibronic structure dominated by C=C stretching and C–H in-plane bending at close to the ground-state frequencies. On the other hand, the calculations predict the $(b_g\pi_2, 3p)$ manifold to lie within the interval 52 700–54 700 cm^{-1} (*6.53–6.78 eV*), and on this basis the 57 009-cm^{-1} feature has been assigned to the forbidden N → V_2 valence promotion! Perhaps the best argument against such a valence assignment is the high-resolution photoelectron spectrum of the $(b_g\pi_2)^{-1}$ ionization in butadiene, Fig. XVII.A-1, which closely mirrors the vibronic envelope of the 57 009-cm^{-1} band [W17]. Such strong similarities occur only between a photoelectron band and the allowed Rydberg absorption bands leading to it [I, 73]. The N → V_2 band calculated to fall in this region more likely corresponds to the broad, forbidden feature centered at 60 000 cm^{-1} (*7.4 eV*).

Comparison of the photoelectron and absorption spectra in Fig. XVII.A-1 leads to the conclusion that *all* the sharp absorption

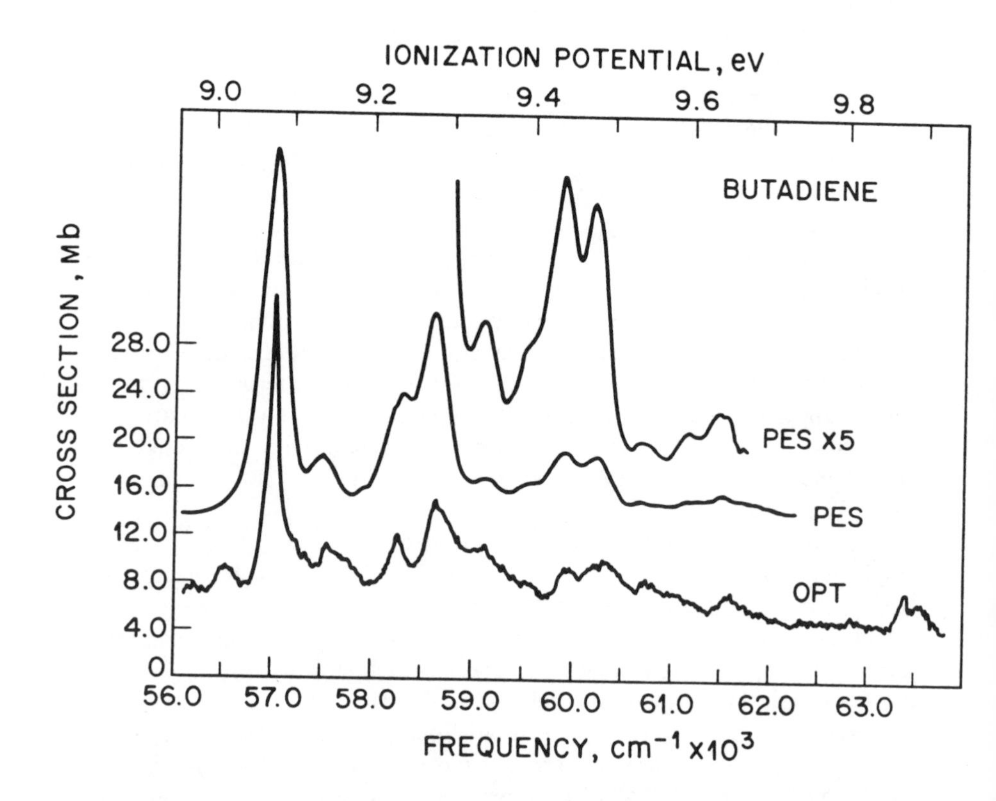

Fig. XVII.A-1. Comparison of the absorption spectrum (*OPT, lower frequency scale*) and $(\pi_2)^{-1}$ Franck-Condon envelope of the photoelectron spectrum of butadiene (*PES, upper energy scale*) [W17]. The intensity scale refers only to the absorption curve.

features in the 56 000–62 000 cm^{-1} (*6.9–7.7 eV*) region of butadiene are due to a single electronic origin ($b_g\pi_2 \rightarrow$ 3p) and associated a_g vibrations. In contrast, McDiarmid [M29] has deduced three origins in this spectral region and Nascimento and Goddard [N16] assign five, with none of them terminating at 3p based on their calculation. The *ab initio* calculations [B77, N16] do offer a solution to one problem, however. Formally, transitions from $b_g\pi_2$ to all three components of 3p are allowed, whereas only two transitions are observed. The calculations do agree that the transition to the lowest frequency component of 3p (3pσ) has an oscillator strength approximately 1 % of that to the other two components, and so will be difficult to find. The calculations place the 3pπ level between 3pσ

and $3p\sigma'$, in contrast to the usual splitting pattern found for planar molecules (Section I.B).

In many molecules, the originating MO in a Rydberg excitation has a nodal pattern which is essentially ns-like (diimide, for example) or np-like (ethylene and acetone, for example). In the first instance, the pseudo-atomic selection rule will allow transitions to np orbitals and in the second instance, to ns and nd orbitals. In the unusual case of butadiene, the uppermost MO ($b_g\pi_2$) is 3d-like, and Rydberg excitations therefore are allowed to np and nf orbitals. Two Rydberg series terminating at np have been observed already. An excellent candidate for the rarely-seen series terminating at nf orbitals is described by McDiarmid [M29] as having $\delta = 0.087$ and $3 \leqslant n \leqslant 13$. The discordant "$n = 3$" feature of the McDiarmid series has been shown by Wiberg *et al.* [W17] to be a vibronic band of the $b_g\pi_2 \rightarrow 3p$ transition, and so the $b_g\pi_2 \rightarrow nf$ series begins properly with $n = 4$ at 65 998 cm^{-1} (*advert., 8.1825 eV*) in butadiene-h_6 [D32]. The transition to the $n = 9$ member of this series at 71 790 cm^{-1} (*8.900 eV*) has a C-type rotational contour [R3] thereby demonstrating an out-of-plane polarization for members of this series. Transitions to four of the seven components of the $b_g\pi_2 \rightarrow nf$ manifold will have out-of-plane polarization. Though the calculation of Buenker *et al.* [B77] assigns the $\delta = 0.087$ series members as forbidden transitions to nd, McDiarmid's experimental work argues for allowed transitions to nf instead. A further nf series having $\delta = 0.04$ is observed as a three-photon resonance in the MPI spectrum [J11, R46].

In ethylene, the (π, π^*) valence configuration is mixed with its Rydberg conjugate (π, 3dxz) and through this, the resulting (π, 3dxz) state is lifted far above the other components of the (π, 3d) complex. In butadiene, the ($b_g\pi_2$, $a_u\pi_3^*$) valence configuration is similarly mixed, albeit more weakly, with its Rydberg conjugate so as to produce a diffuse π^* orbital. As a consequence of the symmetry of the $a_u\pi_3^*$ MO, the conjugate configuration is ($b_g\pi_2$, 4fπ), and one thereby expects the conjugate 4fπ Rydberg orbital to lie considerably above the other 4f components, and in fact, it may have a negative quantum defect. The orbital ordering in the core-split nf manifold of butadiene has yet to be determined either experimentally or theoretically.

The one-photon forbidden excitations to nd orbitals in butadiene are difficult to identify. According to the calculations of Buenker *et al.* [B77], the core-split components of the transitions to 3d are 58 800–62 750 cm^{-1} (*7.3–7.8 eV*) above the ground state. Rothberg *et al.* [R46] report a two-photon resonance to 3d at 59 380 cm^{-1}

(*7.362 eV*), and a four-photon resonance† with an origin at 61 406 cm^{-1} (*7.6198 eV*) [M37] also would appear to terminate at 3d. (The transition to 4s has its origin at 61 436 cm^{-1} (*advert., 7.6169 eV*) [M3]). A short Rydberg series having $\delta = 0.21$ could be $b_g\pi_2 \rightarrow nd$ [M29], if it proves to be real. On the other hand, Mallard *et al.* [M3] have scanned the $b_g\pi_2 \rightarrow nd$ region of butadiene using the MPI technique but failed to find any such two-photon resonances. This paper is particularly recommended to the reader wanting a critical overview of the Rydberg scene in butadiene.

Going beyond the first ionization potential of butadiene, Flicker *et al.* [F17] report electron impact energy losses centered at 76 900 and 89 000 cm^{-1} (*vert., 9.53 and 11.0 eV*); it is not known whether these are Rydberg excitations to higher ionization potentials or involve $N \rightarrow V_{3,4}$ excitations (*cf.* 1,4-cyclohexadiene, Section XVII.C).

The temporary-negative-ion spectrum of butadiene [J16] shows π^* valence TNI resonances at 5000 (^2A_u) and 22 600 cm^{-1} (^2B_g) (*vert., 0.62 and 2.80 eV*), the first of which displays a 1600-cm^{-1} vibrational progression. As expected from simple MO theory, these π^* resonances of butadiene nicely straddle that of ethylene (14 400 cm^{-1} *vert., 1.78 eV*), Fig. II.B-1, and are split by an amount (17 600 cm^{-1}) very close to that separating the two π MO ionization potentials (19 800 cm^{-1}) [K18].

In contrast to the 17 600-cm^{-1} splitting of the π^* MO's observed in the TNI resonances, the two $1s_C \rightarrow \pi^*$ inner-shell transitions of butadiene are said to be separated by only 4000 cm^{-1} [H66]! Excitations from $1s_C$ to π^* are followed by slightly exhalted transitions to 3p and 4p Rydberg orbitals and by two shape resonances [52 000 and 110 000 cm^{-1} (*vert., 6.4 and 13.6 eV*) beyond the $(1s_C)^{-1}$ ionization potential] assumed to terminate at σ^*(C–C) and σ^*(C=C) MO's.

XVII.B. Alkyl Dienes

The question of a low-lying $^1A_g(\pi, \pi^*)$ state in the diene chromophore [II, 170] remains the object of strong interest. Studying variously alkylated *s-cis-* and *s-trans-*dienes, McDiarmid and Doering [M39] conclude that transitions to the lowest detectable ^1A_g state (V_2) fall in a

† This four-photon resonance is peculiar because it is preceded by a three-photon resonance to V_1, and V_1 is itself too short-lived to give an MPI resonance [R46].

range lying above 50 500 cm^{-1} (*6.26 eV*), whereas transitions to the $^1B_uV_1$ states are in the range 38 000−47 500 cm^{-1} (*4.7−5.9 eV*). For a given pattern of alkylation, the transition to ^1B_u is *ca.* 4000 cm^{-1} lower in the *s-cis* isomer than in the *s-trans*. Symmetry designations appropriate to butadiene are used in this Section for dienes of formally lower symmetry in order to stress the similarities of electronic structure and spectra.

Being in the *s-cis* configuration, the 1,3-cyclodienes [II, 170] have $N \rightarrow V_2$ transitions which are intense and centered in the range 63 600−66 500 cm^{-1} (*vert., 7.88−8.24 eV*), while the transitions to V_1 appear as weaker bands at 39 800−42 400 cm^{-1} (*vert., 4.93−5.26 eV*) [F43]. Killat [K15] reports that in 1,3-cyclohexadiene, the oscillator strengths from N to V_1 and V_2 are 0.13 and 0.72, respectively. Comparison of the vapor and solid spectra of 1,3-cyclohexadiene [K15] reveals that the Rydberg excitations near 50 000 cm^{-1} (*6.2 eV*) are suppressed in the solid, that the transition to V_2 is at 1300 cm^{-1} higher frequency in the solid than in the vapor, and that a very intense energy loss at *ca.* 105 000 cm^{-1} (*vert., 13.0 eV*) in the free molecule is not damped in the condensed phase. Because alkylation of butadiene does not shift the T_1 and T_2 levels [M66], it is not surprising to find them at nearly their butadiene frequencies in the 1,3-cyclodienes [F43].

The three transitions of 1,3-cycloheptadiene in the vapor-phase energy loss spectrum [F43] (24 100, 41 000 and 66 500 cm^{-1} *vert., 2.99, 5.08 and 8.25 eV*) appear at identical frequencies as electron transmission maxima in the solid-film spectrum [S10], thereby signifying valence assignments: $N \rightarrow T_1$, $N \rightarrow V_1$ and $N \rightarrow V_2$.

The MPI spectra of methylated butadienes [R46] has led to the delineation of several new Rydberg series. Though the one-, two- and three-photon spectra agree with one another in the asymmetrically-substituted dienes, the multiphoton technique still is advantageous because of its generally superior resolution and freedom from overlapping $N \rightarrow V$ absorption. As in butadiene, the derivatives *trans*-piperylene, isoprene and 2,3-dimethyl butadiene each have $b_g\pi_2 \rightarrow n$p series with $\delta = 0.4$ and 0.6. Additionally, in isoprene and 2,3-dimethyl butadiene, *n*f series are seen in which the high-*n* members are very prominent. The electron impact spectrum of 2,3-dimethyl butadiene cooled in a jet expansion is sharpened, but reveals little that is new [D36, M41], while the optical spectrum of the cooled vapor [S3] reveals sequence structure in the $b_g\pi_2 \rightarrow 3$p band involving methyl-group torsion.

The circular dichroism (CD) spectra of the alkylated dienes are readily interpreted using the assignments given above, provided one assumption holds true. When a molecule displaying sharp Rydberg excitations is placed in solution, the absorption profile usually is so distorted it would seem that the transition has "disappeared." In fact, the transition has broaden so much (and shifted somewhat to higher frequency) that it often cannot be observed in absorption. However, CD is more efficacious at resolving overlapped transitions, and what has "disappeared" in absorption may still be present as a broad but distinct feature in the CD spectrum. Some authors have observed this phenomenon and misinterpreted it as showing a broad valence excitation lurking beneath each of the Rydberg excitations. As demonstrated by Mason and coworkers [B69, D45], the final test in this regard rests in lowering the temperature of the solution, for Rydberg CD bands will show a rapid shift to yet higher frequencies (Fig. II.D-1, for example), whereas valence excitations show a small or zero shift to low frequencies. The CD spectra of optically active dienes in solution can be interpreted readily once it is recognized that though Rydberg excitations usually do not appear in the absorption spectra of such solutions, they do appear in the CD spectra.

The deuterium- and methyl-substituted system 7,7-dimethyl bicyclo-[4.1.1]octa-2,4-diene 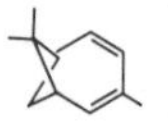 is nominally a planar *cis*-diene and as such displays a relatively weak $N \rightarrow V_1$ transition (ϵ is approximately 3000) centered at 36 000 cm^{-1} (*4.46 eV*) in the vapor [B69] and slightly shifted to lower frequency in 3-methyl pentane solution, Fig. XVII.B-1. Though transitions at 43 100 (π_2, 3s), 46 900 (π_2, 3p) and 51 900 cm^{-1} (π_2, 3d) (*vert., 5.34, 5.81 and 6.43 eV*) in the vapor phase are no longer evident in the solution absorption, they persist as broad features in the CD spectra of the solutions. Lowering the temperature of the solutions, Fig. XVII.B-1, results in dramatic shifts of the Rydberg transitions to higher frequencies. No doubt, there are two or three excitations to (π_2, 3p) in the 47 000−50 000 cm^{-1} (*5.8−6.2 eV*) region. The intense transition to V_2 (ϵ of approximately 15 000 is expected) will be found in this compound near 65 000 cm^{-1} (*vert., 8.0 eV*).

Absorption and CD spectra of α- and β-phellandrene as presented by Gross and Schnepp [G43] would seem to follow the example of the octadiene given above. In the *cisoid* diene α-phellandrene, the pattern of transitions from N to $V_1(\pi_2, \pi_3^*)$ (38 800 cm^{-1} *vert., 4.81 eV*), to (π_2, 3s) (42 500 cm^{-1} *vert., 5.27 eV*), to (π_2, 3p) (47 200 cm^{-1} *vert., 5.85 eV*), to (π_2, 3d) (51 300 cm^{-1} *vert., 6.36 eV*), and to $V_2(\pi_2, \pi_4^*)$ (64 000 cm^{-1} *vert., 7.93 eV*) is evident in the vapor-phase absorption and CD spectra. However, in perfluoro-*n*-heptane solution at room temperature, the CD spectrum of α-phellandrene shows several broad features in the Rydberg region which Gross and Schnepp interpret as underlying valence excitations, but which we assign instead to perturbed Rydberg transitions. The effects on these transitions of lowering the solvent temperature would

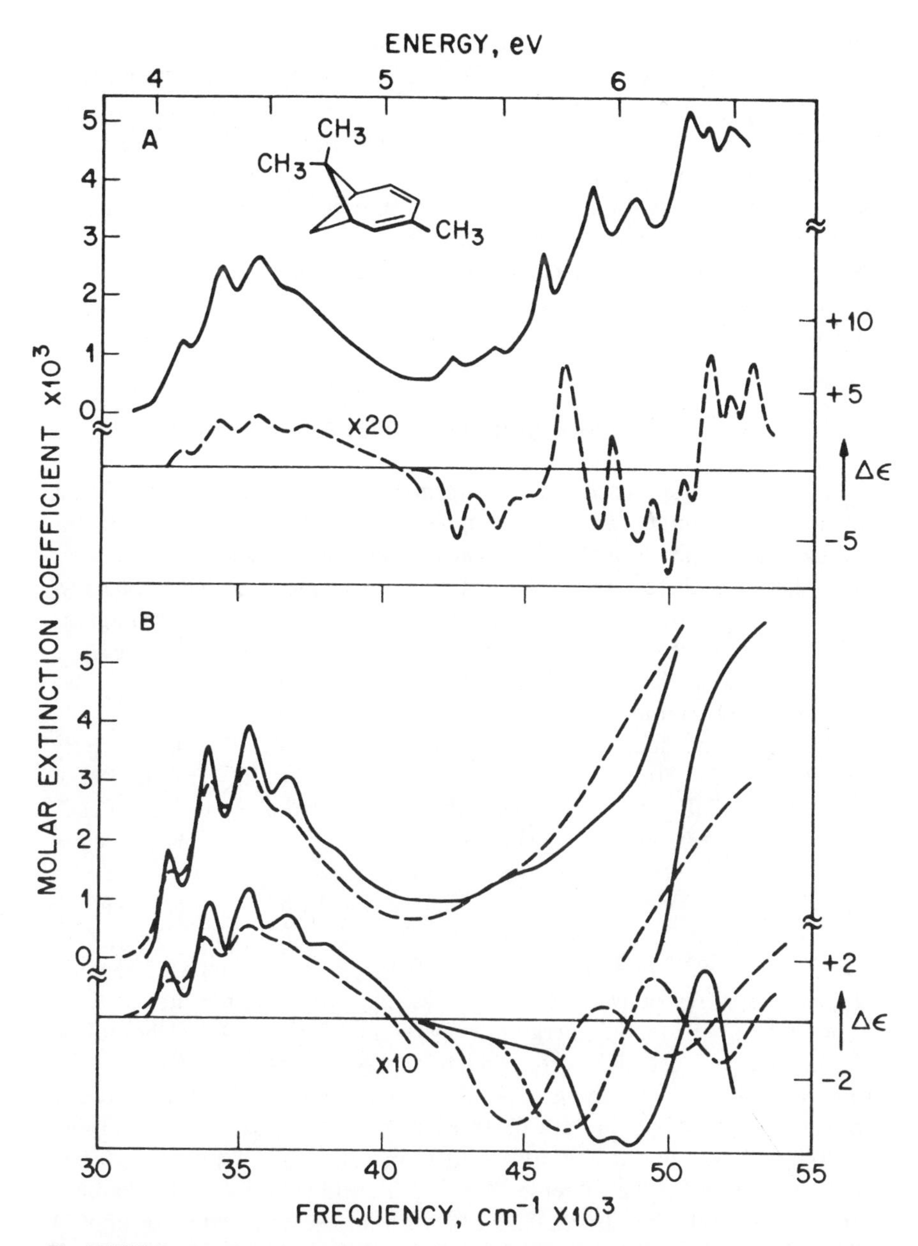

Fig. XVII.B-1 *Panel A.* Absorption (—) and CD spectra (- -) of 4,7,7-trimethyl bicyclo[4.1.1]octa-2,4-diene in the vapor phase. *Panel B.* Absorption spectra of 4,7,7-trimethyl bicyclo[4.1.1]octa-2,4-diene in 3-methyl pentane solution at 93 K (—) and at 293 K (- -), upper two curves, and CD spectra of the 3-methyl pentane solutions at 93 K (—), at 173 K (- · -) and at 293 K (- -), lower three curves [B69].

allow one unambiguously to choose between Rydberg and valence assignments. Lacking this, it is better to assume that these CD features are strongly perturbed Rydberg excitations. Interestingly, the V_1-V_2 split of 26 000 cm^{-1} in α-phellandrene shows that full coupling between the double bonds is maintained even though the axes of the double bonds are said to be skewed [G43]. Assignments in *transoid* β-phellandrene parallel those in the α isomer, however with drastically different $N \rightarrow V_1$ and $N \rightarrow V_2$ intensities, as expected from the relative orientations of their double bonds. The $\pi_2 \rightarrow 3s$ excitation in the β isomer is seen as a weak shoulder at 45 500 cm^{-1} (*vert., 5.64 eV*).

XVII.C. Nonconjugated Dienes and Polyenes

In the nonconjugated dienes, the formally noninteracting double bonds in fact can interact and split appreciably, using either or both the through-space and through-bond routes. This is nicely shown in the work of Jordan *et al.* [J15], Fig. II.B-1, in which photoelectron and temporary-negative-ion spectral data are combined. One sees that the nonconjugated 1-4 interaction can be as large as half the size of the conjugated 1-3 interaction, and that this decreases to a much smaller value for 1-5 nonconjugated interaction.

The electronic spectrum of 1,4-cyclohexadiene [**II**, 171] is relatively broad; only a few states can be identified with any certainty and in most of these the possibility exists that more than one transition is involved in each [F42, K15]. A triplet upper-state energy loss observed at 34 600 cm^{-1} (*vert., 4.29 eV*) contains one or more excitations to states of the sort $^3(\pi, \pi^*)$ [F42]. Transitions from the highest filled MO $b_{1u}\pi_2$ to 3s and 3p are observed by electron impact at 49 600 and 51 800 cm^{-1} (*vert., 6.15 and 6.42 eV*) with term values of 21 500 ($b_{1u}\pi_2 \rightarrow 3s$) and 19 300 cm^{-1} ($b_{1u}\pi_2 \rightarrow 3p$), respectively [F42, K18, M38]. However, the first of these is present as well in the energy-loss spectrum of solid 1,4-cyclohexadiene [K15], and so must also correspond to the symmetry forbidden $N \rightarrow V_1$ valence excitation. Rydberg excitations to 3s and 3p originating at the second highest MO $b_{3g}\pi_1$ are observed in electron impact at 57 300 and 58 900 cm^{-1} (*7.10 and 7.30 eV*) with term values of 22 400 and 20 800 cm^{-1}, respectively. The intense $N \rightarrow V_2$ transition of 1,4-cyclohexadiene comes at 64 100 cm^{-1} (*vert., 7.95 eV*) with an oscillator strength of 1.34 [F42, K15, M38]. As is common with parallel oscillators, the center-of-gravity of $N \rightarrow V$ intensity has moved to higher frequencies in the diene compared to the monoolefin {cyclohexene, 55 200 cm^{-1} (*vert., 6.84 eV*) [F42]}. Many other peaks are observed in the 70 000–105 000-cm^{-1}

(*8.7–13-eV*) region of 1,4-cyclohexadiene, with good agreement between vapor-phase and solid-phase frequencies. Assignment of these valence excitations is vague, but they are likely to involve both $\pi \rightarrow \pi^*$ and $\sigma \rightarrow \pi^*$ promotions.

Rydberg excitations to 3p also have been observed in the MPI spectrum of 1,4-cyclohexadiene as a two-photon resonance (51 773 cm^{-1} *adiabat., 6.4189 eV*, $b_{1u}\pi_2 \rightarrow$ 3p) and as a three-photon resonance (58 872 cm^{-1} *adiabat., 7.2990 eV*, $b_{3g}\pi_1 \rightarrow$ 3p) [S2]. A second ($b_{1u}\pi_2$, 3p) origin is observed at 54 640 cm^{-1} (*6.774 eV*) in the MPI spectrum [R35]. The fact that the lower and higher transitions to 3p are respectively two-photon and three-photon allowed proves that the lower ionization involves a u level ($b_{1u}\pi_2$) and the higher ionization involves a g level ($b_{3g}\pi_1$) [S2].

The splitting among the π MO's of 1,5-cyclooctadiene is very small, Fig. II.B-1, and consequently very little structure can be resolved in the absorption spectrum [M38]. A slight shoulder at 48 400 cm^{-1} (*vert., 5.95 eV*) is either N $\rightarrow$ V_1 or $\pi_2 \rightarrow$ 3s (or both), while N $\rightarrow$ V_2 peaks at 57 000 cm^{-1} (*7.1 eV*). Considering the lack of π-electron conjugation in 1,5-cyclooctadiene, a splitting of 9000 cm^{-1} between the N $\rightarrow$ V_1 and N $\rightarrow$ V_2 transitions seems unlikely, and so the shoulder at 48 400 cm^{-1} is best assigned as $\pi_2 \rightarrow$ 3s, with the transitions to V_1 and V_2 unresolved. Strangely, though there is little or no evidence supporting $\pi - \pi$ interaction in 1,5-cyclooctadiene, its N $\rightarrow$ V_2 band is once again at substantially higher frequency (57 300 cm^{-1} *vert., 7.10 eV*) than that of the monoolefin (cyclooctene 50 800 cm^{-1} *vert., 6.30 eV*).

The nonconjugated hexaene squalene (2,6,10,15,19,23-hexamethyl-2,6,10,14,18,22-tetracosahexaene) has a reflection spectrum in the liquid phase exhibiting a $\pi \rightarrow \pi^*$ band at 50 000 cm^{-1} (*vert., 6.2 eV*) and a plasmon peak at 186 000 cm^{-1} (*23 eV*) in the energy-loss function Im $(-1/\epsilon)$ [P4]. Since all adjacent double bonds are separated by two or more methylene groups in squalene there is no detectable splitting of the $\pi \rightarrow \pi^*$ manifold in this molecule.

Two electron-impact studies on norbornadiene  [D33, F42] are in essential agreement with one another, while substantially altering the old interpretation [II, 176]. The lowest singlet excitation in norbornadiene, at 42 300 cm^{-1} (*vert., 4.25 eV*), corresponds to the forbidden N $\rightarrow$ V_1 promotion, whereas the two overlapping bands at *ca.* 48 000 cm^{-1} (*6 eV*) are assigned as $\pi_2 \rightarrow$ 3s and N $\rightarrow$ V_2. The presence of three bands in the 40 000–50 000-cm^{-1} (*5.0–6.2-eV*) region of norbornadiene is readily apparent in the CD spectra of its optically active derivatives [L25]. In the MPI spectrum of the jet-cooled molecule, only the two-photon $\pi_2 \rightarrow$ 3s excitation appears; the origin is at 44 380 cm^{-1} (*5.502 eV*) [G12],

Fig. XVII.C-1. The intense one-photon transition to V_3 peaks at 55 200 cm^{-1} (*6.84 eV*) with $\pi_2 \rightarrow 3p$ overlapping but centered at 51 500 cm^{-1} (*6.38 eV*). In the two-photon resonant MPI spectrum, a $\pi_2 \rightarrow 3p$ origin is observed at 50 790 cm^{-1} (*6.297 eV*), and the corresponding $\pi_2 \rightarrow 3d$ band origin is observed at 55 250 cm^{-1} (*6.850 eV*). The term values for these Rydberg excitations are in the expected ranges. The final $N \rightarrow V$ excitation ($N \rightarrow V_4$) is assigned by Doering and McDiarmid [D33] to an energy loss at 60 500 cm^{-1} (*vert., 7.50 eV*) based on a one-electron picture.

The four valence excitations to the $V_n(\pi, \pi^*)$ states of norbornadiene also are apparent in the electron transmission spectrum of the solid, each with an appropriate shift to lower frequencies of a few 1000 cm^{-1} compared with the vapor phase [S10]. Additional transmission maxima at 71 000, 95 000 and 107 000 cm^{-1} (*vert., 8.8, 11.8 and 13.3 eV*) correspond to valence excitations of uncertain assignment involving σ/σ^* MO's [F42]. The presence of four excitations to $^1(\pi, \pi^*)$ levels in norbornadiene is expected from MO theory but exceeds by two the number predicted by the independent-systems model. Note too that in the latter, all the transitions to $^3(\pi, \pi^*)$ have the same frequency in norbornadiene and norbornene, whereas the lowest triplets in norbornadiene are not degenerate (27 000 and 31 000 cm^{-1} *vert., 3.4 and 3.9 eV*) and lie below that of norbornene (33 000 cm^{-1} *vert., 4.1 eV*) [F42].

All the Rydberg excitations in norbornadiene originating at π_2 show long progressions of 390 cm^{-1} in the optical and MPI spectra, which progressions in turn are identical to the $(\pi_2)^{-1}$ vibronic profile in the photoelectron spectrum, Fig. XVII.C-1. There are two vibrations which might logically be assigned to this interval: ν_{12} (425 cm^{-1} in the ground state, $HC{-}CH_2{-}CH$ bridgehead deformation) and ν_{11} (500 cm^{-1} in the ground state, HC—CH—CH wing flapping). *Ab initio* calculations on norbornadiene and its ground state cation [G12] show that the geometric change on ionization is almost totally due to a decrease in the dihedral angle between wings, and so the 390-cm^{-1} spectroscopic interval must correspond to ν_{11}'.

One expects the spectrum of Dewar benzene, Fig. X.B-1, to resemble that of norbornadiene, and such is the case [G40, H16, H17]. What is apparently the forbidden $N \rightarrow V_1$ ($^1A_1 \rightarrow {}^1A_2$) band of Dewar benzene has a smooth profile peaking at 48 800 cm^{-1} (*6.05 eV*; $f = 0.023$). Though confirmed to be a valence excitation by external perturbation [H16], the assignment of this band is nontheless somewhat uncertain in that $\sigma \rightarrow \pi^*$ excitations in highly strained unsaturated systems such as this also will fall at low frequency. On the high-frequency wing of the valence transition rests the $\pi_2 \rightarrow 3s$ band [52 600 cm^{-1} (*vert., 6.52 eV*); $f \geqslant 0.04$;

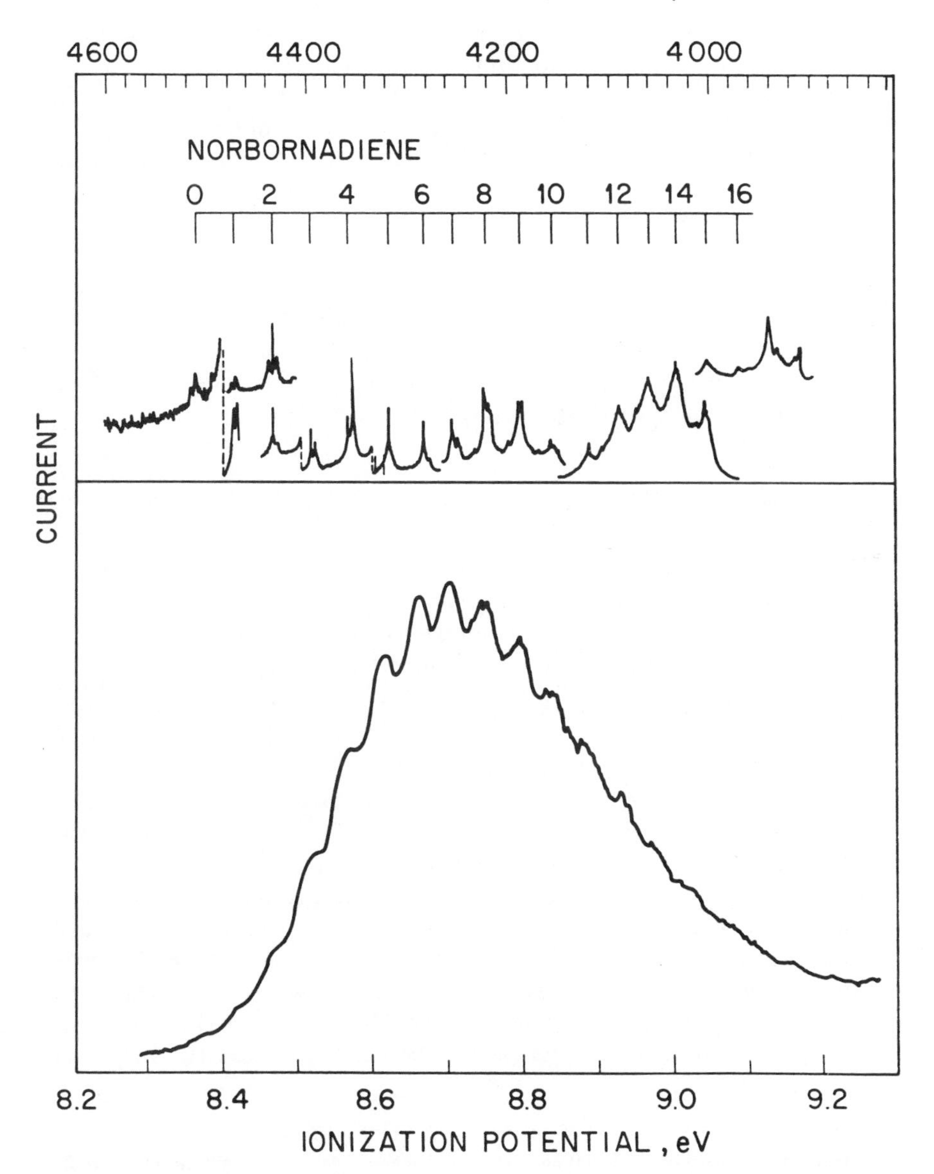

Fig. XVII.C-1. Comparison of the $(\pi_2)^{-1}$ Franck-Condon envelope in the photoelectron spectrum of norbornadiene, and that of the $\pi_2 \rightarrow 3s$ Rydberg transition in the jet-cooled molecule [G12].

23 200-cm^{-1} term value] and a $\pi_2 \rightarrow$ 3p origin appears as a sharp feature at 54 000 cm^{-1} (*6.7 eV*, 21 800-cm^{-1} term value). Vibrational structure in the $\pi_2 \rightarrow$ 3s band of Dewar benzene consists of a long progression (356 ± 40 cm^{-1}) of the totally symmetric wing flapping motion ν_9', which is 382 cm^{-1} in the ground state. This closely parallels the situation in the $\pi_2 \rightarrow$ 3s band of norbornadiene, Fig. XVII.C-1. Continuous absorption (N $\rightarrow$ V$_2$) rises from 54 000 to 59 000 cm^{-1} (*6.7 to 7.3 eV*) in Dewar benzene.

The absorption and CD spectra of a chiral methyl derivative of triquinacene in hydrocarbon solution [P9] reveals oppositely rotating N $\rightarrow$ V$_1$ and N $\rightarrow$ V$_2$ bands at 48 000 and 55 000 cm^{-1} (*vert., 6 and 7 eV*), in accord with the independent-systems picture of three equivalent but weakly interacting double bonds. Cooling the solution resulted in no significant change of the CD spectrum, thereby showing that the bands in question do not have Rydberg upper states, Section II.D.

XVII.D. Heterocyclic Dienes

Recent experimental and theoretical work not only has confirmed much of the earlier supposition concerning the assignment of the spectral bands of furan [**II**, 181], but also has uncovered several new ones. Valence temporary-negative-ion resonances are reported for furan [V9] in which the incident electron is trapped in the π_4^* MO (14 500 cm^{-1} *vert., 1.8 eV*) or the π_5^* MO (25 300 cm^{-1} *vert., 3.14 eV*).† The two lowest excitations to $^3(\pi, \pi^*)$ appear as electron energy losses at 32 200 and 42 100 cm^{-1} (*vert., 3.99 and 5.22 eV*) [F13, F14, V9], and have been assigned upper-state symmetries $^3B_2$ and $^3A_1$ [A29, T12], as in butadiene (Section XVII.A). Some confusion still remains in the singlet valence manifold of furan. Roebber *et al.* [R38] studied the absorption spectrum of jet-cooled furan in the 45 000–51 000-cm^{-1} (*5.6–6.3-eV*) region, and though no sharpening was observed, they were able to resolve the weak feature at *ca.* 47 000 cm^{-1} (*vert., 5.8 eV*) from vibronic components of the more intense N $\rightarrow$ V$_1$ band centered at 48 700 cm^{-1} (*6.04 eV*). The low-frequency feature had been assigned earlier as $1a_2\pi_3 \rightarrow$ 3s [**II**, 181], and

† The π_5^* binding energy of 25 300 cm^{-1} in furan is close to that of π_4^* in butadiene (22 600 cm^{-1}) since π_5^* of furan has a node through the O atom and otherwise resembles π_4^* of butadiene.

theory has since confirmed this [T12]. However Roebber *et al.* claim that this band is affected neither by clustering in the beam nor by solution and hence must be assigned instead as a forbidden N $\rightarrow$ V transition. On the other hand, Roebber *et al.* also report that the MPI spectrum of furan vapor shows a single sharp line at 47 700 cm^{-1} (*5.91 eV*) which they assign as $1a_2\pi_3 \rightarrow 3s$. Though theory places no forbidden N $\rightarrow$ V excitation below the intense N $\rightarrow$ V$_1$ band in furan, it is interesting to note that theory does assign the 62 500-cm^{-1} (*7.75-eV*) band as N $\rightarrow$ V$_2$, and also concludes that there is no diffuse orbital admixture in the upper state of the intense N $\rightarrow$ V$_1$ transition [T12]. Using semiempirical methods, Åsbrink *et al.* [A29] predict two other N $\rightarrow$ V excitations in furan to lie just in the high-frequency wings of the N $\rightarrow$ V$_1$ and N $\rightarrow$ V$_2$ bands discussed above. Applying the electron-transmission technique to solid films of furan, Sanche observes transitions to V$_1$ and V$_2$ at 51 600 and 54 400 cm^{-1} (*vert., 6.40 and 6.75 eV*) and yet a third valence excitation at 65 700 cm^{-1} (*vert., 8.15 eV*) [S8].

In a comprehensive MPI study of furan, Cooper *et al.* [C45] report a Rydberg spectrum which is complete except for the small region covered by Roebber *et al.*! In addition to the Rydberg series already known for furan, Cooper *et al.* find two, three, and four-photon resonances corresponding to a one-photon-forbidden/two-photon-allowed ns series ($n = 4-7$) and an nd series ($n = 3-5$), both converging upon the $(1a_2\pi_3)^{-1}$ ionization potential. Presumably the $n = 3$ member of the ns series is that found by Roebber *et al.* at 47 700 cm^{-1}, however, we note that this feature is reported to consist of only a single line whereas the higher members of the $(1a_2\pi_3, ns)$ series show considerably more vibrational structure. Another series is found in the MPI spectrum of furan has $\delta = 0.035$; because there is no $n = 3$ member, it has been assigned as terminating at nf ($n = 4-7$). A sharp feature in the MPI spectrum at 54 509 cm^{-1} (*advert., 6.7581 eV*) was assigned as $\pi_3 \rightarrow \pi_5^*$, however a valence excitation such as this is not expected in the MPI spectrum, and indeed the 17 140-cm^{-1} term value of this feature suggests instead that it is a component of the $(1a_2\pi_3, 3p)$ Rydberg complex. An MPI resonance to another component of $(1a_2\pi_3, 3p)$ appears as a three-photon resonance at 52 184 cm^{-1} (*advert., 6.4698 eV*) [C45].

Features at 64 900 and 68 700 cm^{-1} (*8.05 and 8.50 eV*) [F14] in furan have term values of 18 800 and 15 000 cm^{-1} with respect to the $(2b_1\pi_2)^{-1}$ ionization potential and are assigned by virtue of their term values and by calculation [T12] as having $^1(2b_1, 3p)$ and $^1(2b_1, 3d)$ upper states, respectively. Energy losses to superexcited states at 88 700 and 96 400 cm^{-1} (*vert., 11.0 and 11.9 eV*) [F14] possibly correspond to $6b_2\sigma \rightarrow 3s$ and $8a_1\sigma \rightarrow 3p$ transitions as determined by their term values.

The general features of the pyrrole spectrum [II, 184] resemble those of furan. Valence TNI resonances corresponding to population of π_4^* and π_5^* in pyrrole appear at 20 200 and 27 700 cm^{-1} (*vert., 2.50 and 3.43 eV*), respectively [V9], and the two lowest $\pi \rightarrow \pi^*$ triplets are found at 33 900 and 42 100 cm^{-1} (*vert., 4.20 and 5.22 eV*) [F13, F14, V9]. The lowest of these triplets is calculated [B87] to have a $^3B_2$ upper state as in furan. The electron energy loss at 42 100 cm^{-1} (*vert., 5.22 eV*) in pyrrole [F13] appears to consist of an unresolved triplet state and a dipole-forbidden singlet state, which the *ab initio* calculations assign to $^3(1a_2\pi_3$, 3s) and $^1(1a_2\pi_3$, 3s) configurations [B87].

The crystal spectrum of pyrrole has been compared with that of the vapor in order to test absorption in the 42 000−54 000-cm^{-1} (*5.2−6.7-eV*) region for Rydberg character [B24]. The very weak feature at 42 500 cm^{-1} (*vert., 5.27 eV*) previously assigned as $1a_2\pi_3 \rightarrow 3s$ [II, 184] is appropriately missing from the crystal spectrum, as is the sharp structure at *ca.* 47 400 cm^{-1} (*5.88 eV*, $1a_2\pi_3 \rightarrow 3p$). The crystal spectrum shows clearly that the $1a_2\pi_3 \rightarrow 3p$ Rydberg transition rests upon a valence $N \rightarrow V_1$ band centered at 48 300 cm^{-1} (*5.99 eV*); the corresponding band in the vapor is centered at 47 500 cm^{-1} (*5.89 eV*) [F14]. The $N \rightarrow V_2$ band of pyrrole peaks at *ca.* 54 000 cm^{-1} (*6.7 eV*) in the crystal, and at 56 000 cm^{-1} (*6.9 eV*) in the vapor [F14]. Electron-transmission spectra of solid pyrrole reveal valence excitations at 44 800, 51 600 and 62 100 cm^{-1} (*vert.*, 5.55, *6.40 and 7.70 eV*) [S8]. As with furan, four $N \rightarrow V_n$ excitations also are calculated to fall in the 48 000−72 000 cm^{-1} (*6−9 eV*) region of pyrrole [A29].

Bavia *et al.* [B24] have constructed four Rydberg series from the one-photon absorption spectrum of pyrrole: $1a_2\pi_3 \rightarrow n$s ($\delta = 0.896$, $n = 3-8$); $1a_2\pi_3 \rightarrow n$p ($\delta = 0.505$, $n = 3-7$); $1a_2\pi_3 \rightarrow n$p$'$ ($\delta = 0.473$, $n = 4-7$) and $1a_2\pi_3 \rightarrow n$d ($\delta = 0.032$, $n = 3-8$). In the nd series, the $n = 3$ member is described as "tentative," and so the series may terminate at nf ($n = 4-8$) instead. The MPI study of pyrrole [C45] failed to find the $n = 3$ member of the ns series, however three np series were constructed ($\delta = 0.53, 0.50$ and 0.46). Another series in the MPI spectrum having $\delta = 0.02$ lacks an $n = 3$ member and therefore is assigned as $1a_2\pi_3 \rightarrow n$f. Superexcited states of pyrrole [F14] appear as broad features in the electron impact spectrum and can be assigned as $n = 3$ Rydberg transitions originating at deeper MO's, as in Table XVII.C-I.

The MPI spectrum of N-methyl pyrrole [C45] is surprisingly different from that of pyrrole itself. Cooper *et al.* report a clear two-photon resonance to the $^1(1a_2\pi$, 3s) state at 41 193 cm^{-1} (*advert., 5.1072 eV*, 22 880-cm^{-1} term value) and a four-member ns series having $\delta = 0.90$.

TABLE XVII.C-I

SUPEREXCITED STATES IN PYRROLE[a]

Frequency, cm^{-1}	(*eV*)	Ionization Potential,[b] cm^{-1}	(*eV*)	Term Value, cm^{-1}	Assignment
77 700	(*9.63*)	101 700	(*12.61*)	24 000	$9a_1 \rightarrow 3s$
82 100	(*10.2*)	101 700	(*12.61*)	19 600	$9a_1 \rightarrow 3p$
89 200	(*11.0*)	110 000	(*13.64*)	20 800	$6b_2 \rightarrow 3p$
94 800	(*11.7*)	115 000	(*14.26*)	20 200	$8a_1 \rightarrow 3p$

[a] Reference [F14].

[b] Reference [**II**, D6].

The $1a_2\pi \rightarrow 3s$ transition in N-methyl pyrrole is one-photon forbidden/two-photon allowed, and the specification of the two-photon origin by MPI then permits a rational assignment of the one-photon vibronic spectrum. A possible np series ($\delta = 0.43$) also might be present in the MPI spectrum.

There is little new to add to the stories on thiophene and selenophene [**II**, 186]. As in the other heterocyclic dienes, the π_4^* and π_5^* MO's of thiophene are involved in valence TNI resonances [9400 and 26 600 cm^{-1} (*vert., 1.2 and 3.30 eV*), respectively] [V9]. Additionally, a resonance is observed at 54 600 cm^{-1} (*vert., 6.77 eV*); since the nearest Rydberg state in thiophene which could function as a parent state for a Feshbach resonance is at 64 560 cm^{-1} (*advert., 7.997 eV*), and since Feshbach decrements (Section II.B) rarely are as large as 10 000 cm^{-1}, it is concluded that this is a valence TNI resonance involving either a σ^* MO or a core excitation. Low-lying triplets are found at 30 200 and 37 300 cm^{-1} (*vert., 3.74 and 4.62 eV*) in thiophene [F13, F14, V9]. Re-examination of the thiophene spectrum has uncovered another Rydberg series ($\delta = 0.896$) terminating at ns with $n = 4-8$ in thiophene-d_4 [B23]. As in the other dienes, there are several superexcited states in thiophene [F14] which can be assigned as Rydberg excitations originating at deeper MO's.

Two new Rydberg series are reported in selenophene converging upon the $(3a_2\pi_5)^{-1}$ ionization limit at 70 780 cm^{-1} (*8.775 eV*): an np series ($n = 6-8$) with $\delta = 0.421$ and an nd series (n = 5–9) with $\delta = 0.043$ [B23]. As in the Rydberg excitations of thiophene, those of selenophene actively involve the ν_4', ν_5', ν_7' and ν_8' vibrational modes. In contrast, a discrete system in selenophene with (0,0) at 47 800 cm^{-1} (*5.93 eV*) has a different vibronic pattern, and so has been assigned instead as $N \rightarrow V_1$ [B23]. Since this band may correspond to a Rydberg excitation originating at $b_1\pi$, it should be tested experimentally for its Rydberg/valence character (Section II.D).

XVII.E. Higher Conjugated Polyenes

The *cis/trans* isomeric mixture of the 1,3,5-hexatrienes [**II**, 189, 336] can be separated chromatographically and spectra of the separate isomers are now available. In electron transmission, it is reported that the π_5^* electron affinity in both isomers is 28 500 cm^{-1} (*vert., 3.53 eV*), whereas π_4^* is at 12 700 cm^{-1} (*1.58 eV*) in the *cis* isomer but at 17 200 cm^{-1} (*2.13 eV*) in the *trans* isomer [B84]. This difference is attributable to the $C_2 - C_5$ interaction in the π_4^* MO of the *cis* isomer.

Triplet excited states in *trans*-1,3,5-hexatriene at 21 000 and 33 100 cm^{-1} (*vert., 2.60 and 4.10 eV*) have $^3B_u(\pi, \pi^*)$ and $^3A_g(\pi, \pi^*)$ symmetries, respectively, and so are closely related to the two lowest triplet states in butadiene [F15, F40, N14, N15, P34, S10]. A weak energy loss at 35 500 cm^{-1} (*vert., 4.40 eV*) reported by Post *et al.* [P34] could not be reproduced under more controlled conditions [F15]. An intense excitation to $V_1(\pi, \pi^*)$ has its origin at 39 670 cm^{-1} (*4.918 eV*) and extends to about 47 000 cm^{-1} (*5.8 eV*) [G3]; the Franck-Condon envelope of this $\pi \rightarrow \pi^*$ excitation is far better resolved than the corresponding band of butadiene (Section XVII.A), with a long progression of C=C stretch (ν_3') clearly predominating [F15].

The transition from the uppermost pi MO $2a_u\pi_3$ [B25] to 3s is allowed in *trans*-1,3,5-hexatriene, and is expected at *ca.* 43 000 cm^{-1} (*5.3 eV*) based on an estimated term value of 23 000 cm^{-1}; it is undoubtedly covered by the intense $N \rightarrow V_1$ excitation having an oscillator strength of 0.6−1.1 [P34]. Following this, there is a feature centered at 50 300 cm^{-1} (*advert., 6.24 eV*) which appears weakly in both the optical [G4] and electron-impact spectra [F15, P34]. Nascimento and Goddard [N14] calculate that this feature has a ($2a_u\pi_3$, 3d) upper state, but the experimental term value of 16 550 cm^{-1} is better suited to a (π_3, 3p) assignment, which also would account for the low intensity. In agreement with the (π_3, 3p) Rydberg assignment, the 50 300-cm^{-1} band appears with high intensity as a two-photon resonance in the MPI spectrum of the vapor-phase *trans* isomer [P10], and is washed out of the spectrum of the neat liquid as determined by thermal blooming [T27]. On the other hand, it is said to appear in the electron transmission spectrum of a solid film of the mixed isomers of hexatriene [S10]. Parker *et al.* [P12] attempted to determine the symmetry of the (π_3, 3p) upper state by measuring the MPI spectra using linearly- and circularly-polarized light. They conclude that the terminating orbital is 3pπ, however our experience with such measurements [H31] is that they cannot be trusted to yield unambiguous symmetries. Interestingly, the lowest component of the (π_3, 3p) manifold of *trans*-1,3,5-hexatriene is computed to be (π_3, 3pσ)[†] [N14], just as in ethylene (Section X.A) and butadiene (Section XVII.A).

The differing symmetries of *cis*-1,3,5-hexatriene (C_{2v}) and its *trans* isomer (C_{2h}) have been used to good advantage by Doering *et al.* [D37] within the Rydberg manifolds. Thus the weak energy loss at 50 300 cm^{-1} (*vert., 6.24 eV*) in the *trans* compound becomes an intense feature at the same frequency in the *cis* compound, leading to a 3pσ assignment of the upper level. Similarly, an intense energy loss at 52 600 cm^{-1} (*advert., 6.52 eV*) in the *trans* spectrum is missing in the *cis* spectrum under optical conditions, implying a (π_3, 3d$a_2\pi$) upper state. The transition to 3pπ peaks at 47 200 cm^{-1} (*5.88 eV*) according to Doering *et al.*

Just beyond the weak $\pi_3 \rightarrow$ 3p excitation of *trans*-1,3,5-hexatriene there appears a set of sharp absorptions of moderate intensity [52 685 cm^{-1} *advert., 6.5320 eV*, f approximately 0.1], assigned by all students of this molecule as valence $\pi \rightarrow \pi^*$ [F15, G4, K6, N14, P34]. However, the optical absorption-band profile [G4] does resemble that of the first band in the photoelectron spectrum [B25] and has the proper intensity and term value (14 180 cm^{-1}) to qualify as an allowed $2a_u\pi_3 \rightarrow$ 3d Rydberg transition. This transition is an excellent candidate for high-pressure or condensed-phase tests [I, 76]. A Rydberg series stretching from 62 225 to 67 000 cm^{-1} (*advert., 7.7147 to 8.3 eV*) is assigned by Gavin and Rice [G4] as terminating at np in spite of its quantum defect of 0.054. Nascimento and Goddard later correct this to nd [N14]. A broad peak at 65 000 cm^{-1} (*vert., 8.0 eV*) in the absorption of the *cis* isomer [G4] is likely an $N \rightarrow V_n$ valence excitation overlaid by weaker but sharper Rydberg transitions.

Once beyond the first ionization potential, our general experience has been that the broad features encountered there in absorption or electron impact most often are symmetry-allowed excitations to $n = 3$ Rydberg levels going to successively higher ionization potentials. This approach

† The 3pσ orbital is aligned short-axis, in-plane.

works again in the case of *trans*-1,3,5-hexatriene. Thus the energy losses at 73 000, 78 000 and 84 700 cm^{-1} (*vert., 9.1, 9.7 and 10.5 eV*) [F15, P34] have the proper term values with respect to the ionization potentials at 93 500 $(10a_g)^{-1}$, 101 600 $(9b_u)^{-1}$ and 96 000 cm^{-1} $(1a_u)^{-1}$ (*11.6, 12.6 and 11.9 eV*) [B25] to terminate at 3p, 3s and 3d, respectively; each is symmetry allowed.

In *cis*-1,3,5-hexatriene in an argon matrix, all the Rydberg absorptions are broadened beyond recognition, and only an intense peak at 39 500 cm^{-1} (*vert., 4.90 eV*, $N \rightarrow V_1$) and a weaker, smooth peak at 50 400 cm^{-1} (*vert., 6.25 eV*, $N \rightarrow V_2$) remain [K15]. In the *cis* isomer, the $N \rightarrow V_2$ transition remains forbidden even though the molecule no longer has a center of symmetry provided that all single bonds are still *trans*. If the $N \rightarrow V_2$ band of the *cis* isomer corresponds to that at *ca.* 53 000 cm^{-1} (*6.6 eV*) in the *trans* isomer, then the matrix spectrum of the *cis* isomer proves that the Rydberg conjecture offered above for the band in the *trans* isomer is incorrect. Two apparent valence excitations in the thin-film spectrum of mixed isomers of 1,3,5-hexatriene are centered at 61 700 and 77 000 cm^{-1} (*7.65 and 9.6 eV*) [S10]. The latter excitation corresponds to that at 78 000 cm^{-1} in the vapor of the *trans* isomer.

As one would expect, the electron energy-loss spectrum of cycloheptatriene [II, 189] closely resembles that of 1,3,5-hexatriene discussed above [F39]. Thus the two singlet-triplet $\pi \rightarrow \pi^*$ energy losses at 24 600 and 31 900 cm^{-1} (*vert., 3.05 and 3.95 eV*) of cycloheptatriene have their counterparts at 21 000 and 33 100 cm^{-1} (*vert., 2.60 and 2.90 eV*) in the open-chain hexatriene. The two $\pi \rightarrow \pi^*$ singlet excitations of hexatriene at 43 000 and 52 700 cm^{-1} (*vert., 5.33 and 6.53 eV*) appear as intense features centered at 39 100 and 51 600 cm^{-1} (*4.85 and 6.40 eV*) in cycloheptatriene. The latter band persists as an intense feature in heptane solution, thus documenting its valence nature in cycloheptatriene, and in 1,3,5-hexatriene by analogy. The $\pi_3 \rightarrow 3p$ band of hexatriene (50 300 cm^{-1}, *6.23 eV*) appears in cycloheptatriene as an inflection at 50 000 cm^{-1} (*vert., 6.2 eV*) with a term value of 19 500 cm^{-1}, while a sharp peak at 55 900 cm^{-1} (*vert., 6.93 eV*) has a 13 600-cm^{-1} term value as appropriate for a $\pi_3 \rightarrow 3d$ assignment. The valence excitations to both the singlet and triplet (π, π^*) levels of cycloheptatriene appear as well in the electron transmission spectrum of the solid film, albeit with depressed frequencies [S10].

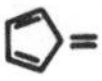

Vacuum-ultraviolet spectral data is available for three conjugated triene isomers of benzene:- trimethylene cyclopropane (TMCP), fulvene, and 3,4-dimethylene cyclobutene (DMCB). Though working with substances such as these is difficult due to their thermal and photochemical instabilities, the most recent spectra appear to be reliable. The N → V_1 band of TMCP lies quite low: 34 700 cm^{-1} (*vert., 4.30 eV*) [D42] with a molar extinction coefficient of 9390 ± 1170 at this frequency [B9]. *Ab initio* theory [R2] assigns this band as an allowed $\pi \rightarrow \pi^*$ transition, $^1A_1 \rightarrow {}^1E'$. A Rydberg series ($\delta = 0.58$; $n = 3-7$) has been deduced, and assigned as $e''\pi \rightarrow n$p [D42].

It is interesting to compare the vibronic envelopes of the $e''\pi \rightarrow$ 3p band of the TMCP absorption spectrum [(0,0) at 53 850 cm^{-1}, *6.676 eV*] with the $(e'')^{-1}$ ionization of the photoelectron spectrum [(0,0) at 72 110 cm^{-1}, *8.940 eV*]. The bands outwardly look much alike, as expected, however Bally and Haselbach [B9] point out that the first vibrational interval in the optical spectrum is 1644 cm^{-1}, whereas it is 1810 cm^{-1} in the photoelectron spectrum. They argue from this that the 1644-cm^{-1} vibration corresponds to ν_2' (C=C stretch, a_1') whereas the 1810-cm^{-1} vibration corresponds to ν_{13}' (C=C stretch, e'). We are reluctant to accept this inasmuch as it is not unusual to have such frequency differences in the same vibration of the Rydberg state and the ion, and there is no evidence otherwise showing the observed vibrations to have different symmetries. As discussed in Section I.D, vibronic bandshape differences between Rydberg and PES bands originating at the same MO can be caused by the preferential mixing of a valence conjugate with the Rydberg MO.

The forbidden $e''\pi \rightarrow$ 3s ($^1A_1' \rightarrow {}^1E''$) transition of TMCP is expected to have a term value much like that observed for the first Rydberg excitation in benzene, *i.e.*, 23 200 cm^{-1}. Though a weak feature is observed at 45 045 cm^{-1} (*advert., 5.5847 eV*) [B9], its term value of 27 100 cm^{-1} is too large for an ($e''\pi$, 3s) Rydberg assignment, and N → V_2 ($^1A_1' \rightarrow {}^1A_1'$) is preferred [R2]. The origin of the ($e''\pi$, 3s) band will fall close to 49 000 cm^{-1} (*6 eV*). Interestingly, the $^2E''$ ion shows no measurable Jahn-Teller splitting and so the corresponding $e''\pi \rightarrow n$s transitions also will be free of this complication.

The $\pi \rightarrow \pi^*$ parade in fulvene begins with two weak features, the first being structured [origin at 19 660 cm^{-1} (*2.437 eV*)], and the second being a broad continuum centered at 27 800 cm^{-1} (*3.45 eV*) [B67, D41]. Taken together, these two transitions have an oscillator strength of 4.2×10^{-3}. *Ab initio* theory [R2] predicts that these are $^1A_1 \rightarrow {}^1B_2(\pi, \pi^*)$ and $^1A_1 \rightarrow {}^1A_1(\pi, \pi^*)$ transitions of modest oscillator strength. There then follows an intense feature ($f = 0.34$) at 45 550 cm^{-1} (*vert., 5.647 eV*) displaying twenty quanta of 495 cm^{-1}, Fig. X.B-1, assigned as the b_1 nontotally symmetric out-of-plane CH_2 deformation [B67]. Theory adequately accounts for this band as N → V_3($^1A_1 \rightarrow {}^1A_1$). All these bands of fulvene have been tested in matrices and under high pressure, and their response confirms their valence $\pi \rightarrow \pi^*$ nature [H16].

The N → V spectra of fulvene and butadiene are obviously very different, and there is little reason to expect them to be otherwise. Consequently, it is a surprise to see how similar their Rydberg spectra are. This similarity arises because the highest occupied MO of fulvene (and furan) ($1a_2''$) has the same pseudo-atomic 3d orbital pattern as that which characterizes the highest occupied MO of butadiene. The optical spectrum of fulvene exhibits two sharp and intense origins at 49 570 and 56 400 cm^{-1} (*advert., 6.146 and 6.992 eV*) with very similar vibrational structure [B67, H16]. The first of these was totally obliterated when tested in an argon matrix and otherwise was strongly perturbed in high-pressure He gas, Figure X.B-1. Clearly, these are Rydberg excitations. With an adiabatic first ionization potential of 67 430 cm^{-1} (*8.36 eV*) [H34], the term values of the two sharp origins in fulvene are 17 860 and 11 030 cm^{-1}. As in butadiene, Rydberg transitions from the $1a_2\pi_3$ MO of fulvene will be allowed to np (n = 3, 4,..) and nf (n = 4, 5,..) orbitals, whereas transitions to ns remain forbidden even though there is no longer a center of symmetry in fulvene. The observed term values for the 49 570- and 56 400-cm^{-1} bands agree with expectations for transitions to 3p and 4p, respectively.

Absorption in dimethylene cyclobutene (DMCB) is broad and poorly resolved [G38, K13], Fig. X.B-1. N → V valence excitations peak at 41 700 cm^{-1} (*5.17 eV*; f = 0.1) and 48 500 cm^{-1} (*6.01 eV*; $f > 1$). Theory predicts an N → V ordering of $^1A_1$, $^1B_2$, $^1A_1$, $^1B_1$, with a monotonic increase of oscillator strength on going from $^1A_1$ to $^1B_1$ [R2]. With an ionization potential of 70 980 cm^{-1} [*advert., 8.80 eV*, $(2b_1\pi_3)^{-1}$] [H34], an allowed $2b_1\pi_3 \rightarrow$ 3s excitation is expected in DMCB at *ca.* 48 000 cm^{-1} (*5.95 eV*) and probably corresponds to one or more of the sharp features resting upon the intense N → V band in this region. The $2b_1\pi_3 \rightarrow$ 3d Rydberg band (13 180-cm^{-1} term value) appears at 57 800 cm^{-1} (*7.17 eV*). Owing to differences of C_2-C_5 pi-electron interactions, the spectra of DMCB and *cis*-1,3,5-hexatriene (*vide supra*) differ noticeably: the N → V_n excitations are *ca.* 7000 cm^{-1} lower in DMCD with N → V_1 the weakest of these, whereas in *cis*-1,3,5-hexatriene N → V_1 is the most intense of the $\pi \rightarrow \pi^*$ excitations.

Interpretation of the electron energy-loss spectrum of cyclooctatetraene [F41, **II**, 189] is very difficult without a sophisticated *ab initio* calculation to lead the way. Lacking that, we can still take the first step by sorting the transitions into valence and Rydberg upper states on the basis of their term values and the photoelectron spectrum [B20], Table XVII.E-I. In making these assignments, it is assumed that all the observed features are symmetry allowed, and that term values of 22 000−23 000 cm^{-1} imply

TABLE XVII.E-I

SPECTROSCOPIC ASSIGNMENTS IN CYCLOOCTATETRAENE[a]

Frequency, cm^{-1}	(*eV*)	Ionization Potential,[b] cm^{-1}	(*eV*)	Term Value, cm^{-1}	Assignment
24 600	(*3.05*)	67 900	(*8.42*)		$^3(\pi, \pi^*)$
32 700	(*4.05*)				$^3(\pi, \pi^*)$
39 000	(*4.84*)				$^3(\pi, \pi^*)$
35 700	(*4.43*)				$^1(\pi, \pi^*)$
48 600	(*6.02*)			19 300	$^1(5a_1\pi_4, 3p)$
51 800	(*6.42*)			16 100	$^1(\pi, \pi^*)$
56 400	(*6.99*)	78 900	(*9.78*)	22 500	$^1(7e\pi_3, 3s)$
67 800	(*8.41*)	78 900	(*9.78*)	11 100	$^1(7e\pi_3, 3d)$
		89 900	(*11.1*)	22 100	$^1(4b_2\pi_2, 3s)$
73 000	(*9.05*)	93 200	(*11.6*)	20 200	$(3b_1\sigma, 3p)$

[a] Reference [F41].

[b] Taken from reference [B20].

transitions terminating at 3s, 17 000–20 000 cm^{-1} imply transitions terminating at 3p, and 11 000–14 000 cm^{-1} imply transitions terminating at 3d. The transition at 51 800 cm^{-1} (*vert., 6.42 eV*) is too intense to be a Rydberg band, and so is assigned as $^1A_1 \rightarrow {}^1E$ ($\pi \rightarrow \pi^*$) [F41]. Little else can be said until MPI spectra and *ab initio* calculations are reported.

The spectrum of spiro[4.4]nonatetraene, , is interesting both for what it shows and what it does not show [B21]. With a first ionization potential of 64 400 cm^{-1} (*advert., 7.98 eV*) from the $1a_2\pi_4$ MO (D_{2d}), Rydberg excitations to 3s (forbidden) and to 3p (allowed) are expected at 42 000 and 44 500 cm^{-1} (*5.21 and 5.52 eV*), respectively, whereas neither is reported. Instead, four valence excitations are observed, the upper three of which appear to be allowed and the lowest forbidden, as judged by their relative intensities. These bands are most readily interpreted using the independent-systems picture, with the fundamental oscillators being those of the cyclopentadiene pi-electron system. Coupling of the $\pi \rightarrow \pi^*$ triplets in cyclopentadiene at 25 000 cm^{-1} (*vert., 3.1 eV*) [F43] lead to a forbidden singlet excited state in spiro[4.4]nonatetraene, corresponding to the weak band observed at 36 500 cm^{-1} (*vert., 4.52 eV*) [C47]. The intense $N \rightarrow V_1$ transitions of cyclopentadiene (42 400 cm^{-1} *vert., 5.26 eV*) are noninteracting in the spiro compound owing to symmetry, and so form a doubly degenerate pair of allowed transitions, assigned to the intense bands at 48 800 cm^{-1} (*vert., 6.05 eV*). In contrast, the $N \rightarrow V_2$ transitions (62 000 cm^{-1} *vert., 7.69 eV* in cyclopentadiene) interact strongly in the tetraene, leading to the intensely allowed transition at 54 600 cm^{-1} (*vert., 6.77 eV*) with polarization predicted to be along the S_4 axis. Since the second component of the V_2/V_2 interaction is forbidden and so does not serve to explain the intense transition reported at 60 000 cm^{-1} (*vert., 7.44 eV*), it is likely that the 60 000-cm^{-1} band is derived from a higher $\pi \rightarrow \pi^*$ excitation of cyclopentadiene. The level structure sketched above for spirononatetraene is closely paralleled by the results of an MO-CI calculation of its spectrum [C47].

The low-resolution electron impact spectrum of solid β-carotene [O8] reveals an $N \rightarrow V_1$ band at 24 200 cm^{-1} (*vert., 3.0 eV*) followed by two broad features at 84 700 and 166 000 cm^{-1} (*vert., 10.5 and 20.6 eV*) which are remarkably like those of polyethylene and the longer alkanes at the same frequencies (Section III.D). The second of these broad bands is most likely the volume plasmon involving the σ(C–C) electrons. In the α,ω-diphenyl polyenes [O8] the two broad bands beyond 80 000 cm^{-1} (*10 eV*) maintain frequencies very close to those in β-carotene.

Spectra of conjugated polyenes in which one or two of the =CH groups are replaced by =NR or =NOR are discussed in papers from Sandorfy's laboratory. Such replacements in butadienes to produce either RCH=CH–CH=NR′ or RCH=N–N=CHR′ lead to low-lying $n_N \rightarrow \pi^*$ excitations [V29]. Because the $\pi \rightarrow \pi^*$ transitions of the C=C and C=N chromophores are separated by only 3000 cm^{-1} (Section XI.B), the $N \rightarrow V_1$ bands of the monoaza and diaza dienes (50 000 cm^{-1} *vert., 6.2 eV*) are only slightly above those of the corresponding all-carbon dienes (45 000 cm^{-1} *vert., 5.6 eV*). The only other identifiable features in these spectra are possible $N \rightarrow V_2$ transitions at 55 000–60 000 cm^{-1} (*vert., 6.8–7.4 eV*); the spectra are broad and no Rydberg excitations can be spotted. In the diaza compound having $R = R' = CH_3$, the $N \rightarrow V_2$ band is more intense than $N \rightarrow V_1$, and they are of approximately equal intensity when $R = R' = C_2H_5$, suggesting that these compounds are not necessarily *trans*-planar.

Vacuum-ultraviolet spectra in the series of azapolyene compounds $R-(HC{=}CH)_xCH{=}NOCH_3$ [D56] with $x = 0$, 1 and 2 have $N \rightarrow V_1$

excitations which are a few thousand cm^{-1} below those of the corresponding all-carbon polyenes. As in simple oximes, the Rydberg excitations are buried by the broad $N \rightarrow V_1$ bands, however weak features in the 50 000−55 000-cm^{-1} (*6.2−6.8-eV*) region for $x = 1$ and 2 are probably a mix of overlapping $N \rightarrow V_2$ and Rydberg excitations.

XVII.F. Addendum

M. Allan, L. Neuhaus and E. Haselbach, (All-E)-1,3,5,7 octatetraene: electron-energy-loss and electron-transmission spectra, *Helv. Chim. Acta* **67**, 1776 (1984).

G. De Alti, P. Decleva and A. Lisini, An ab initio study of the satellite structure in the heteroatom core ionization of furan, pyrrole and thiophene, *Chem. Phys.* **90**, 231 (1984).

P. Dupuis, C. Sandorfy and D. Vocelle, Photoelectron and ultraviolet absorption spectra of the enamines of a series of simple polyenes, *Photochem. Photobiol.* **39**, 391 (1984).

P. Mercier, C. Sandorfy, O. Pilet and P. Vogel, Vapor phase UV spectra of 2,3,5,6,7,8-hexakis(methylene)bicyclo[2.2.2]octane ("[2.2.2] Hericene") and related s-cis-butadiene derivatives, *Can. J. Spectrosc.* **28**, 184 (1983).

R. N. S. Sodhi and C. E. Brion, High resolution carbon 1s and valence shell electronic excitation spectra of allene and trans-1,3-butadiene studied by electron energy loss spectroscopy, submitted to *J. Electron Spectrosc. Related Phenomena.*

E. Lindholm, Interpretation of core-excitation ($1s-\pi^*$) processes, submitted to *J. Chem. Phys.*

CHAPTER XVIII

Cumulenes

XVIII.A. Allene and Ketene

Work on unraveling the complexities of the allene spectrum [**II**, 199] has progressed significantly in the past ten years. The spectral data has been expanded by an electron-impact energy loss study [M73], by absorption and MCD studies in the vapor and solution phases [F50], by CD studies of chiral derivatives [R4] and by *ab initio* calculations of allene's spectral properties [D19, R4]. Let us see how this new work builds upon the base established in [**II**, 199]. The lowest valence configuration in allene, $(2e\pi, 3e\pi^*)$, results in the four singlet excited states $^1A_1$, $^1A_2$, $^1B_1$ and $^1B_2$, with transitions from the ground state allowed only to $^1B_2$. Excitations to $^3A_1$ and $^3B_2$ are observed as energy losses at 34 500 and 39 400 cm^{-1} (*vert., 4.28 and 4.89 eV*) [M73]. A broad region of singlet → singlet energy loss in the 40 000−50 000-cm^{-1} region (*5.0−6.2-eV*) is resolved into two transitions, while the MCD spectrum of allene in n-C_6F_{14} solution (Fig. XVIII.A-1) reveals singlet valence levels at 46 500, 48 800 and 52 600 cm^{-1} (*vert., 5.76, 6.05 and 6.52 eV*) [F50] in a region of very low absorption coefficient. The calculations [D19, R4] predict forbidden transitions to $^1A_2$ and $^1B_1$ in the vicinity of 50 000 cm^{-1}.

The structured absorption of moderate intensity at 54 200 cm^{-1} (*vert., 6.72 eV*) in allene earlier had been deduced to be the $2e\pi \rightarrow 3s$ Rydberg transition based on its term value [**II**, 200], whereas others treat it as a valence excitation [M73]. In the vapor-phase MCD spectrum, the 54 200-cm^{-1} band of allene shows a B term which is totally demolished in solution, Fig. XVIII.A-1, thereby confirming its Rydberg upper-state character. Theory [D19, R4] concurs in this assignment, and further states that the magnetic moment of the ($2e\pi$, 3s) state in the vapor phase is so small that the MCD band is predominantly B-term-like even though the state is formally degenerate and an A term might otherwise be expected. Note however, that if the ($2e\pi$, 3s) Rydberg state is statically Jahn-Teller split as is the $(2e\pi)^{-1}$ band in the photoelectron spectrum, then a B term is expected in the MCD spectrum [F50].

The intense absorption at 55 000−63 000 cm^{-1} (*6.8−7.8 eV*) in the allene molecule was earlier assigned as an overlapping of the allowed $^1A_1 \rightarrow {}^1B_2(2e\pi, 3e\pi^*)$ valence and $2e\pi \rightarrow 3p$ Rydberg excitations [**II**, 201]. In the solution absorption and MCD spectra, Fig. XVIII.A-1, a single broad peak survives centered at 59 500 cm^{-1} (*7.38 eV*) while another appears at 63 300 cm^{-1} (*7.85 eV*) in the MCD. This response to solution indeed is consistent with the presence of overlapped $2e\pi \rightarrow 3e\pi^*$ and $2e\pi \rightarrow 3p$ excitations (Section II.D). In spite of the intensity and solution-perturbation arguments for an allowed $2e\pi \rightarrow 3e\pi^*$ valence excitation in this region, the theory of Diamond and Segal [D19] leads them to assign the two features at 59 500 and 63 300 cm^{-1} to components of the $2e\pi \rightarrow 3p$ Rydberg excitation. On the other hand, according to the calculation of Rauk *et al.*, this is the region of allowed $2e\pi \rightarrow 3e\pi^*$ valence and $2e\pi \rightarrow 3p$ Rydberg absorption, with the ($2e\pi$, $3e\pi^*$) valence configuration strongly mixed with appropriate Rydberg configurations of the same symmetry [R4]. Though assignment of the 63 300-cm^{-1} band to a specific valence excited state is plausible (Fuke and Schnepp propose a $2e\pi \rightarrow \sigma^*$ promotion [F50]), we lean toward a Rydberg assignment instead, for such bands are inclined to appear in MCD solution spectra (Section II.D) and the term value is appropriate for such a transition.

A tight cluster of six $2e\pi \rightarrow 3d$ Rydberg bands are predicted for the 64 700−67 600-cm^{-1} (*8.02−8.38-eV*) region of allene [D19], however the term values for the bands observed in this region are somewhat larger than normally associated with 3d, whereas the assignment of these bands as $2e\pi \rightarrow 3p$ converging upon the *second* Jahn-Teller component of the ^2E ionic state fits nicely. The absorption spectrum of allene-d_4 would be useful in enumerating the number of electronic origins and hence the assignments in this region. The transition which we earlier assigned as

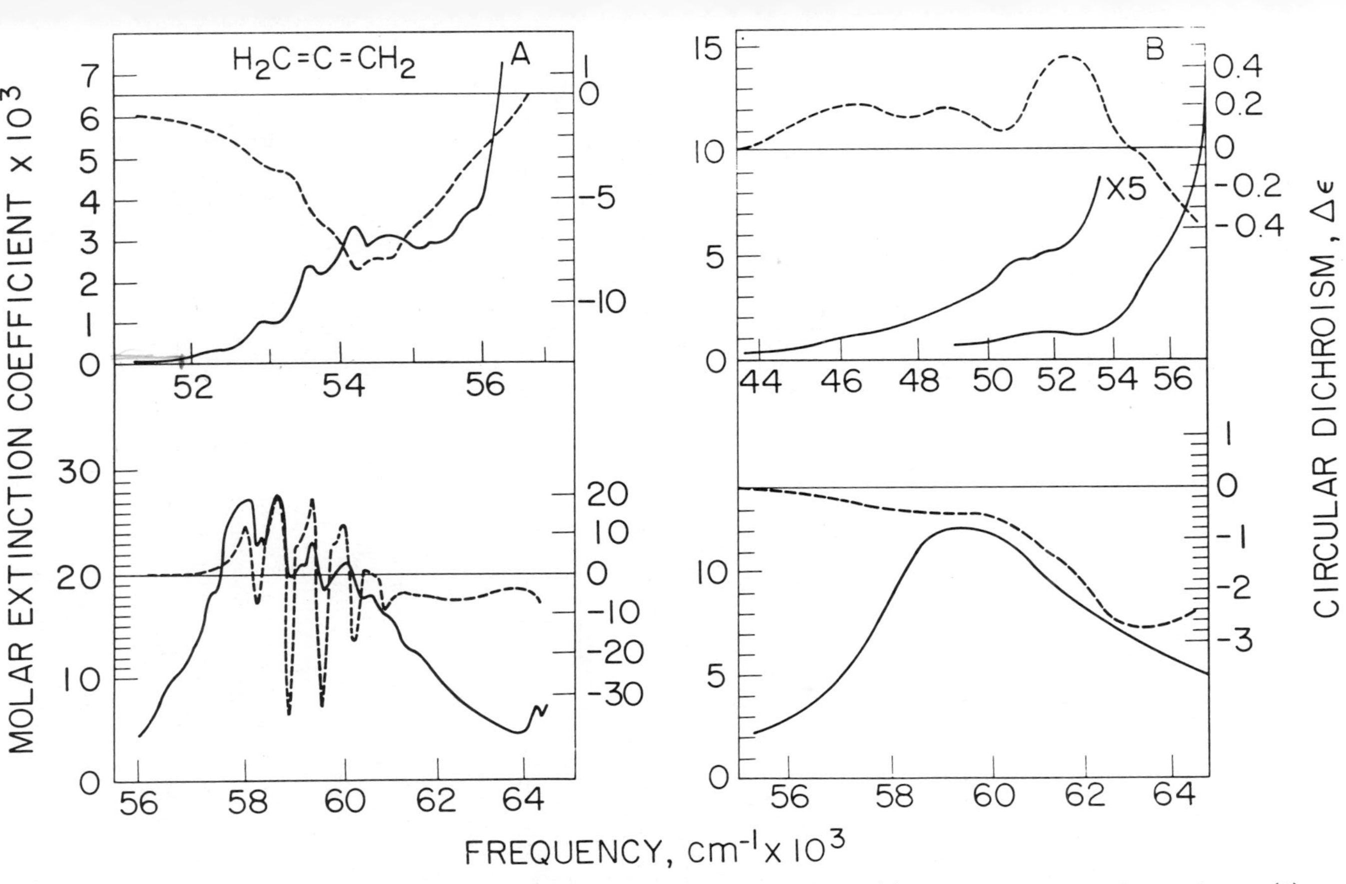

Fig. XVIII.A-1. Absorption (—) and MCD (- -) spectra of allene in the vapor phase **(A)** and in solution in perfluoro-*n*-hexane **(B)** [F50].

$2e\pi \rightarrow 3d$[†] (69 150 cm^{-1} *advert., 8.573 eV*) is calculated by Diamond and Segal [D19] to be the allowed valence excitation to $^1B_2$. Though this *ab initio* calculation adds considerable understanding and is most welcome, it seems likely that it suffers from placing the $2e\pi \rightarrow 3e\pi^*$ valence excitations too high, as do most *ab initio* calculations. An energy loss to a superexcited state at 90 740 cm^{-1} (*vert., 11.25 eV*) [M73] in allene has a term value of 25 400 cm^{-1} with respect to the ionization potential of the $1e\pi$ MO at 116 000 cm^{-1} (*vert., 14.4 eV*) and so at first sight would seem to have a $(1e\pi, 3s)$ assignment. However, the photoionization curve for allene shows considerable autoionization fine structure [P15], with the most intense feature being at 86 200 cm^{-1} (*advert., 10.7 eV*). Thus we assign this feature as $1e\pi \rightarrow 3s$, with a 29 800-cm^{-1} term value, and tentatively identify the 90 740-cm^{-1} feature as a valence $1e\pi \rightarrow 3e\pi^*$ transition. The autoionizing line corresponding to $1e\pi \rightarrow 3p$ is prominent at 97 400 cm^{-1} (*advert., 12.08 eV*) and has a term value of 18 600 cm^{-1}. The absorptive and CD properties of the chiral cumulenes 1,3-dimethyl allene and 1,2-cyclononadiene [R4] out to 55 000 cm^{-1} (*6.8 eV*) are consistent with the allene assignments given above.

Earlier work on ketene [**II**, 203] raised the question of where the lowest transition to 3s might be, and suggested the band at 46 900 cm^{-1} (*vert., 5.81 eV*) as an answer. This suggestion is supported by electron-impact energy loss spectra in which a singlet → singlet band at 47 300 cm^{-1} (*vert., 5.86 eV*) is assigned as $n_O \rightarrow 3s$ [F37], and a threshold electron spectrum showing a singlet → triplet excitation at 46 800 cm^{-1} (*vert., 5.8 eV*) which is calculated to terminate at 3s [V30]. That the $n_O \rightarrow 3s$ transition truly is at 47 300 cm^{-1} in ketene can be checked, for if it is the lowest Rydberg excitation in the molecule, then there will be a Feshbach resonance to the $^2(n_O, 3s^2)$ negative-ion state at *ca.* 43 000 cm^{-1} (*5.3 eV*) in the TNI spectrum (Section II.B). This experiment remains to be done. Another singlet → triplet excitation in the threshold spectrum of ketene at 56 500 cm^{-1} (*vert., 7.0 eV*) is possibly the complement of the singlet → singlet $n_O \rightarrow 3p$ excitation observed at 54 700 cm^{-1} (*6.78 eV*), but the calculation of Harding and Goddard [H13] suggests a $^3(\pi, \pi^*)$ upper-state assignment instead. Two other valence singlet → triplet excitations appear in ketene at 30 600 and 40 300 cm^{-1} (*vert., 3.8 and 5.0 eV*) [F37, V30], which theory assigns as $^3(n_O, \pi^*)$ and $^3(\pi, \pi^*)$, respectively [H13].

[†] An obvious typographical error in [**II**, 202] reports this as $2e\pi \rightarrow 3p$.

XVIII.B. Addendum

See Sodhi and Brion, Section XVII.F.

CHAPTER XIX

Phenyl Compounds

Though benzene and its derivatives remain the most popular systems among molecular spectroscopists, and though considerable progress has been made in the past ten years, still many puzzles remain. Among those of prime concern, one can list:

a) Confusion still exists concerning the identification and assignment of *n*p, *n*d and *n*f Rydberg series going to the first ionization potential. Where are the lower members of these series and how many series are expected to be seen?

b) Only about half of the possible singlet and triplet $\pi \rightarrow \pi^*$ excitations have been observed. Even though the long-sought transition to $^1E_{2g}(\pi, \pi^*)$ appears to have been found, where are the remainder?

c) There are several suggestions from theory that $\sigma \rightarrow \pi^*$ excitations are low-lying. Where are they, and where are the complementary $\pi \rightarrow \sigma^*$ bands?

d) What is the nature of the high-lying temporary-negative-ion resonances in benzene which are not Feshbach resonances? Do these involve $\pi \rightarrow \pi^*$ core excitation? Where are the valence TNI resonances to σ^*?

e) Why is it that in the inner-shell spectra of the halobenzenes, there seems to be no spectral evidence for transitions involving the $\sigma^*(C-X)$ orbital, whereas this orbital is very active in outer-shell spectra?

f) What is the nature of the valence excitations beyond 65 000 cm^{-1} (*8 eV*) in benzene?

g) Why is there no splitting of the $^1E_{1u}(\pi, \pi^*)$ upper state in benzene derivatives of low symmetry?

h) In rare-gas matrices containing benzene, how molecular are the excitonic impurity absorptions? Does benzene form a bubble state (Section II.D) in rare-gas matrices?

i) Where are the σ^* shape resonances in benzene and its derivatives, and how can these be distinguished from peaks in the photoionization cross sections?

In the discussions which follow, most of these questions will be addressed, if not solved. In this, the benzene parent orbital and state nomenclature of the D_{6h} point group (Section XIX.A) will be retained in large part, even though the molecules under discussion are substituted so as to have formally lower symmetries (Sections XIX.B, C, D and E). Spectra of various isomers of benzene (fulvene, dewar benzene, *etc*) are discussed in Section XVII.C.

XIX.A. Benzene

Careful absorption studies on benzene vapor [**II**, 210, 336] with due regard for accurate intensity measurement and spectral bandwidth have been made in the 37 000–290 000-cm^{-1} (*4.6–36-eV*) region [P7]. Over the stated frequency range, the integrated oscillator strength is 17.5, with specific transitions contributing 0.0013 ($^1A_{1g} \rightarrow {}^1B_{2u}$), 0.090 ($^1A_{1g} \rightarrow {}^1B_{1u}$), 0.90–0.95 ($^1A_{1g} \rightarrow {}^1E_{1u}$) and 0.075 (the sum of all discrete Rydberg transitions below the first ionization potential). This last figure (0.075) is surprisingly low in view of the Rydberg oscillator strengths observed in other systems [**I**, 29].

New data is available on the long-sought transition to the $^1E_{2g}(\pi, \pi^*)$ valence state. Nakashima *et al.* [N8] have recorded the $S_1(^1B_{2u}) \rightarrow S_n$ spectrum of benzene vapor and uncovered three peaks. The first, 56 000 cm^{-1} (*vert., 7.0 eV*) above the $^1A_{1g}$ ground state, is a vibronically-induced transition from $^1B_{2u}$ to $^1E_{1u}$. Transitions from $^1B_{2u}$ are electric-dipole allowed only to $^1B_{1g}$ and $^1E_{2g}$. The weak $^1B_{2u} \rightarrow {}^1E_{1u}$ band is followed by two stronger transitions centered at 63 000 and 76 000 cm^{-1} (*vert., 7.8* *and 9.4 eV*) above the $^1A_{1g}$ ground state. Comparisons with excited-state spectra recorded in cyclohexane solution show that these transitions terminate at valence orbitals [N7]. According to both semiempirical [B41] and *ab initio* calculations [H25], the 63 000-cm^{-1} transition from the ground state corresponds to a two-electron $\pi \rightarrow \pi^*$ promotion to $^1E_{2g}$, however, π^* is a 3d Rydberg orbital according to the *ab initio* work, whereas the semiempirical calculations and the experimental work conclude that it is valence. Assignment of the 76 000-cm^{-1} transition is secure, for all workers describe it as $^1A_{1g} \rightarrow {}^1E_{2g}(\pi, \pi^*)$ valence. A singlet $\rightarrow$ triplet energy loss at 51 300 cm^{-1} (*vert., 6.36 eV*) in benzene has been assigned as terminating at $^3E_{2g}(\pi, \pi^*)$ [W25], however, this may be a Rydberg excitation instead (*vide infra*).

In benzene and several of its alkylated derivatives, it has been computed by CNDO methods [B41, B42, R23, R43] that $\sigma \rightarrow \pi^*$ valence excitations will fall in the vicinity of the $\pi \rightarrow \pi^*$ transition to $^1B_{1u}$. Though not *ab initio*, these calculations are difficult to ignore since they do an excellent job of predicting the $\pi \rightarrow \pi^*$ spectrum of benzene. Confident experimental identification of the $\sigma \rightarrow \pi^*$ promotion is difficult since the spectral region in question already is known to contain transitions to $^1B_{1u}(\pi, \pi^*)$, $^3(\pi, 3s)$, $^1(\pi, 3s)$ and $^3E_{2g}(\pi, \pi^*)$. The complementary $\pi \rightarrow \sigma^*$ excitations are calculated to fall on the high-frequency side of the transition to $^1E_{1u}$, in a region already occupied by transitions to $^3(\pi, 3p)$, $^1(\pi, 3p)$ and $^1E_{2g}(\pi, \pi^*)$. MCD and ORD measurements may prove effective in locating such valence states of the phenyl chromophore [R43].

Almost all the vapor-phase benzene transitions have been investigated one way or another in condensed phases, many at low temperature. Low-frequency solvent shifts of the benzene valence transitions to $^1B_{2u}$, $^1B_{1u}$ and $^1E_{1u}$ in liquified rare gases are cataloged by Zelikina and Meister [Z4]. *Ab initio* calculations on the (π, π^*) excited states of benzene [H25] show that whereas $^1B_{2u}$ and $^1B_{1u}$ are quite compact, $^1E_{1u}$ is very diffuse, as expected from its ionic character. Nonetheless, the solvent shift of the $^1E_{1u}$ level in the rare gases is fully in accord with a valence upper state, just as with the V states of ethylene (Section X.A) and butadiene (Section XVII.A). The crystal-induced (0,0) component of the transition to $^1B_{1u}$ is found at 46 500 cm^{-1} (*5.76 eV*) in benzene thin films at 10 K [P6]. Dramatic differences are apparent in the absorption spectra of neat benzene films deposited at different temperatures [S48, S50]. With deposition at high temperature (170 K) and presumably with better crystal quality, the absorption to $^1E_{1u}$ at 90 K is strongly repressed while significant absorption appears in the 68 000–89 000-cm^{-1} (*8.5–11-eV*) region. Certain of these latter features correlate with electron energy losses reported by Killat [K14] for benzene films deposited at 77 K; seven peaks, most likely valence excitations, are enumerated in the 64 000–137 000-cm^{-1} (*8–17-eV*) region. Killat's data on solid benzene also shows an immense peak at 173 000 cm^{-1} (*21.5 eV*) in the spectrum of Im $(-1/\epsilon)$ which is totally missing in the spectrum of ϵ_2 and therefore is assigned as a collective excitation (or plasmon, Section II.D). Electron-transmission spectra of neat benzene films at low impact energy (14 eV) reveal four consecutive triplet excited states ($^3B_{1u}$, $^3E_{1u}$, $^3B_{2u}$ and $^3E_{2g}$) preceding $^1E_{1u}$ [S9]. Surprisingly, this work reports that the $^1A_{1g} \rightarrow {}^1E_{1u}$ band shifts to high frequency on going from a 2 % solution in xenon to the neat benzene crystal.

In a high-symmetry system such as benzene, one expects that MPI spectroscopy will reveal several new Rydberg states (Section II.A), and one is not disappointed in this regard. Thus Johnson [J10] reports the $e_{1g}\pi \rightarrow 3s$ transition in benzene as a (2 + 1) MPI resonance with an origin at 51 085 cm^{-1} (*advert., 6.3336 eV*) followed by a display of a_{1g} and e_{2g} vibrations, Fig. II.A-3. The assignment of this two-photon resonance to the $(e_{1g}\pi, 3s)$ upper state rather than to $^1E_{2g}(\pi, \pi^*)$ is based upon its term value and vibronic structure [J10], upon the condensed-phase effect [S34], and upon the similarities to the $\pi \rightarrow 3s$ transitions in benzene derivatives [K39]. Doering [D29, D31] has shown convincingly that the intensity anomaly in the electron-impact excitation of the $^1A_{1g} \rightarrow {}^1B_{1u}$ band of benzene is due to the presence of the $e_{1g}\pi \rightarrow 3s$ transition [**II**, 214] at exactly the frequency determined by the MPI spectrum. More highly-refined electron impact studies by Wilden and Comer [W25] and by Allan [A8] show yet a third transition among those to $^1B_{1u}$ and $^1E_{1g}$. Though some workers attribute this third transition to the valence excitation to $^3E_{2g}(\pi, \pi^*)$, it is perfectly located for assignment as $^3(e_{1g}\pi, 3s)$, with a singlet-triplet split at the origin of 1700 cm^{-1}. Working at a constant residual energy of 0.20 eV, Allan observes energy losses to higher $(e_{1g}\pi, ns)$ members ($n = 4-7$) in both the triplet (T) and singlet (S) manifolds, leading to quantum defects of $\delta_T = 0.86$ and $\delta_S = 0.84$.

Studies in neat liquid benzene and in cyclohexane solution by Scott and Albrecht [S34] show what is apparently the two-photon $e_{1g}\pi \rightarrow 3s$ Rydberg excitation in spite of the external perturbation. The sharp 51 085-cm^{-1} transition of the vapor phase, Fig. II.A-3, appears as a broad plateau at *ca.* 55 000 cm^{-1} (*6.8 eV*) in the fluorescence excitation spectrum of the neat liquid, with an upper state said to have mixed Rydberg and charge-transfer character. On dilution in cyclohexane, a clear peak develops at 52 600 cm^{-1} (*vert., 6.52 eV*). This behavior on dilution and a strong temperature effect on the peak frequency (Section II.D) argue for a $(\pi, 3s)$ Rydberg upper-state assignment rather than $^1E_{2g}(\pi, \pi^*)$.

Going beyond the transition to 3s, Johnson [J10] observes a plethora of three-photon-resonant MPI features in the 66 000–72 000-cm^{-1} (*8.2–8.9-eV*) region of benzene vapor with frequencies which match exactly the higher Rydberg transitions [**II**, 217] found by Wilkinson [**II**, W26] in the one-photon spectrum. As Johnson states, no other two-photon MPI resonances are found on the low-frequency side of 50 000 cm^{-1} (*6.2 eV*), suggesting that the transition to $^1E_{2g}(\pi, \pi^*)$ does not lie below this frequency; however, one must keep in mind the very strong bias which MPI spectroscopy holds against valence-state resonances even if two-photon allowed (Section II.A).

Whetten *et al.* [W13] describe a 5-member Rydberg series in the MPI spectrum of benzene in the region of 63 000–75 000 cm^{-1} (*7.8–9.3 eV*). These (2 + 1) resonances have intense electronic origins displaying the expected isotope shifts. Their frequencies can be fit as either the $n = 4-8$ members of the $e_{1g}\pi \rightarrow n$s series for which Johnson's band at 51 085 cm^{-1} is the $n = 3$ member, or excluding the transition found by Johnson, as the $n = 3-7$ members of an $e_{1g}\pi \rightarrow n$d series. Note that the lowest member of the series is too low to allow an $e_{1g}\pi \rightarrow n$f assignment. Vibronic analyses of these bands imply that each upper-state is Jahn-Teller stabilized by *ca.* 1000 cm^{-1} via ν_6 and ν_9 vibrational modes. Wiesenfeld and Greene [W18] populated a level at 63 935 cm^{-1} (*7.9267 eV*) in benzene using a sub-picosecond two-photon absorption, and then ionized this level using a second sub-picosecond laser pulse. By varying the time between the two pulses, they directly determined a lifetime of 0.070 psec for this Rydberg level.

Excitation of an electron from $e_{1g}\pi$ into npxye_{1u} and npza_{2u} orbitals results in four singlet states: $^1A_{1u}$, $^1A_{2u}$ and $^1E_{2u}$ from $e_{1g} \rightarrow e_{1u}$, and $^1E_{1u}$ from $e_{1g} \rightarrow a_{2u}$. The transitions to the $n = 3$ members of the $^1A_{2u}$ and $^1E_{1u}$ series have been observed in the one-photon absorption spectrum as sharp-line origins at 55 881 and 59 795 cm^{-1} (*6.9282 and 7.4135 eV*), respectively. The specific assignment of the $(e_{1g}\pi, n\text{p}xy)$ $^1A_{2u}$ level as lying below $(e_{1g}\pi, n\text{p}z)$ $^1E_{1u}$ comes from a vibronic study by Vitenberg *et al.* [V28] of the $\nu_{20}(e_{2u})$ sequence structures surrounding the $n = 4$ origins. In the $n = 4$ and 5 members of the series having the larger δ value, the sequence structure is quite regular as regards both frequency spacing and intensity, implying a nondegenerate electronic upper state, *i.e.*, $^1A_{2u}$. In contrast, the origins of the second series show the irregular sequence spacing and intensity pattern characteristic of a quadratic Jahn-Teller effect in an electronically degenerate state, *i.e.*, $^1E_{1u}$. In partially deuterated benzenes of appropriately low symmetry, it is expected that the degeneracy inherent in the (0,0) bands of the $e_{1g}\pi \rightarrow n$pza_{2u} ($^1A_{1g} \rightarrow {}^1E_{1u}$) transitions of benzene-h_6 will be lifted, whereas this is irrelevant in the $e_{1g}\pi \rightarrow npxye_{1u}$ ($^1A_{1g} \rightarrow {}^1A_{2u}$) transition to a nondegenerate upper state. Indeed, Scharf *et al.* [S24] do find in the one-photon spectra of selectively deuterated species that the (0,0) bands of the $^1A_{1g} \rightarrow {}^1E_{1u}$ Rydberg transitions are noticeably broader in D_{2h} and C_{2v} symmetries than in D_{3h} and D_{6h}, whereas there is no effect of this sort among the members of the $^1A_{1g} \rightarrow {}^1A_{2u}$ series. Further confirmation of the upper-state symmetry assignments of the bands in question as $^1A_{2u}$ and $^1E_{1u}$ comes from the MCD spectrum of benzene-h_6 [S57] in which an A-term is observed for the transition into the degenerate upper state whereas no MCD signal at all is observed for the $^1A_{1g} \rightarrow {}^1A_{2u}$ transition. In regard 3p core splitting, the above assignments support the earlier work of Jonsson and Lindholm [II, J14], and the theoretical work of Hay and Shavitt [H25].

In an MPI investigation of the ($e_{1g}\pi$, 3p) region, Johnson and Korenowski [J13] find the three-photon resonance to the $^1A_{2u}$ level in benzene intermixed with another having its origin at 56 080 cm^{-1} (*advert., 6.953 eV*) which they assign to the $^1E_{2u}$ component. The vibronics of these badly overlapped MPI transitions are readily separated using circularly and linearly polarized light. Whetten *et al.* [W13] also have seen the MPI transition to $^1A_{2u}(e_{1g}\pi$, 3p) as a two-photon resonance vibronically induced by ν_{16}.

Higher Rydberg bands of benzene in rare-gas matrices [H23] have been assembled into two series of 5−6 members each, which closely parallel the ($e_{1g}\pi$, np) and ($e_{1g}\pi$, nd) series of the vapor phase. The authors of this work, Hasnain *et al.* [H23], argue that such impurity transitions in rare gases are not equivalent to the intrinsic Wannier excitons of the host matrix, but instead retain considerable molecular character. Similar arguments are given by Resca and Resta [R10] in regard the spectra of rare-gas alloys.

One final MPI band of benzene remains to be discussed. Whetten *et al.* [W13] report a peculiar two-photon resonance in benzene spanning the 59 000−63 000 cm^{-1} (*7.3−7.8 eV*) region. Since the vibronic analysis of this band implicates single quanta of the Jahn-Teller-active ν_6 and ν_9 modes it is concluded that the electronic upper-state symmetry must be E_g. The origin is at 60 800 cm^{-1} (*advert., 7.54 eV*) and is said not to be related by vibronic borrowing to the one-photon Rydberg transitions in this region. Because the isotope shift of the 60 800-cm^{-1} band is larger than is observed for the other Rydberg states of benzene, and because no higher members of a series can be seen, it is tentatively concluded that the band in question is $^1A_{1g} \rightarrow {}^1E_{2g}(\pi, \pi^*)$ with the π^* MO admixed with Rydberg 3d character. Such an assignment receives considerable support from the work of Nakashima *et al.*, who find the vertical frequency for excitation to

$^1E_{2g}$ to be 63 000 cm^{-1} [N8]. Still, it must be said that it is counter to all our experience that a valence excitation can appear as a resonance at such high frequencies in the MPI spectrum (Section II.A), and we note again that the theoretical work of Hay and Shavitt [H25] assigns the $^1E_{2g}$ upper state in this region as Rydberg. A $\pi \rightarrow \sigma^*$ valence assignment also is a possibility here according to CNDO calculations [B41, R43], while yet another possibility is that the band is vibronically induced ($\pi \rightarrow$ 3d) in the two-photon spectrum, and is either vibronically induced or does not appear at all in the one-photon spectrum.

Wilden and Comer [W25] report yet another $e_{1g}\pi \rightarrow n\text{d}$ quadrupole-allowed Rydberg series in benzene, uncovered by angular-dispersed electron scattering. With the $n = 3$ member at 59 970 cm^{-1} (*vert., 7.435 eV*) and $\delta = 0.266$ for the series, it is assigned as $e_{1g}\pi \rightarrow n\text{d}a_{1g}$ as earlier suggested by the calculation of Betts and McKoy [**I**, B27]. This transition of Wilden and Comer may be vibronically induced by an 800-cm^{-1} vibration in the two-photon spectrum of Whetten *et al.*, thereby appearing at 60 800 cm^{-1} in that spectrum [W13]. An analogous assignment ($3e_{2g}\sigma \rightarrow n\text{d}a_{1g}$) is appropriate for a series of quadrupole-allowed energy losses detected by Wilden and Comer converging upon the $(3e_{2g}\sigma)^{-1}$ ionization potential with $\delta = 0.226$. Two longer series, allowed and converging upon the $(3e_{2g}\sigma)^{-1}$ ionization potential [($\delta = 0.47$, $3e_{2g}\sigma \rightarrow n\text{p}e_{1u}$; $\delta = 0.28$, $3e_{2g}\sigma \rightarrow n\text{f}$ (b_{1u}, b_{2u} or e_{1u})] [W25] also are reported. For the latter series, clearly resolved (0,0) band splittings are observed in the partially deuterated benzenes of D_{2h} symmetry, showing that the upper state must be degenerate in benzene-h_6 [I8, I9, S24].† Experiments of this sort, plus detailed vibronic analysis clearly show that the second ionization potential of benzene corresponds to removal of an electron from the $3e_{2g}\sigma$ MO [I7, I8].

An MCD spectrum of the electronic origins at 67 575 and 67 660 cm^{-1} (*advert., 8.3780 and 8.3886 eV*) in benzene, identified as 3R″ and 3R‴ by Wilkinson [**II**, W26], reveals A-term profiles for each, implying $^1E_{1u}$ upper states as appropriate for the symmetry-allowed $e_{1g}\pi \rightarrow$ 4f excitations [S57]. Quantitative absorption and MCD studies of the transition to $^1E_{1u}$ in benzene lead to an upper-state magnetic moment of -0.14 Bohr magnetons [A15].

† Just as the $^1E_{1u}$ (π, π^*) state of benzene is not Jahn-Teller split due to cancellation effects, the $^1E_{1u}$ state of the ($3e_{2g}\sigma$, nfe_{1u}) configuration also is not Jahn-Teller split in benzene-h_6 [I7].

The $1s_C$ spectrum of benzene [E1, H56, H58] commences with a relatively intense $1s_C \rightarrow \pi^*$ transition with a 41 000-cm^{-1} term value, followed by transitions to 3s, 3p and 3d with respective term values of 25 000, 18 600 and 11 300 cm^{-1}. As expected for a $1s_C \rightarrow \pi^*$ assignment, this transition has out-of-plane polarization [J5]. In accord with the antishielding principal (Section I.B), the ($1s_C$, 3s) term value of 25 000 cm^{-1} is a few thousand cm^{-1} larger than that of the valence configuration ($e_{1g}\pi$, 3s), 23 470 cm^{-1}. Following the $(1s_C)^{-1}$ ionization potential of benzene in the ESCA spectrum, shake-up bands appear at 33 900, 47 600, 56 500, 67 000 and 86 300 cm^{-1} (*vert., 4.2, 5.9, 7.0, 8.3 and 10.7 eV*) higher frequencies, while a strongly asymmetric band suggests a sixth component at 72 600 cm^{-1} (*vert., 9.0 eV*). These transitions are calculated to be $\pi \rightarrow \pi^*$ excitations in the $(1s_C)^{-1}$ positive ion in which the inner-shell hole is localized so as to reduce the benzene symmetry to C_{2v} [B44, L30, O3]. Related $\pi \rightarrow \pi^*$ shake-up transitions also are observed in pyridine and pentafluoropyridine (Section XIX.D). Bigelow and Freund [B44] show that many of these shake-up bands of benzene have their parentage in the $\pi \rightarrow \pi^*$ spectrum of the neutral molecule. Most interestingly, the shake-up excitations in the 73 000–89 000-cm^{-1} (*9–11-eV*) region are derived from the two transitions to $^1E_{2g}$ in the neutral molecule, which are themselves calculated to come at 64 000–81 000 cm^{-1} (*8–10 eV*) in the neutral molecule. These figures agree closely with the experimental results of Nakashima *et al.* [N8] on excitations to $^1E_{2g}$. The lowest $\sigma \rightarrow \sigma^*$ transition in the $(1s_C)^{-1}$ benzene ion is calculated to lie beyond 113 000 cm^{-1} (*14 eV*) with respect to the $(1s_C)^{-1}$ threshold.

Johnson *et al.* [J5] find two shake-up-like features at 26 000 and 77 000 cm^{-1} (*vert., 3.2 and 9.6 eV*) beyond the vapor-phase $(1s_C)^{-1}$ ionization potential of benzene when the benzene is adsorbed onto the (111) plane of Pt; these they assign however to $1s_C \rightarrow \sigma^*$ shape resonances rather than to $\pi \rightarrow \pi^*$ shake-up excitations. According to the relationship between C–C bond distance and σ^* shape-resonance frequency proposed by Hitchcock *et al.* [H66], the feature at higher frequency in the adsorbed-benzene spectrum indeed is a σ^* shape resonance. Note that a band 26 000 cm^{-1} beyond the $(1s_C)^{-1}$ ionization potential of benzene in the electron-impact spectrum which was called a shake-up excitation [H58] does not correlate with any of the shake-up bands of the ESCA spectrum.

Negative-ion formation in benzene involves a long series of electron resonances. Mathur and Hasted [M14] find valence TNI resonances (Section II.B) involving $e_{2u}\pi^*$ and $b_{2g}\pi^*$ MO's [W34] at 8760 and 39 800

cm^{-1} *advert., 1.086 and vert., 4.93 eV*), respectively.[†] Transitions to these levels are observed in most if not all benzene derivatives, with the transition to the $e_{2u}\pi^*$ MO being split into two components in derivatives such as phenol and nitrobenzene [J14, M14].[§] Allan [A8] reports a narrow Feshbach resonance at 47 300 cm^{-1} (*vert., 5.87 eV*) in benzene having the ($e_{1g}\pi$, 3s) parent configuration and the expected 4000-cm^{-1} shift to lower frequency (Section II.B). This Feshbach resonance of benzene moves to 46 000 cm^{-1} (*vert., 5.7 eV*) in toluene [S54], to 45 200 cm^{-1} (*vert., 5.6 eV*) in aniline [S54] and to 44 400 cm^{-1} (*vert., 5.5 eV*) in pyridine [A38] as it tracks the first ionization potential. Additionally, five other TNI features are observed in benzene between 40 000 and 80 000 cm^{-1} (*5 and 10 eV*), which Allan assigns as $\pi \rightarrow \pi^*$ core-excited resonances. Several of these bands also are reported by Smyth *et al.* [S54] and Azria and Schulz [A38]. One of the resonances, at 71 400 cm^{-1} (*vert., 8.85 eV*), is positioned properly for assignment to the $^2(3e_{2g}\sigma, 3p^2)$ Feshbach level.

XIX.B. Alkyl Benzenes

The orbital and state symmetries of the molecules discussed in this Section will be phrased in terms of the D_{6h} point group of benzene unless there is a significant reason to descend to lower symmetry. A complete set of survey spectra of the methyl benzenes [**II**, 223, 338] reporting one-photon absorption frequencies, vibrational increments and oscillator strengths has been published by Bolovinos *et al.* [B51, B52]. The oscillator strengths of the transitions to the $^1B_{1u}$ and $^1E_{1u}$ states generally tend to increase with methylation in an irregular way. A weak band appears at 42 000−44 000 cm^{-1} (*vert., 5.2−5.4 eV*) in 1,2,4,5-tetramethyl benzene, in pentamethyl benzene and in hexamethyl benzene which is not otherwise seen in benzene itself. This band of hexamethyl benzene previously was assigned as $e_{1g}\pi \rightarrow 3s$ [**II**, 226], and we extend this

[†] Burrow *et al.* [B82] point out that according to pi-electron theory, the sum of the $(1a_{2u}\pi)^{-1}$ ionization potential and the $b_{2g}\pi^*$ electron affinity should be very close to 65 000 cm^{-1} (*8.1 eV*), and that taking the $b_{2g}\pi^*$ electron affinity as 39 800 cm^{-1} yields an anomalously low value for this sum of 59 400 cm^{-1} (*7.37 eV*). Note however, that Mathur and Hasted [M14] find yet another TNI resonance in benzene at 33 900 cm^{-1} (*vert., 4.21 eV*); if we assign it to $b_{2g}\pi^*$, then the sum in question becomes 65 200 cm^{-1} (*8.09 eV*).

[§] However, see [F7] for a contrary interpretation.

assignment to the corresponding bands in the partially methylated derivatives. Though the $e_{1g}\pi \rightarrow 3s$ transition can be located in the electron-impact spectrum of benzene, it is too broad to be seen in toluene [D31]. X-ray absorption spectra of adsorbed toluene demonstrate that the $\mathscr{A}(1s_C, \pi^*)$ band is polarized out-of-plane [J5].

Measurement of the optical functions of several of the liquified methyl benzenes by reflection reveals the $^1A_{1g} \rightarrow {}^1E_{1u}$ transition peaked at *ca.* 52 000 cm^{-1} (*6.4 eV*) [A35]. Comparison of the absorption and MCD profiles of the $^1A_{1g} \rightarrow {}^1E_{1u}$ bands of benzene and the methyl benzenes in perfluoro-*n*-hexane solution allows the estimation of the $^1E_{1u}$ magnetic moments [F49], which are found to vary between − 0.05 and − 0.21 Bohr magnetons depending upon the pattern of methylation (See also Section XIX.C for related data on the halobenzenes).

Though the absorption spectrum of *sec*-butyl benzene vapor is rather featureless, its CD spectrum is more revealing [A13]. In particular, two oppositely-rotating peaks appear between the transitions to $^1B_{1u}$ and $^1E_{1u}$, centered at 52 490 and 53 480 cm^{-1} (*6.508 and 6.630 eV*). Though Allen and Schnepp [A13] claim these are derived from the $^1A_{1g} \rightarrow {}^1E_{2g}(\pi, \pi^*)$ transition of benzene, the observed frequencies are too low for this (Section XIX.A), whereas their term values of *ca.* 18 000−20 000 cm^{-1} suggest $\pi \rightarrow 3p$ assignments. In a related case, that of 1-methyl indan, similar CD bands again are seen between the transitions to $^1B_{1u}$ and $^1E_{1u}$ [II, AD2]. A CNDO calculation of the CD spectrum of 1-methyl indan assigns the extraneous bands as $\sigma \rightarrow \pi^*$ [B8], and indeed, CNDO studies of alkyl benzene spectra similarly predict such low-lying $\sigma \rightarrow \pi^*$ valence bands [B41, B42, R23, R43]. It would be most interesting to see if these transitions maintain their low frequencies in a high-quality *ab initio* calculation, and also to see how these bands of *sec*-butyl benzene behave experimentally when put into a condensed phase. The reflection spectrum of [2,2]-paracyclophane [F46] reveals a band at 48 000 cm^{-1} (*vert., 5.9 eV*) said to be derived from the $^1A_{1g} \rightarrow {}^1E_{2g}$ band of benzene, however, as with *sec*-butyl benzene, the frequency is just too low for the $^1E_{2g}$ assignment.

In the Rydberg region of the methyl benzene spectra, Bolovinos *et al.* [B51] report many Rydberg series converging upon the lowest ionization potentials, with series lengths which decrease with increasing methylation. There are no series discernible in hexamethyl benzene. Most compounds in this class show an *n*s series ($\delta = 0.9$), from one to three *n*p series ($\delta = 0.44-0.68$), and several series with δ in the 0.0−0.1 range. In the more highly methylated benzenes, this latter δ value may get as large as 0.25. It is still a point of contention as to whether the $\delta = 0.0-0.25$ series terminate at *n*d or *n*f upper orbitals.

Once again, the MPI spectra prove useful in locating the lower-*n* Rydberg promotions in the alkyl benzenes. Thus in toluene, MPI origins terminating at 3p are observed as two-photon resonances at 52 839 and 54 734 cm^{-1} (*advert., 6.5510 and 6.7860 eV*), only the first of which is at all distinguishable in the one-photon absorption spectrum [U2]. Appropriately, these bands broaden to the high-frequency side when the sample is pressurized with He. The equivalent transitions in *p*-xylene are observed in the MPI spectrum at *higher* frequencies than in toluene, but these are thought to be false origins. In *o*-xylene, the MPI spectrum again reveals two electronic origins, with term values pointing to 3p terminating orbitals. It is surprising to note that the corresponding Rydberg excitations to 3s do not appear in the MPI spectra of toluene or *o*-xylene though $\pi \rightarrow 3s$ is intense in the MPI spectrum of benzene. A lifetime of only 0.17 psec is deduced for the $\pi \rightarrow 4d$ origin of toluene at 64 060 cm^{-1} (*adiabat., 7.9422 eV*) reached by two-photon absorption [W18].

The low-frequency wing (70 000–140 000 cm^{-1}, *9–18 eV*) of the plasmon resonance at 170 000 cm^{-1} (*vert., 21 eV*) in polystyrene [**II**, 256] has been shown to contain several valence transitions [I3, R23, R24]. Two of these, at 77 700 and 98 800 cm^{-1} (*vert., 9.65 and 12.15 eV*), in polystyrene appear to have their analogs in the electron-transmission spectrum of solid benzene at 77 800 and 98 000 cm^{-1} (*vert., 9.65 and 12.15 eV*) [K14, S7, S8]. The first of these is most likely the $^1A_{1g} \rightarrow {}^1E_{2g}(\pi, \pi^*)$ excitation, while the second probably involves σ and/or σ^* MO's. It is also possible that the 77 800-cm^{-1} band of polystyrene is derived from the 72 600-cm^{-1} band of polyethylene, in which case it has strong Rydberg character. Combining data on polystyrene from 0.6 to 8050 eV, Inagaki *et al.* [I3] find that the *f*-sum converges only at several keV even though the highest discrete absorption is at *ca.* 300 eV. Moreover, the *f*-sums of the inner-shell ($1s_C$) and outer-shell (2s and 2p) absorption regions suggest an interaction between these which acts to redistribute the oscillator strengths somewhat. The four bands of the $1s_C$ inner-shell spectrum of polystyrene have been assigned as terminating at π^* and σ^* valence MO's using a calculation which nicely reproduces the experimental spectrum [R23].

A factor yet to be faced in the interpretation of alkyl benzene spectra is stressed by Bigelow [B42] who calculates that such systems will show $\sigma \rightarrow \pi^*$ valence excitations at approximately 52 000 cm^{-1} (*6.5 eV*) with appreciable charge-transfer character. The possibility that such $\sigma \rightarrow \pi^*$ transitions are significant in the spectra of systems such as *sec*-butyl benzene or [2,2]-paracyclophane cannot be discounted.

XIX.C. Halobenzenes

A large number of variously fluorinated benzenes [**II**, 233] have been studied both optically [P29] and by electron impact [F38, F44]. While these compounds clearly display the benzene-like transitions to $^1B_{2u}$, $^1B_{1u}$ and $^1E_{1u}$, Figs. XIX.C-1 and XIX.C-2, several "nonbenzenoid" excitations are reported as well in the more heavily fluorinated species. Working with eleven different fluorobenzenes, Frueholz *et al.* [F38, F44] find quite small shifts in the fluorobenzene series for the $\pi \rightarrow \pi^*$ transitions to $^3B_{1u}$ (31 000 cm^{-1} *vert., 3.85 eV*), $^3B_{2u}$ (45 600 cm^{-1} *vert., 5.65 eV*), $^1B_{2u}$ (37 900 cm^{-1} *vert., 4.70 eV*), $^1B_{1u}$ (50 400 cm^{-1} *vert., 6.25 eV*) and $^1E_{1u}$ (56 500 cm^{-1} *vert., 7.0 eV*). It is reported that no splitting of the $^1E_{1u}$ states of the halobenzenes of low symmetry can be observed [O11], in contradiction to our earlier statement [**II**, 236]. Values of the magnetic moment in the $^1E_{1u}(\pi, \pi^*)$ states of these halobenzenes are $-$ 0.2 to $-$ 0.3 Bohr magnetons, with the 1,3,5-substitution pattern at the lower end of the range [K1], as in the methyl benzenes (Section XIX.B). Once these more common valence items in the fluorobenzene spectra are accounted for, one then finds further features of great interest; these are marked **F** in Fig. XIX.C-2. Focusing on the heavily fluorinated species, previously unreported transitions appear at the frequencies shown in Table XIX.C-I. Because the frequencies of the $\pi \rightarrow \pi^*$ excitations are so static in the fluorobenzenes, one first argues that the **F** bands cannot be of this type. Next, note that in the more heavily fluorinated benzenes the term values of the transitions in question are considerably larger than those expected for Rydberg excitations, whereas in the less heavily fluorinated compounds the term values are solidly in the range expected for Rydberg excitations to 3s and/or 3p.

The appearance of the valence **F** bands in the heavily fluorinated benzenes is strongly reminiscent of the situation in the fluoroethylenes [**II**, 55] where "extraneous" low-frequency valence bands also appear in the more heavily fluorinated systems. In that case, it was argued that fluorination depresses the $\sigma - \sigma^*$ manifold with respect to the $\pi - \pi^*$ manifold so that $\pi \rightarrow \sigma^*$ excitations are low-lying. We believe the same general situation holds for the fluorobenzenes, *i.e.*, the **F** transitions are assignable to $\mathscr{A}(\pi, \sigma^*)$ bands (Section I.C), where σ^* is localized in the C$-$F bond. In the less heavily fluorinated systems, the $\mathscr{A}(\pi, \sigma^*)$ transitions most likely are hidden beneath the more intense transition to $^1E_{1u}(\pi, \pi^*)$.

Triplet $\rightarrow$ triplet excitations have been observed in each of the difluorobenzenes which are 74 000 $\pm$ 500 cm^{-1} (*vert., 9.17 $\pm$ 0.06 eV*) above their respective ground states [A2]. These excitations are obviously

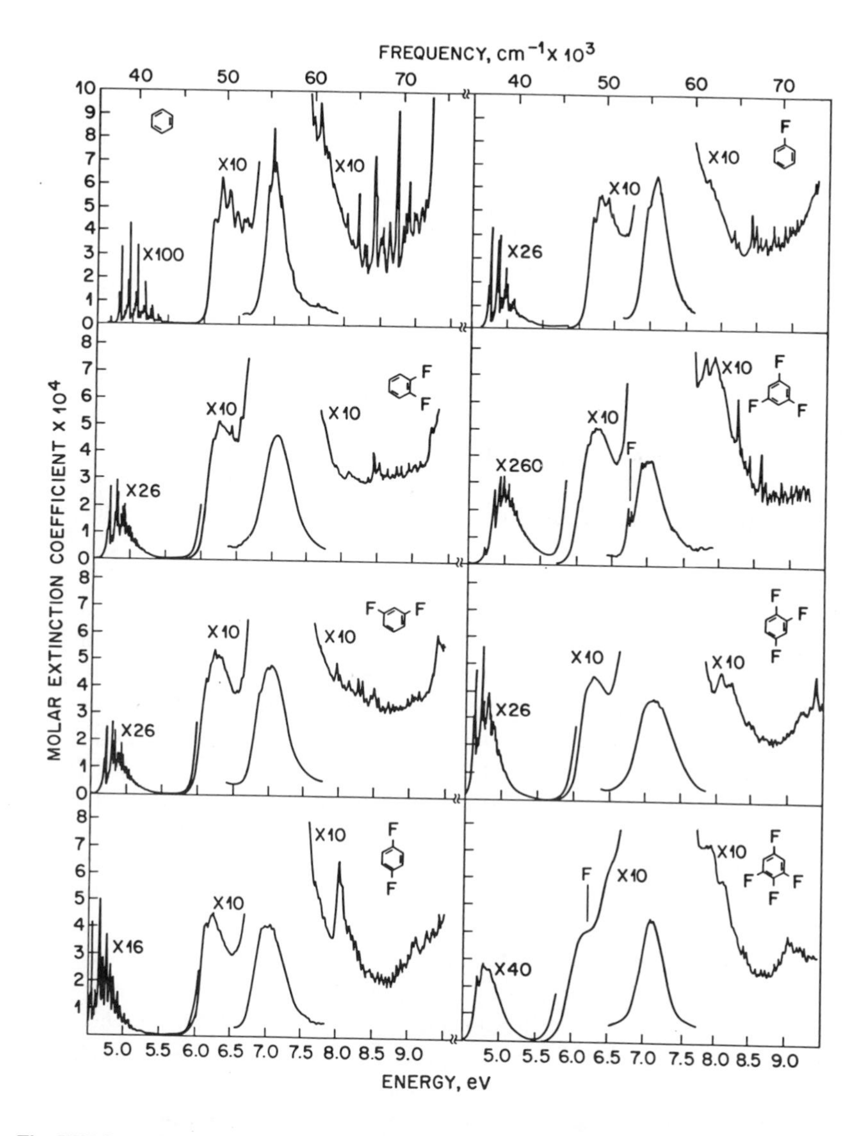

Fig. XIX.C-1. Optical spectra of benzene and the fluorobenzenes in the vapor phase [P29].

related to the $T_1 \rightarrow T_n$ valence bands of benzene and toluene found in the same region, and which have been argued to terminate at $^3E_{2g}(\pi, \pi^*)$. This assignment results in a very small $^3E_{2g}$ - $^1E_{2g}$ split.

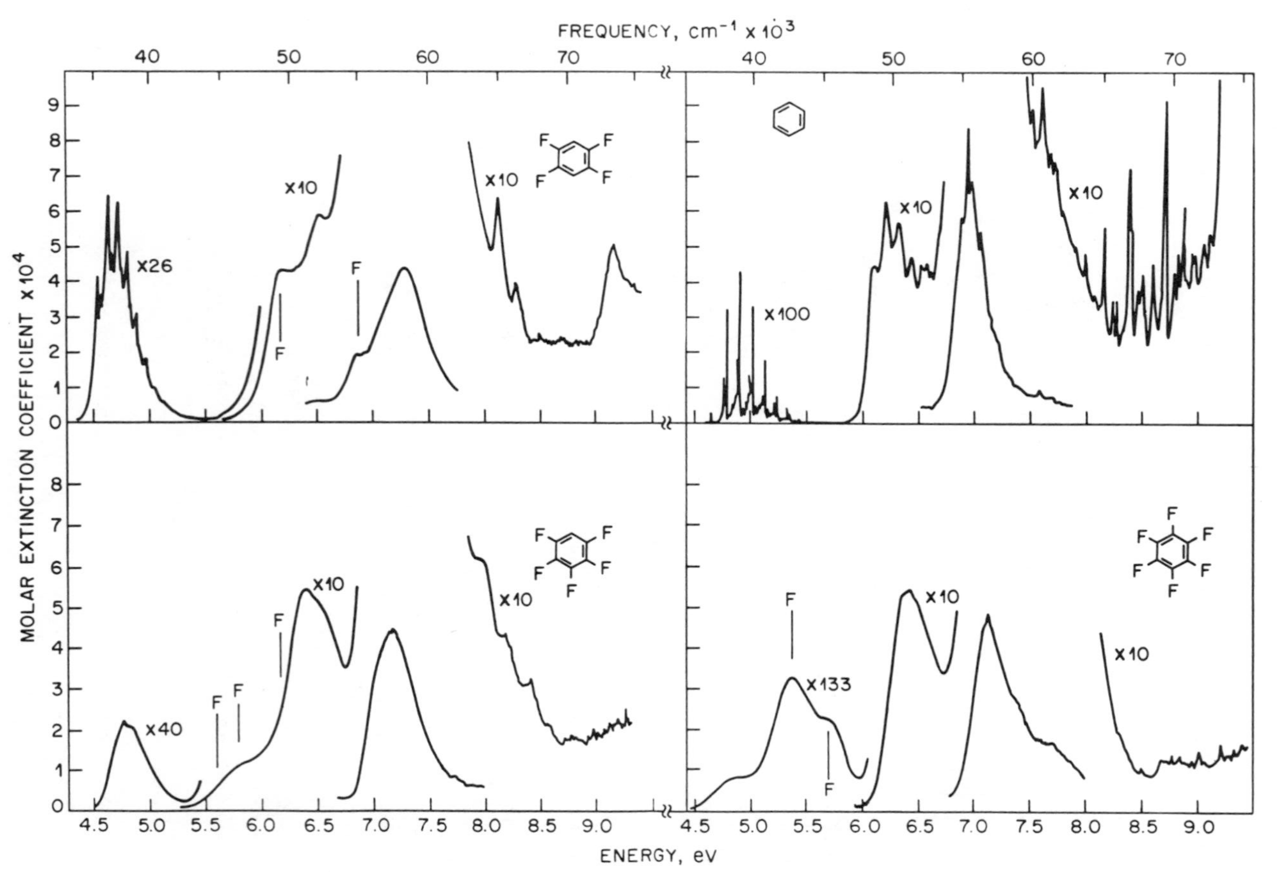

Fig. XIX.C-2. Optical spectra of benzene and the more highly fluorinated benzenes [P29]. Transitions in the fluorobenzenes labelled **F** do not appear in benzene.

TABLE XIX.C-I

"NONBENZENOID" BANDS PRECEDING THE TRANSITION TO $^1E_{1u}$ IN THE FLUOROBENZENES

Compound	Band Position, cm^{-1} (*eV*)[a]	Ionization Potential, cm^{-1} (*eV*)[b]	Term Value, cm^{-1}
1,3,5-Trifluorobenzene	53 900 (*6.68*)	77 600 (*9.62*)	23 700
1,2,3,5-Tetrafluorobenzene	49 400 (*6.13*)	77 100 (*9.56*)	27 700
1,2,4,5-Tetrafluorobenzene	49 800 (*6.18*)	75 500 (*9.36*)	25 700
	55 300 (*6.85*)	75 500 (*9.36*)	20 200
	55 300 (*6.85*)	81 000 (*10.04*)	25 700
Pentafluorobenzene	44 400 (*5.50*)	77 800 (*9.64*)	33 400
	46 000 (*5.71*)	82 300 (*10.2*)	36 300
	48 600 (*6.02*)	77 800 (*9.64*)	29 200
Hexafluorobenzene	43 200 (*5.35*)	79 800 (*9.90*)	36 600
	45 600 (*5.65*)	79 800 (*9.90*)	34 200

[a] From references [F44] and [P29].

[b] Adiabatic values from reference [P29].

As in benzene itself, the $b_1\pi \rightarrow 3s$ Rydberg excitation of fluorobenzene was first detected as a two-photon resonance in the MPI spectrum [K39, K40]. The $b_1\pi \rightarrow 3s$ origin is observed at 50 914 cm^{-1} (*6.3123 eV*) followed by a short vibrational display closely resembling that in benzene; higher-*n* members have not been reported to date. Once the position of the $b_1\pi \rightarrow 3s$ band of fluorobenzene was documented by MPI, Philis *et al.* [P28] returned to study the one-photon spectrum, and indeed found clear evidence for this transition at 50 900 cm^{-1} (*advert., 6.31 eV*), lying just between the valence transitions to $^1B_{1u}$ and $^1E_{1u}$. In an N_2 matrix and in the neat crystal, the structure of the transition to $^1B_{1u}(\pi, \pi^*)$ sharpens whereas that to $(b_1\pi, 3s)$ broadens considerably, as expected. This observation is important, for it eliminates the possibility that the excitation in question is either to the $^1E_{2g}(\pi, \pi^*)$ state calculated by some to be in this region or to a component of an $\mathscr{A}(\pi, \sigma^*)$ band. Transitions to *n*p levels appear as three-photon resonances in fluorobenzene ($\delta = 0.5$) [K40] and another series of MPI bands ($\delta = 0.02$) is within 50 cm^{-1} of a series already identified in one-photon absorption.

The N_2-matrix spectrum of fluorobenzene [P28] also shows a sharpening of the feature observed at 55 300 cm^{-1} (*6.86 eV*) in the vapor-phase spectrum. On this basis, the $\pi \rightarrow 3p$ assignment earlier suggested [II, G11] is refuted, while Philis *et al.* [P28] in turn suggest that the band in question is the (0,0) band of the transition to $^1E_{1u}$, followed by a totally symmetric vibrational progression of 850 cm^{-1}. Shoulders much like the (0,0) band of the $^1A_{1g} \rightarrow {}^1E_{1u}$ band of fluorobenzene also appear in 1,3,5-trifluorobenzene (53 900 cm^{-1} *vert., 6.68 eV*) and in 1,2,4,5-tetrafluorobenzene (55 300 cm^{-1} *vert., 6.85 eV*) [F44, P27], Fig. XIX.C-2. Several assignments are possible for these bands: *a*) the first vibronic component of the transition to $^1E_{1u}$ as in fluorobenzene; *b*) $\pi \rightarrow 3s$ or possibly $\pi \rightarrow 3p$ Rydberg transitions; *c*) a component of the $\pi \rightarrow \sigma^*$ valence manifold; *d*) the $^1A_{1g} \rightarrow {}^1E_{2g}(\pi, \pi^*)$ valence excitation; and *e*) a $\sigma \rightarrow \pi^*$ excitation, as has been proposed for benzene (Section XIX.A) and the alkyl benzenes in this region (Section XIX.B). Philis *et al.* [P27] point out that in electron impact the intensity ratio of the 55 200-cm^{-1} and $^1A_{1g} \rightarrow {}^1E_{1u}$ bands at their peaks change with incident electron energy, implying two different transitions. Thus, option *a*) is not viable. In regard choosing between $\pi \rightarrow \sigma^*$ and $\sigma \rightarrow \pi^*$ assignments, extensive studies on the effects of fluorination on pi-electron systems suggest that $\pi \rightarrow \sigma^*$ will be far below $\sigma \rightarrow \pi^*$ excitations in fluorinated systems. Hence it is concluded that if the transitions in question are valence, they have either $\pi \rightarrow \sigma^*$ or $\pi \rightarrow \pi^*$ assignments. Finally, we note that there is now considerable evidence that the lowest transition to $^1E_{2g}(\pi, \pi^*)$ is on the high-frequency side of the transition to $^1E_{1u}$ (*ca.* 63 000 cm^{-1}, *7.8 eV*) whereas the transitions in question are on the low-frequency side (*ca.* 55 000 cm^{-1}, *6.8 eV*). Thus it is concluded that if the features at *ca.* 55 000 cm^{-1} in the fluorobenzenes are valence, they are $\mathscr{A}(\pi, \sigma^*)$ bands. Rydberg assignments are also possible based on the observed term values, and the problem requires a condensed-phase study to distinguish between the final two options.

Optical absorption studies of the three isomers of difluorobenzene [G15] reveal two Rydberg series in each ($\delta = 0.5$, $\pi \rightarrow n$p; $\delta = 0.05$, $\pi \rightarrow n$d or nf), converging upon the lowest ionization potentials. As is often observed, the vibrational structure in these Rydberg absorption bands closely match those in the corresponding photoelectron spectra. Remembering that the originating MO in these series ($e_{1g}\pi$) has the angular symmetry of an atomic d orbital and that d $\rightarrow$ f transitions will be much more intense than d $\rightarrow$ d, we prefer to assign the series having $\delta = 0.05$ as terminating at nf, with n beginning at 4. In the context of the discussion in Section I.D, there is no 4f wavefunction collapse in this system. The threshold photoelectron spectrum of hexafluorobenzene displays a beautiful series of autoionizing lines in the 97 000–113 000 cm^{-1} (*12–14 eV*) region [D52, D53] which can be ordered so as to form a Rydberg series ($n = 3-6$)

with $\delta = -0.02$; the transitions are assigned as $e_{2g}\sigma \rightarrow nd$ assuming the orbital ordering put forth in [P36, **II**, B60]. Several series analogous to this one of hexafluorobenzene are found in benzene as well (Section XIX.A). Other Rydberg series observed in hexafluorobenzene absorption in this region have $\delta = 0.8$ and $\delta = 0.5$ [S52] and converge upon the third ionization potential $(e_{2g}\sigma)^{-1}$ at 113 100 cm^{-1} (*vert., 14.02 eV*). In constructing these series, aspects of the earlier series of Smith and Raymonda [**II**, S40] have been reassigned. Price *et al.* [P36] report that an intense $a_{2u}\pi \rightarrow ns$ Rydberg series in 1,3,5-trifluorobenzene ($\delta = 0.85$) parallels those observed in benzene and hexafluorobenzene ($\delta = 1.15$) [K27]. There is a strong similarity in the spectra of pentafluorobenzene [**II**, S40] and decafluorobiphenyl, Fig. XX-1, [M49] arising from the fact that the latter is twisted by almost 90° about the central C–C bond, thereby uncoupling the pi electrons of the two rings.

A very large number of energy-loss peaks beyond the first ionization potentials of the fluorobenzenes are cataloged without comment by Frueholz *et al.* [F44]. These are most likely autoionizing Rydberg transitions converging upon higher ionization potentials. $\mathscr{A}(2p_F, \sigma^*)$ bands also are expected in the fluorobenzenes in the region above 97 000 cm^{-1} (*12 eV*), but their intensities are probably too low for them to appear as distinct features. A valence excitation in solid fluorobenzene reported at 73 000 cm^{-1} (*vert., 9.05 eV*) [S8] is quite close to the ${}^1A_{1g} \rightarrow {}^1E_{2g}(\pi, \pi^*)$ excitation of benzene at 76 000 cm^{-1} (Section XIX.A) and may be related.

In the chlorobenzenes, it is reasonable to argue that because the occupied $2p\pi_C$ MO's of the carbon framework are *ca.* 16 000 cm^{-1} higher than the $3p\pi_{Cl}$ MO's [K18], the $\mathscr{A}(2p\pi_C, \sigma^*)$ bands should be at a lower frequencies than the $\mathscr{A}(3p\pi_{Cl}, \sigma^*)$ bands, and that though the $2p\pi - 4p\pi$ split shrinks to *ca.* 12 000 cm^{-1} in the bromobenzenes, the $\mathscr{A}(2p\pi, \sigma^*)$-below-$\mathscr{A}(4p\pi, \sigma^*)$ argument still should hold. Thus it is concluded that the lowest "nonbenzenoid" bands in the chloro- and bromobenzenes are $\pi \rightarrow \sigma^*$ transitions, just as in the fluorobenzenes. The absorption spectrum of hexachlorobenzene [K1] in *n*-heptane solution shows a clear $\mathscr{A}(\pi, \sigma^*)$ band at 41 500 cm^{-1} (*vert., 5.14 eV*) lying just between the transitions to ${}^1B_{2u}$ and ${}^1B_{1u}$.† The corresponding peak in hexabromobenzene falls at 42 400 cm^{-1} (*vert., 5.48 eV*), while 1,3,5-trichloro- and 1,3,5-tribromobenzene also show weak structures in the 40 000–43 000 cm^{-1} (*4.96–5.33 eV*) region which are undoubtedly $\mathscr{A}(\pi, \sigma^*)$ bands as well.

† This spectrum differs significantly from that reported earlier [**II**, 231].

The $\mathscr{A}(\pi, \sigma^*)$ band term values in the haloethylenes are quite constant (29 500 ± 1500 cm^{-1}) irrespective of the number of halogens, their identity, or pattern of substitution, Table X.C-I. In the halobenzenes quoted above, the term values of the $\mathscr{A}(\pi, \sigma^*)$ bands again are constant, 33 000 ± 1000 cm^{-1}, but noticeably higher than in the haloethylenes. The higher-lying $\mathscr{A}(n\mathrm{p}, \sigma^*)$ bands of the halobenzenes have not been located, however, from the known separation of the $(\pi)^{-1}$ and $(n\mathrm{p})^{-1}$ ionization potentials [K18] and the assumption of the equality of the $\mathscr{A}(\pi, \sigma^*)$ and $\mathscr{A}(n\mathrm{p}, \sigma^*)$ term values (Section X.C), we estimate that the $\mathscr{A}(n\mathrm{p}, \sigma^*)$ bands will be found near 54 000 cm^{-1} (*vert., 6.7 eV*). These generalizations do not hold for the fluorobenzenes, for which the $\mathscr{A}(\pi, \sigma^*)$ term values are as large as 36 600 cm^{-1} and possibly as small as 23 700 cm^{-1}, Table XIX.C-I.

In benzene and many of its derivatives, the three lowest unoccupied π^* MO's are derived from the $e_{2u}\pi^*$ and $b_{2g}\pi^*$ MO's of the benzene ring. The temporary-negative-ion resonances involving the $e_{2u}\pi^*$ MO's in benzene, 1,3,5-trifluorobenzene and hexafluorobenzene are observed at *ca.* 8000 cm^{-1} (*1 eV*) and those involving $b_{2g}\pi^*$ appear at *ca.* 37 000 cm^{-1} (*vert., 4.6 eV*) [F25]. In fluorobenzenes of lower symmetry and in derivatives such as phenol, toluene, benzaldehyde, *etc.*, the $e_{2u}\pi^*$ level splits to yield two TNI resonances in the 0–12 000 cm^{-1} (*0–1.5 eV*) region [C22, J17]. In the temporary-negative-ion spectrum of chlorobenzene [M14], the transitions into the three vacant π^* MO's come at 7300, 14 000 and 37 700 cm^{-1} (*vert., 0.90, 1.74 and 4.68 eV*). Another broad resonance at 66 300 cm^{-1} (*vert., 8.22 eV*) in chlorobenzene is just 2200 cm^{-1} below the intense D-band of the neutral molecule [**II**, 235], and so is assigned as a Feshbach resonance (Section II.B). Since the D-band is assigned as a $3\mathrm{p}\pi_{\mathrm{Cl}} \rightarrow 4\mathrm{p}$ excitation, with $(3\mathrm{p}\pi)^{-1}$ being the third ionization potential in the molecule, the $^2(3\mathrm{p}\pi, 4\mathrm{p}^2)$ resonance configuration of the negative ion may be looked upon as a superexcited Feshbach state. It is especially peculiar that no valence TNI resonance involving the σ^*(C–Cl) orbital is observed in chlorobenzene; perhaps this is in some way related to the absence of transitions terminating at σ^* in the inner-shell spectra of these compounds.

Inner-shell spectra in the halobenzenes [H58] are complicated by the fact that the C–H and C–X groups in C_6H_5X have resolvably different $(1s_C)^{-1}$ ionization potentials. The difference is largest in fluorobenzene. In keeping with the weak perturbation offered by the halogen to the phenyl group in the monohalobenzenes, the inner-shell excitation energies and term values fall into narrow ranges in these compounds. As in benzene, the inner-shell spectra of the halobenzenes commence with $1s_C \rightarrow \pi^*$ valence excitations, however, differing in that two of these appear in each of the halobenzenes. It is clear from the splitting that the separation is due to differences in the $(1s_C)^{-1}$ ionization potentials rather than to splitting of the π^* degeneracy. In benzene, the $(1s_C, \pi^*)$ term value is 41 100 cm^{-1} while in the halobenzenes values lie in the range 43 000 – 45 000 cm^{-1}. The relatively intense $1s_C \rightarrow \pi^*$ excitations in the halobenzenes are followed by the usual parade of Rydberg excitations: $1s_C \rightarrow 3s$ (27 400–23 400 cm^{-1} term values); $1s_C \rightarrow 3p$ (20 200–18 600 cm^{-1} term values) and $1s_C \rightarrow 3d, 4s$ (12 900–10 500 cm^{-1} term values). In accord with the antishielding argument (Section I.B), the $(1s_C, 3s)$ term values in the inner-shell spectra are significantly larger than the $(\phi_i, 3s)$ term values of the outer-shell spectra. Shake-up bands in the photoelectron spectrum of chlorobenzene occurring 56 000 and 77 000 cm^{-1} (*vert., 7.0 and 9.5 eV*) beyond the $(1s_C)^{-1}$ ionization potential have been assigned as $\pi \rightarrow \pi^*$ transitions in the $(1s_C)^{-1}$ positive ion [O3], as in benzene itself (Section XIX.A). Curiously, there are no signs in the halobenzenes of the $1s_C \rightarrow \sigma^*(C-X)$ valence excitations which are so prominent in other halogen-containing systems (Chapter IV).

Inner-shell spectra originating at the halogen atom orbitals of the halobenzenes might be expected to show prominent transitions to $\sigma^*(C-X)$, however, once again the first bands in these spectra [H58] are assigned as terminating at π^* ($2s_{Cl} \rightarrow \pi^*$, 43 500-$cm^{-1}$ term value and $2p_{Cl} \rightarrow \pi^*$, 41 100-$cm^{-1}$ term value in chlorobenzene, for example). Perhaps the intense bands in the $1s_C$ spectra do terminate at π^* as stated in the original work, whereas the intense bands originating at the inner-shell halogen orbitals terminate at $\sigma^*(C-X)$. The observed term values are compatible with this, and the relative intensities of the transitions to π^* and σ^* might be explained in terms of the overlap between the orbitals involved in the transitions. These valence transitions in the halobenzenes are followed by Rydberg excitations of normal term values [H58].

XIX.D. Azabenzenes

Bolovinos *et al.* [B53] have presented a detailed description of the spectra of pyridine, pyrimidine, pyrazine and *sym*-triazine to 77 000 cm^{-1} (*9.5 eV*). As seen in Fig. XIX.D-1, the benzenoid pattern of $\pi \rightarrow \pi^*$ transitions to $^1B_{2u}$, $^1B_{1u}$ and $^1E_{1u}$ is preserved in the azabenzenes [**II**, 237], the general effects of replacing a C–H group with N being to shift the $\pi \rightarrow \pi^*$ excitations upward and to introduce low-lying $n_N \rightarrow \pi^*$

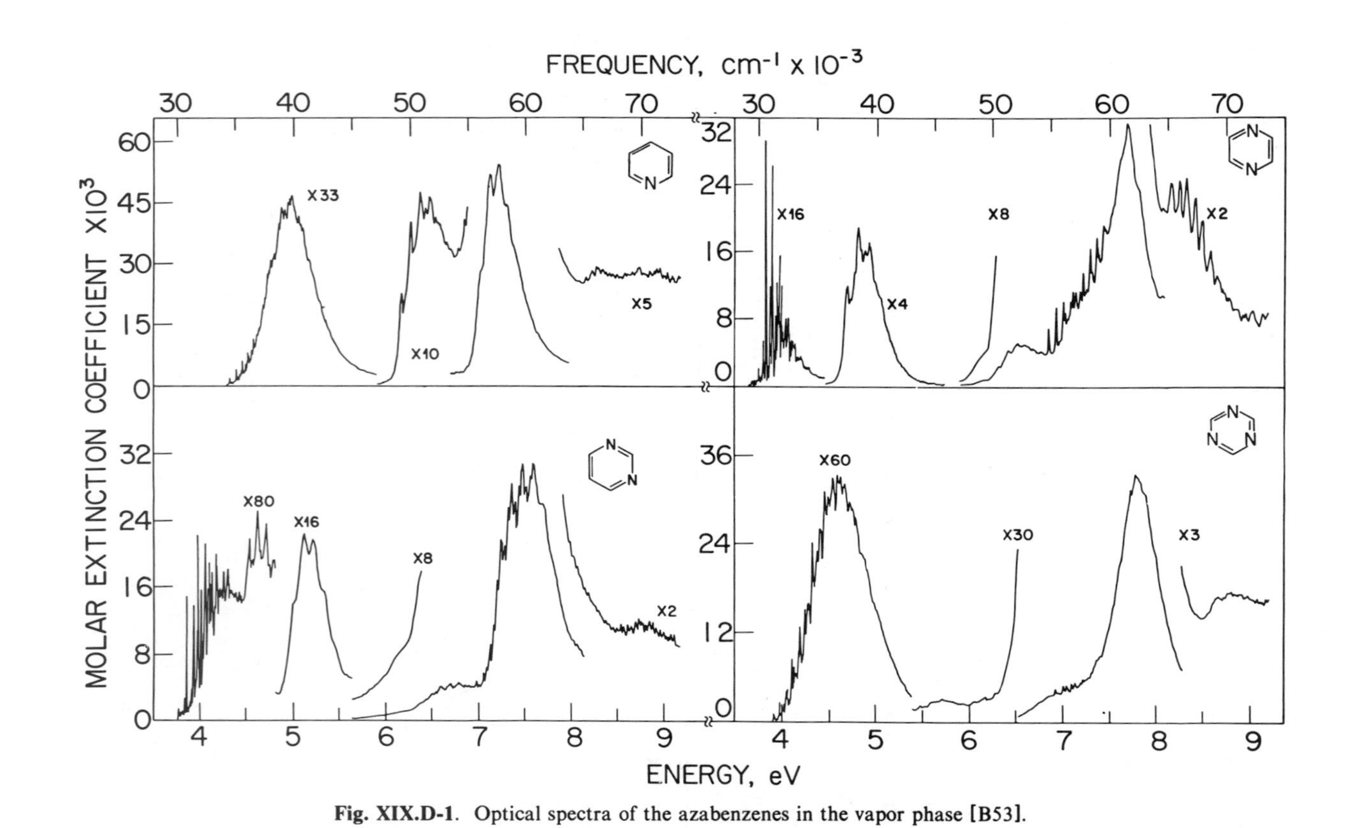

Fig. XIX.D-1. Optical spectra of the azabenzenes in the vapor phase [B53].

transitions [W1].[†] Though there is no obvious splitting of the $^1E_{1u}$ state in low symmetry, one does note that the $^1A_{1g} \rightarrow {}^1E_{1u}$ bandwidth in pyrimidine is approximately twice those in benzene and the other azabenzenes.

In the spectra of pyridine in Ar and Kr matrices, the transitions to $^1B_{1u}$ and $^1E_{1u}$ are seen to consist of well-resolved vibronic progressions of $\nu_2'(a_1)$, and the electronic splitting of the benzene-derived $^1E_{1u}$ state is concluded to be less than the width of a single vibronic line (300 cm^{-1}) [O11]. The vibronic structure in the $^1A_{1g} \rightarrow {}^1E_{1u}$ region of pyridine was earlier interpreted as involving an $n_N \rightarrow 3p$ Rydberg transition overlapping the $\pi \rightarrow \pi^*$ band [II, 239], but further work by Olsher [O12] and Bolovinos *et al.* [B53] shows only the 800 cm^{-1} $\nu_2'(a_1)$ progression in the $^1E_{1u}$ state. In the optically-active pyridine derivative (+)-5,6,7,8-tetrahydro-8,9,9-trimethyl-5,8-methanoquinoline 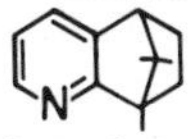 the CD spectrum shows two peaks in the $^1E_{1u}$ region (52 400 and 56 200 cm^{-1} *vert., 6.49 and 6.96 eV*) in hydrocarbon solution [R44]. It is not clear whether these are the symmetry-split components of the transition to $^1E_{1u}$ or if one of them involves a $\pi \leftrightarrow \sigma$ excitation. Inasmuch as Rydberg excitations often appear in solution CD spectra (Section II.D), Rydberg assignments also are possibilities here. The validity of Rydberg assignments would be confirmed if the frequencies of the CD bands increase as the solution is cooled (Section II.D). A band observed in the CD spectrum of the pyridine derivative at 49 000 cm^{-1} (*vert., 6 eV*) is assigned as a second $n_N \rightarrow \pi^*$ excitation, however a Rydberg assignment ($n_N \rightarrow 3s$) also is a possibility.

Several Rydberg series involving nd terminating orbitals ($\delta = 0.22$) and converging upon the lowest $(n_N)^{-1}$ ionization potentials of pyridine and pyrimidine have been assembled [B53]. In pyridine, two np series have δ values of 0.60 and 0.45, whereas in pyrimidine they are 0.62 and 0.40. The unobserved one-photon Rydberg transitions from n_N to 3s in pyridine and pyrazine are prominent as two-photon resonances in the MPI spectra at 50 669 and 50 830 cm^{-1} (*advert., 6.2820 and 6.3020 eV*), respectively [B32, T26]. The lowest transitions to 3s in benzene, pyridine and pyrazine involve a common set of upper-state vibrations, though with different Franck-Condon factors and anharmonicities due to the different originating MO's in these excitations. The $(n_N, 3s)$ term values of

[†] The benzene symmetry notation has been preserved, unless otherwise noted.

pyridine and pyrazine are approximately 1000 cm^{-1} larger than that of ($e_{1g}\pi$, 3s) in benzene due to the increased 3s orbital penetration at the N atom. In *sym*-triazine [**II**, 247], the $n_N(e') \rightarrow$ 3s Rydberg excitation first observed as a one-photon absorption, Fig. XIX.D-1, also is reported as a two-photon resonance in the MPI spectrum of the jet-cooled molecule [W15]. The $^1E'(n_N, 3s)$ upper state of *sym*-triazine is dynamically Jahn-Teller unstable, the active vibration being the ν_6 ring distortion. Three short Rydberg series going to higher ionization potentials are claimed for pyridine [S52].

Several temporary-negative-ion resonances (Section II.B) are recorded in vapors of pyridine and the diazabenzenes [M13, V7]. In pyridine, the three valence TNI transitions to π^* are observed at frequencies of 6400, 9280 and 38 000 cm^{-1} (*vert., 0.79, 1.15 and 4.71 eV*),[†] and these valence resonances are followed by broad features at 58 600 and 63 400 cm^{-1} (*vert., 7.27 and 7.86 eV*). Assignment of these latter resonances is problematic, for they are above the lowest Rydberg excitations of pyridine and hence would not seem to be Feshbach configurations. What appears to be the corresponding resonance in benzene is a single band at 71 400 cm^{-1} (*vert., 8.85 eV*), suggesting that the resonance is degenerate in benzene but split into two components in pyridine. A possibility is that they are valence TNI resonances involving $\pi \rightarrow \pi^*$ core excitations. In the diazabenzenes, the valence TNI resonances to the π_4^* and π_5^* MO's are in the 800–8000-cm^{-1} (*0.1–1-eV*) region.

The electron energy-loss spectrum of poly(2-vinyl pyridine) closely resembles that of polystyrene, with the plasmon loss at 170 000 cm^{-1} (*vert., 21 eV*) dominating the Im$(-1/\epsilon)$ spectrum [R23]. $1s_C$ excitations in poly(2-vinyl pyridine) terminating at π_4^*, π_5^* and π_6^* show a splitting pattern closely resembling that for the π_4^*, π_5^* and π_6^* valence TNI resonances in pyridine [M13, V7]. The inner-shell excitations to π^* are then followed by several excitations which are either $1s_C \rightarrow \sigma^*$ shape resonances or $\pi \rightarrow \pi^*$ shake-up bands. Shake-up excitations in the $(1s_C)^{-1}$ ESCA spectrum of pyridine at 50 000 and 59 700 cm^{-1} (*vert., 6.2 and 7.4 eV*) beyond the $(1s_C)^{-1}$ ionization potential [C27] are reminiscent of those observed in benzene at 47 600 and 56 500 cm^{-1} (*vert., 5.9 and 7.0 eV*), which in turn are derived from the $^1A_{1g} \rightarrow {}^1B_{1u}(\pi, \pi^*)$ and $^1A_{1g} \rightarrow {}^1E_{1u}(\pi, \pi^*)$ transitions of the neutral molecule. Similar features are observed in the shake-up spectrum of pentafluoropyridine [C27].

[†] However, see the footnote on p. 359.

In addition to the transitions to $^1B_{2u}$, $^1B_{1u}$ and $^1E_{1u}$, the absorption spectrum of neat liquid pyridine shows a strongly rising absorption between 73 000 and 85 000 cm^{-1} *(9.0 and 10.6 eV)* which is not present in the vapor spectrum [M2]. Possibly this is an intermolecular charge transfer of the sort postulated for other neat-liquid and solid phases [**II**, 139]. The electron-transmission spectrum of solid pyridine consists of the known $\pi \rightarrow \pi^*$ excitations followed by peaks at 67 700, 80 700 and 111 000 cm^{-1} (*vert., 8.40, 10.05 and 13.75 eV*) which were assigned as Rydberg excitations [S7, S8]. Inasmuch as the 67 700-cm^{-1} peak is observed at just this frequency in the vapor phase [V7], it is more likely that it is a valence excitation, possibly $^1A_{1g} \rightarrow {}^1E_{2g}(\pi, \pi^*)$ as in benzene (Section XIX.A). The electron energy-loss spectra of benzene, pyridine and pyrazine adsorbed onto the (111) face of a Ag crystal show all $\pi \rightarrow \pi^*$ excitations within $\pm$ 2000 cm^{-1} of their vapor-phase values [A37]. Additionally, the azabenzenes show metal $\leftrightarrow$ molecule charge-transfer excitations at low frequencies which are reminiscent of the lowest valence TNI resonances in these molecules.

XIX.E. Other Phenyl Compounds

Comparison of the spectra of aniline and its N,N-dialkyl derivatives [**II**, 253] in the vapor and condensed phases reveal several Rydberg features [F48]. In aniline itself, early members of two *n*p and two *n*d series are assigned, paralleling the situation in benzene, whereas in N,N-diethyl aniline the *n*s series also is seen clearly. Fuke and Nagakura [F48] relate this difference to the fact that the originating orbital in aniline is a ring π MO, whereas in the diethyl derivative, the originating MO is the lone pair on nitrogen, thereby making the Rydberg spectrum resemble that of the alkyl amines (Section VI.A). The lowest Rydberg excitation in N,N-dimethyl aniline (37 000 cm^{-1} *vert., 4.6 eV*) is polarized either out-of-plane or in-plane short axis [D8], and apparently can be seen as well in the polarized absorption spectrum of the neat crystal [B36]. Low-resolution threshold electron impact spectra of various substituted benzenes clearly show the valence TNI resonances involving the lower π^* MO's [C22].

Single-crystal transmission spectra of hydroquinone and its dimethyl ether (*p*-dimethoxybenzene) show the usual parade of benzenoid $\pi \rightarrow \pi^*$ bands with their expected polarizations, however, in hydroquinone, another intense feature is seen at 60 000 cm^{-1} (*vert., 7.4 eV*) with long-axis polarization [K10]. Possibly, this band is the result of a substituent-to-ring charge transfer (OH $\rightarrow \pi^*$).

Spectra of phenyl-containing polymers in the vacuum-ultraviolet region generally offer very little insight. Thus interpretation of the spectrum of poly(vinyl cinnamate) [O4] is intractable because it has present simultaneously the bands of the alkane backbone, the vinyl, the carboxyl and the phenyl groups. Similar complications hold for the spectra of Kapton [A21] and Lexan [R26], but that of polystyrene is much more manageable [**II**, 256 and Section XIX.B].

XIX.F. Addendum

R. L. Whetten and E. R. Grant, Strong vibronic coupling in molecular Rydberg states, *J. Chem. Phys.* **80**, 5999 (1984).

S. G. Grubb, R. L. Whetten, A. C. Albrecht and E. R. Grant, A precise determination of the first ionization potential of benzene, *Chem. Phys. Lett.* **108**, 420 (1984).

R. L. Whetten, K.-J. Fu and E. R. Grant, High Rydberg states of jet-cooled toluene observed by ultraviolet two-photon absorption spectroscopy: ultrafast radiationless decay and pseudo-Jahn-Teller effects, *Chem. Phys.* **90**, 155 (1984).

H.-P. Fenzlaff and E. Illenberger, Low energy electron impact on benzene and the fluorobenzenes. Formation and dissociation of negative ions, *Int. J. Mass Spectrom. Ion Phys.* **59**, 185 (1984).

CHAPTER XX

Higher Aromatics

Vacuum-ultraviolet spectra of aromatics other than benzene and its derivatives are most disappointing. Once beyond the six pi electrons of benzene, the numbers of $\pi \rightarrow \pi^*$ transitions become uncomfortably large, the $\pi \leftrightarrow \sigma$ excitations also intervene in great numbers, and sharp Rydberg excitations are uncommonly sparse considering how sharp the $(\pi)^{-1}$ bands are in the photoelectron spectra. Even MPI spectroscopy (Section II.A) has been of little use in finding the Rydberg excited states of molecules such as naphthalene, anthracene, *etc.* A large number of aromatic crystal spectra are in the literature, several with polarization determinations, but orbital assignments are usually uncertain or nonexistent.

Spectral studies on biphenyl in the vapor and crystalline phases present data which is difficult to understand because the spectra are not totally consistent with one another, and one is not certain how to correlate the transitions of the two phases. This latter arises owing to unknown crystal shifts and to a drastic change of geometry on going from the vapor phase (45° dihedral angle between phenyl planes) to the crystal (0° dihedral angle between phenyl planes). It is agreed [M49, S49] that the biphenyl spectra in both phases involve two weak $\pi \rightarrow \pi^*$ transitions (37 000–42 000 cm^{-1} *vert., 4.6–5.2 eV*), followed by an intense peak at *ca.* 52 000 cm^{-1} (*6.4 eV*) which may consist of more than one transition.

These excitations of biphenyl probably correlate directly with the $\pi \rightarrow \pi^*$ bands of benzene (Section XIX.A) terminating at $^1B_{2u}$, $^1B_{1u}$ and $^1E_{1u}$ states, and with those of fluorene in the same spectral region [V18]. Further, it seems that the $^1E_{1u}$ transitions in the two rings of biphenyl interact strongly, resulting in a second component observed at 61 800 cm^{-1} (*vert., 7.66 eV*) with short-axis polarization. As biphenyl is substituted in the o,o' positions with fluorine [M49] or methyl groups [S49] so as to drive the rings into a nearly perpendicular conformation, the $^1E_{1u}$ splitting of 10 000 cm^{-1} in biphenyl itself subsides to *ca.* 4000 cm^{-1}, Fig. XX-1. Though McLaughlin and Clark [M49] derived polarization directions for the biphenyl crystal transitions, it is not clear how bands of the vapor are related to bands of the crystal. Thus, for example, it was concluded that the weak band at 57 300 cm^{-1} (*vert., 7.10 eV*) in the vapor spectrum has long-axis polarization since that is the polarization at 57 300 cm^{-1} in the crystal. However, judging by relative intensities it appears that it is the 61 800-cm^{-1} band of the vapor phase which is centered at 57 300 cm^{-1} in the crystal. There are no obvious Rydberg excitations in the optical spectrum of biphenyl.

In decafluorobiphenyl [M49], the steric requirements of the *ortho* fluorine atoms force the phenyl rings into a near-perpendicular geometry, and with this the spectrum becomes much more like that of pentafluorobenzene (Section XIX.C) than biphenyl. As in pentafluorobenzene, Fig. XIX.C-2, decafluorobiphenyl should show a number of valence $\mathscr{A}(\pi, \sigma^*)$ bands in the 40 000–55 000-cm^{-1} (*5–7-eV*) region. Tentatively, this is the assignment of choice for the apparently extraneous bands in the spectra of decafluorobiphenyl at 51 800 cm^{-1} (*vert., 6.42 eV*) and of o,o'-difluorobiphenyl at 50 000 cm^{-1} (*vert., 6.2 eV*), Fig. XX-1. Note, however, that the same band seems also to be present in the spectrum of o,o'-dimethyl biphenyl [S49].

On going from benzene to biphenyl, the two largest spectral changes are the disappearance of sharp-line Rydberg excitations and the splitting of the local $^1E_{1u}$ absorptions into two intense peaks. In p-terphenyl, the region of local $^1E_{1u}$ absorption shows three peaks [H44, H48, S36, V19], each of which has short-axis polarization according to the single-crystal work of Venghaus and Hinz [V19]. The spectra must be far more complicated than this however, for the independent-systems approach places 6 transitions in the $^1A_{1g} \rightarrow {}^1E_{1u}$ region and charge transfer will increase this number greatly. Venghaus and Hinz [V19] list a number of other p-terphenyl transitions in the 65 000–145 000-cm^{-1} (*8–18-eV*) region without comment. Again, as in biphenyl, there are no obvious Rydberg excitations in p-terphenyl.

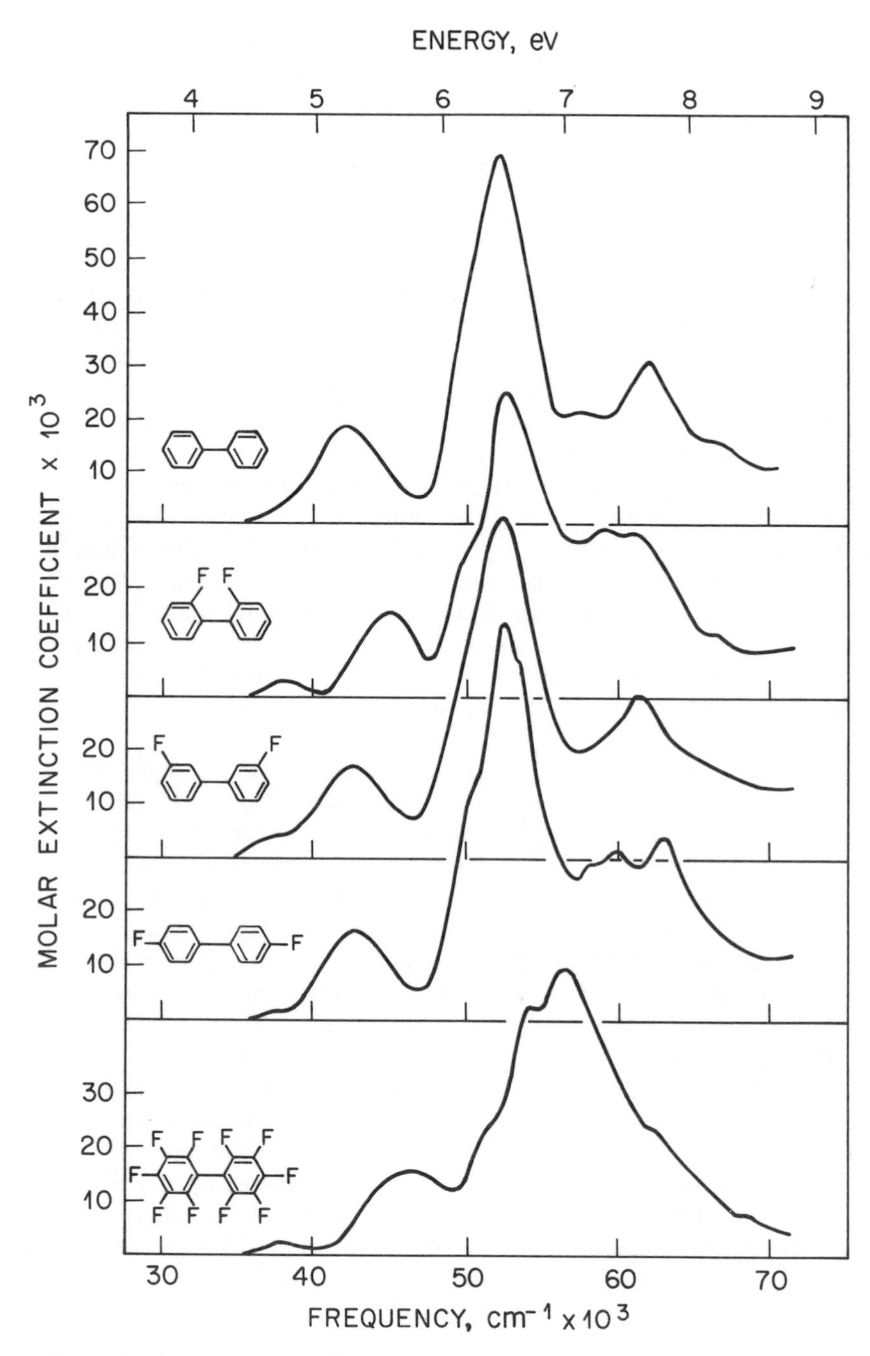

Fig. XX-1. Absorption spectra of various fluorinated biphenyls in the vapor phase [M49].

The electronic origins in naphthalene vapor [**II**, 339] at 45 070 and 45 390 cm^{-1} (*5.588 and 5.628 eV*) earlier were declared to correspond to valence excitations [**II**, 259], however, it is now clear that these prominent features of the vapor-phase spectrum do not appear either in the solution spectrum [K35] or the rare-gas matrix spectra [H22], and therefore a $1a_u\pi \rightarrow 3p$ Rydberg assignment is now to be preferred for this core-split spectral complex. (See below as well for confirmation of this assignment from TNI spectra.) Solvent perturbation of the Rydberg excitations of naphthalene in two instances reveal underlying $\pi \rightarrow \pi^*$ excitations. Thus solution of naphthalene in perfluoro-*n*-heptane demolishes the Rydberg bands in the 55 000–65 000-cm^{-1} (*6.8–8.0-eV*) region to uncover a valence transition at 62 000 cm^{-1} (*vert., 7.7 eV*) [K35], while in rare-gas matrices the Rydberg transitions in the 49 000–57 000-cm^{-1} (*6.1–7.1-eV*) region are blotted out to reveal a structured $\pi \rightarrow \pi^*$ excitation at 54 000 cm^{-1} (*vert., 6.7 eV*) [H22]. According to the semiempirical calculation of Åsbrink *et al.* [A27], the upper states of these two $\pi \rightarrow \pi^*$ transitions have $^1B_{2u}$ and $^1B_{3u}$ symmetries, respectively. The reflection spectrum of polycrystalline anthracene [**II**, 340] to 75 000 cm^{-1} (*9.3 eV*) [P2] is consistent with the single-crystal work of Kunstreich and Otto [**II**, K40]. Polarized reflection spectra of the lowest six $\pi \rightarrow \pi^*$ excitations of single-crystal anthracene reveal the directional dispersion of the excitonic bands [K28]. The crystal spectrum of anthracene is reviewed by Koch and Otto [K30].

The resonant scattering of slow electrons from naphthalene vapor [M15] is readily interpretable. Resonances found at 1600, 6100 and 10 900 cm^{-1} (*vert., 0.20, 0.76 and 1.35 eV*) correspond to residence of the incident electron in the three lowest π^* MO's ($2b_{2g}$, $2b_{3g}$ and $3b_{1u}$), respectively [A27, P31]. Yet another resonance in naphthalene at 27 000 cm^{-1} (*vert., 3.4 eV*) [V22] possibly involves occupation of a σ^* MO. These valence TNI resonances are followed by two others at 42 700 and 60 900 cm^{-1} (*vert., 5.29 and 7.55 eV*) which are most likely Feshbach resonances involving the lower Rydberg states of naphthalene as parents, Section II.B. In [**II**, 259], the Rydberg nature of electronic origins at 45 070 and 45 390 cm^{-1} (*5.588 and 5.628 eV*) was denied, but we now see that assigning them as terminating at 3p (*vide supra*) then leads naturally to a $^2(1a_u\pi, 3p^2)$ assignment for the TNI resonance at 42 700 cm^{-1}. As for the higher resonance at 60 900 cm^{-1}, it is at the expected frequency for a $^2(1b_{3g}\pi, 3d^2)$ Feshbach resonance where $1b_{3g}\pi$ is the third $(\pi)^{-1}$ ionization potential of naphthalene and the parent state $(1b_{3g}\pi, 3d)$ is at 66 900 cm^{-1} (*vert., 8.29 eV*) [**II**, 260].

Assignments in the spectra of azulene and its derivatives remain somewhat cloudy. Electron impact on azulene at low energy reveals a triplet state at 19 000 cm^{-1} (*vert., 2.4 eV*), possibly a second triplet at 24 000 cm^{-1} (*vert., 3.0 eV*), and three $\pi \rightarrow \pi^*$ singlet states corresponding to those known optically in the 32 000−44 000-cm^{-1} (*4.0−5.5-eV*) region [K35]. McGlynn and coworkers [B57, L24] have searched the ultraviolet spectra of azulene and several derivatives for Rydberg excitations. Accepting the $2a_2\pi \rightarrow 3d$ assignment of the sharp-line structure beginning at 47 000 cm^{-1} (*adiabat., 5.8 eV*) in azulene [**II**, 267], McGlynn and coworkers trace this transition through several derivatives in the 45 000−50 000 cm^{-1} (*5.6−6.2 eV*) region, and point out how similar are the vibronic envelopes of the $2a_2\pi \rightarrow 3d$ optical and $(2a_2\pi)^{-1}$ photoelectron bands. Solution of azulene in *n*-pentane and in perfluoro-*n*-hexane [K35] acts to remove all the sharp Rydberg features in the 45 000−60 000-cm^{-1} (*5.6−7.4-eV*) region of the vapor-phase spectrum, exposing three broad bands at 48 200, 52 500 and 56 300 cm^{-1} (*vert., 5.98, 6.51 and 6.98 eV*). Apparently, these are $\pi \rightarrow \pi^*$ valence excitations. The $2a_2\pi \rightarrow 3p$ Rydberg transition of the azulene chromophore is said by McGlynn *et al.* to rest upon the high-frequency wing of the $\pi \rightarrow \pi^*$ band at *ca.* 35 000 cm^{-1} (*vert., 4.3 eV*), expressing itself spectrally as a widening of the former in the vapor-phase spectrum, while the $2a_2\pi \rightarrow 3s$ transition is totally engulfed by the $\pi \rightarrow \pi^*$ absorption in the 25 000−35 000 cm^{-1} (*3.1−4.3 eV*) region. No Rydberg absorptions going to higher ionization potentials can be assigned in the azulene spectrum out to 90 000 cm^{-1} (*11 eV*) [B57].

Reflection spectra of liquified 2-ethyl naphthalene and quinoline [M2] out to 86 000 cm^{-1} (*10.7 eV*) are qualitatively very similar, with the bands of quinoline shifted somewhat to higher frequencies with respect to those of 2-ethyl naphthalene. This is also the pattern of frequency shifts observed between the corresponding bands of benzene and pyridine, Sections XIX.A and XIX.D. Comparison of the threshold electron-impact spectrum of quinoline vapor [P31] and the reflection spectrum of the liquid shows a band-for-band correspondence in the 24 000−73 000-cm^{-1} (*3−9-eV*) region, thereby identifying all bands as valence. As expected, the electron-impact spectra of quinoline and isoquinoline look very much alike [P31].

Comparison of the vapor and solution spectra of indole strongly suggests the presence of Rydberg excitations at 39 500, 41 100 and 44 200 cm^{-1} (*vert., 4.90, 5.10 and 5.48 eV*) having term values of 22 700, 21 100 and 18 000 cm^{-1} [L2]. These transitions originate at the lone-pair orbital n_N and terminate at 3s, 3p and 3p′ orbitals. A CNDO study of poly-*N*-vinyl carbazole [B43] concludes that the intense transitions below 69 000 cm^{-1} (*8.5 eV*) involve $\pi \rightarrow \pi^*$ promotions, whereas alkyl backbone absorption (Section III.D) begins to dominate at higher frequencies [F22].

Dupuis *et al.* [D55] point out a most interesting possibility. Because porphyrins have unusually low ionization potentials (*ca.* 48 000 cm^{-1} *vert., 6 eV*), the lowest frequency $\pi \rightarrow 3s$ Rydberg excitation is expected at *ca.* 26 000 cm^{-1} (*3.2 eV*), assuming a (π, 3s) term value of 22 000 cm^{-1}. Though blotted out by the Soret band in absorption and severely perturbed by the condensed-phase effect, this low-lying Rydberg state still may be of consequence in the photophysics of porphyrins and in photosynthesis.

There are spectra without end of crystalline catacondensed hydrocarbons in the vacuum ultraviolet (see for example [S36]), however there are so many electrons and orbitals in these large molecules that it is presently an impossible task to assign the spectra with any confidence. We will mention a few noteworthy points in these spectra, but otherwise make no attempt to assign bands.

The optical spectrum of a thin film of naphthacene [**II**, 266] in the vacuum ultraviolet shows broad peaks at 60 000 and 66 000 cm^{-1} (*7.4 and 8.2 eV*) [K2]. Similar features are observed in the energy loss spectrum of the naphthacene crystal [U1]. Spectra to 80 000 cm^{-1} (*10 eV*) of the three isomers of naphthacene: chrysene, benzanthracene and triphenylene [H48] are very different from one another and from naphthacene. Moreover, comparison of the solid-state spectra with solution spectra shows that there are strong crystal effects operating. In pentacene [R11, S36], the thin-film spectra differ greatly depending upon the temperature at which the film was deposited [K2]. The polarized reflection spectrum of single-crystal perylene [**II**, 341] to 62 000 cm^{-1} (*7.7 eV*) [F47] is consistent with pi-electron calculations on this molecule. Spectra of pyrene crystal and vapor [V17, V20, **II**, 266, 341] display six sharp $\pi \rightarrow \pi^*$ transitions in the region up to 56 000 cm^{-1} (*7 eV*), followed by a slowly rising absorption consisting of many overlapping bands. All six of the $\pi \rightarrow \pi^*$ bands are unaffected by solution in perfluoro-*n*-heptane [K35]. This general absorption pattern in pyrene appears again in fluorene [H49, V18] where sharp $\pi \rightarrow \pi^*$ excitations fill the region up to approximately 65 000 cm^{-1} (*8 eV*), yielding then to a slowly rising absorption peaking at 130 000 cm^{-1} (*16 eV*). Intense, broad excitations in these pi-electron molecules at high frequencies may be related to the pi-electron plasmon in graphite at 218 000 cm^{-1} (*vert., 27 eV*), Section II.D, or simply may be due to maxima in the photoionization cross sections, Section II.C.

Spectra of crystalline tetracyanoquinodimethane (TCNQ) and its complexes with large-molecule charge transfer donors have been reported.

Four bands of TCNQ in the region up to 65 000 cm^{-1} (*8 eV*) [R20, R21] are explained semiquantitatively as $\pi \rightarrow \pi^*$ transitions [A26, R21], though there is some controversy as to whether the 46 000-cm^{-1} (*vert., 5.7-eV*) band is allowed or forbidden. Spectra of TCNQ complexes with tetrathiafulvalene [R21] and with decamethyl ferrocene [T6] are available.

CHAPTER XXI

Inorganic Systems

The discussion in this Chapter is limited to discrete, polyatomic species, either neutral or charged, in the vapor phase, in solution or in the crystalline phase. Omitted are the extended solids in which discrete polyatomic species do not appear.

XXI.A. Halides

The X-ray absorption spectrum of the PF_6^- ion is most interesting, for it is isoelectronic with SF_6, (Section VIII.B), on which there has been so much work. Because anions do not show Rydberg transitions (Section I.B), the PF_6^- spectrum will consist only of valence excitations to $6a_{1g}\sigma^*$ and $6t_{1u}\sigma^*$ MO's. Spectra of NH_4PF_6 and KPF_6 [V27] originating at $1s_P$, $2p_P$ and $1s_F$ inner-shell orbitals show valence transitions to $6a_{1g}\sigma^*$(P–F) with term values of 47 000 cm^{-1}, followed by valence transitions to $6t_{1u}\sigma^*$(P–F) with term values of *ca.* 17 000 cm^{-1}. Appropriately, the intensity patterns of these two bands in the $1s_p$, $2p_p$ and $1s_F$ spectra are forbidden/allowed ($1s_P \rightarrow \sigma^*$), allowed/forbidden ($2p_P \rightarrow \sigma^*$) and allowed/allowed ($1s_F \rightarrow \sigma^*$) respectively. The valence (ϕ_i, $6a_{1g}\sigma^*$) term value of 47 000 cm^{-1} in PF_6^- matches closely the

(ϕ_i, $6a_{1g}\sigma^*$) term value of 48 000 cm^{-1} in SF_6 (Table VIII.B-I), however this agreement does not extend to the (ϕ_i, $6t_{1u}\sigma^*$) term values (17 000 cm^{-1}, PF_6^-; 32 000 cm^{-1}, SF_6). As in SF_6, Fig. VIII.B-1, there are two broad excitations beyond each of the inner-shell ionization potentials in the PF_6^- ion, which here are taken to be SU1 and SU2 shake-up bands, but which Vinogradov *et al.* [V27] assign as terminating at components of 3d. Note that though 3d is a Rydberg orbital in SF_6, in the isoelectronic PF_6^- anion there are no Rydberg levels possible (Section I.B), and so "3d" does not exist as a bound level of the system. Calculations on the $1s_{As}$ spectrum of the AsF_6^- ion predict that all valence excitations to the σ^*(As–F) orbitals will have negative term values, and so will appear as very broad shape resonances [M67], in this way resembling the $1s_B$ spectrum of BF_3 (Section V.C).

Ab initio calculations of the term values in the $1s_{As}$ spectrum of AsF_3 [M67] do rather well. The $\mathscr{A}(1s_{As}, \sigma^*)$ bands of upper-state symmetries E and A_1 are calculated to have term values of 84 300 and 64 500 cm^{-1} respectively, and are followed by the Rydberg excitation to 5s with a 40 800-cm^{-1} term value. The first and third of these predicted bands have been observed. In AsF_5, there are again two valence excitations from $1s_{As}$ to σ^* (e' and a'') with term values of only 40 500 and 17 500 cm^{-1}, however the third component of the σ^*(As–F) manifold is beyond the $(1s_{As})^{-1}$ ionization limit and so appears as a shape resonance with a negative term value of − 56 000 cm^{-1}. The inner-shell spectrum of the inorganic anion BF_4^- is discussed in Section V.C.

The gaseous hexafluoride MoF_6 [II, 278] shows a parade of six electronic transitions in the region 47 600–81 000 cm^{-1} (*5.9–10.0 eV*), whereas the corresponding bands of WF_6 lie between 58 300 and 89 100 cm^{-1} (*7.23 and 11.1 eV*). In both compounds, several of the transitions display vibrational progressions of ν_1' (M–F stretch, *ca.* 650 cm^{-1}) [M27]. As regards electronic structure, the 7 uppermost filled MO's of MoF_6 and WF_6 are derived from the $2p\pi$ and $2p\sigma$ orbitals of the fluorine atoms interacting with the appropriate central-atom orbitals, whereas the two lowest vacant valence MO's are the t_{2g} and e_g components of the central-atom nd orbital sets. Clearly, the lowest excitations in these molecules will be ligand → metal charge transfer involving the $2p\pi$ and $2p\sigma$ orbitals on fluorine and the $t_{2g}\pi^*$ valence d orbitals of the central atom. McDiarmid [M27, M36] has listed the term values of the first 6 transitions in MoF_6 and WF_6 with respect to the first 6 ionization potentials in the photoelectron spectra and as can be seen in Table XXI.A-I, they are remarkably constant. This constancy of the term value argues strongly for a common terminating orbital in all these transitions, which we take as $t_{2g}\pi^*$ [D5]. Note that the $t_{2g}\pi^*$ term values

TABLE XXI.A-I

SPECTRAL TRANSITIONS IN TRANSITION-METAL COMPOUNDS[a]

	Transition Frequency, cm^{-1} (*eV*)	Ionization Potential, cm^{-1} (*eV*)	Term Value (cm^{-1})	Assignment	References
MoF_6					[M27,M36]
	47 600 (*5.9*)	121 600 (*15.07*)	74 000	$\mathscr{A}(2p\pi, t_{2g}\pi^*)$	
	52 700 (*6.54*)	125 400 (*15.55*)	72 700	$\mathscr{A}(2p\pi, t_{2g}\pi^*)$	
	57 400 (*7.12*)	127 400 (*15.80*)	70 000	$\mathscr{A}(2p\pi, t_{2g}\pi^*)$	
	69 400 (*8.62*)	142 100 (*17.62*)	72 700	$\mathscr{A}(2p\pi, t_{2g}\pi^*)$	
	74 400 (*9.22*)	149 400 (*18.53*)	75 000	$\mathscr{A}(2p\sigma, t_{2g}\pi^*)$	
	80 980 (*10.04*)	154 000 (*19.09*)	73 000	$\mathscr{A}(2p\sigma, t_{2g}\pi^*)$	
WF_6					[M27,M36,R13]
	58 300 (*7.23*)	123 800 (*15.35*)	65 500	$\mathscr{A}(2p\pi, t_{2g}\pi^*)$	
	64 900 (*8.05*)	129 600 (*16.07*)	64 700	$\mathscr{A}(2p\pi, t_{2g}\pi^*)$	
	69 400 (*8.60*)	135 700 (*16.83*)	66 300	$\mathscr{A}(2p\pi, t_{2g}\pi^*)$	
	72 100 (*8.94*)	138 900 (*17.22*)	66 800	$\mathscr{A}(2p\pi, t_{2g}\pi^*)$	
	80 900 (*10.03*)	148 800 (*18.45*)	67 900	$\mathscr{A}(2p\sigma, t_{2g}\pi^*)$	
	89 100 (*11.05*)	156 200 (*19.36*)	67 100	$\mathscr{A}(2p\sigma, t_{2g}\pi^*)$	
	94 770 (*11.75*)	123 800 (*15.35*)	29 000	$(2p\pi, 6s)$[b]	
	101 600 (*12.6*)	123 800 (*15.35*)	22 200	$(2p\pi, 6p)$[b]	
	108 900 (*13.5*)	123 800 (*15.35*)	14 900	$(2p\pi, 5d)$[b]	
	113 000 (*14.0*)	135 700 (*16.83*)	22 700	$(2p\pi, 6p)$[b]	
$TiCl_4$					[**II**, 276]
	36 000 (*4.5*)	95 000 (*11.8*)	59 000	$\mathscr{A}(3p\pi, e\pi^*)$	
	43 000 (*5.3*)	103 100 (*12.78*)	60 100	$\mathscr{A}(3p\pi, e\pi^*)$	
	56 000 (*7.0*)	95 000 (*11.8*)	38 500	$\mathscr{A}(3p\pi, t_2\sigma^*)$	
CrO_2Cl_2					[J3; L16]
	19 000 (*2.36*)	{ 95 200 (*11.8*) 96 200 (*11.93*)	76 700	$\mathscr{A}(np, 3d\pi^* a_1)$	
	24 500 (*3.04*)	102 700 (*12.73*)	78 200	$\mathscr{A}(np, 3d\pi^* a_1)$	
	34 000 (*4.21*)	111 300 (*13.8*)	77 300	$\mathscr{A}(np, 3d\pi^* a_1)$	

TABLE XXI.A-I (continued)

	Transition Frequency, cm^{-1} (*eV*)	Ionization Potential, cm^{-1} (*eV*)	Term Value (cm^{-1})	Assignment	References
	37 000 (*4.59*)	114 500 (*14.2*)	77 500	$\mathscr{A}(np, 3d\pi^* a_1)$	
	44 000 (*5.46*)	119 400 (*14.8*)	75 400	$\mathscr{A}(np, 3d\pi^* a_1)$	
	50 000 (*6.20*) }	{ 95 200 (*11.8*)	45 200	$\mathscr{A}(np, 3d\sigma^*)$	
	53 500 (*6.63*) }	{ 96 200 (*11.93*)	42 700	$\mathscr{A}(np, 3d\sigma^*)$	
	60 000 (*7.44*)	102 700 (*12.73*)	42 700	$\mathscr{A}(np, 3d\sigma^*)$	
RuO_4					[II, 281]
	25 500 (*3.16*)	97 500 (*12.1*)	72 000	$\mathscr{A}(2p\pi, e\pi^*)$	
	33 500 (*4.15*)	104 000 (*12.9*)	70 500	$\mathscr{A}(2p\pi, e\pi^*)$	
	38 000 (*4.7*)	111 400 (*13.8*)	73 400	$\mathscr{A}(2p\pi, e\pi^*)$	
OsO_4					[II, 281]
	35 000 (*4.3*)	99 400 (*12.3*)	64 400	$\mathscr{A}(2p\pi, e\pi^*)$	
	42 000 (*5.2*)	106 000 (*13.1*)	64 000	$\mathscr{A}(2p\pi, e\pi^*)$	
	49 000 (*6.1*)	108 900 (*13.5*)	59 900	$\mathscr{A}(2p\pi, e\pi^*)$	

[a] See Table VIII.B-I for related data on SF_6.

[b] These are Rydberg excitations.

in Table XXI.A-I are approximately 30 000 cm^{-1} larger than is expected for Rydberg excitations, in accord with our claim that the lowest *n*d virtual orbitals in these compounds are part of the valence shell. The charge-transfer excitations are not fully assigned however, for there is still uncertainty as to the symmetries of the originating MO's [D5, M8, M36], and no account has been taken of the accompanying Rydberg excitations. Two intense excitations in WF_6 at 101 600 and 108 900 cm^{-1} (*vert., 12.6 and 13.5 eV*) are said to terminate at the $e_g\sigma^*$ component of 5d [R13], but we note in Table XXI.A-I that the term values for these peaks and the two peaks beyond them are compatible as well with Rydberg assignments.

The situation in UF_6 is much like that in MoF_6 and WF_6, for all the transitions in UF_6 up to 80 000 cm^{-1} (*10 eV*) seem to be ligand → metal charge transfer. However, the situation is much more complicated in UF_6, for the lowest empty U orbitals, 5f, are split into three components (a_{2u}, t_{2u}, t_{1u}) with an overall splitting of only 12 000 cm^{-1} while the t_{2g} and e_g components of the 6d orbital set are higher but close by. Additionally, relativistic and spin-orbit effects will be very important in this molecule. The UF_6 spectrum commences with a large number of weak excitations between 26 000 and 38 000 cm^{-1} (*3.2 and 4.7 eV*) [C7, C26, M30, R13, S67] which have been assigned by Hay [H27] as dipole-forbidden transitions terminating at the 5f components. Hay assigns the lowest frequency of these as originating at a spin-orbit component of t_{2u}, however Martensson *et al.* [M8, M9] assign the highest filled MO of UF_6 as t_{1g} on the basis of a study of relative photoionization intensities. Further, Rianda *et al.* [R13] assign the more intense transitions from 46 800 to 80 000 cm^{-1} (*5.8 to 10 eV*) as terminating at 6d whereas Hay calculates that they terminate at 5f. Interestingly, though the 5f orbitals might be thought to have little overlap with the fluorine orbitals, Hay calculates very large oscillator strengths for the allowed ligand → 5f charge-transfer transitions. Further consideration [R36] shows that the strongest of these transitions at 97 000−113 000 cm^{-1} (*12−14 eV*) [S67] is the molecular analog of the d → f giant resonance observed in the lanthanide and actinide atoms, Section I.D. Though this transition [e_g(d) → $3t_{1u}$(f)] at first sight would appear to have a Rydberg upper state, it is a valence excitation owing to 5f collapse (Section I.D). With a first ionization potential of 114 000 cm^{-1} (*vert., 14.14 eV*), the Rydberg excitations in UF_6 are above 75 000 cm^{-1} (*9.3 eV*), so that all excitations below this frequency will be $\mathscr{A}$(2p, 5f) bands.

The $\mathscr{A}$-band term value approach used so successfully in understanding the spectra of MoF_6 and WF_6 can be applied as well to the spectrum of $TiCl_4$ with its virtual 3d orbitals of $e\pi^*$ and $t_2\sigma^*$ character. In fact, the first two bands in $TiCl_4$ [**II**, 276, 343] at 36 000 and 43 000 cm^{-1} (*vert., 4.5 and 5.3 eV*) [**II**, 278] have term values of 59 000 and 60 100 cm^{-1} with respect to the two lowest ionization potentials, Table XXI.A-I. The constancy of these term values strongly suggests we are dealing with $\mathscr{A}(3p, e\pi^*)$ charge-transfer transitions. The following set of bands at 56 500−59 000 cm^{-1} (*vert., 7.00−7.30 eV*) however, have term values of only 48 000−50 000 cm^{-1} with respect to the third ionization potential. In this case, we assign these bands as $1t_1 \rightarrow t_2\sigma^*$, leading to a $e\pi^* - t_2\sigma^*$ split of 20 500 cm^{-1}, which is about a factor of two larger than is normally seen for $\frac{40}{9}$ Dq in tetrahedral species [M22]. The lowest Rydberg excitation in $TiCl_4$ falls at 70 000 cm^{-1} (*vert., 8.7 eV*) [**II**, 277].

The approach used above for $TiCl_4$ is now applied to the spectrum of CrO_2Cl_2, Table XXI.A-I, again with success. The first five absorption bands of CrO_2Cl_2 [J3] have term values of 76 800 ± 1500 cm^{-1} with respect to the ionization potentials given by Lee and Rabalais [L16]; the constancy of the term values suggests that all are $\mathscr{A}(\phi_i, 3d\pi^*)$ bands, where ϕ_i is $2p_O$ or $3p_{Cl}$. Interestingly, there are no PES bands between

14.8 and 21.2 eV in CrO_2Cl_2, so the highest frequency band terminating at $3da_1\pi^*$ must be that at 44 000 cm^{-1} (*vert., 5.46 eV*). The bands at yet higher frequencies are assumed to terminate at $3d\sigma^*$, with term values of *ca.* 43 000 cm^{-1}. According to this analysis, the $3d\pi^* - 3d\sigma^*$ splitting is approximately 34 000 cm^{-1} in CrO_2Cl_2.

Nielsen *et al.* [N23] present a marvelous set of absorption spectra of the fluorides XeF_2, XeF_4 and XeF_6 [II, 269, 341] determined using synchrotron radiation in the 50 000–280 000-cm^{-1} (*6–35-eV*) region, Fig. XXI.A-1. The Rydberg transitions of XeF_2 below 105 000 cm^{-1} (*13 eV*) show considerable vibronic structure involving ν_1' and ν_2'. The Rydberg assignments detailed in [II, 271] are confirmed in the work of Nielsen *et al.*, and the series extended. One also sees in Fig. XXI.A-1, the electronically forbidden transition from $\pi_u(3/2)$ to 6p as a small hump at 76 600 cm^{-1} (*vert., 9.50 eV*), and an electronically allowed $\sigma_g \rightarrow n$p Rydberg series beginning at 87 350 cm^{-1} (*advert., 10.83 eV*). Relevant term values for Rydberg orbitals in XeF_2 are (ϕ_i, 6s), 30 400 cm^{-1}; (ϕ_i, 6p), 22 980–21 940 cm^{-1}; and (ϕ_i, 6d), 18 390 cm^{-1}. The first two of these are very close to what is normally found in molecular spectra, whereas a term value for the lowest nd orbital usually is 13 000 cm^{-1}, rather than 18 000 cm^{-1}. However, note that the 5d term value in the Xe atom itself is 16 500 cm^{-1}, reflecting significant penetration via the 3d and 4d precursors in the core.

Nielsen *et al.* assign the broad feature at 75 200 cm^{-1} (*vert., 9.32 eV*) in XeF_4 as $a_{2u} \rightarrow$ 6s, where they have reversed the earlier ordering of the highest a_{2u} and a_{1g} MO's [II, 272]. Regardless of the orbital symmetries involved, the assignment of this feature as Rydberg is unacceptable in view of its experimental oscillator strength of 0.8 [II, 272]. However, the Rydberg pattern of Fig. XXI.A-1 does indeed suggest that the uppermost occupied MO has u symmetry rather than g. With this orbital assignment, the relevant term values become (ϕ_i, 6s), 30 500 cm^{-1}; (ϕ_i, 6p), 21 800–20 300 cm^{-1}; and (ϕ_i, 5d), 17 900 cm^{-1}, in good agreement with those of XeF_2. In XeF_6, only an $a_{1g} \rightarrow n$p Rydberg series is discernible, with the 6p orbital having a term value of 19 150 cm^{-1}.

Both XeF_2 and XeF_4 show prominent peaks at approximately 113 000 cm^{-1} (*vert., 14.0 eV*) which are assigned as maxima in the cross sections for photoionization of the uppermost MO's, while peaks at *ca.* 185 000 cm^{-1} (*23 eV*) in all three xenon fluorides correspond to maxima in the photoionization cross sections of fluorine 2p electrons. In Xe, those Rydberg transitions converging upon $^2P_{1/2}$ which are above the $^2P_{3/2}$ threshold show extremely prominent Fano antiresonances. The spectrum of XeF_2 shows a remarkably similar set of antiresonances in the 103 000–108 000 cm^{-1} (*12.8–13.4 eV*) region, Fig. XXI.A-1, however,

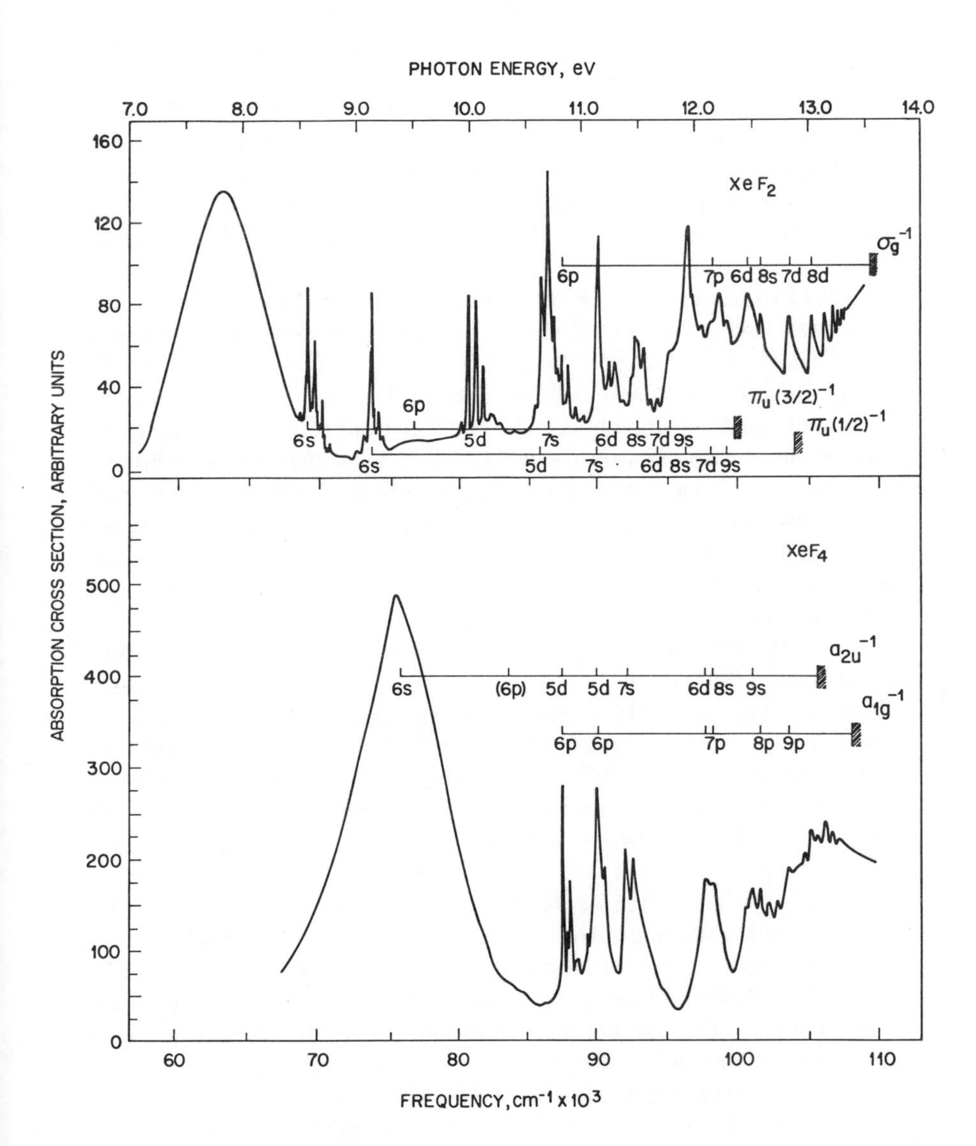

Fig. XXI.A-1. Optical absorption spectra and assignments in XeF_2 (*upper*) and XeF_4 (*lower*) in the vapor phase [N23].

we note that in this case it is the Rydberg transitions converging upon the $(10\sigma_g)^{-1}$ ionization potential which are mixed with the continuum of the $^2\Pi_{1/2}$ ion.

An inner-shell transition in XeF_6 at *ca.* 532 000 cm^{-1} (*vert., 66 eV*) [N22] has a term value of approximately 80 000 cm^{-1} with respect to the $(4d)^{-1}$ ionization potential at 613 000 cm^{-1} (*vert., 76 eV*), which clearly labels it as an inner-shell $\mathscr{A}(4d, \sigma^*)$ band. In fact Nielsen *et al.* assign it as 4d → 5pσ^*.† The 5p level in XeF_6 is within the valence shell, since 5s is doubly occupied in the ground state. In a parallel situation, the 4d → 5p inner-shell excitation of TeF_4 is reported in the 355 000–400 000-cm^{-1} (*44–50-eV*) region [B29]. Since the Te atom in this complex also has the 5s orbital doubly occupied, one is inclined to consider the (4d, 5p) configuration to be inner-shell valence, as in XeF_6. Indeed, the core splitting of the 5p orbitals by 24 000 cm^{-1} and the 55 000-cm^{-1} separation between the transitions to 5p and 6p clearly show that the 5p orbitals in TeF_4 are part of the valence shell while 6s and 6p are Rydberg. It would be most interesting to see how the transitions in question in TeF_4 behave in a matrix, and to have the $(4d)^{-1}$ ionization potential in order to calculate term values. Each of these considerations leads directly to an experimental distinction between Rydberg and valence configurations.

Though the reflectivity of BeF_2 glass suggests an intrinsic energy gap of approximately 90 000 cm^{-1} (*11 eV*), the transmission spectrum reveals an absorption edge at 67 000 cm^{-1} (*8.3 eV*) [W30]. Purification of the sample should result in transmission up to or beyond the LiF cutoff. The extreme transparency of BeF_2 is not surprising considering that its very large bond dissociation energy (152.5 kcal/mole) implies that the $\mathscr{A}(2p, \sigma^*)$ absorption bands will lie so high as to have negative term values.

As with the alkyl halides (Chapter IV), the transition-metal halides show inner-shell $\mathscr{A}$-band transitions, with the σ^* terminating orbital being the e_g component of the nd orbital set in octahedral symmetry [S78, S79]. It would be most interesting to calculate term values for these excited states, however the appropriate ionization potentials are not available. With due account of the large effect of antishielding, the inner-shell $\mathscr{A}$ bands of the transition metal fluorides may well have term values approaching 100 000 cm^{-1}. An especially clear example of an $\mathscr{A}(1s, \sigma^*)$

† Taking the inner-shell $\mathscr{A}(4d, \sigma^*)$-band term value as 80 000 cm^{-1} in XeF_6 and using it then in the outer-shell spectrum leads one to predict $\mathscr{A}(2p, \sigma^*)$ bands at *ca.* 25 000 cm^{-1} (*3.1 eV*) in this molecule! Indeed, weak transitions are reported in this region [N23].

band in a metal halide is that of $CuCl_4^{2-}$ at *ca.* 8980 eV [H2], wherein the absorption is largely electric quadrupole in nature and the polarization is that appropriate for a $1s_{Cu} \rightarrow 3d\sigma^*(x^2 - y^2)$ transition.

XXI.B. Oxides

The oxyanions MO_4^{x-} of the nonmetallic elements are generally quite transparent in the vacuum-ultraviolet region, both as ions in solution and as constituents of ionic crystals. Such transparency is not surprising considering that they all have closed-shell electronic structures resembling that of quartz. The spectrum of such an ion in solution is especially interesting, for in this case Rydberg excitations do not exist as such (Section I.B), however the spectrum is complicated by the presence of intermolecular charge-transfer-to-solvent (CTTS) excitations. General aspects of CTTS transitions are reviewed by Blandamer and Fox [B49].

Spectra of $H_2PO_4^-$ and other phosphate anions in aqueous solution [H7] show a sharply rising absorption near 50 000 cm^{-1} (*6.2 eV*) which has not peaked by 55 500 cm^{-1} (*6.9 eV*). The broad transition has been assigned as CTTS [B49, H7]. In the crystalline state, the $H_2PO_4^-$ anion is responsible for an absorption edge at approximately 55 000 cm^{-1} (*6.9 eV*) in crystals a few mm. thick [S53]. Reflection studies [O13, S4] place the peak of this band at 65 000 cm^{-1} (*8 eV*) in the K^+ and NH_4^+ salts, and the close similarity in the two crystals is used to argue for transitions localized within the $H_2PO_4^-$ ions. It is quite possible that the band beyond 55 500 cm^{-1} in aqueous solutions of $H_2PO_4^-$ in fact is related to the 65 000-cm^{-1} band of the anhydrous salt, in which case the CTTS assignment would be incorrect. Note also that this region of $H_2PO_4^-$ solution absorption is quite close to the 58 000 cm^{-1} (*7.2 eV*) absorption edge found for thin layers of neat $(CH_3O)_3PO$ [**I**, 96].

The sulfate and perchlorate anions are isoelectronic with the phosphate anion and share its optical properties. In aqueous solutions, the SO_4^{2-} ion shows an absorption peak at approximately 58 000 cm^{-1} (*7.2 eV*), while the corresponding feature in HSO_4^- peaks beyond 60 000 cm^{-1} (*7.4 eV*) [B49]. The electron-impact energy loss spectrum of the sulfate ion in a crystal of $CaSO_4$ has its first peak at 62 000 cm^{-1} (*7.7 eV*) [H79], and at 66 000 cm^{-1} (*8.2 eV*) in K_2SO_4 [P33]. The absorption of the perchlorate ion in water has not peaked by 55 500 cm^{-1} (*6.9 eV*) [B49] and no spectral data on crystals is available. Since the ionization potentials increase in the order PO_4^{3-}, SO_4^{2-}, ClO_4^- [N18], the CTTS frequencies will increase in the same order.

It seems most likely that the transitions at *ca.* 60 000 cm^{-1} in the tetrahedral oxyanions $MO_4H_y^{x-}$ have closely related outer-shell valence assignments, however there are no clear suggestions as to which orbitals are involved. Almost certainly the transitions in question originate with the $1t_1$ set of lone-pair orbitals on the oxygen atoms, these being the highest-occupied MO's in all species, and terminate either at valence MO's or possibly at charge-transfer-to-solvent configurations. However, the latter seems unlikely in view of the slight shifts induced in these bands on going from aqueous solutions to the anhydrous crystals.

In the oxyanions and corresponding neutral molecules containing a transition-metal central atom, the spectral transparency disappears completely as the low-lying nd shell offers manifold opportunities for ligand-to-metal charge transfer bands. Thus in both CrO_2F_2 and CrO_2Cl_2 [J3] there are at least five transitions below 50 000 cm^{-1} (*6.2 eV*), while in the latter, "very intense" excitations fall at 53 500 and 60 000 cm^{-1} (*vert., 6.63 and 7.44 eV*), Table XXI.A-I. These relate to those of MnO_4^- (52 900 cm^{-1} *vert., 6.56 eV*) and CrO_4^{2-} (55 560 cm^{-1} *vert., 6.888 eV*) in the same region [**II**, 280]. The ions MoO_4^{2-} and WO_4^{2-} will have spectra related to that of CrO_4^{2-}. The reflection spectra of the Ca salts of the MoO_4^{2-} and WO_4^{2-} ions [G33] show several peaks in the 40 000–65 000-cm^{-1} (*5–8-eV*) region but it is not clear which correlates with the intense vacuum-ultraviolet band of the CrO_4^{2-} ion. Intensity measurements would help here. Excitations of the TiO_4^{2-} ion up to 260 000 cm^{-1} (*32 eV*) have been explained in terms of ligand-to-metal and cation-to-anion charge transfer excitations [B22].

The ions MoO_4^{2-} and WO_4^{2-} in turn are isoelectronic with the neutral species RuO_4 and OsO_4 [**II**, 280], the spectra of which have been redetermined and extended [R37]. The spectra of the tetroxy anions [G33] and their vapor-phase counterpart tetroxides however appear to bear no relationship to one another, reflecting the facts that the vapor-phase spectra show a number of Rydberg excitations which do not exist for anions, and that the crystal spectra will show strong effects due to the Madelung potential and crystal band formation which are irrelevant for vapor-phase spectra. Nonetheless, as shown in Table XXI.A-I, the lower bands of RuO_4 and OsO_4 each show constant term values, indicating that the terminating orbital in each case is the $e\pi^*$ MO derived from the nd central-atom orbitals. As is also the case in the MoF_6/WF_6 pair, the $\mathscr{A}(2p, e\pi^*)$-band term values in the second-row complex RuO_4 are considerably larger than those in the third-row complex OsO_4.

The first-row anion $B_5O_{10}H_x^{x-5}$ (x has not been determined definitely) in the $KB_5O_8 \cdot 4H_2O$ crystal transmits to approximately 60 000 cm^{-1} (*7.5 eV*) in an 8 mm. crystal [P5], and at least part if not all of the absorption at this frequency is due to water absorption (Section VII.A). It is likely that boric acid and its alkyl esters also will have high transmission in the vacuum ultraviolet.

XXI.C. Miscellaneous

The two-photon resonant MPI spectrum of dimethyl mercury in a jet-cooled molecular beam uncovers a well-resolved Rydberg excitation in the 51 000−56 000-cm^{-1} *(6.3−6.9-eV)* region [G9]. Its Rydberg nature is revealed by clustering the $Hg(CH_3)_2$ with N_2 in the beam and observing a drastic "solvent" perturbation. The term value of the transition (23 550 cm^{-1}) and the vibronic analysis lead to an upper-state ($6p\sigma_+$, $6p\pi$) configuration of E" symmetry. Other spin-orbit components of the $6p\sigma_+ \rightarrow 6p\pi$ transition have term values as large as 32 500 cm^{-1}. The electron impact spectrum of $Hg(CH_3)_2$ out to 240 000 cm^{-1} *(30 eV)* energy loss [G9] was interpreted in terms of allowed Rydberg excitations to the lowest Rydberg orbital ($6p\pi$) with term values in the range 23 000−34 900 cm^{-1}. Identification of the terminating orbital as "$6p\pi$" is uncertain, since $6p\pi$ should be part of the valence shell rather than Rydberg, whereas 7p is too high to give term values of 23 000−35 000 cm^{-1}. We need more experience with heavy-metal alkyl spectra before anything further can be said.

Spectra of the thiocyanate ion NCS^- in a variety of vacuum-ultraviolet solvents illustrate the extreme solvent shifts and intensity changes which characterize charge-transfer-to-solvent (CTTS) transitions [F24]. Thus, a band at 40 500 cm^{-1} *(vert., 5.02 eV)* in an aqueous solution containing NCS^- appears at 52 450 cm^{-1} *(vert., 6.503 eV)* in hexafluoroisopropanol; a band at 49 380 cm^{-1} *(vert., 6.122 eV)* in acetone solution is shifted to 56 980 cm^{-1} *(vert., 7.064 eV)* in hexafluoroisopropanol, and a band at 56 060 cm^{-1} *(vert., 6.95 eV)* in acetonitrile solution appears at 56 880 cm^{-1} *(vert., 7.05 eV)* in triethyl phosphate. We must conclude that there is more than one acceptor state in a given solvent, for the different CTTS transition frequencies of NCS^- in a given solvent do not reflect the different valence-MO binding energies of the anion [W33].

Lussier *et al.* [L31] have studied the vacuum-ultraviolet spectra of several heavy-metal hexafluoroacetyl acetonates (HFAA) in an effort to find Rydberg excitations in coordination complexes. The intense $\pi \rightarrow \pi^*$ band of the free ligand at 63 700 cm^{-1} *(vert., 7.90 eV)*, Fig. XII.D-3, appears as well at nearly the same frequency in the $M(HFAA)_3$ complexes where M = Al and Sc. The $\pi \rightarrow$ 3s, 3p Rydberg bands of the ligands appear prominently in the 50 000−60 000-cm^{-1} *(6.2−7.4-eV)* region of the Sc complex, but not at all in the Al complex. Introduction of 3d orbital occupation (M = V, Cr, Mn, Fe, Cu) lowers the first ionization potential so that possible 3d → 4s, 4p Rydberg transitions occur at low frequencies, *i.e.*, below 50 000 cm^{-1} (*6.2 eV*) when M = Mn and Fe. Comparison of vapor and solution or matrix spectra would be most useful here in distinguishing between valence charge transfer and Rydberg transitions.

XXI.D. Addendum

A. M. Mamedov, M. A. Osman and L. C. Hajieva, VUV reflectivity of $LiNbO_3$ and $LiTaO_3$ single crystals, *Appl. Phys.* **34A**, 189 (1984).

C. J. Chen and R. M. Osgood, A spectroscopic study of the excited states of dimethylzinc, dimethylcadmium and dimethylmercury, *J. Chem. Phys.* **81**, 327 (1984).

CHAPTER XXII

Biological Molecules

The prognosis for understanding the vacuum-ultraviolet spectra of biological materials is not an optimistic one. These molecules in general are of very high molecular weight, low symmetry, contain a variety of chromophores and are studied in solution where Rydberg excitations are severely perturbed but may not disappear entirely. Though there is considerable activity in the vacuum-ultraviolet spectroscopy of biological materials, it is almost totally in the area of circular dichroism, and is aimed at investigating molecular conformation rather than electronic structure. From the molecular orbital point of view, there is little that one can say regarding such experiments. Consequently, these studies are simply listed in Table XXII-I without comment. The magnitude of the problem here is easily illustrated by the spectrum of the nucleic acid uracil in which there are possible 15 $\pi \rightarrow \pi^*$ and 6 n $\rightarrow \pi^*$ transitions in addition to 63 transitions originating at n and π MO's and terminating at $n = 3$ Rydberg orbitals. In the region up to 72 000 cm^{-1} (*9 eV*), theory predicts one n $\rightarrow \pi^*$ band followed by 7 $\pi \rightarrow \pi^*$ transitions [A25] in fair agreement with electron-impact and optical [H70] studies of uracil. Note however that the theory ignores Rydberg excitations.

TABLE XXII-I

VACUUM ULTRAVIOLET DATA ON BIOLOGICAL MATERIALS

Material	Measurement	Range, cm^{-1}	References
Galactomannans	CD	52 000–72 000	*a*
Poly(galacturonic acid)	CD	42 000–72 000	*b*
Oligosaccharides	CD	42 000–59 000	*c*
Alginates	CD	40 000–72 000	*d*
Dextran	CD	43 000–67 000	*e*
Carageenan	CD	50 000–67 000	*f*
Chitin, chitan	CD	40 000–62 000	*g*
Agarose	CD	48 000–69 000	*h*
Aldopyranosides	CD	50 000–61 000	*i*
Acetylated glucans	CD	40 000–72 000	*j,k*
Hyaluronate saccharides	CD	42 000–56 000	*l*
Glycopeptides	CD	40 000–67 000	*m,n,o*
Oligopeptides	CD	42 000–72 000	*p,q,r,s,t,u,v,w, x,y, z,aa,ab*
Enzymes	Abs.; PE	43 000–91 000	*u,v,ac*
DNA	Abs.; Refl.; CD	16 000–660 000	*ad,ae,af*
Bovine plasma albumin	Abs.; Refl.	16 000–660 000	*ag*
Chloroplasts	Abs.	13 000–177 000	*ah*
Red blood cells	Abs.	17 000–89 000	*ah*

[a]L. A. Buffington, E. S. Stevens, E. R. Morris and D. A. Rees, *Intern. J. Biol. Macromol.* **2**, 199 (1980). [b]J. N. Liang and E. S. Stevens, *Intern. J. Biol. Macromol.* **4**, 316 (1982). [c]C. A. Bush, *in* "Excited States in Organic Chemistry and Biochemistry", B. Pullman and N. Goldblum *eds.*, D. Reidel, Dordrecht, 1977, p. 209. [d]J. N. Liang, E. S. Stevens, S. A. Frangou, E. R. Morris and D. A. Rees, *Intern. J. Biol. Macromol.* **2**, 204 (1980). [e]A. J. Stipanovic, E. S. Stevens and K. Gekko, *Macromol.* **13**, 1471 (1980). [f]J. S. Balcerski, E. S. Pysh, G. C. Chen and J. T. Yang, *J. Amer. Chem. Soc.* **97**, 6274 (1975). [g]L. A. Buffington and E. S. Stevens, *J. Amer. Chem. Soc.* **101**, 5159 (1979). [h]J. N. Liang, E. S. Stevens, E. R. Morris and D. A. Rees, *Biopolymers* **18**, 327 (1979). [i]R. G. Nelson and W. C. Johnson, Jr., *J. Amer. Chem. Soc.* **98**, 4296 (1976). [j]C. A. Bush and S. Ralapati, *ACS Symp. Series* **150**, 293 (1981). [k]A. J. Stipanovic and E. S. Stevens, *Biopolymers* **20**, 1183 (1981). [l]M. K. Cowman, C. A. Bush and E. A. Balazs, *Biopolymers* **22**, 1319 (1983). [m]C. A. Bush, V. K. Dua, S. Ralapati, C. D. Warren, G. Spik, G. Strecker and J. Montreuil, *J. Biol. Chem.* **257**, 8199 (1982). [n]D. Herschlag, E. S. Stevens and

J. E. Gander, *Intern. J. Peptide Protein Res.* **22**, 16 (1983). [o]C. A. Bush, R. E. Feeney, D. T. Osuga, S. Ralapati and Y. Yeh, *Intern. J. Peptide Protein Res.* **17**, 125 (1981). [p]R. T. Coffey, E. S. Stevens, C. Toniolo, and G. M. Bonora, *Makromol. Chem.* **182**, 941 (1981). [q]A. J. Stipanovic and E. S. Stevens, *Biopolymers* **20**, 1565 (1981). [r]D. D. Jenness, C. Sprecker and W. C. Johnson, Jr., *Biopolymers* **15**, 513 (1976). [s]R. T. Coffey, E. S. Stevens, A. Cosani and E. Peggion, *Macromol.* **16**, 1243 (1983). [t]M. A. Young and E. S. Pysh, *J. Amer. Chem. Soc.* **97**, 5100 (1975). [u]S. Brahms and J. Brahms, *J. Mol. Biol.* **138**, 149 (1980). [v]S. Brahms and J. G. Brahms, *J. chim. Phys.* **76**, 841 (1979). [w]J. S. Balcerski, E. S. Pysh, G. M. Bonora and C. Toniolo, *J. Amer. Chem. Soc.* **98**, 3470 (1976). [x]S. K. Brahmachari, V. S. Ananthanarayanan, S. Brahms, J. Brahms, R. S. Rapaka and R. S. Bhatnagar, *Biochem. Biophys. Res. Commun.* **86**, 605 (1979). [y]C. Toniolo, G. M. Bonora, M. Palumbo, E. Peggion and E. Stevens, *Biopolymers* **17**, 1713 (1978). [z]M. M. Kelly, E. S. Pysh, G. M. Bonora and C. Toniolo, *J. Amer. Chem. Soc.* **99**, 3264 (1977). [aa] D. J. Paskowski, E. S. Stevens, G. M. Bonora and C. Toniolo, *Biochem. Biophys. Acta* **535**, 188 (1978). [ab] M. Bakir and E. S. Stevens, *Intern. J. Peptide Protein Res.* **19**, 133 (1982). [ac] S. Iwanami and N. Oda, *Radiat. Res.* **90**, 466 (1982). [ad] I. Földvári, A. Fekete and G. Corradi, *J. Biochem. Biophys. Methods* **5**, 319 (1982). [ae] C. A. Sprecher, W. A. Baase and W. C. Johnson, Jr., *Biopolymers* **18**, 1009 (1979). [af] T. Inagaki, R. N. Hamm, E. T. Arakawa and L. R. Painter, *J. Chem. Phys.* **61**, 4246 (1974). [ag] T. Inagaki, R. N. Hamm, E. T. Arakawa and R. D. Birkhoff, *Biopolymers* **14**, 839 (1975). [ah] M. W. Williams, E. T. Arakawa, R. D. Birkhoff, R. N. Hamm, H. C. Schweinler and R. A. MacRae, *Rad. Research* **61**, 185 (1975).

References

[A1] I. Abbati, L. Braicovich and B. De Michelis, Investigating ultraviolet photoelectron spectroscopy of ice, *Solid State Commun.* **29**, 511 (1979).

[A2] G. D. Abbott, J. Dyke, D. Phillips and M. E. Sime, Triplet-triplet absorption spectra of some benzenoid hydrocarbons in the vapour phase, *J. Chem. Soc., Faraday Trans. II* **78**, 1971 (1982).

[A3] T. Abuain, I. C. Walker and D. F. Dance, The lowest triplet state in ammonia and methylamine detected by electron-impact excitation, *J. Chem. Soc., Faraday Trans. II* **80**, 641 (1984).

[A4] Y. Achiba, K. Sato, K. Shobatake and K. Kimura, The mechanism for photofragmentation of H_2S revealed by multiphoton ionization photoelectron spectroscopy, *J. Chem. Phys.* **77**, 2709 (1982).

[A5] D. B. Adams, Ab initio calculations concerning core Auger shifts in some silicon compounds, *J. Chem. Soc., Faraday Trans. II* **73**, 991 (1977).

[A6] O. V. Agashkin, V. I. Pesterev and V. Z. Gabdrakipov, Spectra of invertomers of piperidine in the far ultraviolet, *Dokl. Akad. Nauk SSSR* **222**, 373 (1975).

[A7] V. N. Akimov, A. S. Vinogradov and T. M. Zimkina, X-ray absorption spectra of water and ammonia molecules, *Opt. Spectrosc.* **53**, 280 (1982).

[A8] M. Allan, Forward electron scattering in benzene: forbidden transitions and excitation functions, *Helv. Chim. Acta* **65**, 2008 (1982).

[A9] M. Allan, Excited states of 1,3-butadiyne by electron energy loss spectroscopy, *J. Chem. Phys.* **80**, 6020 (1984).

[A10] M. Allan, Vibrational and electronic excitation in *p*-benzoquinone by electron impact, *Chem. Phys.* **84**, 311 (1984).

[A11] M. Allan, Electronic structure of the 1,3-butadiyne anion studied by electron transmission and vibrational excitation spectra in the gas phase, *Chem. Phys.* **86**, 303 (1984).

[A12] M. Allan, M. B. Robin and P. A. Snyder, to be published.

[A13] S. D. Allen and O. Schnepp, The circular dichroism spectrum of *S*(+)-*sec*-butyl benzene, *Chem. Phys. Lett.* **29**, 210 (1974).

[A14] S. D. Allen and O. Schnepp, Circular dichroism spectrum of a saturated hydrocarbon, (–) (3*S*:5*S*)-2,2,3,5-tetramethylheptane, *J. Chem. Soc., Chem. Commun.*, 904 (1974).

[A15] S. D. Allen, M. G. Mason, O. Schnepp and P. J. Stevens, The magnetic circular dichroism spectrum of benzene and toluene and the magnetic moment of the $^1E_{1u}$ state, *Chem. Phys. Lett.* **30**, 140 (1975).

[A16] D. D. Altenloh and B. R. Russell, Electric dichroism spectroscopy in the vacuum ultraviolet. Dimethyl sulfide and thiirane, *Chem. Phys. Lett.* **77**, 217 (1981).

[A17] D. D. Altenloh and B. R. Russell, Electric dichroism spectroscopy in the vacuum ultraviolet. Thietane, tetrahydrothiophene and tetrahydrothiopyran, *J. Phys. Chem.* **86**, 1960 (1982).

[A18] D. D. Altenloh, L. R. Ashworth and B. R. Russell, Excited-state dipole moment and polarizability for the 3s Rydberg of ethylene oxide, *J. Phys. Chem.* **87**, 4348 (1983).

[A19] M. Andersen, S. Nir, J. M. Heller, Jr. and L. R. Painter, Optical constants and dispersion equations of lecithin, cholesterol, fucose and chloroform: measurements in vacuum ultraviolet to visible wavelength regions, *Radiat. Res.* **76**, 493 (1978).

[A20] J. Appell, J. Durup, F. C. Fehsenfeld and P. Fournier, Doubly ionized states of some polyatomic molecules studied by double charge transfer spectroscopy, *J. Phys. B: At. Mol. Phys.* **7**, 406 (1974).

[A21] E. T. Arakawa, M. W. Williams, J. C. Ashley and L. R. Painter, The optical properties of Kapton: measurements and applications, *J. Appl. Phys.* **52**, 3579 (1981).

[A22] T. Ari and J. B. Hasted, Electron energy-loss spectra of acetaldehyde and formic acid, *Chem. Phys. Lett.* **85**, 153 (1982).

[A23] D. R. Armstrong, J. Jamieson and P. G. Perkins, Systematic *ab initio* calculations on one- and three-dimensional polymers, *Theoret. Chim. Acta* **50**, 193 (1978).

[A24] L. Armstrong, Jr., Numerology, hydrogenic levels and the ordering of excited states in one-electron atoms, *Phys. Rev.* **25A**, 1794 (1982).

[A25] L. Åsbrink, C. Fridh and E. Lindholm, The semiempirical method HAM/3, applied to uracil, *Tetrahedron Lett.* **52**, 4627 (1977).

[A26] L. Åsbrink, C. Fridh and E. Lindholm, Electronic structure of TCNQ, studied with HAM/3, *Int. J. Quantum Chem.* **13**, 331 (1978).

[A27] L. Åsbrink, C. Fridh and E. Lindholm, Ionization, excitation and electron affinity of naphthalene, studied with HAM/3, *Z. Naturforsch.* **33A**, 172 (1978).

[A28] L. Åsbrink, C. Fridh and E. Lindholm, Valence excitation of linear molecules. I. Excitation and UV spectra of N_2, CO, acetylene and HCN, *Chem. Phys.* **27**, 159 (1978).

[A29] L. Åsbrink, C. Fridh and E. Lindholm, Interpretation of electronic spectra. I. Spectra of furan, pyrrole and cyclopentadiene, studied with HAM/3, *J. Electron Spectrosc. Related Phenomena* **16**, 65 (1979).

[A30] M. N. R. Ashfold and J. P. Simons, Vacuum ultraviolet photodissociation spectroscopy of CH_3CN, CD_3CN, CF_3CN and CH_3NC, *J. Chem. Soc., Faraday II* **7**, 1263 (1978).

[A31] M. N. R. Ashfold, M. T. Macpherson and J. P. Simons, Photochemistry and spectroscopy of simple polyatomic molecules in the vacuum ultraviolet, *Top. Current Chem.* **86**, 1 (1979).

[A32] M. N. R. Ashfold and R. N. Dixon, Multiphoton ionization spectroscopy of H_2S: A reinvestigation of the $^1B_1 - {}^1A_1$ band at 139.1 nm, *Chem. Phys. Lett.* **93**, 5 (1982).

[A33] M. N. R. Ashfold, J. M. Bayley and R. N. Dixon, A new electronic state of water revealed by gas phase multiphoton ionization spectroscopy, *J. Chem. Phys.* **79**, 4080 (1983).

[A34] M. N. R. Ashfold, J. M. Bayley and R. N. Dixon, Molecular predissociation dynamics revealed through multiphoton ionization spectroscopy. I. The $\tilde{C}$ $^1B_1$ states of H_2O and D_2O, *Chem. Phys.* **84**, 35 (1984).

[A35] J. S. Attrey, R. D. Birkhoff and L. R. Painter, Optical and dielectric functions of liquid methylated benzenes in the 2- to 10-eV spectral region, *J. Appl. Phys.* **53**, 9005 (1982).

[A36] P. Avouris, A. R. Rossi and A. C. Albrecht, Electronic band-shape calculations on ammonia, *J. Chem. Phys.* **74**, 5516 (1981).

[A37] P. Avouris and J. E. Demuth, Electronic excitations of benzene, pyridine and pyrazine adsorbed on Ag(111), *J. Chem. Phys.* **75**, 4783 (1981).

[A38] R. Azria and G. J. Schulz, Vibrational and triplet excitation by electron impact in benzene, *J. Chem. Phys.* **62**, 573 (1975).

[A39] R. Azria, Y. Le Coat, G. Lefevre and D. Simon, Dissociative electron attachment on H_2S: energy and angular distributions of H^- ions, *J. Phys. B: At. Mol. Phys.* **12**, 679 (1979).

[B1] C. Backx, G. R. Wight, R. R. Tol and M. J. van der Wiel, Electron-electron coincidence measurements of CH_4, *J. Phys. B: At. Mol. Phys.* **8**, 3007 (1975).

[B2] C. Backx and M. J. van der Wiel, Electron-ion coincidence measurements of CH_4, *J. Phys. B: At. Mol. Phys.* **8**, 3020 (1975).

[B3] G. Bader, L. Caron and L. Sanche, Electron-exciton complex formation in organic solids, *Solid State Commun.* **38**, 849 (1981).

[B4] M. A. Baig, J. P. Connerade, J. Dagata and S. P. McGlynn, Quasi-atomic Rydberg states of a complex molecule: CH_3I, *J. Phys. B: At. Mol. Phys.* **14**, L25 (1981).

[B5] M. A. Baig, J. Hormes, J. P. Connerade and S. P. McGlynn, Rydberg transitions in the H_2S molecule, *J. Phys. B: At. Mol. Phys.* **14**, L725 (1981).

[B6] M. A. Baig, J. P. Connerade and J. Hormes, Autoionization resonances in the $4p\pi$ spectrum of methyl bromide, *J. Phys. B: At. Mol. Phys.* **15**, L5 (1982).

[B7] N. L. Baker and B. R. Russell, The first s-Rydberg transitions of ethyl bromide and ethyl iodide as related to the corresponding methyl halides, *J. Mol. Spectrosc.* **69**, 211 (1978).

[B8] S. Baldwin-Boisclair and D. D. Shillady, An improved CNDO/S study of the circular dichroism of 1-methyl indan in the 275-170 nm region, *Chem. Phys. Lett.* **58**, 405 (1978).

[B9] T. Bally and E. Haselbach, *Tris*-(methylidene)-cyclopropane ("[3] radialene") Part 2. Electronic states of the molecular cation and revised UV absorption spectrum of the parent neutral, *Helv. Chim. Acta* **61**, 754 (1978).

[B10] M. S. Banna and D. A. Shirley, Molecular photoelectron spectroscopy at 132.3 eV: N_2, CO, C_2H_4 and O_2, *J. Electron Spectrosc. Related Phenomena* **8**, 255 (1976).

[B11] M. K. Barbarez, D. K. Das Gupta and D. Hayward, On the mechanisms of characteristic electron energy loss spectra of polymers, *J. Phys. D: Appl. Phys.* **10**, 1789 (1977).

[B12] R. L. Barinskii and I. M. Kulikova, Rydberg constants in the K-absorption spectrum of sulfur in the SF_6 molecule, *J. Struct. Chem.* **14**, 335 (1973).

[B13] R. L. Barinskii and I. M. Kulikova, Investigation of the excited states of gas molecules using K-absorption spectra, *Bull. Acad. Sci. USSR, Phys. Ser.* **38**, 444 (1974).

[B14] B. Baron, D. Hoover and F. Williams, Vacuum ultraviolet photoelectric emission from amorphous ice, *J. Chem. Phys.* **68**, 1197 (1978).

[B15] A. Barth, R. J. Buenker, S. D. Peyerimhoff and W. Butscher, Theoretical study of the core-ionized and various core-excited and shake-up states of acetylene and ethylene by ab initio MRD-CI methods, *Chem. Phys.* **46**, 149 (1980).

[B16] H. Basch, Satellite bands in the x-ray photoelectron spectrum of formaldehyde, *Chem. Phys.* **10**, 157 (1975).

[B17] H. Basch, Satellite bands and the ion states of formamide, *Chem. Phys. Lett.* **37**, 447 (1976).

[B18] E. Bastian, P. Potzinger, A. Ritter, H.-P. Schuchmann, C. von Sonntag and G. Weddle, The direct photolysis of tetramethylsilane in the gas and liquid phases, *Ber. Bunsenges. Phys. Chem.* **84**, 56 (1980).

[B19] P. K. Basu, U. Chandra Singh, K. N. Tantry, V. Ramamurthy and C. N. R. Rao, Nonbonded interactions in 2,2,4,4-tetramethyl-1,3-cyclobutanedithione and 2,2,4,4-tetramethyl-3-thio-1,3-cyclobutanedione, *J. Mol. Struct. (Theochem)* **76**, 237 (1981).

[B20] C. Batich, P. Bischof and E. Heilbronner, The photoelectron spectra of cyclooctatetraene and its hydrogenated derivatives, *J. Electron Spectrosc. Related Phenomena* **1**, 333 (1972/73).

[B21] C. Batich, E. Heilbronner, E. Rommel, M. F. Semmelhack and J. S. Foos, Equivalence of the energy gaps $\Delta I(1,2)$ and $\Delta E(1,2)$ between corresponding bands in the photoelectron (I) and electronic absorption (E) spectra of spiro[4.4]nonatetraene. An amusing consequence of spiro conjugation, *J. Amer. Chem. Soc.* **96**, 7662 (1974).

[B22] D. Bäuerle and W. Braun, Vacuum ultraviolet reflectivity and band structure of $SrTiO_3$ and $BaTiO_3$, *Z. Phys.* **B29**, 179 (1978).

[B23] M. Bavia, C. Zauli and L. Fusina, Rydberg states in selenophene, *Mol. Phys.* **30**, 1289 (1975).

[B24] M. Bavia, F. Bertinelli, C. Taliani and C. Zauli, The electronic spectrum of pyrrole in the vapor and crystal, *Mol. Phys.* **31**, 479 (1976).

[B25] M. Beez, G. Bieri, H. Bock and E. Heilbronner, The ionization potentials of butadiene, hexatriene and their methyl derivatives: evidence for through space interaction between double bond π-orbitals and nonbonded pseudo-π orbitals of methyl groups?, *Helv. Chim. Acta* **56**, 1028 (1973).

[B26] S. Bell, T. L. Ng and A. D. Walsh, Vacuum ultraviolet spectra of formic and acetic acids, *J. Chem. Soc., Faraday Trans. II* **71**, 393 (1975).

[B27] S. Bell, Electronic structure and spectrum of cyanogen, *Chem. Phys. Lett.* **67**, 498 (1979).

[B28] T. N. Bell, K. A. Perkins and P. G. Perkins, Heats of formation and dissociation of methylsilanes and chlorosilanes and derived radicals, *J. Chem. Soc., Faraday Trans. I* **77**, 1779 (1981).

[B29] H. G. Bennewitz, W. H. E. Schwarz and K. H. Thunemann, Electronic excitation in reactive collisions of Cs atoms with nonmetal fluorides, *Chem. Phys.* **52**, 227 (1980).

[B30] C. Benoit and J. A. Horsley, Electronic states of $C_2H_4^{++}$ from double charge transfer spectroscopy and from SCF-CI calculations, *Mol. Phys.* **30**, 557 (1975).

[B31] J. O. Berg and G. W. Robinson, An approach to the understanding of radiation chemistry in the condensed phase, *Chem. Phys. Lett.* **34**, 211 (1975).

[B32] J. O. Berg, D. H. Parker and M. A. El-Sayed, Assignment of the lowest ionization potentials in pyridine and pyrazine by multiphoton ionization spectroscopy, *Chem. Phys. Lett.* **56**, 411 (1978).

[B33] H. Bergmann and H. Bock, Photoelectron spectra and molecular properties XLVI. Nitroso compounds – electron-rich molecules, *Z. Naturforsch.* **30B**, 629 (1975).

[B34] J. Berkowitz, "Photoabsorption, Photoionization and Photoelectron Spectroscopy," Academic Press, New York, 1979.

[B35] M. J. Berry, Chloroethylene photochemical lasers: vibrational energy content of the HCl molecular elimination products, *J. Chem. Phys.* **61**, 3114 (1974).

[B36] F. Bertinelli, A. Brillante, P. Palmieri and C. Taliani, Valence and Rydberg excited states of aniline. An MO/CI study and the UV crystal spectra of aniline and N,N-dimethylaniline, *Gazz. Chim. Ital.* **110**, 321 (1980).

[B37] C. Bertucci, R. Lazzaroni, P. Salvadori and W. C. Johnson, Jr., Far U.V. circular dichroism spectra of (*S*)-(+)-1,2,2-trimethylpropyl ethyl ether: solvent effects, *J. Chem. Soc., Chem. Commun.*, 590 (1981).

[B38] C. Bertucci, E. Chiellini, P. Salvadori and W. C. Johnson, Jr., Far-ultraviolet circular dichroism spectra of isotactic optically active poly(alkyl vinyl ethers) and their low molecular weight models, *Macromol.* **16**, 507 (1983).

[B39] C. Bertucci, C. Rosini, R. Lazzaroni, D. Pini and P. Salvadori, Far ultraviolet circular dichroism spectra of a simple vinyl ether chromophore, *J. Chem. Res. (S)*, 236 (1983).

[B40] M. Bettendorf, R. J. Buenker, S. D. Peyerimhoff and J. Römelt, Ab initio CI calculation of the effects of Rydberg-valence mixing in the electronic spectrum of the HF molecule, *Z. Phys.* **304A**, 125 (1982).

[B41] R. W. Bigelow, Further application of the CNDO/S method to the photoelectron and optical spectra of benzene, *J. Chem. Phys.* **66**, 4241 (1977).

[B42] R. W. Bigelow, An assessment of the electronic structure of alkylbenzenes, *J. Chem. Phys.* **70**, 2315 (1979).

[B43] R. W. Bigelow, Higher-lying singlet states of poly-*N*-vinylcarbazole: a CNDO/S study, *J. Chem. Phys.* **71**, 1037 (1979).

[B44] R. W. Bigelow and H.-J. Freund, The core-hole excitation spectrum of benzene: a symmetry-adapted CNDO/S equivalent-core study including "spin-symmetry breaking" configurations, *J. Chem. Phys.* **77**, 5552 (1982).

[B45] R. D. Birkhoff, R. N. Hamm, M. W. Williams, E. T. Arakawa and L. R. Painter, Optical properties of liquids in the vacuum UV, *in* "Chemical Spectroscopy and Photochemistry in the Vacuum-Ultraviolet," C. Sandorfy, P. J. Ausloos and M. B. Robin, *eds.*, D. Reidel, Dordrecht, 129 (1974).

[B46] R. D. Birkhoff, L. R. Painter and J. M. Heller, Jr., Optical and dielectric functions of liquid glycerol from gas photoionization measurements, *J. Chem. Phys.* **69**, 4185 (1978).

[B47] R. D. Birkhoff, J. M. Heller, Jr., H. H. Hubbell, Jr. and L. R. Painter, Optical and dielectric functions of liquid hexamethylphosphoric triamide between 2 and 25 eV, *J. Chem. Phys.* **74**, 200 (1981).

[B48] R. D. Birkhoff, J. M. Heller, Jr., L. R. Painter, J. C. Ashley and H. H. Hubbell, Jr., Photoemission and electron mean free paths in liquid formamide in the vacuum UV, *J. Chem. Phys.* **76**, 5208 (1982).

[B49] M. J. Blandamer and M. F. Fox, Theory and applications of charge-transfer-to-solvent spectra, *Chem. Rev.* **70**, 59 (1969).

[B50] H. Bock, K. Wittel, M. Veith and N. Wiberg, Photoelectron spectra and molecular properties. XLIV. On the blue color of bis(trimethylsilyl) diimine, *J. Amer. Chem. Soc.* **98**, 109 (1976).

[B51] A. Bolovinos, J. Philis, E. Pantos, P. Tsekeris and G. Andritsopoulos, The methylbenzenes *vis-a-vis* benzene. Comparison of their spectra in the Rydberg series region, *J. Chem. Phys.* **75**, 4343 (1981).

[B52] A. Bolovinos, J. Philis, E. Pantos, P. Tsekeris and G. Andritsopoulos, The methylbenzenes *vis-a-vis* benzene. Comparison of their spectra in the valence-shell transitions region, *J. Mol. Spectrosc.* **94**, 55 (1982).

[B53] A. Bolovinos, P. Tsekeris, J. Philis, E. Pantos and G. Andritsopoulos, Absolute VUV absorption spectra of some gaseous azabenzenes, *J. Mol. Spectrosc.* **103**, 240 (1984).

[B54] M. J. W. Boness, I. W. Larkin, J. B. Hasted and L. Moore, Virtual negative ion spectra of hydrocarbons, *Chem. Phys. Lett.* **1**, 292 (1967).

[B55] J. Bordas, A. J. Grant, H. P. Hughes, A. Jacobsson, H. Kamimura, F. A. Levy, K. Nakao, Y. Natsume and A. D. Yoffe, The band structure and optical properties of sulfur nitrogen polymer: II. Optical properties from 0.5 to 27 eV, *J. Phys. C: Solid State Phys.* **9**, L277 (1976).

[B56] F. K. Botz and R. E. Glick, Methane temporary negative ion resonances, *Chem. Phys. Lett.* **33**, 279 (1975).

[B57] D. Bouler, W. S. Felps, J. Lewis, R. V. Nauman and S. P. McGlynn, Molecular Rydberg states. Azulene and 4,6,8-trimethylazulene, *Chem. Phys. Lett.* **67** 420 (1979).

[B58] T. D. Bouman and A. E. Hansen, *Ab initio* calculations of oscillator and rotatory strengths in the random-phase approximation: twisted mono-olefins, *J. Chem. Phys.* *66*, 3460 (1977).

[B59] T. D. Bouman and A. E. Hansen, Ab initio calculations of oscillator and rotatory strengths in the random-phase approximation: planar and twisted butadiene, *Chem. Phys. Lett.* **53**, 160 (1978).

[B60] E. Boursey and J.-Y. Roncin, High resolution spectroscopy of the electronic excitation of NO trapped in rare gas matrices. The B $^2\Pi \leftarrow$ X $^2\Pi$ system, *J. Mol. Spectrosc.* *55*, 31 (1975).

[B61] A. G. Briggs, Rydberg term table - a computer version, *Spectrochim. Acta* **37B,** 1079 (1982).

[B62] P. Brint, K. Wittel, P. Hochmann, W. S. Felps and S. P. McGlynn, Molecular Rydberg transitions. 3. A linear combination of Rydberg orbitals (LCRO) model for the two-chromophoric system 2,2,4,4-tetramethylcyclobutane-1,3-dione (TMCBD), *J. Amer. Chem. Soc.* **98**, 7980 (1976).

[B63] C. E. Brion, S. Daviel, R. Sodhi and A. P. Hitchcock, Recent advances in inner-shell excitation of free molecules by electron energy loss spectroscopy, *Conference Proceedings, X ray and Atomic Inner-Shell Physics* **94**, 429 (1982).

[B64] M. Brith-Lindner and S. D. Allen, The near vacuum UV magnetic circular dichroism of ethylene in the gas phase, *Chem. Phys. Lett.* **47**, 32 (1977).

[B65] E. Broclawik, J. Mrozek and V. H. Smith, Jr., A quantum chemical investigation of the ammonium radical. I. SCF-Xα-SW calculations of the electronic structure and Rydberg spectra, *Chem. Phys.* **66**, 417 (1982).

[B66] F. C. Brown, R. Z. Bachrach and A. Bianconi, Fine structure above the carbon K edge in methane and in the fluoromethanes, *Chem. Phys. Lett.* **54**, 425 (1978).

[B67] R. D. Brown, P. J. Domaille and J. E. Kent, The experimental electronic and vibrational spectra of fulvene, *Aust. J. Chem.* **23**, 1707 (1970).

[B68] R. S. Brown, A photoelectron investigation of the peroxide bond, *Can. J. Chem.*, **53**, 3439 (1975).

[B69] A. R. Browne, A. F. Drake, F. R. Kearney, S. F. Mason and L. A. Paquette, Electronic absorption and circular dichroism spectra of the perturbed coplanar *cis*-diene chromophore in deuterium and methyl substituted 7,7-dimethylbicyclo[8.1.1]octa-2,4-dienes, *J. Amer. Chem. Soc.* **105**, 6123 (1983).

[B70] P. J. Bruna, S. D. Peyerimhoff, R. J. Buenker and P. Rosmus, Non-empirical SCF and CI study of the ground and excited states of thioformaldehyde, *Chem. Phys.* **3**, 35 (1974).

[B71] P. J. Bruna, R. J. Buenker and S. D. Peyerimhoff, Theoretical prediction of the electronic spectrum of thioacetone, *Chem. Phys.* **22**, 375 (1977).

[B72] P. J. Bruna, Comparison between some simple carbonyl and thiocarbonyl compounds using combined *ab initio* methods, *Prog. Theoret. Org. Chem.* **2**, 401 (1977).

[B73] A. E. Bruno, D. J. Clouthier, P. G. Mezey and R. P. Steer, Rydberg and valence-shell transitions in the quartz ultraviolet spectra of aliphatic thiones, *Can. J. Chem.* **59**, 952 (1981).

[B74] R. J. Buenker and S. D. Peyerimhoff, Calculations on the electronic spectrum of water, *Chem. Phys. Lett.* **29**, 253 (1974).

[B75] R. J. Buenker and S. D. Peyerimhoff, Mixed valence-Rydberg states, *Chem. Phys. Lett.* **36**, 415 (1975).

[B76] R. J. Buenker and S. D. Peyerimhoff, Ab initio calculations on the electronic spectrum of ethane, *Chem. Phys.* **8**, 56 (1975).

[B77] R. J. Buenker, S. Shih and S. D. Peyerimhoff, All-valence-electron CI treatment of the electronic spectrum of *trans*-butadiene, *Chem. Phys. Lett.* **44**, 385 (1976).

[B78] R. J. Buenker and S. D. Peyerimhoff, All valence-electron configuration mixing calculations for the characterization of the $^1(\pi, \pi^*)$ states of ethylene, *Chem. Phys.* **9**, 75 (1976).

[B79] R. J. Buenker, S. D. Peyerimhoff and S.-K. Shih, Comment on the role of configuration selection methods in describing the V state of ethylene, *J. Chem. Phys.* **69**, 3882 (1978).

[B80] R. J. Buenker, S.-K. Shih and S. D. Peyerimhoff, An MRD-CI study of the vertical $^1(\pi, \pi^*)$ V-N transition of ethylene using an AO basis with optimized Rydberg $nd\pi$ species and two separate carbon d polarization functions, *Chem. Phys.* **36**, 97 (1979).

[B81] R. J. Buenker, S. D. Peyerimhoff and S.-K. Shih, Ab initio study of the spatial extension of the ethylene V state, *Chem. Phys. Lett.* **69**, 7 (1980).

[B82] P. D. Burrow, J. A. Michejda and K. D. Jordan, Experimental study of the negative ion states of styrene. A test of the pairing theorem, *J. Amer. Chem. Soc.* **98**, 6392 (1976).

[B83] P. D. Burrow, A. Modelli, N. S. Chiu and K. D. Jordan, Temporary Σ and Π anions of the chloroethylenes and chlorofluoroethylenes, *Chem. Phys. Lett.* **82**, 270 (1981).

[B84] P. D. Burrow and K. D. Jordan, Electron transmission spectroscopy of 1,3,5 hexatriene: isomeric differences in π^* orbital energies, *J. Amer. Chem. Soc.* **104**, 5247 (1982).

[B85] P. D. Burrow, A. Modelli, N. S. Chiu and K. D. Jordan, Temporary negative ions in the chloromethanes, $CHCl_2F$ and CCl_2F_2: characterization of the σ^* orbitals, *J. Chem. Phys.* **77**, 2699 (1982).

[B86] P. G. Burton, S. D. Peyerimhoff and R. J. Buenker, Theoretical studies of the electronic spectrum of thioformaldehyde, *Chem. Phys.* **73**, 83 (1982).

[B87] W. Butscher and K.-H. Thunemann, A nonempirical SCF and CI study of the electronic spectrum of pyrrole, *Chem. Phys. Lett.* **57**, 224 (1978).

[C1] J. W. Caldwell and M. S. Gordon, Excited states and photochemistry of saturated molecules. Minimal plus Rydberg basis set calculations on the vertical spectra of CH_4, C_2H_6, C_3H_8 and n-C_4H_{10}, *Chem. Phys. Lett.* **59**, 403 (1978).

[C2] J. W. Caldwell and M. S. Gordon, Excited states and photochemistry of saturated molecules. The vibrational structure in the electronic spectrum of ethane, *J. Mol. Spectrosc.* **96**, 383 (1982).

[C3] J. W. Caldwell and M. S. Gordon, Excited states and photochemistry of saturated molecules. 11. Potential energy surfaces in low-lying states of ethane, *J. Phys. Chem.* **86**, 4307 (1982).

[C4] L. S. Caputi and L. Papagno, Plasmon excitation in graphite by electron energy loss, *Phys. Lett.* **93A**, 417 (1983).

[C5] T. A. Carlson, W. B. Dress, F. A. Grimm and J. S. Haggerty, Study of satellite structure found in the photoelectron spectra of gaseous alkenes, *J. Electron Spectrosc. Related Phenomena* **10,** 147 (1977).

[C6] T. A. Carlson, M. O. Krause, F. A. Grimm, P. Keller and J. W. Taylor, Angle-resolved photoelectron spectroscopy of CCl_4: the Cooper minimum in molecules, *J. Chem. Phys.* **77,** 5340 (1982).

[C7] D. C. Cartwright, S. Trajmar, A. Chutjian and S. Srivastava, Cross sections for electron impact excitation of electronic states in UF_6 at incident electron energies of 10, 20 and 40 eV, *J. Chem. Phys.* **79,** 5483 (1983).

[C8] A. J. Caruso, Mass absorption coefficients for polypropylene and paralene C between 8.34 Å and 452 Å, *Appl. Opt.* **13,** 1744 (1974).

[C9] G. C. Causley and B. R. Russell, Vacuum ultraviolet absorption spectra of the bromomethanes, *J. Chem. Phys.* **62,** 848 (1975).

[C10] G. C. Causley and B. R. Russell, Vacuum ultraviolet absorption spectra of dichlorosilane, dichloromethylsilane and dichlorodimethylsilane, *J. Electron Spectrosc. Related Phemonena* **8,** 71 (1976).

[C11] G. C. Causley, J. B. Clark and B. R. Russell, The vacuum ultraviolet spectrum of bromosilane, *Chem. Phys. Lett.* **38,** 602 (1976).

[C12] G. C. Causley and B. R. Russell, The vacuum ultraviolet absorption spectra of the Group IVA tetrachlorides, *J. Electron Spectrosc. Related Phenomena* **11,** 383 (1977).

[C13] G. C. Causley and B. R. Russell, Electric dipole moment and polarizability of the 1749 Å, $^1B_2$ excited state of formaldehyde, *J. Chem. Phys.* **68,** 3797 (1978).

[C14] G. C. Causley and B. R. Russell, Electric dichroism spectroscopy in the vacuum ultraviolet. 2. Formaldehyde, acetaldehyde and acetone, *J. Amer. Chem. Soc.* **101,** 5573 (1979).

[C15] G. C. Causley and B. R. Russell, Electric dichroism spectroscopy in the vacuum ultraviolet. I. Cyclobutanone, cyclopentanone, cyclohexanone and cycloheptanone, *J. Chem. Phys.* **72,** 2623 (1980).

[C16] L. S. Cederbaum, J. Schirmer, W. Domcke and W. von Niessen, Complete breakdown of the quasiparticle picture for inner valence electrons, *J. Phys. B: At. Mol. Phys.* **10,** L549 (1977).

[C17] R. J. Celotta and R. H. Huebner, Electron impact spectroscopy: an overview of the low-energy aspects, *in* "Electron Spectroscopy: Theory, Techniques and Applications," Vol. 3, C. R. Brundle and A. D. Baker, *eds.*, Academic, New York, 41 (1979).

[C18] D. Chadwick and A. Katrib, Photoelectron spectra of acetaldehyde and acetyl halides, *J. Electron Spectrosc. Related Phenomena* **3,** 39 (1974).

[C19] K. W. Chang and W. R. M. Graham, Vacuum UV spectra of photolyzed C_2H_2 in solid Ar at 8 K, *J. Chem. Phys.* **76,** 5238 (1982).

[C20] F. T. Chau and L. Karlsson, Vibronic interaction in molecules and ions. I. The first-order Jahn-Teller interaction in doubly degenerate electronic states of C_{3v} type molecules, *Phys. Scr.* **16,** 248 (1977).

[C21] N. S. Chiu, P. D. Burrow and K. D. Jordan, Temporary anions of the fluoroethylenes, *Chem. Phys. Lett.* **68,** 121 (1979).

[C22] L. G. Christophorou, D. L. McCorkle and J. G. Carter, Compound-negative-ion-resonant states and threshold-electron excitation spectra of monosubstituted benzene derivatives, *J. Chem. Phys.* **60,** 3779 (1974).

[C23] L. G. Christophorou, Negative ions of polyatomic molecules, *Environ. Health Perspectives* **36,** 3 (1980).

[C24] A. Chutjian, Electron impact excitation of the low-lying electronic states of formaldehyde, *J. Chem. Phys.* **61,** 4279 (1974).

[C25] A. Chutjian, R. I. Hall and S. Trajmar, Electron-impact excitation of H_2O and D_2O at various scattering angles and impact energies in the energy-loss range 4.2 – 12 eV, *J. Chem. Phys.* **63**, 892 (1975).

[C26] A. Chutjian, S. K. Srivastava, S. Trajmar, W. Williams and D. C. Cartwright, Electron-impact excitation of UF_6 at an electron energy of 20 eV in the energy-loss range of 0–10 eV, *J. Chem. Phys.* **64**, 4791 (1976).

[C27] D. T. Clark and D. B. Adams, Some aspects of shake-up phenomena in pyridine and perfluoropyridine, *J. Electron Spectrosc. Related Phenomena* **7**, 401 (1975).

[C28] J. B. Clark and B. R. Russell, Vacuum ultraviolet absorption spectra of several Group IVA tetrabromides, *J. Electron Spectrosc. Related Phenomena* **11**, 371 (1977).

[C29] P. A. Clark and L. W. Pickett, The vacuum ultraviolet spectra of ethyleneimine and trimethyleneimine, *J. Chem. Phys.* **64**, 2062 (1976).

[C30] J. T. Clarke, H. W. Moos and P. D. Feldman, The far-ultraviolet spectra and geometric albedos of Jupiter and Saturn, *Astrophys. J.* **255**, 806 (1982).

[C31] D. J. Clouthier, A. R. Knight, R. P. Steer, R. H. Judge and D. C. Moule, The $\tilde{B}\,(^1A') \leftarrow \tilde{X}\,(^1A')$ spectrum of ClFCS, *J. Mol. Spectrosc.* **83**, 148 (1980).

[C32] M. J. Coggiola, O. A. Mosher, W. M. Flicker and A. Kuppermann, Electronic spectroscopy of the fluoroethylenes by electron impact, *Chem. Phys. Lett.* **27**, 14 (1974).

[C33] M. J. Coggiola, W. M. Flicker, O. A. Mosher and A. Kuppermann, Electron-impact spectroscopy of the fluoroethylenes, *J. Chem. Phys.* **65**, 2655 (1976).

[C34] D. Cohen, M. Levi, B. S. Green, R. Arad-Yellin, H. Basch and A. Gedanken, Excited electronic states of ethylene sulfide: CD study of (+)-(*R*,*R*)-2,3-dimethylthiirane, *J. Phys. Chem.* **87**, 4585 (1983).

[C35] D. Cohen, M. Levi, H. Basch and A. Gedanken, Excited electronic states of optically active substituted ethylene oxides: (–)-*S*-2-methyloxirane and (–)(*S*,*S*)-2,3-dimethyloxirane, *J. Amer. Chem. Soc.* **105**, 1738 (1983).

[C36] B. E. Cole and R. N. Dexter, Photo-absorption cross sections for chlorinated methanes and ethanes between 46 and 100 Å, *J. Quant. Spectrosc. Radiat. Transfer* **19**, 303 (1978).

[C37] R. Colin, M. Herman and I. Kopp, Renner-Teller interactions in the acetylene molecule, *Mol. Phys.* **37**, 1397 (1979).

[C38] R. Colle, R. Montagnani, P. Riani and O. Salvetti, Calculation of some electronic excited states of formaldehyde; remarks on the $\pi \rightarrow \pi^*$ transition, *Theoret. Chim. Acta* **49**, 37 (1978).

[C39] J. P. Connerade, The non-Rydberg spectroscopy of atoms, *Contemp. Phys.* **19**, 415 (1978).

[C40] J. P. Connerade, M. A. Baig, S. P. McGlynn and W. R. S. Garton, Rovibronic structure of Rydberg and non-Rydberg states of H_2O and D_2O, *J. Phys. B: At. Mol. Phys.* **13**, L705 (1980).

[C41] J. P. Connerade, M. A. Baig and S. P. McGlynn, Autoionization in polyatomic molecules, *J. Phys. B: At. Mol. Phys.* **14**, L67 (1981).

[C42] J. P. Connerade, Synchrotron spectroscopy of the giant resonances in the lanthanides and actinides and its relevance to valence changes, *J. Less Common Metals* **93**, 171 (1983).

[C43] R. E. Connors, J. L. Roebber and K. Weiss, Vacuum ultraviolet spectroscopy of cyanogen and cyanoacetylenes, *J. Chem. Phys.* **60**, 5011 (1974).

[C44] J. B. Coon, C. E. Jones and C. E. Blount, Ultraviolet absorption spectra of ammonia and deuterated ammonia in solid argon, *Spectrochim. Acta* **36A**, 439 (1980).

[C45] C. D. Cooper, A. D. Williamson, J. C. Miller and R. N. Compton, Resonantly enhanced multiphoton ionization of pyrrole, N-methyl pyrrole and furan, *J. Chem. Phys.* **73**, 1527 (1980).

[C46] D. O. Cowan, R. Gleiter, J. A. Hashmall, E. Heilbronner and V. Hornung, Interaction between the orbitals of lone pair electrons in dicarbonyl compounds, *Angew. Chem. Int. Ed. Engl.* **10**, 401 (1971).

[C47] D. P. Craig, P. J. Stiles, P. Palmieri and C. Zauli, Electronic circular dichroism of twisted [4.4]-spirononatetraene, *J. Chem. Soc., Faraday Trans. II* **75**, 97 (1979).

[C48] E. M. Custer and W. T. Simpson, Franck-Condon principle as involving torsional oscillations: height of the potential barrier to internal rotation in an electronically excited state of ethane, *J. Chem. Phys.* **60**, 2012 (1974).

[C49] T. Cvitaš, H. Güsten and L. Klasinc, Photoelectron spectra of chlorofluoromethanes, *J. Chem. Phys.* **67**, 2687 (1977).

[D1] J. A. Dagata, G. L. Findley, S. P. McGlynn, J. P. Connerade and M. A. Baig, Molecular Rydberg transitions. Multichannel approaches to electronic states: CH_3I, *Phys. Rev.* **24A**, 2485 (1981).

[D2] D. F. Dance and I. C. Walker, Threshold electron energy-loss spectra for simple alkynes, *J. Chem. Soc. Faraday* **70**, 1426 (1974).

[D3] D. F. Dance and I. C. Walker, Electronic energy loss spectrum of ethylene, *J. Chem. Soc. Faraday Trans. II* **71**, 1903 (1975).

[D4] A. Dargelos and C. Sandorfy, The photoelectron and far-ultraviolet absorption spectra of simple oximes, *J. Chem. Phys.* **67**, 3011 (1977).

[D5] F. E. Darling and J. P. Dahl, The electronic structure of WF_6 elucidated by SCF-Xα-SW calculations, *Chem. Phys.* **20**, 129 (1977).

[D6] P. Dauber, M. Brith, E. Huler and A. Warshel, The vibronic structure of crystalline ethylene, *Chem. Phys.* **7**, 108 (1975).

[D7] P. Dauber and M. Brith, A study of the electronic spectrum of halogen derivatives of ethylene, *Chem. Phys.* **11**, 143 (1975).

[D8] A. Davidsson and B. Nordén, A new electronic transition in the polarized spectrum of dimethyl aniline, *Chem. Phys. Lett.* **28**, 39 (1974).

[D9] J. L. Dehmer and D. Dill, Molecular effects on inner-shell photoabsorption. K-shell spectrum of N_2, *J. Chem. Phys.* **65**, 5327 (1976).

[D10] J. L. Dehmer, J. Siegel and D. Dill, Shape resonances in e-SF_6 scattering, *J. Chem. Phys.* **69**, 5205 (1978).

[D11] J. L. Dehmer, A. C. Parr, S. Wallace and D. Dill, Photoelectron branching ratios and angular distributions for the valence levels of SF_6 in the range $16 \leqslant h\nu \leqslant 30$ eV, *Phys. Rev.* **26A**, 3283 (1982).

[D12] P. M. Dehmer, Rydberg states of van der Waals molecules - a comparison with Rydberg states of atoms and chemically bonded species, *Comments At. Mol. Phys.* **13**, 205 (1983).

[D13] J. Delhalle, J. M. Andre, S. Delhalle, J. J. Pireaux, R. Caudano and J. J. Verbist, Electronic structure of polyethylene: theory and ESCA measurements, *J. Chem. Phys.* **60**, 595 (1974).

[D14] D. Demoulin and M. Jungen, Theoretical assignments of the electronic spectrum of acetylene, *Theoret. Chim. Acta* **34**, 1 (1974).

[D15] D. Demoulin, The shapes of some excited states of acetylene, *Chem. Phys.* **11**, 329 (1975).

[D16] D. Demoulin and M. Jungen, Expanded electronic states of the fluorine molecule, *Chem. Phys.* **16**, 311 (1976).

[D17] D. Demoulin, Empirically adjusted ab initio calculations for the valence and Rydberg excited states of formic acid, *Chem. Phys.* **17**, 471 (1976).

[D18] M. J. S. Dewar, Y. Yamaguchi and S. H. Suck, MO studies of polymers. I. Use of MNDO to calculate geometries, vibrational frequencies and the electronic band structures of polymers; formalism and application to polyethylene, *Chem. Phys.* **43**, 145 (1979).

[D19] J. Diamond and G. A. Segal, The vertical excited states and magnetic circular dichroism spectrum of allene, *J. Amer. Chem. Soc.* **106**, 952 (1984).

[D20] H. R. Dickinson and W. C. Johnson, Jr., Optical properties of sugars. II. Vacuum ultraviolet absorption of model compounds, *J. Amer. Chem. Soc.* **96**, 5050 (1974).

[D21] G. H. F. Diercksen, W. P. Kraemer, T. N. Rescigno, C. F. Bender, B. V. McKoy, S. R. Langhoff and P. W. Langhoff, Theoretical studies of photoexcitation and ionization in H_2O, *J. Chem. Phys.* **76**, 1043 (1982).

[D22] G. H. F. Diercksen and P. W. Langhoff, Theoretical studies of photoexcitation and ionization in H_2S, to be published.

[D23] T. G. DiGiuseppe, J. W. Hudgens and M. C. Lin, Detection of gas-phase methyl radicals using multiphoton ionization, *Chem. Phys. Lett.* **82**, 267 (1981).

[D24] T. G. DiGiuseppe, J. W. Hudgens and M. C. Lin, New electronic states in CH_3 observed using multiphoton ionization, *J. Chem. Phys.* **76**, 3337 (1982).

[D25] T. G. DiGiuseppe, J. W. Hudgens and M. C. Lin, Multiphoton ionization of CH_3 radicals in the gas phase, *J. Phys. Chem.* **86**, 36 (1982).

[D26] M. A. Dillon, R.-G. Wang and D. Spence, Electron impact spectroscopy of methane and methane-d_4, *J. Chem. Phys.* **80**, 5581 (1984).

[D27] M. A. Dillon, personal communication, 1983.

[D28] U. Dinur, R. J. Hemley and M. Karplus, Equilibrium geometry and dynamics of the valence excited states of 1,3-butadiene, *J. Phys. Chem.* **87**, 924 (1983).

[D29] J. P. Doering, Electronic energy levels of benzene below 7 eV, *J. Chem. Phys.* **67**, 4065 (1977).

[D30] J. P. Doering, Electron impact study of the energy levels of *trans*-1,3-butadiene, *J. Chem. Phys.* **70**, 3902 (1979).

[D31] J. P. Doering, Electron impact study of the 50, 000 cm^{-1} band of benzene, *J. Chem. Phys.* **71**, 20 (1979).

[D32] J. P. Doering and R. McDiarmid, Electron impact study of the energy levels of *trans*-1,3-butadiene. II. Detailed analysis of valence and Rydberg transitions, *J. Chem. Phys.* **73**, 3617 (1980); Erratum, *J. Chem. Phys.* **75**, 500 (1981).

[D33] J. P. Doering and R. McDiarmid, An electron impact investigation of the forbidden and allowed transitions of norbornadiene, *J. Chem. Phys.* **75**, 87 (1981).

[D34] J. P. Doering and R. McDiarmid, 100 eV electron impact study of 1,3-butadiene, *J. Chem. Phys.* **75**, 2477 (1981).

[D35] J. P. Doering and R. McDiarmid, An electron impact investigation of the 3p-Rydberg transitions in acetone, *J. Chem. Phys.* **76**, 1838 (1982).

[D36] J. P. Doering, Electron scattering from supersonic jet-cooled molecules, *J. Chem. Phys.* **79**, 2083 (1983).

[D37] J. P. Doering, A. Sabljić and R. McDiarmid, Rydberg states of hexatriene. An electron impact investigation, *J. Phys. Chem.* **88**, 835 (1984).

[D38] J. P. Doering, M. B. Robin, N. A. Kuebler, A. Gedanken and K. Ragavachari, to be published.

[D39] C. E. Doiron, M. E. Macbeath and T. B. McMahon, Triplet states of borazine and N-trimethylborazine determined by trapped electron spectroscopy, *Chem. Phys. Lett.* **59**, 90 (1978).

[D40] C. E. Doiron and T. B. McMahon, Triplet states of fluorocarbonyl compounds determined by trapped electron spectroscopy, *Chem. Phys. Lett.* **63**, 204 (1979).

[D41] P. J. Domaille, J. E. Kent and M. F. O'Dwyer, The visible spectrum of fulvene, *Chem. Phys.* **6**, 66 (1974).

[D42] E. A. Dorko, R. Scheps and S. A. Rice, Comments on the ultraviolet spectrum and photophysical properties of trimethylenecyclopropane, *J. Phys. Chem.* **78**, 568 (1974).

[D43] J. Doucet, P. Sauvageau and C. Sandorfy, Photoelectron and far-ultraviolet absorption spectra of chlorofluoro derivatives of ethane, *J. Chem. Phys.* **62**, 355 (1975).

[D44] J. Doucet, R. Gilbert, P. Sauvageau and C. Sandorfy, Photoelectron and far-ultraviolet spectra of CF_3Br, CF_2BrCl and CF_2Br_2, *J. Chem. Phys.* **62**, 366 (1975).

[D45] A. F. Drake and S. F. Mason, The absorption and circular dichroism spectra of chiral olefins, *Tetrahedron* **33**, 937 (1977).

[D46] S. D. Druger, Resonance lineshapes in the Rydberg spectra of large molecules, *Chem. Phys.* **67**, 3238 (1977).

[D47] C. R. Drury-Lessard and D. C. Moule, The higher Rydberg states of formaldehyde, *Chem. Phys. Lett.* **47**, 300 (1977).

[D48] C. R. Drury-Lessard and D. C. Moule, Ring puckering in the $^1B_2$ (n, 3s) Rydberg electronic state of cyclobutanone, *J. Chem. Phys.* **68**, 5392 (1978).

[D49] C. R. Drury and D. C. Moule, The ultraviolet absorption spectrum of thioformaldehyde, *J. Mol. Spectrosc.* **92**, 469 (1982).

[D50] C. R. Drury, J. Y. K. Lai and D. C. Moule, The $\tilde{B}\ ^1A_2 \leftarrow \tilde{X}\ ^1A_1$ absorption spectrum of thioformaldehyde, *Chem. Phys. Lett.* **87**, 520 (1982).

[D51] W. Duch and G. A. Segal, Theoretical calculation of the absorption and magnetic circular dichroism spectrum of a Jahn-Teller distorted excited state: the $1E'$ excited state of cyclopropane, *J. Chem. Phys.* **79**, 2951 (1983).

[D52] G. Dujardin and S. Leach, Rydberg series converging to the $\tilde{C}$ state of $C_6F_6^+$ in threshold photoelectron and VUV absorption spectra of hexafluorobenzene, *Chem. Phys. Lett.* **96**, 337 (1983).

[D53] G. Dujardin, S. Leach, O. Dutuit, T. Govers and P. M. Guyon, Autoionization processes in *sym*-trifluorobenzene and hexafluorobenzene: studies involving threshold photoelectrons and ion fluorescence, *J. Chem. Phys.* **79**, 644 (1983).

[D54] A. Yu. Dukhnyakov and A. S. Vinogradov, Fine structure of x-ray absorption spectra of the SiF_4 molecule and the Na_2SiF_6 coordination compound near the K edge of the fluorine atom, *Opt. Spectrosc. (USSR)* **53**, 502 (1982).

[D55] P. Dupuis, R. Roberge and C. Sandorfy, The very low ionization potentials of porphyrins and the possible role of Rydberg states in photosynthesis, *Chem. Phys. Lett.* **75**, 434 (1980).

[D56] P. Dupuis, R. Roberge, C. Sandorfy and D. Vocelle, Ultraviolet absorption and photoelectron spectra of the oximes and O-methyloximes of a series of polyenic Schiff bases. Relation to visual pigments, *J. Chem. Phys.* **74**, 256 (1981).

[D57] G. Durand and F. Volatron, An improved basis set for H_2O. Test on Rydberg excited states, *J. Chem. Phys.* **80**, 1937 (1984).

[E1] W. Eberhardt, R.-P. Haelbich, M. Iwan, E. E. Koch and C. Kunz, Fine structure at the carbon 1s K edge in vapors of simple hydrocarbons, *Chem. Phys. Lett.* **40**, 180 (1976).

[E2] v. H. Ebinghaus, K. Kraus, W. Müller-Duysing and H. Neuert, Negative ionen durch elektronenresonanzeinfang in PH_3, AsH_3 and SiH_4, *Z. Naturforsch.* **19A**, 732 (1964).

[E3] D. A. Edmonson, J. S. Lee and J. P. Doering, Inelastic scattering of positive ions and electrons from water: the 4–6 eV energy loss region, *J. Chem. Phys.* **69**, 1445 (1978).

[E4] R. F. Egerton and M. J. Whelan, The electron energy loss spectrum and band structure of diamond, *Phil. Mag.* **30**, 739 (1974).

[E5] T. S. Eichelberger, IV and G. J. Fisanick, Multiphoton ionization spectroscopy of the 3s Rydberg state in the deuterated acetaldehydes, *J. Chem. Phys.* **74**, 5962 (1981).

[E6] S. T. Elbert and E. R. Davidson, *Ab initio* calculations on urea, *Int. J. Quantum Chem.* **8**, 857 (1974).

[E7] S. T. Elbert, S. D. Peyerimhoff and R. J. Buenker, All-valence-electron CI calculations on the electronic spectrum of diborane, *Chem. Phys.* **11**, 25 (1975).

[E8] W. C. Ermler and R. S. Mulliken, Energies and orbital sizes for some Rydberg and valence states of the nitrogen molecule, *J. Mol. Spectrosc.* **61**, 100 (1976).

[E9] K. Evans, R. Scheps, S. A. Rice and D. Heller, Primary photochemical and photophysical processes in chloro- and bromo-acetylene, *J. Chem. Soc., Faraday II* **69**, 856 (1973).

[E10] E. M. Evleth and E. Kassab, Theoretical analysis of the role of Rydberg states in the photochemistry of some small molecules, *in* "Quantum Theory of Chemical Reactions" Vol. II, R. Daudel, A. Pullman, L. Salem, and A. Veillard, *eds.*, D. Riedel, Dordrecht, 261 (1980).

[E11] E. M. Evleth and E. Kassab, Theoretical aspects of small molecule Rydberg photochemistry, *in* "Computational Theoretical Organic Chemistry," I. G. Csizmadia and R. Daudel, *eds.*, D. Reidel, Dordrecht, 379 (1981).

[E12] E. M. Evleth, J. T. Gleghorn and E. Kassab, Clarification of the concept of avoided crossings in the NH-bond-rupture surfaces of excited ammonia, *Chem. Phys. Lett.* **80**, 558 (1981).

[E13] E. M. Evleth and J. T. Gleghorn, Solvent effects on molecular Rydberg states: the B $^1\Sigma_u$ state of molecular hydrogen in helium, *Chem. Phys. Lett.* **94**, 373 (1983).

[E14] E. M. Evleth, H. Z. Cao, E. Kassab and A. Sevin, Computation of the direct adiabatic channel for the relaxation of electronically excited C_2H_5 radical to give $H + C_2H_4$, *Chem. Phys. Lett.* **109**, 45 (1984).

[E15] H. Eyring, J. Walter and G. E. Kimball, "Quantum Chemistry", J. Wiley, New York, 163 (1944).

[F1] J. E. Falk and R. J. Fleming, Calculation of the electronic energy band structure of polyethylene, *J. Phys. C: Solid State Phys.* **6**, 2954 (1973).

[F2] S. Felps, P. Hochmann, P. Brint and S. P. McGlynn, Molecular Rydberg transitions. The lowest-energy Rydberg transitions of s-type in CH_3X and CD_3X, X = Cl, Br and I, *J. Mol. Spectrosc.* **59**, 355 (1976).

[F3] W. S. Felps, K. Wittel and S. P. McGlynn, Molecular Rydberg transitions. *Cis*-dibromoethylene, *J. Mol. Spectrosc.* **71**, 101 (1978).

[F4] W. S. Felps, J. D. Scott, G. L. Findley and S. P. McGlynn, Molecular Rydberg transitions. XX. Vibronic doubling in alkyl bromides, *J. Chem. Phys.* **74**, 4832 (1981).

[F5] W. S. Felps, S. P. McGlynn, and G. L. Findley, Molecular Rydberg transitions. Cyanogen chloride, *J. Mol. Spectrosc.* **86**, 71 (1981).

[F6] W. S. Felps, G. L. Findley and S. P. McGlynn, Molecular Rydberg transitions. Intermediate coupling in simple bromides, *Chem. Phys. Lett.* **81**, 490 (1981).

[F7] E. P. Fesenko and L. V. Iogansen, Resonances at 1.1 and 1.7 eV in scattering of slow electrons from benzene molecule, *Chem. Phys. Lett.* **48**, 22 (1977).

[F8] G. L. Findley, K. Wittel, W. S. Felps and S. P. McGlynn, Molecular Rydberg transitions. VIII. The geometry of ethylene in the R_{1s} state, *Int. J. Quantum Chem.* **11**, 229 (1977).

[F9] G. J. Fisanick, T. S. Eichelberger, IV, M. B. Robin and N. A. Kuebler, Two-color multiphoton spectroscopy: detection of $n = 12-41$ Rydberg states in diazabicyclooctane, *J. Phys. Chem.* **87**, 2240 (1983).

[F10] U. Fischbach, R. J. Buenker and S. D. Peyerimhoff, Non-empirical calculations on the Rydberg states of ethylene, *Chem. Phys.* **5**, 265 (1974).

[F11] L. Flamigni, F. Barigelletti, S. Dellonte and G. Orlandi, Temperature dependence of fluorescence lifetime of cyclic alkanes: Mechanism of S_1 deactivation, *Chem. Phys. Lett.* **89**, 13 (1982).

[F12] W. M. Flicker, O. A. Mosher and A. Kuppermann, Singlet → triplet transitions in methyl-substituted ethylenes, *Chem. Phys. Lett.* **36**, 56 (1975).

[F13] W. M. Flicker, O. A. Mosher and A. Kuppermann, Triplet states of furan, thiophene and pyrrole, *Chem. Phys. Lett.* **38**, 489 (1976).

[F14] W. M. Flicker, O. A. Mosher and A. Kuppermann, Electron impact investigation of electronic excitations in furan, thiophene and pyrrole, *J. Chem. Phys.* **64**, 1315 (1976).

[F15] W. M. Flicker, O. A. Mosher and A. Kuppermann, Low energy, variable angle electron-impact excitation of 1,3,5-hexatriene, *Chem. Phys. Lett.* **45**, 492 (1977).

[F16] W. M. Flicker, O. A. Mosher and A. Kuppermann, Electron impact spectroscopy of the alkynes: A comparison of propyne and 1-butyne with acetylene, *J. Chem. Phys.* **69**, 3311 (1978).

[F17] W. M. Flicker, O. A. Mosher and A. Kuppermann, Electron-impact investigation of excited singlet states in 1,3-butadiene, *Chem. Phys.* **30**, 307 (1978).

[F18] J.-H. Fock, P. Gürtler and E. E. Koch, Molecular Rydberg transitions in carbon monoxide: term value/ionization energy correlation of BF, CO and N_2, *Chem. Phys.* **47**, 87 (1980).

[F19] J.-H. Fock and E. E. Koch, Partial cross sections and autoionization resonances in the valence shell photoemission from solid acetylene, *Chem. Phys. Lett.* **105**, 38 (1984).

[F20] H. M. Foley, On the ordering of atomic energy levels in excited states, *Phys. Rev.* **19A**, 2134 (1979).

[F21] P. D. Foo and K. K. Innes, New experimental tests of existing interpretations of electronic transitions of ethylene, *J. Chem. Phys.* **60**, 4582 (1974).

[F22] R. L. Ford, Vacuum-ultraviolet absorption spectra and optical constants of poly(N-vinyl carbazole) films, *J. Opt. Soc. Amer.* **64**, 952 (1974).

[F23] M. S. Foster, A. D. Williamson and J. L. Beauchamp, Photoionization mass spectrometry of trans-azomethane, *Int. J. Mass Spectrom. Ion Phys.* **15**, 429 (1974).

[F24] M. F. Fox, C. B. Smith and E. Hayon, Far-ultraviolet solution spectroscopy of thiocyanate, *J. Chem. Soc., Faraday I* **77**, 1497 (1981).

[F25] J. R. Frazier, L. G. Christophorou, J. G. Carter and H. C. Schweinler, Low-energy electron interactions with organic molecules: negative ion states of fluorobenzene, *J. Chem. Phys.* **69**, 3807 (1978).

[F26] R. R. Freeman, R. M. Jobson, J. Bokor and W. E. Cooke, Planetary atoms, *Opt. Sci.* **40**, 220 (1983).

[F27] R. S. Freund, High-Rydberg molecules, *in* "Rydberg States of Atoms and Molecules," R. F. Stebbings and F. B. Dunning, *eds.*, Cambridge University Press, 355 (1982).

[F28] C. Fridh, L. and E. Lindholm, Valence excitation of linear molecules. II. Excitation and UV spectra of C_2N_2, CO_2 and N_2O, *Chem. Phys.* **27**, 169 (1978).

[F29] C. Fridh, Electronic excitation of formic acid, *J. Chem. Soc., Faraday Trans. II* **74**, 190 (1978).

[F30] C. Fridh, Electronic excitation of acetonitrile, methylisonitrile and propyne, *J. Chem. Soc. Faraday Trans. II* **74**, 2193 (1978).

[F31] C. Fridh, L. and E. Lindholm, Acrolein: excitation, ionization and electron affinity studied with HAM/3, *Phys. Scr.* **20**, 603 (1979).

[F32] C. Fridh, Interpretation of electronic spectra. Part 4. Spectra of cyclopropane, ethylenimine and ethylene oxide, studied with HAM/3, *J. Chem. Soc., Faraday Trans. II* **75**, 993 (1979).

[F33] H. Friedrich, B. Sonntag, P. Rabe, W. Butscher and W. H. E. Schwarz, Term values and valence-Rydberg mixing in core-excited states. SiH_4 and PH_3, *Chem. Phys. Lett.* **64**, 360 (1979).

[F34] H. Friedrich, B. Pittel, P. Rabe, W. H. E. Schwarz and B. Sonntag, Core to valence/Rydberg excitations in SiF_4, *J. Phys. B: At. Mol. Phys.* **13**, 25 (1980).

[F35] D. C. Frost, S. T. Lee, C. A. McDonnell and N. P. C. Westwood, Photoelectron spectroscopic studies of some nitrosyl and nitryl halides and nitric acid, *J. Electron Spectrosc. Related Phenomena* **7**, 331 (1975).

[F36] D. C. Frost, S. T. Lee, C. A. McDowell and N. P. C. Westwood, The photoelectron spectra of diazene, diazene-d_2, and *trans*-methyldiazine, *J. Chem. Phys.* **64**, 4719 (1976).

[F37] R. P. Frueholz, W. M. Flicker and A. Kuppermann, Excited electronic states of ketene, *Chem. Phys. Lett.* **38**, 57 (1976).

[F38] R. P. Frueholz, W. M. Flicker, O. A. Mosher and A. Kuppermann, Excited electronic states of the fluorobenzenes by variable angle electron impact spectroscopy, *Chem. Phys. Lett.* **52**, 86 (1977).

[F39] R. P. Frueholz, R. Rianda and A. Kuppermann, Excited electronic states of 1,3,5-cycloheptatriene, *Chem. Phys. Lett.* **55**, 88 (1978).

[F40] R. P. Frueholz and A. Kuppermann, Vibronic structure of the second triplet state of 1,3,5-hexatriene, *J. Chem. Phys.* **69**, 3433 (1978).

[F41] R. P. Frueholz and A. Kuppermann, Electron spectroscopy of 1,3,5,7-cyclooctatetraene by low-energy, variable-angle electron impact, *J. Chem. Phys.* **69**, 3614 (1978).

[F42] R. P. Frueholz, W. M. Flicker, O. A. Mosher and A. Kuppermann, Excited electronic states of cyclohexene, 1,4-cyclohexadiene, norbornene, and norbornadiene as studied by electron-impact spectroscopy, *J. Chem. Phys.* **70**, 1986 (1979).

[F43] R. P. Frueholz, W. M. Flicker, O. A. Mosher and A. Kuppermann, Electronic spectroscopy of 1,3-cyclopentadiene, 1,3-cyclohexadiene and 1,3-cycloheptadiene by electron impact, *J. Chem. Phys.* **70**, 2003 (1979).

[F44] R. P. Frueholz, W. M. Flicker, O. A. Mosher and A. Kuppermann, Electronic spectroscopy of benzene and the fluorobenzenes by variable angle electron impact, *J. Chem. Phys.* **70**, 3057 (1979).

[F45] M. Fujii, T. Ebata, N. Mikami and M. Ito, Two-color multiphoton ionization of diazabicyclooctane in a supersonic free jet, *Chem. Phys. Lett.* **101**, 578 (1983).

[F46] K. Fuke and S. Nagakura, Polarized reflection and absorption spectra of the [2.2]-paracyclophane crystal, *Bull. Chem. Soc. Jpn.* **48**, 46 (1975).

[F47] K. Fuke, K. Kaya, T. Kajiwara and S. Nagakura, The polarized reflection and absorption spectra of perylene crystals in monomeric and dimeric forms, *J. Mol. Spectrosc.* **63**, 98 (1976).

[F48] K. Fuke and S. Nagakura, Rydberg transitions of aniline and N,N-diethylaniline, *J. Mol. Spectrosc.* **64**, 139 (1977).

[F49] K. Fuke, A. Gedanken and O. Schnepp, The magnetic moment of the $^1E_{1u}$ N-V state of benzene and its methyl derivatives, *Chem. Phys. Lett.* **67**, 483 (1979).

[F50] K. Fuke and O. Schnepp, Absorption and magnetic circular dichroism spectra of allene, *Chem. Phys.* **38**, 211 (1979).

[F51] W. Fuss, UV spectra of aminoboranes: $\pi \rightarrow \pi^*$ and Rydberg transitions, *Z. Naturforsch.* **29B**, 514 (1974).

[F52] W. Fuss and H. Bock, Photoelectron spectra and molecular properties. XXXVI. $(H_3C)_3B$, $(H_3C)_2BF$, $(H_3C)_2BN(CH_3)_2$, and $(H_3C)_2CC(CH_3)_2$: The use of ionization potentials in assigning UV spectra, *J. Chem. Phys.* **61**, 1613 (1974).

[G1] A. R. Gallo and K. K. Innes, A $^1B_1$ Rydberg state of the H_2S molecule, *J. Mol. Spectrosc.* **54**, 472 (1975).

[G2] R. A. Gangi and R. F. W. Bader, Theoretical confirmation of the stability of the H_3O radical, *Chem. Phys. Lett.* **11**, 216 (1971).

[G3] R. M. Gavin, Jr., S. Risemberg and S. A. Rice, Spectroscopic properties of polyenes. I. The lowest energy allowed singlet-singlet transition for *cis*- and *trans*- 1,3,5-hexatriene, *J. Chem. Phys.* **58**, 3160 (1973).

[G4] R. M. Gavin, Jr. and S. A. Rice, Spectroscopic properties of polyenes. II. The vacuum ultraviolet spectra of *cis*- and *trans*-1,3,5-hexatriene, *J. Chem. Phys.* **60**, 3231 (1974).

[G5] A. Gedanken and M. D. Rowe, Magnetic circular dichroism spectra of the methyl halides. Resolution of the $n \rightarrow \sigma^*$ continuum, *Chem. Phys. Lett.* **34**, 39 (1975).

[G6] A. Gedanken and O. Schnepp, The magnetic circular dichroism spectrum of acetylene, *Chem. Phys. Lett.* **37**, 373 (1976).

[G7] A. Gedanken and O. Schnepp, The excited states of cyclopropane. MCD spectrum, and CD spectrum of an optically active derivative, *Chem. Phys.* **12**, 341 (1976).

[G8] A. Gedanken and M. Levy, New instrument for circular dichroism measurements in the vacuum ultraviolet, *Rev. Sci. Instrum.* **48**, 1661 (1977).

[G9] A. Gedanken, M. B. Robin and N. A. Kuebler, Excitation, ionization and fragmentation in dimethyl mercury, *Inorg. Chem.* **20**, 3340 (1981).

[G10] A. Gedanken, N. A. Kuebler and M. B. Robin, An MPI search for the $\pi \rightarrow 3p$ Rydberg states of ethylene, *J. Chem. Phys.* **76**, 46 (1982).

[G11] A. Gedanken, M. B. Robin and Y. Yafet, The methyl iodide multiphoton ionization spectrum with intermediate resonance in the A-band region, *J. Chem. Phys.* **76**, 4798 (1982).

[G12] A. Gedanken, G. J. Fisanick, K. Raghavachari, T. S. Eichelberger, IV, M. B. Robin and N. A. Kuebler, Multiphoton spectroscopy and photochemistry in norbornadiene and related systems, to be published.

[G13] F. A. Gianturco and D. G. Thompson, The scattering of slow electrons by polyatomic molecules. A model study for CH_4, H_2O and H_2S, *J. Phys. B: At. Mol. Phys.* **13**, 613 (1980).

[G14] E. Gilberg, M. J. Hanus and B. Foltz, Investigation of the electronic structure of ice by high resolution x-ray spectroscopy, *J. Chem. Phys.* **76**, 5093 (1982).

[G15] R. Gilbert and C. Sandorfy, Rydberg transitions in the ultraviolet spectra of difluorobenzenes, *Chem. Phys. Lett.* **27**, 457 (1974).

[G16] R. Gilbert, P. Sauvageau and C. Sandorfy, Vacuum ultraviolet absorption spectra of chlorofluoromethanes from 120 to 65 nm, *J. Chem. Phys.* **60**, 4820 (1974).

[G17] J. C. Giordan, Negative ions: effect of α- vs. β-silyl substitution on the negative ion states of π systems, *J. Amer. Chem. Soc.* **105**, 6544 (1983).

[G18] J. H. Glownia, S. J. Riley and S. D. Colson, MPI spectroscopy of expansion-cooled ammonia: structure and dynamics at 50 K, *J. Chem. Phys.* **72**, 5998 (1980).

[G19] J. H. Glownia, S. J. Riley and S. D. Colson, The MPI spectrum of expansion-cooled ammonia: photophysics and new assignments of electronic excited states, *J. Chem. Phys.* **73**, 4296 (1980).

[G20] W. A. Goddard, III and W. J. Hunt, The Rydberg nature and assignments of excited states of the water molecule, *Chem. Phys. Lett.* **24**, 464 (1974).

[G21] W. Goetz, D. C. Moule and D. A. Ramsay, The electric dipole moment of the $\tilde{c}\ ^1B_2$ (3s ← n) state of thioformaldehyde, *Can. J. Phys.* **59**, 1635 (1981).

[G22] E. Goldstein, S. Vijaya and G. A. Segal, Theoretical study of the optical absorption and magnetic circular dichroism spectra of cyclopropane, *J. Amer. Chem. Soc.* **102**, 6198 (1980).

[G23] N. Gonohe, N. Yatsuda, N. Mikami and M. Ito, One-photon and two-photon electronic spectra of two caged amines, *Bull. Chem. Soc. Jpn.* **55**, 2796 (1982).

[G24] J. Goodman and L. E. Brus, Rydberg states in condensed phases: Evidence for small "bubble" formation around NO 3sσ (A $^2\Sigma^+$) in solid rare gases, *J. Chem. Phys.* **67**, 933 (1977).

[G25] J. Goodman and L. E. Brus, Vibrational relaxation and small "bubble" spectroscopy of the NO 3s (A $^2\Sigma^1$) Rydberg state in solid rare gases, *J. Chem. Phys.* **69**, 4083 (1978).

[G26] M. S. Gordon, Excited states and photochemistry of saturated molecules. Extended basis calculations on the 1 B_1 ($^1T_2$) state of methane, *Chem. Phys. Lett.* **52**, 161 (1977).

[G27] M. S. Gordon and J. W. Caldwell, Excited states and photochemistry of saturated molecules. VII. Potential energy surfaces in excited singlet states of methane, *J. Chem. Phys.* **70**, 5503 (1979).

[G28] M. S. Gordon, Excited states and photochemistry of saturated molecules. The $^1B_1$ ($^1T_2$) surface in silane, *Chem. Phys. Lett.* **70**, 343 (1980).

[G29] M. S. Gordon, Potential energy surfaces in excited states of saturated molecules, *in* "Potential Energy Surfaces and Dynamics Calculations," D. G. Truhlar, *ed.*, Plenum, New York, 185 (1981).

[G30] R. D. Gordon, The conformation of acetaldehyde in its n → 3s Rydberg state, *J. Chem. Phys.* **73**, 5907 (1980).

[G31] O. Goscinski, J. Müller, E. Poulain and H. Siegbahn, Fluorine core ESCA linewidths in CH_3F and CF_4, *Chem. Phys. Lett.* **55**, 407 (1978).

[G32] G. Gottarelli, B. Samori, G. L. Bendazzoli, P. Palmieri and A. F. Drake, Low-lying optically active Rydberg transitions in propyleneimine, *Chem. Phys. Lett.* **45**, 318 (1977).

[G33] R. Grasser, E. Pitt, A. Scharmann and G. Zimmerer, Optical properties of $CaWO_4$ and $CaMoO_4$ crystals in the 4 to 25 eV region, *Phys. Status Solidi* **69B**, 359 (1975).

[G34] R. G. Green and R. P. Wayne, Vacuum ultraviolet absorption spectra of halogenated methanes and ethanes, *J. Photochem.* **6**, 375 (1976/77).

[G35] F. R. Greening, A united-atom treatment of the Rydberg states of linear molecules with $^2\Pi$ cores, *Chem. Phys. Lett.* **34**, 581 (1975).

[G36] F. R. Greening and G. W. King, Rydberg states of carbon dioxide and carbon disulfide, *J. Mol. Spectrosc.* **59**, 312 (1976).

[G37] T. A. Gregory and S. Lipsky, Determination of the Einstein A coefficient for the A → X transition of ammonia-d_3, *J. Chem. Phys.* **65**, 5469 (1976).

[G38] D. W. T. Griffith, J. E. Kent and M. F. O'Dwyer, The vibrational and electronic spectra of 3,4-dimethylenecyclobutene, *Aust. J. Chem.* **25**, 241 (1972).

[G39] D. W. T. Griffith, J. E. Kent and M. F. O'Dwyer, The electronic spectrum of benzvalene, *J. Mol. Spectrosc.* **58**, 427 (1975).

[G40] D. W. T. Griffith, J. E. Kent and M. F. O'Dwyer, The electronic spectrum of Dewar benzene, *J. Mol. Spectrosc.* **58**, 436 (1975).

[G41] D. F. Grosjean and P. Bletzinger, Photoionization and photoabsorption characteristics of laser seed compounds, *IEEE J. Quant. Electronics* **QE13**, 898 (1977).

[G42] K. P. Gross and O. Schnepp, Circular dichroism spectra of straight chain mono-olefins. Assignment of ethylene transitions, *Chem. Phys. Lett.* **36**, 531 (1975).

[G43] K. P. Gross and O. Schnepp, Absorption and circular dichroism spectra of the *cis*- and *trans*-butadiene chromophores α- and β-phellandrene, *J. Chem. Phys.* **68**, 2647 (1978).

[G44] P. Gürtler, V. Saile and E. E. Koch, Rydberg series in the absorption spectra of H_2O and D_2O in the vacuum ultraviolet, *Chem. Phys. Lett.* **51**, 386 (1977).

[G45] T. Gustafsson, Partial photoionization cross sections of SF_6 between 20 and 54 eV: an interpretation of the photoelectron spectrum, *Phys. Rev.* **18A**, 1481 (1978).

[H1] S. Habousha and A. Gedanken, The nature of the n $\rightarrow$ 3p transition in DABCO: A magnetic circular dichroism study, *J. Phys. Chem.* **86**, 3920 (1982).

[H2] J. E. Hahn and K. O. Hodgson, Polarized x-ray absorption spectroscopy, *in* "Inorganic Chemistry: Toward the 21st Century," M. H. Chisholm, *ed.*, ACS Symp. Series 211, Washington, D.C., 431 (1983).

[H3] H.-J. Haink, E. Heilbronner, V. Hornung and E. Kloster-Jensen, Die photoelektron-spektren der monohalogenacetylene, *Helv. Chim. Acta* **53,** 1073 (1970).

[H4] H.-J. Haink and M. Jungen, Excited states of the polyacetylenes. Analysis of the near ultraviolet spectra of diacetylene and triacetylene, *Chem. Phys. Lett.* **61**, 319 (1979).

[H5] E. Haller, H. Köppel and L. S. Cederbaum, Multimode Jahn-Teller and pseudo-Jahn-Teller effects in BF_3^+, *J. Chem. Phys.* **78**, 1359 (1983).

[H6] K.-H. Hallmeier, R. Szargau, A. Meisel, E. Hartmann and E. S. Gluskin, Investigation of core-excited quantum yield spectra of high-symmetric boron compounds, *Spectrochim. Acta* **37A,** 1049 (1981).

[H7] M. Halmann and I. Platzner, The photochemistry of phosphorus compounds. Part II. Far ultraviolet absorption spectra of some phosphorus oxyanions in aqueous solution, *J. Chem. Soc.,* 1440 (1965).

[H8] A. M. Halpern, Structural effects on photophysical processes in saturated amines. II., *J. Amer. Chem. Soc.* **96**, 7655 (1974).

√ [H9] S. Hamai and F. Hirayama, Fluorescence of acetylenic hydrocarbons, *J. Chem. Phys.* **71**, 2934 (1979).

[H10] M. J. Hanus and E. Gilberg, The K absorption spectra of chlorine in gaseous CH_3Cl, C_2H_5Cl, CF_2Cl_2 and C_2H_3Cl, *J. Phys. B: At. Mol. Phys.* **9**, 137 (1976).

[H11] A. H. Hardin and C. Sandorfy, Photoelectron and vacuum ultraviolet spectra of a series of fluoroethers, *J. Fluorine Chem.* **5**, 435 (1975).

[H12] L. B. Harding and W. A. Goddard, III, The generalized valence band description of the valence states of formamide, *J. Amer. Chem. Soc.* **97**, 6300 (1975).

[H13] L. B. Harding and W. A. Goddard, III, The generalized valence bond description of the low-lying states of ketene, *J. Amer. Chem. Soc.* **98,** 6093 (1976).

[H14] L. B. Harding and W. A. Goddard, III, Ab initio theoretical studies of the Rydberg states of formaldehyde, *J. Amer. Chem. Soc.* **99**, 677 (1977).

[H15] P. J. Harman, J. E. Kent, T. H. Gan, J. B. Peel and G. D. Willett, The photoelectron spectrum of benzvalene, *J. Amer. Chem. Soc.* **99**, 943 (1977).

[H16] P. J. Harman, J. E. Kent, M. F. O'Dwyer and M. H. Smith, Quartz ultraviolet Rydberg transitions in benzene isomers, *Aust. J. Chem.* **32**, 2579 (1979).

[H17] P. J. Harman, J. E. Kent, M. F. O'Dwyer and D. W. T. Griffith, Photochemistry of benzene isomers. 2. Benzvalene and Dewar benzene, *J. Phys. Chem.* **85**, 2731 (1981).

[H18] L. E. Harris, The lower electronic states of nitrite and nitrate ion, nitromethane, nitramide, nitric acid and nitrate esters, *J. Chem. Phys.* **58**, 5615 (1973).

[H19] S. Hashimoto, K. Seki and H. Inokuchi, Anisotropic absorption of oriented syndiotactic-1,2-polybutadiene thin films in the vacuum ultraviolet region, *Rpts. Prog. Polymer Phys. Jpn.* **22**, 439 (1979).

[H20] S. Hashimoto, K. Seki, N. Sato and H. Inokuchi, Anisotropic vacuum UV absorption spectra of oriented polyacrylonitrile thin films, *Rept. Prog. Polym. Phys. Jpn.* **24**, 455 (1981).

[H21] S. Hashimoto, K. Seki, N. Sato and H. Inokuchi, Electronic properties of polymers. Anisotropic light absorption and photoelectron emission of oriented polyethylene films in the vacuum ultraviolet, *J. Chem. Phys.* **76**, 163 (1982).

[H22] S. S. Hasnain, P. Print, T. D. S. Hamilton and I. H. Munro, Absorption spectrum of naphthalene in rare-gas matrices, *J. Mol. Spectrosc.* **72**, 349 (1978).

[H23] S. S. Hasnain, T. D. S. Hamilton, I. H. Munro and E. Pantos, Spectroscopic study of solid benzene and benzene isolated in rare-gas matrices, *J. Mol. Spectrosc.* **72**, 406 (1978).

[H24] S. Havriliak and H. F. King, Rydberg radicals. I. Frozen-core model for Rydberg levels of the ammonium radical, *J. Amer. Chem. Soc.* **105**, 4 (1983).

[H25] P. J. Hay and I. Shavitt, Ab initio configuration interaction studies of the π-electron states of benzene, *J. Chem. Phys.* **60**, 2865 (1974).

[H26] P. J. Hay, The excited states and positive ions of SF_6, *J. Amer. Chem. Soc.* **99**, 1013 (1977).

[H27] P. J. Hay, *Ab initio* studies of excited states of polyatomic molecules including spin orbit and multiplet effects: the electronic states of UF_6, *J. Chem. Phys.* **79**, 5469 (1983).

[H28] T. Hayaishi, S. Iwata, M. Sasanuma, E. Ishiguro, Y. Morioka, Y. Iida and M. Nakamura, Photoionization mass spectrometric study of acetylene in the VUV region, *J. Phys. B: At. Mol. Phys.* **15**, 79 (1982).

[H29] B. A. Heath, N. A. Kuebler and M. B. Robin, Multiphoton ionization spectra of polycyclic alkanes, *J. Chem. Phys.* **70**, 3362 (1979).

[H30] B. A. Heath, M. B. Robin, N. A. Kuebler, G. J. Fisanick and T. S. Eichelberger, IV, Multiphoton ionization spectroscopy of acetaldehyde in its lowest Rydberg state, *J. Chem. Phys.* **72**, 5565 (1980).

[H31] B. A. Heath, G. J. Fisanick, M. B. Robin and T. S. Eichelberger, IV, Extracting the two-photon polarization ratio of the $n \rightarrow 3s$ transition in acetaldehyde from MPI measurements, *J. Chem. Phys.* **72**, 5991 (1980).

[H32] B. A. Heath and M. B. Robin, Torsional modes in the multiphoton ionization spectrum of tetrachloroethylene, *J. Amer. Chem. Soc.* **102**, 1796 (1980).

[H33] D. W. O. Heddle, Resonances in optical excitation functions, *Contemp. Phys.* **17**, 443 (1976).

[H34] E. Heilbronner, R. Gleiter, H. Hopf, V. Hornung and A. de Meijere, Photoelectron-spectroscopic evidence for the orbital sequence in fulvene and 3,4-dimethylene-cyclobutene, *Helv. Chim. Acta* **54**, 783 (1971).

[H35] E. Heilbronner, T. B. Jones, E. Kloster-Jensen and J. P. Maier, Electronic states of di-*t*-butylpolyacetylene radical cations, *Helv. Chim. Acta* **61**, 2040 (1978).

[H36] J. M. Heller, Jr., R. N. Hamm, R. D. Birkhoff and L. R. Painter, Collective oscillation in liquid water, *J. Chem. Phys.* **60**, 3483 (1974).

[H37] J. M. Heller, Jr., R. D. Birkhoff and L. R. Painter, Collective oscillation in liquid normal tetradecane and normal heptadecane, *J. Chem. Phys.* **62**, 4121 (1975).

[H38] J. M. Heller, Jr., R. D. Birkhoff and L. R. Painter, Isotopic effects on the electronic properties of H_2O and D_2O in the vacuum ultraviolet, *J. Chem. Phys.* **67**, 1858 (1977).

[H39] J. M. Heller, Jr., H. H. Hubbell, Jr., L. R. Painter and R. D. Birkhoff, Optical studies of liquid formamide in the vacuum ultraviolet, *J. Chem. Phys.* **71**, 4641 (1979).

[H40] R. J. Hemley, J. I. Dawson and V. Vaida, Franck-Condon analysis of the 1 $^1A_g^- \rightarrow 1\ ^1B_u^+$ transition of 1,3-butadiene from absorption and Raman intensities, *J. Chem. Phys.* **78**, 2915 (1983).

[H41] M. Herman and R. Colin, The geometrical structure of the acetylene molecule in the $\tilde{X}$, $\tilde{G}$ and $\tilde{I}$ states, *Bull. Soc. Chim. Belg.* **89**, 335 (1980).

[H42] M. Herman and R. Colin, The absorption spectra of C_2H_2, C_2D_2, and C_2HD in the region 1260 to 1370 Å, *J. Mol. Spectrosc.* **85**, 449 (1981).

[H43] M. Herman and R. Colin, High resolution spectroscopic study of the Rydberg series of the acetylene isotopic molecules, *Phys. Scr.* **25**, 275 (1982).

[H44] H. W. Hermann and I. V. Hertel, Electron energy loss spectra of dye vapors, *Appl. Phys.* **15**, 185 (1978).

[H45] G. Herzberg, Rydberg spectra of triatomic hydrogen and of the ammonium radical, *Discuss. Faraday Soc.* **71**, 165 (1981).

[H46] G. Herzberg and J. T. Hougen, Spectra of the ammonium radical: the Schuster band of ND_4, *J. Mol. Spectrosc.* **97**, 440 (1983).

[H47] B. Hess, P. J. Bruna, R. J. Buenker and S. D. Peyerimhoff, Ab initio study of the electronic spectrum of acetone, *Chem. Phys.* **18**, 267 (1976).

[H48] S. Hino, T. Veszpremi, K. Ohno, H. Inokuchi and K. Seki, Absorption spectra of volatile aromatic hydrocarbon films in the vacuum ultraviolet region, *Chem. Phys.* **71**, 135 (1982).

[H49] H. J. Hinz and H. Venghaus, Electron energy loss measurements on fluorene ($C_{13}H_{10}$) and *p*-terphenyl ($C_{18}H_{14}$) single crystals and vapors, *in* "Vacuum Ultraviolet Radiation Physics," E.-E. Koch, R. Haensel and C. Kunz, *eds*., Pergamon, 359 (1974).

[H50] T. Hirano, A. Sato, T. Tsuruta and W. C. Johnson, Jr., Optical rotatory dispersion and vacuum ultraviolet circular dichroism of poly [(*R*)-oxypropylene], *J. Polymer Sci.* **17**, 1601 (1979).

[H51] K. Hirao, Direct cluster expansion method. Application to glyoxal, *J. Chem. Phys.* **79**, 5000 (1983).

[H52] K. Hiraoka and W. H. Hamill, Characteristic energy losses by slow electron impact on thin-film alkanes at 77 °K, *J. Chem. Phys.* **59**, 5749 (1973).

[H53] F. Hirayama and S. Lipsky, Fluorescence of mono-olefinic hydrocarbons, *J. Chem. Phys.* **62**, 576 (1975).

[H54] F. Hirota, N. Kakuta and S. Shibata, High-energy electron scattering by diborane, *J. Phys. B: At. Mol. Phys.* **14**, 3299 (1981).

[H55] A. P. Hitchcock, M. Pocock and C. E. Brion, Isotope effects on intensities in the K-shell excitation spectra of CH_4 and CD_4 studied by 2.5 KeV electron impact, *Chem. Phys. Lett.* **48**, 125 (1977).

[H56] A. P. Hitchcock and C. E. Brion, Carbon K-shell excitation of C_2H_2, C_2H_4, C_2H_6 and C_6H_6 by 2.5 KeV electron impact, *J. Electron Spectrosc. Related Phenomena* **10**, 317 (1977).

[H57] A. P. Hitchcock and C. E. Brion, Inner shell excitation of CH_3F, CH_3Cl, CH_3Br and CH_3I by 2.5 KeV electron impact, *J. Electron Spectrosc. Related Phenomena* **13**, 193 (1978).

[H58] A. P. Hitchcock, M. Pocock, C. E. Brion, M. S. Banna, D. C. Frost, C. A. McDowell and B. Wallbank, Inner shell excitation and ionization of the monohalobenzenes, *J. Electron Spectrosc. Related Phenomena* **13**, 345 (1978).

[H59] A. P. Hitchcock and C. E. Brion, Inner-shell excitation and EXAFS-type phenomena in the chloromethanes, *J. Electron Spectrosc. Related Phenomena* **14**, 417 (1978).

[H60] A. P. Hitchcock and C. E. Brion, Inner shell excitation of SF_6 by 2.5 keV electron impact, *Chem. Phys.* **33**, 55 (1978).

[H61] A. P. Hitchcock and C. E. Brion, On the assignment of the carbon K-shell spectra of the methyl halides, *J. Electron Spectrosc. Related Phenomena* **17**, 139 (1979).

[H62] A. P. Hitchcock and M. J. van der Wiel, Absolute oscillator strengths (5–63 eV) for photoabsorption and ionic fragmentation of SF_6, *J. Phys. B: At. Mol. Phys.* **12**, 2153 (1979).

[H63] A. P. Hitchcock and C. E. Brion, Inner shell electron energy loss studies of HCN and C_2N_2, *Chem. Phys.* **37**, 319 (1979).

[H64] A. P. Hitchcock and C. E. Brion, Inner-shell excitation of formaldehyde, acetaldehyde and acetone studied by electron impact, *J. Electron Spectrosc. Related Phenomena* **19**, 231 (1980).

[H65] A. P. Hitchcock, Bibliography of atomic and molecular inner-shell excitation studies, *J. Electron Spectrosc. and Related Phenomena* **25**, 245 (1982).

[H66] A. P. Hitchcock, S. Beaulieu, T. Steel, J. Stöhr and F. Sette, Carbon K-shell electron energy loss spectra of 1- and 2-butenes, *trans*-1,3-butadiene and perfluoro-2-butene. Carbon-carbon bond lengths from continuum shape resonances, *J. Chem. Phys.* **80**, 3927 (1984).

[H67] P. Hochmann, P. H. Templet, H.-t. Wang and S. P. McGlynn, Molecular Rydberg transitions. I. Low-energy Rydberg transitions in methyl halides, *J. Chem. Phys.* **62**, 2588 (1975).

[H68] D. M. P. Holland, J. B. West, A. C. Parr, D. L. Ederer, R. Stockbauer, R. D. Buff and J. L. Dehmer, Constant photoelectron energy spectroscopy of acetylene, *J. Chem. Phys.* **78**, 124 (1983).

[H69] J. M. Hollas and Z. R. Lemanczyk, Rydberg transitions of H_2Se: a rotational contour analysis of the 148.3 nm band, *J. Mol. Spectrosc.* **66**, 79 (1977).

[H70] P.-W. Hou and R. P. Rampling, Far-vacuum ultraviolet transmission spectrum of uracil thin films, *J. Appl. Phys.* **47**, 4572 (1976).

[H71] T. Huang and W. H. Hamill, Characteristic energy loss, luminescence and luminescence excitation spectra of methane and other alkane solids under low-energy electron impact, *J. Phys. Chem.* **78**, 2077 (1974).

[H72] C. Hubrich, C. Zetzsch and F. Stuhl, Absorption spectra of halogenated methanes in the region of 275 to 160 nm at temperatures of 298 and 208° K, *Chem. Ber.* **81**, 437 (1977).

[H73] C. Hubrich and F. Stuhl, The ultraviolet absorption of some halogenated methanes and ethanes of atmospheric interest, *J. Photochem.* **12**, 93 (1980).

[H74] J. W. Hudgens, T. G. DiGuiseppe and M. C. Lin, Two photon resonance enhanced multiphoton ionization spectroscopy and state assignments of the methyl radical, *J. Chem. Phys.* **79**, 571 (1983).

[H75] R. H. Huebner, C. H. Ferguson, R. J. Celotta and S. R. Mielczarek, Apparent oscillator strength distributions derived from electron energy-loss measurements: methane and *n*-hexane, *in* "Vacuum Ultraviolet Radiation Physics," E.-E. Koch, R. Haensel and C. Kunz, *eds*., Pergamon, 181 (1974).

[H76] R. H. Huebner, D. L. Bushnell, Jr., R. J. Celotta, S. R. Mielczarek and C. E. Kuyatt, Ultraviolet photoabsorption by halocarbons 11 and 12 from electron impact measurements, *Nature* **257**, 376 (1975).

[H77] L. M. Hunter, D. Lewis and W. H. Hamill, Low-energy electron reflection spectrometry for thin films of *n*-hexane, benzene and ice at 77°K, *J. Chem. Phys.* **52**, 1733 (1970).

[H78] S. R. Hunter and L. G. Christophorou, Electron attachment to the perfluoroalkanes n-C_nF_{2n+2} ($n = 1 - 6$) using high pressure swarm techniques, *J. Chem. Phys.* **80**, 6150 (1984).

[H79] R. Huzimura and C. Sumida, An electron energy-loss spectroscopic study of $CaSO_4$ crystals, *Jpn. J. Appl. Phys.* **21**, L458 (1982).

[I1] E. Illenberger, H. Baumgärtel and S. Süzer, Electron-attachment spectroscopy: formation and dissociation of negative ions in the fluorochloroethylenes, *J. Electron Spectrosc. Related Phenomena* **33**, 123 (1984).

[I2] T. Inagaki, Optical absorption of aliphatic amino acids in the far ultraviolet, *Biopolymers* **12**, 1353 (1973).

[I3] T. Inagaki, E. T. Arakawa, R. N. Hamm and M. W. Williams, Optical properties of polystyrene from the near-infrared to the X-ray region and convergence of optical sum rules, *Phys. Rev.* **B15**, 3243 (1977).

[I4] K. K. Innes, B. P. Stoicheff and S. C. Wallace, Four-wave sum mixing (130–180 nm) in molecular vapors, *Appl. Phys. Lett.* **29**, 715 (1976).

[I5] E. Ishiguro, M. Sasanuma, H. Masuko, Y. Morioka and M. Nakamura, Absorption spectra of H_2O and D_2O molecules in the vacuum ultraviolet region, *J. Phys. B: At. Mol. Phys.* **11**, 993 (1978).

[I6] E. Ishiguro, S. Iwata, Y. Suzuki, A. Mikuni and T. Sasaki, The boron K photoabsorption spectra of BF_3, BCl_3 and BBr_3, *J. Phys. B: At. Mol. Phys.* **15**, 1841 (1982).

[I7] J. Itah, B. Katz and B. Scharf, The assignment of the second ionization process in benzene, *Chem. Phys. Lett.* **47**, 245 (1977).

[I8] J. Itah, B. Katz and B. Scharf, Direct observation of the splitting of the vibrationless level in a Jahn-Teller active state in deuterobenzenes of D_{2h} symmetry, *Chem. Phys. Lett.* **48**, 111 (1977).

[I9] J. Itah, B. Katz and B. Scharf, The effect of vibrational perturbation on the vibronic structure in Jahn-Teller active states, *Chem. Phys. Lett.* **52**, 92 (1977).

[I10] S. Iwanami and N. Oda, Determination of ionization potentials of poly(glycine), poly(L-alanine) and poly-(L-valine) films by vacuum ultraviolet photoelectron spectroscopy, *Biopolymers* **19**, 1919 (1980).

[I11] S. Iwanami and N. Oda, Photoabsorption and photoelectron-yield spectra of polypeptides in the vacuum-UV region, *Radiat. Res.* **95**, 24 (1983).

[J1] J. K. Jacques and R. F. Barrow, The transition V $^1\Sigma^+ - X\ ^1\Sigma^+$ in hydrogen chloride, *Proc. Phys. Soc. London* **73**, 538 (1959).

[J2] A. Jain and D. G. Thompson, Low-energy electron scattering by H_2S molecules; elastic, rotational and vibrational excitation, *J. Phys. B: At. Mol. Phys.* **17**, 443 (1983).

[J3] J. P. Jasinski, S. L. Holt, J. H. Wood and L. B. Asprey, Experimental and calculated electronic structure of gaseous CrO_2F_2 and CrO_2Cl_2, *J. Chem. Phys.* **63**, 757 (1975).

[J4] H. W. Jochims, W. Lohr and H. Baumgärtel, Photoreactions of small organic molecules. V. Absorption, photoion and resonance photoelectron spectra of CF_3Cl, CF_2Cl_2, $CFCl_3$ in the energy range 10–25 eV, *Ber. Bunsen Gesellschaft* **80**, 130 (1976).

[J5] A. L. Johnson, E. L. Muetterties and J. Stöhr, Orientation of complex molecules chemisorbed on metal surfaces: near-edge X-ray absorption studies, *J. Amer. Chem. Soc.* **105**, 7183 (1983).

[J6] K. E. Johnson and S. Lipsky, On the location of the lowest triplet excitation in ammonia, *J. Chem. Phys.* **66**, 4719 (1977).

[J7] K. E. Johnson, K. Kim, D. B. Johnston and S. Lipsky, Electron impact spectra of methane, ethane and neopentane, *J. Chem. Phys.* **70**, 2189 (1979).

[J8] K. E. Johnson, D. B. Johnston and S. Lipsky, A re-examination of the intensity distribution in the electron energy-loss spectrum of ethylene, *J. Chem. Phys.* **70**, 3144 (1979).

[J9] K. E. Johnson, D. B. Johnston and S. Lipsky, The electron impact spectra of some mono-olefinic hydrocarbons, *J. Chem. Phys.* **70**, 3844 (1979).

[J10] P. M. Johnson, The multiphoton ionization spectrum of benzene, *J. Chem. Phys.* **64**, 4143 (1976).

[J11] P. M. Johnson, The multiphoton ionization spectrum of *trans*-1,3-butadiene, *J. Chem. Phys.* **64**, 4638 (1976).

[J12] P. M. Johnson, Molecular multiphoton ionization, *Acc. Chem. Res.* **13**, 20 (1980).

[J13] P. M. Johnson and G. M. Korenowski, The discovery of a 3p Rydberg state in benzene by three-photon resonant multiphoton ionization spectroscopy, *Chem. Phys. Lett.* **97**, 53 (1983).

[J14] K. D. Jordon, J. A. Michejda and P. D. Burrow, Electron transmission studies of the negative ion states of substituted benzenes in the gas phase, *J. Amer. Chem. Soc.* **98**, 7189 (1976).

[J15] K. D. Jordan, J. A. Michejda and P. D. Burrow, A study of the negative ion states of selected cyclodienes by electron transmission spectroscopy, *Chem. Phys. Lett.* **42**, 227 (1976).

[J16] K. D. Jordan and P. D. Burrow, Studies of the temporary anion states of unsaturated hydrocarbons by electron transmission spectroscopy, *Acc. Chem. Res.* **11**, 341 (1978).

[J17] K. D. Jordan and P. D. Burrow, Comment on the negative ion states of fluorobenzenes, *J. Chem. Phys.* **71**,, 5384 (1979).

[J18] J. Jortner and A. Gaathon, Effects of phase density on ionization processes and electron localization in fluids, *Can. J. Chem.* **55**, 1801 (1977).

[J19] R. H. Judge, C. R. Drury-Lessard and D. C. Moule, The far-ultraviolet spectrum of thioformaldehyde, *Chem. Phys. Lett.* **53**, 82 (1978).

[J20] R. H. Judge and D. C. Moule, A vibronic analysis of the $\widetilde{B}\ ^1A_1\ (\pi\pi^*) \leftarrow \widetilde{X}\ ^1A_1$ electronic transition in thiophosgene, *J. Mol. Spectrosc.* **80**, 363 (1980).

[J21] R. H. Judge, D. C. Moule, A. E. Bruno and R. P. Steer, Thioketone spectroscopy: an analysis of the lower electronic transitions in thioacetone and thioacetaldehyde, *Chem. Phys. Lett.* **102**, 385 (1983).

[J22] M. Jungen, J. Vogt and V. Staemmler, Feshbach resonances and dissociative electron attachment of H_2O, *Chem. Phys.* **37**, 49 (1979).

[K1] A. Kaito, A. Tajiri and M. Hatano, Magnetic circular dichroism of some halogenobenzenes, *Chem. Phys. Lett.* **25**, 548 (1974).

[K2] Y. Kamura, K. Seki and H. Inokuchi, Near and vacuum ultraviolet absorption spectra of polycrystalline and amorphous evaporated films of naphthacene, pentacene, perylene and coronene, *Chem. Phys. Lett.* **30**, 35 (1975).

[K3] C.-M. Kao and P. E. Cade, The electronic/positronic structure of positron/pseudohalide systems: [OH^-; e^+], [SH^-; e^+], [CN^-; e^+] and [N_3^-; e^+], *J. Chem. Phys.* **80,** 3234 (1984).

[K4] L. Karlsson, L. Mattsson, R. Jadrny, T. Bergmark and K. Siegbahn, Vibrational and vibronic structure in the valence electron spectrum of H_2S, *Phys. Scr.* **13**, 229 (1976).

[K5] A. Karpfen, *Ab initio* studies on polymers. V. All-*trans*-polyethylene, *J. Chem. Phys.* **75**, 238 (1981).

[K6] M. Karplus, R. M. Gavin, Jr. and S. A. Rice, Reinterpretation of hexatriene spectrum and comparison with theory, *J. Chem. Phys.* **63,** 5507 (1975).

[K7] E. Kassab, J. T. Gleghorn and E. M. Evleth, Model calculations of intermolecular interactions in the lowest energy Rydberg state of ammonia, *Chem. Phys. Lett.* **70**, 151 (1980).

[K8] S. Katsumata, H. Wakabayashi and K. Kimura, Effect of alkylation on the iodine 6s Rydberg states of iodoalkanes, *Bull. Chem. Soc. Jpn.* **48**, 2223 (1975).

[K9] R. Kaufel, E. Illenberger and H. Baumgärtel, Formation and dissociation of the chloroethylene anions, *Chem. Phys. Lett.* **106,** 342 (1984).

[K10] K. Kaya, K. Fuke and S. Nagakura, Polarized absorption spectra of the hydroquinone and 1,4-dimethoxybenzene crystals in the near and vacuum ultraviolet regions, *Bull. Chem. Soc. Jpn.* **47**, 438 (1974).

[K11] P. R. Keller, J. W. Taylor, T. A. Carlson and F. A. Grimm, Angle-resolved photoelectron spectroscopy of the chloro-substituted methanes, *Chem. Phys.* **79,** 269 (1983).

[K12] R. E. Kennerly, R. A. Bonham and M. McMillan, The total absolute electron scattering cross sections for SF_6 for incident energies between 0.5 and 100 eV including resonance structure, *J. Chem. Phys.* **70,** 2039 (1979).

[K13] J. E. Kent, P. J. Harman and M. F. O'Dwyer, Photochemistry of benzene isomers. 1. Fulvene and 3,4-dimethylenecyclobutene, *J. Phys. Chem.* **85**, 2726 (1981).

[K14] U. Killat, Optical properties of solid benzene derived from electron energy losses, *Z. Phys.* **263,** 83 (1973).

[K15] U. Killat, Electron energy-loss measurements on 1,3-cyclohexadiene and 1,4-cyclohexadiene, *Z. Phys.* **270,** 169 (1974).

[K16] U. Killat, Optical properties of C_6H_{12}, C_6H_{10}, C_6H_8, C_6H_6, C_7H_8, C_6H_5Cl and C_5H_5N in the solid and gaseous state derived from electron energy losses, *J. Phys. C: Solid State Phys.* **7,** 2396 (1974).

[K17] Y.-K Kim, M. Inokuti, G. E. Chamberlain and S. R. Mielczarek, Minima of generalized oscillator strengths, *Phys. Rev. Lett.* **21,** 1146 (1968).

[K18] K. Kimura, S. Katsumata, Y. Achiba, T. Yamazaki and S. Iwata, "Handbook of He I Photoelectron Spectra of Fundamental Organic Molecules," Japan Scientific Societies Press, Tokyo (1981).

[K19] K. Kimura and J. Hormes, Roles of the Rydberg transitions in fast excitation transfer studied in cyclohexane and *n*-heptane using synchrotron radiation, *J. Chem. Phys.* **79**, 2756 (1983).

[K20] G. C. King and J. W. McConkey, Electron energy-loss spectroscopy in the chloro-fluoro-methanes, *J. Phys. B: At. Mol. Phys.* **11**, 1861 (1978).

[K21] G. C. King, J. W. McConkey, F. H. Read and B. Dobson, Negative-ion resonances associated with inner-shell excited states of N_2, NO, N_2O, CO and CO_2, *J. Phys. B: Atom. Mol. Phys.* **13**, 4315 (1980).

[K22] G. W. King and A. W. Richardson, The ultraviolet absorption of cyanogen halides. Part I. Identification and correlation, *J. Mol. Spectrosc.* **21**, 339 (1966).

[K23] G. W. King and A. W. Richardson, The ultraviolet absorption of cyanogen halides. Part II. Analysis of the B and C systems, *J. Mol. Spectrosc.* **21**, 353 (1966).

[K24] N. Kizilkilic, H.-P. Schuchmann and C. von Sonntag, The photolysis of tetrahydrofuran and some of its methyl derivatives at 185 nm, *Can. J. Chem.* **58**, 2819 (1980).

[K25] E. Kloster-Jensen, H.-J. Haink and H. Christen, The electronic spectra of unsubstituted mono to penta-acetylene in the gas phase and in solution in the range 1100 to 4000 Å, *Helv. Chim. Acta* **57**, 1731 (1974).

[K26] E. Kloster-Jensen, J. P. Maier, O. Marthaler and M. Mohraz, The $\tilde{A}\ ^2\Pi_u \rightarrow \tilde{X}\ ^2\Pi_g$ emission and the photoelectron spectrum of dicyanodiacetylene radical cation, *J. Chem. Phys.* **71**, 3125 (1979).

[K27] E. E. Koch, A. Otto, V. Saile and N. Schwentner, Absorption spectra of gaseous benzene, perdeuterobenzene, pyridine and fluorinated benzenes in the VUV, *in* "Vacuum Ultraviolet Radiation Physics," E. E. Koch, R. Haensel and C. Kunz, *eds*., Pergamon, 77 (1974).

[K28] E. E. Koch and A. Otto, Optical properties of anthracene single crystals in the excitonic region of the spectrum between 4 and 10.5 eV, *Chem. Phys.* **3**, 370 (1974).

[K29] E. E. Koch, V. Saile and N. Schwentner, Fine structure in the VUV absorption spectrum of neopentane at 16 eV, *Chem. Phys. Lett.* **33**, 322 (1975).

[K30] E. E. Koch and A. Otto, Vacuum ultraviolet and electron energy loss spectroscopy of gaseous and solid organic compounds, *Int. J. Radiat. Phys. Chem.* **8**, 113 (1976).

[K31] E. E. Koch and B. F. Sonntag, Molecular spectroscopy with synchrotron radiation, Chapter 6 *in* "Synchrotron Radiation", C. Kunz, *ed., Topics in Current Physics,* Springer, Heidelberg, 269 (1979).

[K32] S. Kohda-Sudoh and S. Katagiri, MO calculations of the Rydberg transitions of the methane molecule, *Bull. Chem. Soc. Jpn.* **51**, 94 (1978).

[K33] G. Köhler, Medium effects on the lowest Rydberg transitions of saturated amines, *J. Mol. Struct.* **114**, 191 (1984).

[K34] H. Köppel, L. S. Cederbaum, W. Domcke and W. von Niessen, The Jahn-Teller effect in NH_3^+, *Mol. Phys.* **35**, 1283 (1978).

[K35] G. A. Kourouklis, K. Siomos and L. G. Christophorou, Vacuum ultraviolet absorption spectra of aromatic molecules in solution, *J. Mol. Spectrosc.* **92**, 127 (1982).

[K36] R. Krässig, D. Reinke and H. Baumgärtel, Photoreactions of small organic molecules. III. Comparison of the VUV absorption, photoelectron and differential photoionization spectra of propylene, cyclopropane and ethylene oxide, *Chem. Ber.* **79**, 116 (1976).

[K37] J. Kreile, A. Schweig and W. Thiel, Shape resonances in the valence-shell photoionization of cyanogen, *Chem. Phys. Lett.* **100**, 351 (1983).

[K38] J. Kreile, A. Schweig and W. Thiel, Shape resonances in photoionization: correlation with STO-3G MO results, *Chem. Phys. Lett.* **108**, 259 (1984).

[K39] K. Krogh-Jespersen, R. P. Rava and L. Goodman, The 51, 000 cm^{-1} state in benzene: E_{1g} Rydberg or E_{2g} valence?, *Chem. Phys. Lett.* **64**, 413 (1979).

[K40] K. Krogh-Jespersen, R. P. Rava and L. Goodman, Multiphoton ionization spectrum of fluorobenzene, *Chem. Phys.* **47**, 321 (1980).

[K41] A. Kuboyama, S. Matsuzaki, H. Tagaki and H. Arano, Studies of the $\pi \rightarrow \pi^*$ absorption bands of *p*-quinones and *o*-benzoquinone, *Bull. Chem. Soc. Jpn.* **47**, 1604 (1974).

[K42] A. Kuboyama, Studies of the $\pi \rightarrow \pi^*$ absorption bands of phenanthrenequinone, *J. Natl. Chem. Lab. Industry (Japan)* **71**, 180 (1976).

[K43] A. Kuppermann, W. M. Flicker and O. A. Mosher, Electronic spectroscopy of polyatomic molecules by low-energy variable angle electron impact, *Chem. Rev.* **79**, 77 (1979).

[K44] S. E. Kupriyanov, A. A. Perov, A. Yu. Zayats and A. N. Stepanov, Electron-impact excitation of molecules to long-lived Rydberg states, *Sov. Tech. Phys. Lett.* **7**, 369 (1981).

[K45] H. A. Kurtz and K. D. Jordan, A comparison of the positron energy levels of [F^-; e^+] and [CN^-; e^+] with the Rydberg levels of Na and NO, *J. Phys. B: At. Mol. Phys.* **12**, L473 (1979).

[K46] G. J. Kutcher and A. E. S. Green, A model for energy deposition in liquid water, *Radiat. Res.* **67**, 408 (1976).

[K47] W. Kutzelnigg, Chemical bonding in higher main group elements, *Angew. Chem. Int. Ed. Engl.* **23**, 272 (1984).

[L1] F. Lahmani, C. Lardeaux, M. Lavollee and D. Solgadi, Photodissociation of methylnitrite in the vacuum ultraviolet: I. Identification and quantum yields of electronically excited NO products, *J. Chem. Phys.* **73**, 1187 (1980).

[L2] H. Lami, Presence of a low-lying "Rydberg" band in the vapor phase absorption spectra of indole and 1-methyl indole, *Chem. Phys. Lett.* **48**, 447 (1977).

[L3] P. W. Langhoff, S. R. Langhoff and C. T. Corcoran, Photoabsorption in formaldehyde, *J. Chem. Phys.* **67**, 1722 (1977).

[L4] P. W. Langhoff, A. E. Orel, T. N. Rescigno and B. V. McKoy, Photoabsorption in formaldehyde: intensities and assignments in the discrete and continuous spectral intervals, *J. Chem. Phys.* **69**, 4689 (1978).

[L5] P. W. Langhoff, B. V. McKoy, R. Unwin and A. M. Bradshaw, Experimental and theoretical studies of the valence-shell photoionization cross sections of acetylene: strong autoionization in the $(1\pi_u^{-1})X\ ^2\Pi_u$ partial channel, *Chem. Phys. Lett.* **83**, 270 (1981).

[L6] S. R. Langhoff, S. T. Elbert, C. F. Jackels and E. R. Davidson, The $^1A_1\ \pi \rightarrow \pi^*$ state of formaldehyde, *Chem. Phys. Lett.* **29**, 247 (1974).

[L7] C. Larrieu, A. Dargelos and M. Chaillet, Theoretical study of nitrous acid electronic spectrum and photofragmentation, *Chem. Phys. Lett.* **91**, 465 (1982).

[L8] E. N. Lassettre and A. Skerbele, Generalized oscillator strengths for 7.4 eV excitation of H_2O at 300, 400 and 500 eV kinetic energy. Singlet-triplet energy differences, *J. Chem. Phys.* **60**, 2464 (1974).

[L9] A. H. Laufer, An excited state of acetylene; photochemical and spectroscopic evidence, *J. Chem. Phys.* **73**, 49 (1980).

[L10] R. E. LaVilla and R. D. Deslattes, K-absorption fine structures of sulfur in gaseous SF_6, *J. Chem. Phys.* **44**, 4399 (1966).

[L11] R. E. LaVilla, The sufur $K\beta$ emission and K-absorption spectra from gaseous H_2S. III., *J. Chem. Phys.* **62**, 2209 (1975).

[L12] J. S. Lee, T. C. Wong and R. A. Bonham, A search for coherent features in the angular dependence of the 10 eV and 14 eV energy loss lines in CCl_4 using 25 keV electrons, *J. Chem. Phys.* **63**, 1609 (1975).

[L13] L. C. Lee, R. W. Carlson, D. L. Judge and M. Ogawa, The absorption cross sections of N_2, O_2, CO, NO, CO_2, N_2O, CH_4, C_2H_4, C_2H_6 and C_4H_{10} from 180 to 700 Å, *J. Quant. Spectrosc. Radiat. Transfer* **13**, 1023 (1973).

[L14] L. C. Lee, E. Phillips and D. L. Judge, Photoabsorption cross sections of CH_4, CF_4, CF_3Cl, SF_6 and C_2F_6 from 175 to 770 Å, *J. Chem. Phys.* **67**, 1237 (1977).

[L15] L. C. Lee and C. C. Chiang, Fluorescence yield from photodissociation of CH_4 at 1060–1420 Å, *J. Chem. Phys.* **78**, 688 (1983).

[L16] T.-H. Lee and J. W. Rabalais, Photoelectron spectrum and ground state electronic structure of chromyl chloride vapor, *Chem. Phys. Lett.* **34**, 135 (1975).

[L17] B. H. Lengsfield, III, P. E. M. Siegbahn and B. Liu, *Ab-initio* assignment of the UV spectra of the ethyl, isopropyl and t-butyl radicals, *J. Chem. Phys.* **81**, 710 (1984).

[L18] C. R. Lessard and D. C. Moule, The $\widetilde{B}\ ^1A_1 \leftarrow \widetilde{X}\ ^1A_1$, $\pi^* \leftarrow \pi$ transition in thiocarbonyldifluoride at 2000 Å, *Spectrochim. Acta* **29A**, 1085 (1973).

[L19] C. R. Lessard and D. C. Moule, The $^1A_2$ (n, 3px) Rydberg state of formaldehyde, *J. Mol. Spectrosc.* **60**, 343 (1976).

[L20] C. R. Lessard and D. C. Moule, The assignment of the Rydberg transitions in the electronic absorption spectrum of formaldehyde, *J. Chem. Phys.* **66**, 3908 (1977).

[L21] M. Levi, D. Cohen, V. Schurig, H. Basch and A. Gedanken, Circular dichroism of an optically active olefin chromophore: (*R*)-3-methylcyclopentene, *J. Amer. Chem. Soc.* **102**, 6972 (1980).

[L22] M. Levi, R. Arad-Yellin, B. S. Green and A. Gedanken, Excited states of ethylene oxide. C.D. spectrum of (*S*,*S*)-(−)-2,3-dimethyl oxiran, *J. Chem. Soc., Chem. Commun.*, 847 (1980).

[L23] D. Lewis, P. B. Merkel and W. H. Hamill, Low-energy electron reflection spectrometry for thin films of aromatic and aliphatic molecules at 77 °K, *J. Chem. Phys.* **53**, 2750 (1970).

[L24] J. W. Lewis, R. V. Nauman, D. B. Bouler, Jr. and S. P. McGlynn, Molecular Rydberg states. Low-energy Rydberg states of azulene, *J. Phys. Chem.* **87**, 3611 (1983).

[L25] D. A. Lightner, J. K. Gawrónski and T. D. Bouman, Electronic structure of symmetric homoconjugated dienes. Circular dichroism of (1*S*)-2-deuterio- and 2-methyl norbornadiene and (1*S*)-2-deuterio- and 2-methyl bicyclo[2.2.2]octadiene, *J. Amer. Chem. Soc.* **102**, 5749 (1980).

[L26] E. Lindholm, G. Bieri, L. and C. Fridh, Interpretation of electron spectra. III. Spectra of formamide studied with HAM/3, *Int. J. Quantum Chem.* **14**, 737 (1978).

[L27] S. Lipsky, Ionization and excitation in non-polar organic liquids, *J. Chem. Ed.* **58**, 93 (1981).

[L28] D. H. Liskow and G. A. Segal, Ab initio theoretical calculations on the circular dichroism spectrum of *trans*-cyclooctene, *J. Amer. Chem. Soc.* **100**, 2945 (1978).

[L29] W. L. Luken, Rydberg-valence mixing in atoms and molecules, *Chem. Phys. Lett.* **61**, 162 (1979).

[L30] S. Lunell, S. Svensson, P. A. Malmqvist, U. Gelius, E. Basilier and K. Siegbahn, A theoretical and experimental study of the carbon 1s shake-up structure of benzene, *Chem. Phys. Lett.* **54**, 420 (1978).

[L31] L. S. Lussier, C. Sandorfy, A. Goursot, E. Pénigault and J. Weber, Search for the Rydberg states of transition metal complexes. Hexafluoroacetylacetonates, *J. Phys. Chem.* **88**, 5492 (1984).

[L32] D. Lynch, M.-T. Lee, R. R. Lucchese and V. McKoy, Studies of the photoionization cross sections of acetylene, *J. Chem. Phys.* **80**, 1907 (1984).

[M1] L. E. Machado, E. P. Leal, G. Csanak, B. V. McKoy and P. W. Langhoff, Photo-excitation and ionization in acetylene, *J. Electron Spectrosc. Related Phenomena* **25**, 1 (1982).

[M2] R. A. MacRae, M. W. Williams and E. T. Arakawa, Optical properties of some aromatic liquids in the vacuum ultraviolet, *J. Chem. Phys.* **61**, 861 (1974).

[M3] W. G. Mallard, J. H. Miller and K. C. Smyth, The *ns* Rydberg series of 1,3-*trans*-butadiene observed using multiphoton ionization, *J. Chem. Phys.* **79**, 5900 (1983).

[M4] J.-P. Malrieu, Comments about the representation of Rydberg and ionic excited states and their photochemistry, *Theoret. Chim. Acta* **59**, 251 (1981).

[M5] A. P. Mamedov, L. Ya. Panova, and L. A. Shabalinskaya, Ultraviolet absorption spectra of methyl-substituted ammonium halides, *Opt. Spectrosc.* **43**, 334 (1977).

[M6] P. Marmet and L. Binette, Excited states of CH_4 and CD_4 between 18 and 22 eV, *J. Phys. B: At. Mol. Phys.* **11**, 3707 (1978).

[M7] G. V. Marr and R. M. Holmes, The angular distribution of photoelectrons from CH_4 as a function of photon energy from near threshold to 30 eV, *J. Phys. B: At. Mol. Phys.* **13**, 939 (1980).

[M8] N. Martensson, P.-Å. Malmquist and S. Svensson, On the assignment of the HOMO in UF_6, *Chem. Phys. Lett.* **100**, 375 (1983).

[M9] N. Martensson, P.-Å. Malmquist, S. Svensson and B. Johansson, The electron spectrum of UF_6 recorded in the gas phase, *J. Chem. Phys.* **80**, 5458 (1984).

[M10] H. S. W. Massey, Dissociative attachment, *Endeavor* **4**, 78 (1980).

[M11] H. Masuko, Y. Morioka, M. Nakamura, E. Ishiguro and M. Sasanuma, Absorption spectrum of the H_2S molecule in the vacuum ultraviolet region, *Can. J. Phys.* **57**, 745 (1979).

[M12] D. Mathur and J. B. Hasted, Electron scattering by water and alcohol molecules, *Chem. Phys. Lett.* **34**, 90 (1975).

[M13] D. Mathur and J. B. Hasted, Temporary negative-ion states in pyridine and diazine molecules, *Chem. Phys.* **16**, 347 (1976).

[M14] D. Mathur and J. B. Hasted, Resonant scattering of slow electrons from benzene and substituted benzene molecules, *J. Phys. B: At. Mol. Phys.* **9**, L31 (1976).

[M15] D. Mathur and J. B. Hasted, Resonant scattering of slow electrons from naphthalene vapor, *Chem. Phys. Lett.* **48**, 50 (1977).

[M16] D. Mathur, Collisions of slow electrons with methane: ionization, fragmentation and resonances, *J. Phys. B: At. Mol. Phys.* **13**, 4703 (1980).

[M17] D. Mathur, Electron-induced proton production by dissociative autoionization in CH_4, *Chem. Phys. Lett.* **81**, 115 (1981).

[M18] D. Mathur, F. A. Rajgara and A. Roy, Inner-shell resonances in elastic scattering of electrons from N_2 and CH_4, *Chem. Phys. Lett.* **104**, 500 (1984).

[M19] D. Mathur, F. A. Rajgara and A. Roy, Differential elastic scattering of electrons of the carbon K-shell resonance in methane, *Chem. Phys. Lett.* **107**, 39 (1984).

[M20] C. Mayhew, M. A. Baig and J. P. Connerade, Rydberg transitions in the H_2Se molecule, *J. Phys. B: At. Mol. Phys.* **16**, L757 (1983).

[M21] L. N. Mazalov, A. P. Sadovskii, V. M. Bertenev, V. V. Murakhtanov, E. A. Gal'tsova and L. I. Chernyavskii, X-ray spectra of the H_2S molecule and its electronic structure, *Teor. Eksp. Khim.* **7**, 46 (1971).

[M22] D. S. McClure, Electronic spectra of molecules and ions in crystals, *in* "Solid State Physics," F. Seitz and D. Turnbull, *eds.*, Academic, New York, 399 (1959).

[M23] W. L. McCubbin and R. Manne, Simple LCAO band calculation for the ideal polyethylene chain, *Chem. Phys. Lett.* **2**, 230 (1968).

[M24] R. McDiarmid, C—O torsional vibrations in the 1900 Å transition of diethyl ether, *J. Chem. Phys.* **60**, 3340 (1974).

[M25] R. McDiarmid, On the 1700 Å transition of dimethyl ether, *J. Chem. Phys.* **60**, 4091 (1974).

[M26] R. McDiarmid, Assignments of Rydberg and valence transitions in the electronic absorption spectrum of dimethyl sulfide, *J. Chem. Phys.* **61**, 274 (1974).

[M27] R. McDiarmid, Assignments in the ultraviolet spectra of MoF_6 and WF_6, *J. Chem. Phys.* **61**, 3333 (1974).

[M28] R. McDiarmid, On the forbidden 2000 Å transition of *trans*-1,3-butadiene, *Chem. Phys. Lett.* **34**, 130 (1975).

[M29] R. McDiarmid, On the ultraviolet spectrum of *trans*-1,3-butadiene, *J. Chem. Phys.* **64**, 514 (1976).

[M30] R. McDiarmid, Assignments in the electronic spectrum of UF_6, *J. Chem. Phys.* **65**, 168 (1976).

[M31] R. McDiarmid, An experimental observation of the origin of the V ← N transition of ethylene, *J. Chem. Phys.* **67**, 3835 (1977).

[M32] R. McDiarmid, Assignments in the ultraviolet spectra of symmetrically substituted sulfides, *J. Chem. Phys.* **68**, 945 (1978).

[M33] R. McDiarmid, Transitions in the electronic absorption spectra of *cis* and *trans*-difluoroethylene, *J. Chem. Phys.* **69**, 2043 (1978).

[M34] R. McDiarmid and J. P. Doering, Anomalous V_1 ← N vibrational band intensities in the low energy electron impact spectra of *trans*-dienes, *J. Chem. Phys.* **73**, 4192 (1980).

[M35] R. McDiarmid, A reinvestigation of the absorption spectrum of ethylene in the vacuum ultraviolet, *J. Phys. Chem.* **84**, 64 (1980).

[M36] R. McDiarmid, On the orbital energies and symmetries of MoF_6, WF_6 and ReF_6, *Chem. Phys. Lett.* **76**, 300 (1980).

[M37] R. McDiarmid and A. Auerbach, Four-photon resonant multiphoton ionization spectroscopy, *Chem. Phys. Lett.* **76**, 520 (1980).

[M38] R. McDiarmid and J. P. Doering, Electron impact investigation of the electronic transitions of 1,4-cyclohexadiene and 1,5-cyclooctadiene, *J. Chem. Phys.* **75**, 2687 (1981).

[M39] R. McDiarmid and J. P. Doering, On the valence excited states of conjugated dienes, *Chem. Phys. Lett.* **88**, 602 (1982).

[M40] R. McDiarmid, Comment on "Nonadiabatic treatment of the intensity distribution in the V-N bands of ethylene," *J. Chem. Phys.* **79**, 3170 (1983).

[M41] R. McDiarmid, A. Sabljić and J. P. Doering, The NV_2 valence transition of 2,3-dimethylbutadiene, *J. Chem. Phys.* **80**, 4561 (1984).

[M42] S. P. McGlynn, S. Chattopadhyay, P. Hochmann and H.-t. Wang, Molecular Rydberg transitions. IX. Term value-ionization energy correlations, *J. Chem. Phys.* **68**, 4738 (1978).

[M43] S. P. McGlynn, J. D. Scott, W. S. Felps and G. L. Findley, Molecular Rydberg transitions. XVI. MCD of CH_3Br, *J. Chem. Phys.* **72**, 421 (1980).

[M44] S. P. McGlynn, W. S. Felps, J. D. Scott and G. L. Findley, Molecular Rydberg transitions. XVIII. Vibronic doubling in methyl iodide, *J. Chem. Phys.* **73**, 4925 (1980).

[M45] S. P. McGlynn and G. L. Findley, Giant molecule interactions, *J. Photochem.* **17**, 461 (1981).

[M46] S. P. McGlynn, W. S. Felps and G. L. Findley, Molecular Rydberg transitions. The $\pi \rightarrow 4s$ transition of ClCN, *Chem. Phys. Lett.* **78**, 89 (1981).

[M47] S. P. McGlynn, J. D. Scott and W. S. Felps, Molecular Rydberg spectroscopy magnetic field effects in alkyl halides, *J. Phys. (Paris), Colloq. C2* **43**, 305 (1982).

[M48] V. McKoy, T. A. Carlson and R. R. Lucchese, Photoelectron dynamics of molecules, *J. Phys. Chem.* **88**, 3188 (1984).

[M49] T. G. McLaughlin and L. B. Clark, The electronic spectrum of biphenyl, *Chem. Phys.* **31**, 11 (1978).

[M50] B. N. McMaster, J. Mrozek and V. H. Smith, Jr., An ab initio molecular orbital study of the ammonium radical, *Chem. Phys.* **73**, 131 (1982).

[M51] C. R. McMillin, W. B. Rippon and A. G. Walton, Vacuum ultraviolet spectroscopy of poly-α-amino acids, *Biopolymers* **12**, 589 (1973).

[M52] L. E. McMurchie and E. R. Davidson, Configuration interaction calculations on the planar $^1(\pi, \pi^*)$ state of ethylene, *J. Chem. Phys.* **66**, 2959 (1977).

[M53] L. E. McMurchie and E. R. Davidson, Singlet Rydberg states of ethylene, *J. Chem. Phys.* **67**, 5613 (1977).

[M54] I. Messing, B. Raz and J. Jortner, Medium effects on the vibrational structure of some molecular Rydberg excitations, *Chem. Phys.* **23**, 351 (1977).

[M55] I. Messing, B. Raz and J. Jortner, Perturbations of molecular extravalence excitations by rare-gas fluids, *Chem. Phys.* **25**, 55 (1977).

[M56] S. R. Mielczarek and K. J. Miller, Dependence of generalized oscillator strengths of H_2O on momentum transfer, *Chem. Phys. Lett.* **10**, 369 (1971).

[M57] F. H. Mies, Perturbed Rydberg series: relationship between quantum-defect and configuration-interaction theory, *Phys. Rev.* **20A**, 1773 (1979).

[M58] M. Miladi, J.-P. Le Falher, J.-Y. Roncin and H. Damany, Pressure effects on the vibronic transitions of NH_3 and NO, *J. Mol. Spectrosc.* **55**, 81 (1975).

[M59] J. C. Miller, R. N. Compton and C. D. Cooper, Vacuum ultraviolet spectroscopy of molecules using third-harmonic generation in rare gases, *J. Chem. Phys.* **76**, 3967 (1982).

[M60] K. J. Miller, Dependence of generalized oscillator strengths of ethylene on momentum transfer, *J. Chem. Phys.* **51**, 5235 (1969).

[M61] K. J. Miller, Study of cross-sections, oscillator strengths, generalized oscillator strengths, and atomic like character of the molecular orbitals of formaldehyde for inelastic transitions to valence and Rydberg states, *J. Chem. Phys.* **62**, 1759 (1975).

[M62] B. E. Mills and D. A. Shirley, K-shell correlation-state spectra in formamide, *Chem. Phys. Lett.* **39**, 236 (1976).

[M63] K. M. Monahan, R. L. Russell and W. C. Walker, Classifying Rydberg and valence transitions in CF_2Cl_2, *J. Chem. Phys.* **64**, 5309 (1976).

[M64] D. Moncrieff, I. H. Hillier, S. A. Pope and M. F. Guest, Satellite structure in the 1s photoelectron spectra of CH_4, NH_3 and H_2O calculated by ab initio methods, *Chem. Phys.* **82**, 139 (1983).

[M65] R. Montagnani, P. Riani and O. Salvetti, *Ab initio* calculation of some low-lying electronic states of methane, *Theoret. Chim. Acta* **32**, 161 (1973).

[M66] J. H. Moore, Y. Sato and S. W. Staley, Energies and relative cross-sections of singlet-triplet transitions in methyl-substituted 1,3-butadienes by ion-impact spectroscopy, *J. Chem. Phys.* **69**, 1092 (1978).

[M67] H. Morawitz and P. Bagus, Excitonic fine structure of the arsenic K-edge in AsF_n ($n = 3, 5, 6$) cage molecules, *Chem. Phys. Lett.* **107**, 59 (1984).

[M68] H. Morita, K. Fuke and S. Nagakura, Electronic structure and spectra of acrylic acid in the vapor and condensed phases, *Bull. Chem. Soc. Jpn.* **49**, 922 (1976).

[M69] H. Morita, K. Fuke and S. Nagakura, Charge transfer character in the intramolecular hydrogen bond: electron structures and spectra of hydrogen maleate anion and related molecules, *Bull. Chem. Soc. Jpn.* **50**, 645 (1977).

[M70] R. A. Morris, C. J. Patrissi, D. J. Sardella, P. Davidovits and D. L. McFadden, Efficient vibrational excitation of gaseous CF_4 and C_2F_6 by low-energy electrons, *Chem. Phys. Lett.* **102**, 41 (1983).

[M71] O. A. Mosher, W. M. Flicker and A. Kuppermann, Triplet states in 1,3-butadiene, *Chem. Phys. Lett.* **19**, 332 (1973).

[M72] O. A. Mosher, M. S. Foster, W. M. Flicker, A. Kuppermann and J. L. Beauchamp, Electron impact spectroscopy of *trans*-azomethane, *Chem. Phys. Lett.* **29**, 236 (1974).

[M73] O. A. Mosher, W. M. Flicker and A. Kuppermann, Electronic spectroscopy of propadiene (allene) by electron impact, *J. Chem. Phys.* **62**, 2600 (1975).

[M74] O. A. Mosher, M. S. Foster, W. M. Flicker, J. L. Beauchamp and A. Kuppermann, Electronic spectroscopy of *trans*-azomethane by electron impact, *J. Chem. Phys.* **62**, 3424 (1975).

[M75] D. C. Moule and A. D. Walsh, Ultraviolet spectra and excited states of formaldehyde, *Chem. Rev.* **75**, 67 (1975).

[M76] J. Müller and S. Canuto, Theoretical studies of photodissociation and Rydbergization in the first triplet state (3s $^3A_2''$) of ammonia, *Chem. Phys. Lett.* **70**, 236 (1980).

[M77] J. Müller, H. Ågren and S. Canuto, *Ab initio* studies of the photodissociation in the first excited states of $\tilde{A}\ ^1A_1$ and $\tilde{a}\ ^3A_1$ of PH_3, *J. Chem. Phys.* **76**, 5060 (1982).

[M78] W. Müller, C. Nager and P. Rosmus, The frozen orbital approximation for calculating ionization energies with application to propane, *Chem. Phys.* **51**, 43 (1980).

[M79] R. S. Mulliken, Mixed V states, *Chem. Phys. Lett.* **25**, 305 (1974).

[M80] R. S. Mulliken, Rydberg states and Rydbergization, *Acc. Chem. Res.* **9**, 7 (1976).

[M81] R. S. Mulliken, Rydberg and valence shell states and their interaction, *Chem. Phys. Lett.* **46**, 197 (1977).

[M82] R. S. Mulliken, The excited states of ethylene, *J. Chem. Phys.* **66**, 2448 (1977).

[M83] R. S. Mulliken, The excited states of ethylene, *J. Chem. Phys.* **71**, 556 (1979).

[N1] Y. Nagahira, H. Fukutome and Y. Jido, Vacuum ultraviolet absorption spectra of stearic acid multilayers, *Chem. Phys. Lett.* **34**, 95 (1975).

[N2] Y. Nagahira, K. Matsuki and H. Fukutome, The electronic absorption spectra of fatty acid mono-, bi- and multilayers with bivalent metal ions, *Bull. Chem. Soc. Jpn.* **54**, 1208 (1981).

[N3] Y. Nagahira, K. Matsuki and H. Fukutome, The vacuum ultraviolet absorption spectra of fatty acid multilayers, *Bull. Chem. Soc. Jpn.* **54**, 1217 (1981).

[N4] H. Nakanishi, H. Morita and S. Nagakura, Charge transfer bands observed with the intermolecular hydrogen-bonded systems between acetic acid and some aliphatic amines, *J. Mol. Spectrosc.* **65**, 295 (1977).

[N5] H. Nakanishi, H. Morita and S. Nagakura, Electronic structures and spectra of the keto and enol forms of acetylacetone, *Bull. Chem. Soc. Jpn.* **50**, 2255 (1977).

[N6] H. Nakanishi, H. Morita and S. Nagakura, Charge-transfer character in the intramolecular hydrogen bond: vacuum ultraviolet spectra of acetylacetone and its fluoro derivative, *Bull. Chem. Soc. Jpn.* **51**, 1723 (1978).

[N7] N. Nakashima, M. Sumitani, I. Ohmine and K. Yoshihara, Nanosecond laser photolysis of the benzene monomer and eximer, *J. Chem. Phys.* **72**, 2226 (1980).

[N8] N. Nakashima, H. Inoue, M. Sumitani and K. Yoshihara, Laser flash photolysis of benzene. III. $S_n \leftarrow S_1$ absorption of gaseous benzene, *J. Chem. Phys.* **73**, 5976 (1980).

[N9] Y. Nakato and H. Tsubomura, Emission from molecular "Rydberg" states in organic media, *J. Lumin.* **12/13**, 845 (1976).

[N10] H. Nakatsuji, K. Ohta and K. Hirao, Cluster expansion of the wave function. Electron correlations in the ground state, valence and Rydberg excited states, ionized states and electron attached states of formaldehyde by SAC and SAC-CI theories, *J. Chem. Phys.* **75**, 2952 (1981).

[N11] H. Nakatsuji, Cluster expansion of the wave function. Valence and Rydberg excitations, ionizations and inner-valence ionizations of CO_2 and N_2O studied by the SAC and SAC-CI theories, *Chem. Phys.* **75**, 425 (1983).

[N12] H. Nakatsuji, Cluster expansion of the wave function. Valence and Rydberg excitations and ionizations of ethylene, *J. Chem. Phys.* **80**, 3703 (1984).

[N13] M. S. Nakhmanson and V. I. Baranovskii, Electronic structure of BF_3 and BCl_3 molecules and interpretation of X-ray absorption spectra of boron in these compounds, *Theoret. Exp. Chem.* **7**, 15 (1971).

[N14] M. A. C. Nascimento and W. A. Goddard, III, The excited electronic states of all-*trans*-1,3,5-hexatriene, *Chem. Phys. Lett.* **60**, 1977 (1979).

[N15] M. A. C. Nascimento and W. A. Goddard, III, The valence electronic excited states of trans-1,3-butadiene and trans, trans,-1,3,5-hexatriene from generalized valence bond and configuration interaction calculations, *Chem. Phys.* **36**, 147 (1979).

[N16] M. A. C. Nascimento and W. A. Goddard, III, The Rydberg states of trans-butadiene from generalized valence bond and configuration interaction calculations, *Chem. Phys.* **53**, 251 (1980).

[N17] C. R. Natoli, Near edge absorption structure in the framework of the multiple scattering model. Potential resonance or barrier effects? *in* "EXAFS and Near Edge Structure," A. Bianconi, L. Incoccia and S. Stipcich, *eds.*, Springer, New York, 43 (1983).

[N18] V. I. Nefedov, Yu. A. Buslaev, N. P. Sergushin, L. Baier, Yu. V. Kokunov and A. A. Kuznetsova, Regularities in the electronic structure of isoelectronic compounds, *Izv. Acad. Nauk USSR Ser. Fiz.* **38**, 448 (1974).

[N19] D. A. Nelson and J. J. Worman, The far ultraviolet spectra of some simple azomethines, *Tetrahedron Lett.*, 507 (1966).

[N20] P. S. Neudorfl, R. A. Back and A. E. Douglas, The absorption spectrum of trans-diimide (N_2H_2) in the vacuum ultraviolet region, *Can. J. Chem.* **59**, 506 (1981).

[N21] L. Ng, V. Balaji and K. D. Jordan, Measurement of the vertical electron affinities of cyanogen and 2,4 hexadiyne, *Chem. Phys. Lett.* **101**, 171 (1983).

[N22] U. Nielsen, R. Haensel and W. H. E. Schwarz, The electronic and geometric structure of the free XeF_6 molecule, *J. Chem. Phys.* **61**, 3581 (1974).

[N23] U. Nielsen and W. H. E. Schwarz, VUV spectra of the xenon fluorides, *Chem. Phys.* **13**, 195 (1976).

[N24] G. C. Nieman and S. D. Colson, A new electronic state of ammonia observed by multiphoton ionization, *J. Chem. Phys.* **68**, 5656 (1978).

[N25] G. C. Nieman and S. D. Colson, Characterization of the $\widetilde{C}'$ state of ammonia observed by three-photon, gas-phase spectroscopy, *J. Chem. Phys.* **71**, 571 (1979).

[N26] C. Nishijima, H. Nakayama, T. Kobayashi and K. Yokota, Photoelectron spectrum of malonaldehyde, *Chem. Lett. (Jpn.)*, 5 (1975).

[N27] L. E. Nitzsche and E. R. Davidson, A perturbation theory calculation on the $^1\pi\pi^*$ state of formamide, *J. Chem. Phys.* **68**, 3103 (1978).

[N28] L. E. Nitzsche and E. R. Davidson, Ab initio calculation of some vertical excitation energies of N-methyl acetamide, *J. Amer. Chem. Soc.* **100**, 7201 (1978).

[N29] J. A. Nuth and S. Glicker, The vacuum ultraviolet spectra of HCN, C_2N_2 and CH_3CN, *J. Quant. Spectrosc. Radiat. Transfer* **28**, 223 (1982).

[O1] H. Ogata, J. Kitayama, M. Koto, S. Kojima, Y. Nihei and H. Kamada, Vacuum ultraviolet absorption and photoelectron spectra of aliphatic ketones, *Bull. Chem. Soc. Jpn.* **47**, 958 (1974).

[O2] I. Ya. Ogurtsov and L. A. Kazantseva, Interpretation of the angular dependence of the electron impact spectrum for the $\widetilde{A} \leftarrow \widetilde{X}$ transition in ammonia, *J. Mol. Struct.* **73,** 85 (1981).

[O3] T. Ohta, T. Fujikawa and H. Kuroda, Satellite bands in the gas-phase X-ray photoelectron spectra of benzene and its mono-substituted derivatives, *Chem. Phys. Lett.* **32,** 369 (1975).

[O4] S. Oikawa, M. Tsuda, N. Ueno and K. Sugita, Excited states of poly(vinyl cinnamate) studied by low-energy electron loss spectroscopy and INDO/S computation, *Chem. Phys. Lett.* **74,** 379 (1980).

[O5] H. Okabe, Photodissociation of acetylene and bromoacetylene in the vacuum ultraviolet: production of electronically excited C_2H and C_2, *J. Chem. Phys.* **62,** 2782 (1975).

[O6] H. Okabe, M. Kawasaki and Y. Tanaka, The photodissociation of CH_2I_2: production of electronically excited I_2, *J. Chem. Phys.* **73,** 6162 (1980).

[O7] H. Okabe, Photochemistry of acetylene at 1470 Å, *J. Chem. Phys.* **75,** 2772 (1981).

[O8] T. Okabe, Characteristic energy loss spectra and optical constants of some solid hydrocarbons, *J. Phys. Soc. Jpn.* **35,** 1496 (1973).

[O9] P. R. Olivato, H. Viertler, B. Wladislaw, P. Sauvageau and C. Sandorfy, The far ultraviolet spectra of $CH_3SCH_2SCH_3$ and $CH_3SCH_2COCH_3$, *J. Chem. Phys.* **70,** 1677 (1979).

[O10] E. Oliveros, M. Riviere, C. Teichteil, and J.-P. Malrieu, CI(CIPSI) calculations of the vertical ionization and excitation energies of the formamide molecule, *Chem. Phys. Lett.* **57,** 220 (1978).

[O11] U. Olsher, R. Lubart and M. Brith, The electronic spectrum of pyridine in rare gas matrices, *Chem. Phys.* **17,** 237 (1976).

[O12] U. Olsher, The electronic absorption spectrum of gaseous pyridine 1800 Å $\pi \rightarrow \pi^*$ transition, *J. Chem. Phys.* **66,** 5242 (1977).

[O13] R. Onaka and H. Ito, Electrooptic effect and vacuum ultraviolet absorption of KDP and ADP, *Izvest. Akad. Nauk. SSSR, Ser. Fiz.* **41,** 530 (1977).

[O14] G. O'Sullivan, The absorption spectrum of CH_3I in the extreme VUV, *J. Phys. B: At. Mol. Phys.* **15,** L327 (1982).

[O15] G. O'Sullivan, Chlorine L-edge absorption in CCl_4 and CCl_2F_2, *J. Phys. B: At. Mol. Phys.* **15,** 2385 (1982).

[P1] M. N. Paddon-Row, N. G. Rondan, K. N. Houk, and K. D. Jordan, Geometries of the radical anions of ethylene, fluoroethylene, 1,1-difluoroethylene and tetrafluoroethylene, *J. Amer. Chem. Soc.* **104,** 1143 (1982).

[P2] L. R. Painter, T. S. Riedinger, R. D. Birkhoff and J. M. Heller, Jr., Optical properties of polycrystalline anthracene in the 3.2–9.3 eV spectral region, *J. Appl. Phys.* **51,** 1747 (1980).

[P3] L. R. Painter, E. T. Arakawa, M. W. Williams and J. C. Ashley, Optical properties of polyethylene: measurement and applications, *Radiat. Res.* **83,** 1 (1980).

[P4] L. R. Painter, J. S. Attrey, H. H. Hubbell, Jr. and R. D. Birkhoff, Vacuum ultraviolet optical properties of squalane and squalene, *J. Appl. Phys.* **55,** 756 (1984).

[P5] J. A. Paisner, M. L. Spaeth, D. C. Gerstenberger and I. W. Ruderman, Generation of tunable radiation below 2000 Å by phase-matched sum-frequency mixing in $KB_5O_8 \cdot 4D_2O$, *Appl. Phys. Lett.* **32,** 476 (1978).

[P6] E. Pantos, A. M. Taleb, T. D. S. Hamilton and I. H. Munro, Low-temperature spectra of the $^1B_{1u} \leftarrow {}^1A_{1g}$ transition of C_6H_6, sym-$C_6H_3D_3$ and C_6D_6, *Mol. Phys.* **28,** 1139 (1974).

[P7] E. Pantos, J. Philis and A. Bolovinos, The extinction coefficient of benzene vapor in the region 4.6 to 36 eV, *J. Mol. Spectrosc.* **72,** 36 (1978).

[P8] L. A. Paquette, C. W. Doecke, F. R. Kearney, A. F. Drake and S. F. Mason, Chiral perturbation of olefins by deuterium substitution. The optical activity and circular dichroism behavior of (1*S*)-[2-2H] norbornene and deuterated apoborenes, *J. Amer. Chem. Soc.* **102,** 7228 (1980).

[P9] L. A. Paquette, F. R. Kearney, A. F. Drake and S. F. Mason, The absorption and circular dichroism spectra of chiral triquinacenes, *J. Amer. Chem. Soc.* **103,** 5064 (1981).

[P10] D. H. Parker, S. J. Sheng and M. A. El-Sayed, Multiphoton ionization spectrum of *trans-* hexatriene in the 6.2 eV region, *J. Chem. Phys.* **65,** 5534 (1976).

[P11] D. H. Parker and P. Avouris, Multiphoton ionization spectra of two caged amines, *Chem. Phys. Lett.* **53,** 515 (1978).

[P12] D. H. Parker, J. O. Berg and M. A. El-Sayed, The symmetry of the 6.2 eV two-photon Rydberg state in hexatriene from the polarization properties of the multiphoton ionization spectrum, *Chem. Phys. Lett.* **56,** 197 (1978).

[P13] D. H. Parker, J. O. Berg and and M. A. El-Sayed, Multiphoton ionization spectroscopy, *in* "Advances in Laser Chemistry", A. H. Zewail, *ed.*, Springer, Berlin, 319 (1978).

[P14] D. H. Parker and P. Avouris, Multiphoton ionization and two-photon fluorescence excitation spectroscopy of triethylenediamine, *J. Chem. Phys.* **71,** 1241 (1979).

[P15] A. C. Parr, A. J. Jason and R. Stockbauer, Photoionization and threshold photoelectron-photoion coincidence study of allene from onset to 20 eV, *Int. J. Mass Spectrom. Ion Phys.* **26**, 23 (1978).

[P16] A. C. Parr, A. J. Jason and R. Stockbauer, Photoionization and threshold photoelectron-photoion coincidence study of cyclopropene from onset to 20 eV, *Int. J. Mass Spectrom. Ion Phys.* **33**, 243 (1980).

[P17] A. C. Parr, D. L. Ederer, J. B. West, D. M. P. Holland and J. L. Dehmer, Triply differential photoelectron studies of non-Franck-Condon behavior in the photoionization of acetylene, *J. Chem. Phys.* **76**, 4349 (1982).

[P18] A. A. Pavlychev, A. S. Vinogradov, D. E. Onopko and S. A. Titov, Structure of continuous X-ray absorption near the $L_{2,3}$ (2p) ionization threshold of silicon in SiF_6^{2-}, *Sov. Phys. Solid State* **20**, 2121 (1978).

[P19] A. A. Pavlychev, A. S. Vinogradov, T. M. Zimkina, D. E. Onopko and S. A. Titov, Structure of the x-ray absorption spectra of a number of tetrahedral molecules, *Opt. Spectrosc.* **47**, 40 (1979).

[P20] A. A. Pavlychev, A. S. Vinogradov, T. M. Zimkina, D. E. Onopko and S. A. Titov, Structure of the X-ray absorption spectra of tetrahedral molecules. The Si $L_{II,III}$ and F K spectra of the SiF_4 molecule, *Opt. Spectrosc.* **52**, 302 (1982).

[P21] J. C. Person and P. P. Nicole, New determinations of the oscillator strength distribution and the photoionization yields for methane and n-hexane, *in* "Vacuum Ultraviolet Radiation Physics," E.-E. Koch, R. Haensel and C. Kunz, *eds.*, Pergamon, 184 (1974).

[P22] J. C. Person and P. P. Nicole, Absorption cross sections at high energies. Neopentane from 10 to 21 eV and ethane from 22 to 54 eV, *in* "Fifth International Conference on Vacuum Ultraviolet Radiation Physics," M. C. Castex, M. Pouey and N. Pouey, *eds.*, CNRS, Meudon, France, 111 (1977).

[P23] C. Petrongolo, R. J. Buenker and S. D. Peyerimhoff, Nonadiabatic treatment of the intensity distribution in the V-N bands of ethylene, *J. Chem. Phys.* **76,** 3655 (1982).

[P24] S. D. Peyerimhoff, Study of the mixing of Rydberg and valence states using the MRD-CI method, *Gazz. Chim. Ital.* **108,** 411 (1978).

[P25] S. D. Peyerimhoff and R. J. Buenker, Potential curves for dissociative electron attachment of $CFCl_3$, *Chem. Phys. Lett.* **65,** 434 (1979).

[P26] S. D. Peyerimhoff and R. J. Buenker, Calculation of electronically excited states in molecules: intensity and vibrational structure of spectra, photochemical implications, *in* "Computational Methods in Chemistry," J. Bargon, *ed.*, Plenum Corp., N.Y., N.Y., 175 (1980).

[P27] J. Philis, E. Pantos, G. Andritsopoulos, A. Bovolinos and A. Ioannidou, Detection of the $\pi\pi^*$ E_{2g} state in sym-tetrafluorobenzene, *Chem. Phys. Lett.* **77,** 623 (1981).

[P28] J. Philis, A. Bovolinos, G. Andritsopoulos, E. Pantos and T. D. S. Hamilton, Comment on confirmation of the 3s Rydberg assignment of fluorobenzene, *Chem. Phys. Lett.* **77,** 627 (1981).

[P29] J. Philis, A. Bovolinos, G. Andritsopoulos, E. Pantos and P. Tsekeris, A comparison of the absorption spectra of the fluorobenzenes and benzene in the region 4.5–9.5 eV, *J. Phys. B: At. Mol. Phys.* **14,** 3621 (1981).

[P30] E. Phillips, L. C. Lee and D. L. Judge, Absolute photoabsorption cross sections for H_2O and D_2O from λ 180–790 Å, *J. Quant. Spectrosc. Radiat. Transfer* **18,** 309 (1977).

[P31] M. N. Pisanias, L. G. Christophorou and J. G. Carter, Compound negative ion resonances and threshold electron excitation spectra of quinoline and isoquinoline, *Chem. Phys. Lett.* **13,** 433 (1972).

[P32] A. A. Plankaert, J. Doucet and C. Sandorfy, Comparative study of the vacuum ultraviolet absorption and photoelectron spectra of some simple ethers and thioethers, *J. Chem. Phys.* **60,** 4846 (1974).

[P33] V. G. Plekhanov and V. S. Osminin, Reflection spectra of alkali-metal sulfates at 78 K, *Opt. Spectrosc.* **39,** 337 (1975).

[P34] D. E. Post, Jr., W. M. Hetherington, III and B. Hudson, 100 eV electron impact excitation spectra of 1,3,5-hexatriene, *Chem. Phys. Lett.* **35,** 259 (1975).

[P35] A. M. Pravilov, F. I. Vilesov, V. A. Elokhin, V. S. Ivanov and A. S. Kozlov, Absorption spectra and primary photolysis processes of some fluorinated organic compounds in the near and vacuum ultraviolet, *Sov. J. Quantum Electron.* **8,** 355 (1978).

[P36] W. C. Price, A. W. Potts and J. A. Williams, The orbital interpretation of the photoelectron spectrum of benzene, 1,3,5-trifluorobenzene and hexafluorobenzene, *Chem. Phys. Lett.* **37,** 17 (1976).

[R1] M. Rabrenović, C. J. Proctor, T. Ast, C. G. Herbert, A. G. Brenton and J. H. Beynon, Charge stripping of hydrocarbon positive ions, *J. Phys. Chem.* **87,** 3305 (1983).

[R2] P. Rancurel, B. Huron, L. Praud, J. P. Malrieu and G. Berthier, The electronic spectra of benzene and its conjugated isomers: A full perturbative CI approach, *J. Mol. Spectrosc.* **60,** 259 (1976).

[R3] B. D. Ransom, K. K. Innes and R. McDiarmid, Polarization of the strongest Rydberg transitions of 1,3-butadiene, *J. Chem. Phys.* **68,** 2007 (1978).

[R4] A. Rauk, A. F. Drake and S. F. Mason, Excited states and optical activity of allenes. Allene, 1,3-dimethyl allene and 1,2-cyclononadiene, *J. Amer. Chem. Soc.* **101,** 2284 (1979).

[R5] A. Rauk and S. Collins, The ground and excited states of hydrogen sulfide, methanethiol and hydrogen selenide, *J. Mol. Spectrosc.* **105**, 438 (1984).

[R6] A. Rauk, The optical activity of the three-membered ring: oxiranes, aziridines, diaziridines and oxaziridines, *J. Amer. Chem. Soc.* **103**, 1023 (1981).

[R7] S. Raynor and D. R. Hershbach, Electronic structure of Rydberg states of H_3, NeH, H_2F, H_3O, NH_4 and CH_5 molecules, *J. Phys. Chem.* **86**, 3592 (1982).

[R8] F. H. Read, A modified Rydberg formula, *J. Phys. B: At. Mol. Phys.* **10**, 449 (1977).

[R9] F. H. Read, A new class of atomic states: the "Wannier-ridge" resonances, *Aust. J. Phys.* **35**, 475 (1982).

[R10] L. Resca and R. Resta, Rydberg states in condensed matter, *Phys. Rev.* **19B**, 1683 (1979).

[R11] M. Rei Vilar, M. Heyman and M. Schott, Spectroscopy of low-energy electrons backscattered from an organic solid surface: pentacene, *Chem. Phys. Lett.* **94**, 522 (1983).

[R12] R. Rianda, R. P. Frueholz and W. A. Goddard, III, The low lying states of ammonia; generalized valence bond and configuration interaction studies, *Chem. Phys.* **19**, 131 (1977).

[R13] R. Rianda, R. P. Frueholz and A. Kuppermann, Electronic spectroscopy of UF_6 and WF_6 by electron impact, *J. Chem. Phys.* **70**, 1056 (1979).

[R14] R. Rianda, R. P. Frueholz and A. Kuppermann, Singlet → triplet transitions in C≡N containing molecules by electron impact, *J. Chem. Phys.* **80**, 4035 (1984).

[R15] A. Richartz, The positive ions of ethane, *Prog. Theoret. Org. Chem.* **2**, 64 (1977).

[R16] A. Richartz, R. J. Buenker, P. J. Bruna and S. D. Peyerimhoff, Stability and structure of the $C_2H_6^+$ ion: investigation of the photoelectron spectrum of ethane below 14 eV using *ab initio* methods, *Mol. Phys.* **33**, 1345 (1977).

[R17] A. Richartz, R. J. Buenker and S. D. Peyerimhoff, Ab initio MRD-CI study of ethane: the 14–25 eV PES region and Rydberg states of positive ions, *Chem. Phys.* **28**, 305 (1978).

[R18] A. Richartz, R. J. Buenker and S. D. Peyerimhoff, Calculation of the vertical electronic spectrum of propane, *Chem. Phys.* **31**, 187 (1978).

[R19] D. M. Rider, G. W. Ray, E. J. Darland and G. E. Leroi, A photoionization mass spectrometric investigation of CH_3CN and CD_3CN, *J. Chem. Phys.* **74**, 1652 (1981).

[R20] J. J. Ritsko, N. O. Lipari, P. C. Gibbons, S. E. Schnatterly, J. R. Fields and R. Devaty, Observation of electric monopole transitions in tetracyanoquinodimethane, *Phys. Rev. Lett.* **36**, 210 (1976).

[R21] J. J. Ritsko, L. J. Brillson and D. J. Sandman, Electron-energy loss spectroscopy of tetracyanoquinodimethane, TCNQ, tetrathiofulvalene, TTF, and the salt TTF-TCNQ, *Solid State Commun.* **24**, 109 (1977).

[R22] J. J. Ritsko, L. J. Brillson, R. W. Bigelow and T. J. Fabish, Electron energy loss spectroscopy and the optical properties of polymethylmethacrylate from 1 to 300 eV, *J. Chem. Phys.* **69**, 3931 (1978).

[R23] J. J. Ritsko and R. W. Bigelow, Core excitons and the dielectric response of polystyrene and poly(2-vinylpyridine) from 1 to 400 eV, *J. Chem. Phys.* **69**, 4162 (1978).

[R24] J. J. Ritsko, Momentum dependent dielectric response of polystyrene, *J. Chem. Phys.* **70**, 4656 (1979).

[R25] J. J. Ritsko, Electron energy loss spectroscopy of pristine and radiation damaged polyethylene, *J. Chem. Phys.* **70**, 5343 (1979).

[R26] J. J. Ritsko, Electronic states and excitations in polymers, *Org. Coat. Plast. Chem.* **42,** 358 (1980).

[R27] J. J. Ritsko, G. Crecelius and J. Fink, Thermal transformation of local to extended electronic states in polymers, *Phys. Rev.* **27B,** 2612 (1983).

[R28] R. Roberge, C. Sandorfy, J. I. Matthews and O. P. Strausz, The far ultraviolet and He I photoelectron spectra of alkyl and fluorine substituted silane derivatives, *J. Chem. Phys.* **69,** 5105 (1978).

[R29] R. Roberge and D. R. Salahub, Valence and Rydberg excited states of H_2S: an SCF-Xα -SW molecular orbital study, *J. Chem. Phys.* **70,** 1177 (1979).

[R30] R. Roberge, C. Sandorfy and O. P. Strausz, The electronic spectra of the alkyl, fluorine and chlorine substituted derivatives of silane, *Theoret. Chim. Acta* **52,** 171 (1979).

[R31] M. B. Robin, Rydberg excitations in X-ray absorption spectra, *Chem. Phys. Lett.* **31,** 140 (1975).

[R32] M. B. Robin and N. A. Kuebler, Electronic structure and spectra of small rings. VI. Multiphoton ionization spectra of the saturated three-membered rings, *J. Chem. Phys.* **69,** 806 (1978).

[R33] M. B. Robin and N. A. Kuebler, Allowed character of the 1900-Å band of borazine, *J. Mol. Spectrosc.* **70,** 472 (1978).

[R34] M. B. Robin and N. A. Kuebler, Ultraviolet hot-band spectra using a CO_2 laser, *Chem. Phys. Lett.* **80,** 512 (1981).

[R35] M. B. Robin and N. A. Kuebler, unpublished results.

[R36] M. B. Robin, Molecular analogs of the 4f wavefunction collapse and giant resonances in atoms, to be published.

[R37] J. L. Roebber, R. N. Wiener and C. A. Russell, Vacuum ultraviolet spectra of osmium tetroxide, *J. Chem. Phys.* **60,** 3166 (1974).

[R38] J. L. Roebber, D. P. Gerrity, R. Hemley and V. Vaida, Electronic spectrum of furan from 2200 to 1950 Å, *Chem. Phys. Lett.* **75,** 104 (1980).

[R39] K. Rohr, Differential scattering experiments for $e - H_2S$ collisions in the low-energy range, *J. Phys. B: At. Mol. Phys.* **11,** 4109 (1978).

[R40] K. Rohr, Cross beam experiment for the scattering of low-energy electrons from methane, *J. Phys. B: At. Mol. Phys.* **13,** 4897 (1980).

[R41] J. Römelt, S. D. Peyerimhoff and R. J. Buenker, Ab initio MRD-CI study of the Rydberg states of methylene, *Chem. Phys.* **54,** 147 (1981).

[R42] J.-Y. Roncin, M. Miladi, H. Damany and B. Vodar, Pressure effect on the electronic spectrum of NO. Relationship with matrix isolated spectrum, *in* "Vacuum Ultraviolet Radiation Physics," E.-E. Koch, R. Haensel and C. Kunz, *eds.*, Pergamon, 67 (1974).

[R43] J. S. Rosenfield, Magnetic rotational strengths of higher states in benzene: evidence for the existence of an out-of-plane transition, *Chem. Phys. Lett.* **39,** 391 (1976).

[R44] C. Rosini, C. Bertucci, D. Pini, P. Salvadori, F. Soccolini and G. Delogu, The vacuum ultraviolet circular dichroism spectrum of an isolated pyridine chromophore, *J. Chem. Soc., Chem. Commun.*, 287 (1983).

[R45] A. R. Rossi and P. Avouris, Theoretical studies of the electronic structure and spectra of low-lying states of NH_3^+, *J. Chem. Phys.* **79,** 3413 (1983).

[R46] L. J. Rothberg, D. P. Gerrity and V. Vaida, Electronic spectra of butadiene and its methyl derivatives: a multiphoton ionization study, *J. Chem. Phys.* **73,** 5508 (1980).

[R47] M. D. Rowe and A. Gedanken, Magnetic circular dichroism of the carbon tetrahalides in the $n_x \rightarrow \sigma^*$ region, *Chem. Phys.* **10,** 1 (1975).

[R48] C. Roxlo and A. Mandl, Vacuum ultraviolet absorption cross-sections for halogen containing molecules, *J. Appl. Phys.* **51,** 2969 (1980).

[R49] R. Runau, S. D. Peyerimhoff and R. J. Buenker, Ab initio study of the photodissociation of ammonia, *J. Mol. Spectrosc.* **68,** 253 (1977).

[S1] P. M. Saatzer, R. D. Koob and M. S. Gordon, Excited states and photochemistry of saturated molecules. Part 3. Structure and bonding in n-alkane excited states using INDO-CI, *J. Chem. Soc., Faraday Trans. II* **73,** 829 (1977).

[S2] A. Sabljic, A. Auerbach and R. McDiarmid, Symmetry of Rydberg transitions in 1,4-cyclohexadiene: multiphoton ionization investigation, *Int. J. Quantum Chem. Symp.* **16,** 357 (1982).

[S3] A. Sabljic and R. McDiarmid, Methyl torsions in the Rydberg states of 2,3-dimethylbutadiene, *Chem. Phys. Lett.* **106,** 132 (1984).

[S4] S. Saito, K. Wada and R. Onaka, Vacuum ultraviolet reflection spectra of KDP and ADP, *J. Phys. Soc. Jpn.* **37,** 711 (1974).

[S5] D. R. Salahub, Semi-empirical MO-CI calculations on excited states IV. Unsaturated molecules, *Theoret. Chim. Acta* **22,** 330 (1971).

[S6] W. R. Salaneck, N. O. Lipari, A. Paton, R. Zallen and K. S. Liang, Electronic structure of S_8, *Phys. Rev.* **12B,** 1493 (1975).

[S7] L. Sanche, Resonance transmission of 0−15 eV electrons in solid benzene and pyridine, *Chem. Phys. Lett.* **65,** 61 (1979).

[S8] L. Sanche, Transmission of 0−15 eV monoenergetic electrons through thin-film molecular solids, *J. Chem. Phys.* **71,** 4860 (1979).

[S9] L. Sanche and M. Michaud, Electron energy-loss vibronic spectroscopy of matrix-isolated benzene and multilayer benzene films, *Chem. Phys. Lett.* **80,** 184 (1981).

[S10] L. Sanche, G. Bader and L. Caron, Transmission of 0−15 eV monoenergetic electrons through aliphatic and alicyclic hydrocarbon flims, *J. Chem. Phys.* **76,** 4016 (1982).

[S11] R. T. Sanderson, "Polar Covalence," Academic Press, New York, 1983.

[S12] C. Sandorfy, Chemical spectroscopy in the vacuum ultraviolet, *J. Mol. Struct.* **19,** 183 (1973).

[S13] C. Sandorfy, Photochemical primary steps in the Rydberg states of organic molecules: the example of ethane, *Chem. Reactions* **1,** 159 (1975).

[S14] C. Sandorfy, Rydberg and valence-shell transitions and the photochemical primary steps in saturated organic molecules, *Z. Phys. Chem.* **101,** 307 (1976).

[S15] C. Sandorfy, The photochemical consequences of the Rydberg-valence shell distinction, *Prog. Theoret. Org. Chem.* **2,** 384 (1977).

[S16] C. Sandorfy, Far-ultraviolet absorption spectra of organic molecules: valence-shell and Rydberg transitions, *Top. Current Chem.* **86,** 92 (1979).

[S17] C. Sandorfy, Spectroscopy of non-aromatic Schiff bases, *J. Photochem.* **17,** 297 (1981).

[S18] C. Sandorfy and L. S. Lussier, Valence shell and Rydberg transitions of larger molecules, *in* "Photophysics and Photochemistry in the Vacuum Ultraviolet", S. P. McGlynn, G. L. Findley and R. H. Huebner, *eds.*, D. Reidel, Dordrecht (1984).

[S19] M. Sasanuma, E. Ishiguro, H. Masuko, Y. Morioka and M. Nakamura, Absorption structures of SF_6 in the VUV region, *J. Phys. B: At. Mol. Phys.* **11,** 3655 (1978).

[S20] M. Sasanuma, E. Ishiguro, T. Hayaisha, H. Masuko, Y. Morioka, T. Nakajima and M. Nakamura, Photoionization of SF_6 in the XUV region, *J. Phys. B: At. Mol. Phys.* **12,** 4057 (1979).

[S21] P. Sauvageau, J. Doucet, R. Gilbert and C. Sandorfy, Vacuum ultraviolet and photoelectron spectra of fluoroethanes, *J. Chem. Phys.* **61,** 391 (1974).

[S22] J. Schander and B. R. Russell, Vacuum ultraviolet spectra of bromoethylene and dibromoethylenes, *J. Amer. Chem. Soc.* **98,** 6900 (1976).

[S23] J. Schander and B. R. Russell, The perfluoro effect on the molecular Rydberg states of bromo- and iodoethylene, *J. Mol. Spectrosc.* **65,** 379 (1977).

[S24] B. Scharf, R. Vitenberg, B. Katz and Y. B. Band, Variations of splittings of vibronic Jahn-Teller states in asymmetrically deuterated benzenes, *J. Chem. Phys.* **77**, 2226 (1982).

[S25] G. J. Schulz, Resonances in electron impact on atoms, *Rev. Mod. Phys.* **45**, 378 (1973).

[S26] G. J. Schulz, Resonances in electron impact on diatomic molecules, *Rev. Mod. Phys.* **45**, 423 (1973).

[S27] W. H. E. Schwarz, Continuous change from valence to Rydberg type states. An example of XUV spectroscopy, *Chem. Phys.* **9,** 157 (1975).

[S28] W. H. E. Schwarz, Interpretation of the core electron excitations spectra of hydride molecules and the properties of hydride radicals, *Chem. Phys.* **11,** 217 (1975).

[S29] W. H. E. Schwarz and R. J. Buenker, Use of the Z + 1 core analogy model: examples from the core-excitation spectra of CO_2 and N_2O, *Chem. Phys.* **13**, 153 (1976).

[S30] W. H. E. Schwarz, L. Mensching, K. H. Hallmeier and R. Szargau, K-shell excitations of BF_3, CF_4 and MBF_4 compounds, *Chem. Phys.* **82**, 57 (1983).

[S31] J. D. Scott and B. R. Russell, Electronic symmetries, dipole moments and polarizabilities of molecules in Rydberg states determined from electric-field effects on absorption spectra, *J. Chem. Phys.* **63,** 3243 (1975).

[S32] J. D. Scott, W. S. Felps and S. P. McGlynn, Molecular Rydberg transitions: field effects in the vacuum ultraviolet, *Nucl. Instrum. Methods* **152,** 231 (1978).

[S33] J. D. Scott, W. S. Felps, G. L. Findley and S. P. McGlynn, Molecular Rydberg transitions. XII. Magnetic circular dichroism of methyl iodide, *J. Chem. Phys.* **68,** 4678 (1978).

[S34] T. W. Scott and A. C. Albrecht, A Rydberg transition in benzene in the condensed phase: two photon fluorescence excitation studies, *J. Chem. Phys.* **74,** 3807 (1981).

[S35] G. A. Segal, K. Wolf and J. J. Diamond, Theoretical study of the circular dichroism of (+)-(*S*)-2-butanol in the vacuum ultraviolet, *J. Amer. Chem. Soc.* **106**, 3175 (1984).

[S36] K. Seki, T. Hirooka, Y. Kamura and H. Inokuchi, Photoemission from polycyclic aromatic crystals in the vacuum ultraviolet region. V. Photoelectron spectroscopy by the rare-gas resonance lines and vacuum-ultraviolet absorption spectra, *Bull. Chem. Soc. Jpn.* **49,** 904 (1976).

[S37] K. Seki, K. Nakagawa, N. Sato, H. Inokuchi, S. Hashimoto, K. Inoue, S. Suga, H. Kanzaki and K. Takagi, Anisotropic carbon 1s XUV absorption spectra of oriented *st*-1,2-polybutadiene films, *Chem. Phys. Lett.* **70,** 220 (1980).

[S38] K. Seki, U. Karlsson, R. Engelhardt and E. E. Koch, Intramolecular energy band dispersion of *n*-$C_{36}H_{74}$ observed by angle-resolved photoemission with synchrotron radiation, *Chem. Phys. Lett.* **103**, 343 (1984).

[S39] M. Seki, K. Kobayashi and J. Nakahara, Optical spectra of hexagonal ice, *J. Phys. Soc. Jpn.* **50,** 2643 (1981).

[S40] C. J. Seliskar and R. E. Hoffman, Electronic spectroscopy of malondialdehyde, *Chem. Phys. Lett.* **43,** 481 (1976).

[S41] G. Seng and F. Linder, Vibrational excitation of polar molecules by electron impact. II. Direct and resonant excitation in H_2O, *J. Phys. B: At. Mol. Phys.* **9,** 2539 (1976).

[S42] E. H. Sharman, O. Schnepp, P. Salvadori, C. Bertucci and L. Lardicci, Circular dichroism spectra of aliphatic ethers, *J. Chem. Soc., Chem. Commun.*, 1000 (1979).

[S43] T. E. Sharp and J. T. Dowell, Isotope effects in dissociative attachment of electrons in methane, *J. Chem. Phys.* **46**, 1530 (1967).

[S44] T. E. Sharp and J. T. Dowell, Dissociative attachments of electrons in ammonia and ammonia-d_3, *J. Chem. Phys.* **50**, 3024 (1969).

[S45] D. A. Shaw, G. C. King, D. Cvejanovic and F. H. Read, Electron impact excitation of inner-shell excited states of CO, *J. Phys. B: At. Mol. Phys.* **17**, 2091 (1984).

[S46] T. Shibaguchi, H. Onuki and R. Onaka, Electronic structures of water and ice, *J. Phys. Soc. Jpn.* **42**, 152 (1977).

[S47] S.-K. Shih, S. D. Peyerimhoff and R. J. Buenker, Ab initio configuration interaction calculations for the electronic spectrum of hydrogen sulfide, *Chem. Phys.* **17**, 391 (1976).

[S48] M. Shiho, The vacuum ultraviolet absorption spectra of solid benzene from 5 eV to 11 eV, *Jpn. J. Appl. Phys.* **12**, 314 (1973).

[S49] M. Shiho, The vacuum ultraviolet absorption spectra of solid biphenyl, *J. Phys. Soc. Jpn.* **36**, 1636 (1974).

[S50] M. Shiho, Observation of a new type of absorption spectrum of solid benzene in the vacuum ultraviolet region, *J. Phys. Soc. Jpn.* **43**, 2105 (1977).

[S51] O. Sinanoglu, Theory of intravalency and Rydberg transitions in molecules, *in* "Chemical Spectroscopy and Photochemistry in the Vacuum-Ultraviolet", C. Sandorfy, P. J. Ausloos and M. B. Robin, *eds.*, D. Reidel, Dordrecht, 337 (1974).

[S52] C. Sluse-Goffart and J. Momigny, The absorption spectra of fluorobenzene, hexafluorobenzene and pyridine from the ionization threshold up to 13.5 eV, *Chem. Phys. Lett.* **25**, 231 (1974).

[S53] W. L. Smith, KDP and ADP transmission in the vacuum ultraviolet, *Appl. Opt.* **16**, 1798 (1977).

[S54] K. C. Smyth, J. A. Schiavone and R. S. Freund, Electron impact excitation of metastable states of benzene, toluene and aniline, *J. Chem. Phys.* **61**, 1789 (1974).

[S55] P. A. Snyder, Circular dichroism of (−)-α-pinene and (−)-β-pinene in the vacuum ultraviolet. A comparison of theoretical and experimental values, *in* "Vacuum Ultraviolet Radiation Physics," E.-E. Koch, R. Haensel and C. Kunz, *eds.*, Pergamon, 97 (1974).

[S56] P. A. Snyder and W. C. Johnson, Jr., Circular dichroism of *l*-borneol, *J. Amer. Chem. Soc.* **100**, 2939 (1978).

[S57] P. A. Snyder, P. A. Lund, P. N. Schatz and E. M. Rowe, Magnetic circular dichroism (MCD) of the Rydberg transitions in benzene using synchrotron radiation, *Chem. Phys. Lett.* **82**, 546 (1981).

[S58] P. A. Snyder, P. N. Schatz and E. M. Rowe, Vacuum ultraviolet magnetic circular dichroism measurements with synchrotron radiation. An initial study and interpretation of Rydberg transitions in ethylene, *Ann. Israel Phys. Soc.* **6**, 144 (1984).

[S59] R. N. S. Sodhi, S. Daviel, C. E. Brion and G. G. B. de Sousa, Electron energy loss spectra of the silicon 2p, 2s and carbon 1s and valence shells of tetramethylsilane, to appear in *J. Electron Spectrosc. Related Phenomena.*

[S60] B. Sonntag, Atomic and molecular effects in the VUV spectra of solids, *J. Phys. (Paris) Colloq.* **39**, C4, 9 (1978).

[S61] D. Spence, Systematics of Feshbach resonances in the molecular halogens, *Phys. Rev.* **A10**, 1045 (1974).

[S62] D. Spence and T. Noguchi, Feshbach resonances associated with Rydberg states of the hydrogen halides, *J. Chem. Phys.* **63**, 505 (1975).

[S63] D. Spence, Classification of Feshbach resonances in electron-molecule scattering, "*IX Int. Conf. Phys. Electronic Atomic Collisions*", J. S. Risely and R. Geballe, *Eds*., U. Wash. Press, Seattle, 241 (1976).

[S64] D. Spence, New aid to the classification of Feshbach resonances. Application to Ne, Kr, Ar and Xe, *Phys. Rev.* **15A,** 883 (1977).

[S65] D. Spence, Prediction of low energy molecular Rydberg states from Feshbach resonance spectra, *J. Chem. Phys.* **66,** 669 (1977).

[S66] S. M. Spyrou, I. Sauers and L. G. Christophorou, Electron attachment to the perfluoroalkanes n-C_NF_{2N+2} ($N = 1-6$) and i-C_4F_{10}, *J. Chem. Phys.* **78,** 7200 (1983).

[S67] S. K. Srivastava, D. C. Cartwright, S. Trajmar, A. Chutjian, and W. X. Williams, Photoabsorption spectrum of UF_6 by electron impact, *J. Chem. Phys.* **65,** 208 (1976).

[S68] V. Staemmler, R. Jaquet and M. Jungen, CEPA calculations on open-shell molecules. IV. Electron correlation effects in B_1 Rydberg states of H_2O, *J. Chem. Phys.* **74,** 1285 (1981).

[S69] R. H. Staley, L. B. Harding, W. A. Goddard, III and J. L. Beauchamp, Triplet states of the amide group. Trapped electron spectra of formamide and related molecules, *Chem. Phys. Lett.* **36,** 589 (1975).

[S70] J. G. Stamper and R. F. Barrow, The $V(^1\Sigma^+)$ - N $(^1\Sigma^+)$ transition of hydrogen bromide, *J. Phys. Chem.* **65,** 250 (1961).

[S71] R. P. Steer, Structure and decay dynamics of electronic excited states of thiocarbonyl compounds, *Rev. Chem. Intermed.* **4,** 1 (1980).

[S72] L. Z. Stenkamp and E. R. Davidson, An ab initio study of formamide, *Theoret. Chim. Acta* **44,** 405 (1977).

[S73] W. M. St. John, III, R. C. Estler and J. P. Doering, Low-energy electron impact study of acetone, *J. Chem. Phys.* **61**, 763 (1974).

[S74] R. Stockbauer and M. G. Inghram, The fragmentation of propane and deuteropropane molecular ions, *J. Chem. Phys.* **65**, 4081 (1976).

[S75] J. A. Stockdale, F. J. Davis, R. N. Compton and C. E. Klots, Production of negative ions from CH_3X molecules (CH_3NO_2, CH_3CN, CH_3I and CH_3Br) by electron impact and by collisions with atoms in excited Rydberg states, *J. Chem. Phys.* **60,** 4279 (1974).

[S76] R. S. Stradling, M. A. Baldwin, A. G. Loudon and A. Maccoll, Low energy electron impact spectra of some simple alkynes, *J. Chem. Soc., Faraday II* **72,** 871 (1976).

[S77] R. S. Stradling and A. G. Loudon, Electron impact spectroscopy of some simple nitriles, *J. Chem. Soc., Faraday Trans. II* **73,** 623 (1977).

[S78] C. Sugiura and T. Suzuki, K x-ray absorption spectra of chlorine in 3d transition-metal complexes, *J. Chem. Phys.* **75,** 4357 (1981).

[S79] C. Sugiura and M. Ohashi, "White line" in the chlorine K x-ray absorption spectra of K_2PdCl_4, K_2PtCl_4, K_2PdCl_6 and K_2PtCl_6, *J. Chem. Phys.* **78,** 88 (1983).

[S80] M. Suto and N. Washida, Emission spectra of CF_3 radicals. II. Analysis of the UV emission spectrum of CF_3 radicals, *J. Chem. Phys.* **78**, 1012 (1983).

[S81] M. Suto and N. Washida, Emission spectra of CF_3 radicals. I. UV and visible emission spectra of CF_3 observed in the VUV photolysis and the metastable argon atom reaction of CF_3H, *J. Chem. Phys.* **78**, 1007 (1983).

[S82] M. Suto and L. C. Lee, Emission spectra of CF_3 radicals. V. Photodissociation of CF_3H, CF_3Cl and CF_3Br by vacuum ultraviolet, *J. Chem. Phys.* **79**, 1127 (1983).

[S83] M. Suto and L. C. Lee, OH ($A\ ^2\Sigma^+ \rightarrow X\ ^2\Pi$) yield from photodissociation of H_2O_2 at 106−193 nm, *Chem. Phys. Lett.* **98**, 152 (1983).

[S84] M. Suto and L. C. Lee, Photodissociation of NH_3 at 106–200 nm, *J. Chem. Phys.* **78**, 4515 (1983).

[S85] M. Suto and L. C. Lee, Quantitative photoexcitation and fluorescence studies of C_2H_2 in vacuum ultraviolet, *J. Chem. Phys.* **80**, 4824 (1984).

[S86] M. Suto and L. C. Lee, Photoabsorption and photodissociation of $HONO_2$ in the 105–220 nm region, *J. Chem. Phys.* **81**, 1294 (1984).

[S87] S. Svensson, H. Ågren and U. I. Wahlgren, SCF and limited CI calculations on the 1s shake-up spectrum of H_2O, *Chem. Phys. Lett.* **38**, 1 (1976).

[S88] J. R. Swanson, D. Dill and J. L. Dehmer, Shape resonance effects in the photoabsorption spectra of BF_3, *J. Chem. Phys.* **75**, 619 (1981).

[T1] W.-C. Tam and C. E. Brion, Electronic spectra of aliphatic carbonyl compounds by electron impact spectroscopy. I. Saturated aldehydes, *J. Electron Spectrosc. Related Phenomena* **3**, 467 (1974).

[T2] W.-C. Tam and C. E. Brion, Electronic spectra of aliphatic carbonyl compounds by electron impact spectroscopy. II. Saturated ketones, *J. Electron Spectrosc. Related Phenomena* **4**, 139 (1974).

[T3] W.-C. Tam and C. E. Brion, Electronic spectra of aliphatic carbonyl compounds by electron impact spectroscopy. III. Unsaturated compounds, *J. Electron Spectrosc. Related Phenomena* **4**, 149 (1974).

[T4] H. Tanaka, T. Okada, L. Boesten, T. Suzuki, T. Yamamoto and M. Kubo, Differential cross-sections for elastic scattering of electrons by CH_4 in the energy range of 3 to 20 eV, *J. Phys. B: At. Mol. Phys.* **15**, 3305 (1982).

[T5] H. Tanaka, M. Kubo, N. Onodera and A. Suzuki, Vibrational excitation of CH_4 by electron impact: 3–20 eV, *J. Phys. B: At. Mol. Phys.* **16**, 2861 (1983).

[T6] D. B. Tanner, J. S. Miller, M. J. Rice and J. J. Ritsko, Optical and electron-energy-loss studies of the monomeric and dimeric phases of decamethyl ferrocinium tetracyanoquino dimethanide, (DMeFc) (TCNQ), *Phys. Rev.* **21B**, 5835 (1980).

[T7] S. Taylor, D. G. Wilden and J. Comer, Electron energy-loss spectroscopy of forbidden transitions to valence and Rydberg states of formaldehyde, *Chem. Phys.* **70**, 291 (1982).

[T8] J. Texter and E. S. Stevens, Random phase circular dichroism calculations of the $\sigma^*/3s \leftarrow n$ transition in chiral alcohols, *J. Chem. Phys.* **70**, 1440 (1979).

[T9] W. Thiel, Characterization of resonances in photoionization, *J. Electron Spectrosc. Related Phenomena* **31**, 151 (1983).

[T10] R. Thomson and P. A. Warsop, Electronic spectrum of the haloacetylenes. Part 1. Chloroacetylene, *Trans. Faraday Soc.* **65**, 2806 (1969).

[T11] R. Thomson and P. A. Warsop, Electronic spectrum of the haloacetylenes. Part 2. Bromoacetylene, *Trans. Faraday Soc.* **66**, 1871 (1970).

[T12] K.-H. Thunemann, R. J. Buenker and W. Butscher, Configuration interaction study of the electronic spectrum of furan, *Chem. Phys.* **47**, 313 (1980).

[T13] I. A. Topol, A. V. Kondratenko and L. N. Mazalov, Study of highly excited states and resonances in the PCl_3 molecule by a method of $X\alpha$-scattered waves, *Opt. Spectrosc.* **50**, 267 (1981).

[T14] K. Toriyama, K. Nunome and M. Iwasaki, ESR evidence for twisted structure of methyl-substituted ethylene radical cations: trimethylethylene and propylene, *Chem. Phys. Lett.* **107**, 86 (1984).

[T15] J. A. Tossell and J. W. Davenport, MX-α calculation of the elastic electron scattering cross sections and x-ray absorption spectra of CX_4 and SiX_4 (X = H, F, Cl), *J. Chem. Phys.* **80**, 813 (1984).

[T16] I. Trabjerg, M. Vala and J. Baiardo, Rydberg transitions in bichromophoric molecules tetramethyl-1,3-cyclobutanedithione (TMCBDT) and tetramethyl-3-thiocyclobutan-1-one (TMTCB), *Mol. Phys.* **50,** 193 (1983).

[T17] S. Trajmar and A. Chutjian, Electron impact excitation of SF_6, *J. Phys. B: At. Mol. Phys.* **10,** 2943 (1977).

[T18] G. Trinquier and J.-P. Malrieu, The zwitterionic singlet excited state of ethylene as a singlet methylene dimer, *Chem. Phys. Lett.* **72,** 328 (1980).

[T19] M. Tronc, G. C. King and F. H. Read, Carbon K-shell excitation in small molecules by high-resolution electron impact, *J. Phys. B: At. Mol. Phys.* **12,** 137 (1979).

[T20] M. Tronc and R. Azria, Differential cross section for CN^- formation from dissociative electron attachment to the cyanogen molecule C_2N_2, *Chem. Phys. Lett.* **85,** 345 (1982).

[T21] D. T. Truch, D. R. Salomon and D. A. Armstrong, The Rydberg spectroscopy of methyl chloride, *J. Mol. Spectrosc.* **78,** 31 (1979).

[T22] G. Trudeau, J.-M. Dumas, P. Dupuis, M. Guerin and C. Sandorfy, Intermolecular interactions and anesthesia: infrared spectroscopic studies, *Top. Current Chem.* **93,** 91 (1981).

[T23] B. P. Tsai and T. Baer, Analysis of autoionizing Rydberg states in HI and CH_3I. Comments on Rydberg electron wave functions, *J. Chem. Phys.* **61,** 2047 (1974).

[T24] M. Tsuboi, A. Y. Hirakawa, T. Hoshino, T. Ishiguro, K. Kimura and S. Katsumata, Vibrational structures of a few electronic bands of thioformamide, *J. Mol. Spectrosc.* **63,** 80 (1976).

[T25] M. Tsuboi and A. Y. Hirakawa, The 240 nm band system of methylamine: a preliminary analysis of the rotational structure of the 0-0 band of CH_3ND_2, *Can. J. Phys.* **60,** 844 (1982).

[T26] R. E. Turner, V. Vaida, C. A. Molini, J. O. Berg and D. H. Parker, The multiphoton ionization spectra of pyridine and pyrazine, *Chem. Phys.* **28,** 47 (1978).

[T27] A. J. Twarowski and D. S. Kliger, A search for a low-lying excited ^1A state in 1,3,5-hexatriene, *Chem. Phys. Lett.* **50,** 36 (1977).

[U1] N. Ueno, A. Okazaki, Y. Hayasi and S. Kiyono, Electron scattering spectra of naphthacene and perylene polycrystals, *Chem. Phys. Lett.* **42,** 119 (1976).

[U2] G. O. Uneberg, P. A. Campo and P. Johnson, A study of the 3p Rydberg transitions of the methylbenzenes by multiphoton ionization spectroscopy, *J. Chem. Phys.* **73,** 1110 (1980).

[U3] R. Unwin, I. Khan, N. V. Richardson, A. M. Bradshaw, L. S. Cederbaum and W. Domcke, The effect of a resonance on vibrational structure in the photoelectron spectrum of acetylene, *Chem. Phys. Lett.* **77,** 242 (1981).

[V1] V. Vaida, R. E. Turner, J. L. Casey and S. D. Colson, Multiphoton transitions in trans-butadiene observed by multiphoton ionization and thermal lensing spectroscopy, *Chem. Phys. Lett.* **54,** 25 (1978).

[V2] V. Vaida and G. M. McClelland, Electronic absorption spectroscopy of cooled supersonic expansions: dynamics of the $^1B_{1u}$ state of *trans*-butadiene, *Chem. Phys. Lett.* **71,** 436 (1980).

[V3] M. Vala, I. Trabjerg and E. N. Svendsen, The vacuum ultraviolet spectrum of tetramethyl-1,3-cyclobutanedione, *Acta Chem. Scand.* **A28,** 37 (1974).

[V4] K. E. Valenta, K. Vasudevan and F. Grein, *Ab initio* studies of the geometry, electronic spectrum and vertical ionization potentials of oxygen difluoride (OF_2), *J. Chem. Phys.* **72,** 2148 (1980).

[V5] M. J. van der Wiel, W. Stoll, A. Hamnett and C. E. Brion, Partial oscillator strengths (25–50 eV) of the one electron states of CH_4^+ measured in an (e, 2e) experiment, *Chem. Phys. Lett.* **37,** 240 (1976).

[V6] M. van Hemert and A. van der Avoird, *Ab initio* calculation of the first order interaction energy in excited dimers. The H_2O-H_2O and H_2O-Ne dimers, *J. Chem. Phys.* **71,** 5310 (1979).

[V7] E. H. van Veen and F. L. Plantenga, Threshold electron-impact excitation spectrum of pyridine, *Chem. Phys. Lett.* **30,** 28 (1975).

[V8] E. H. van Veen and F. L. Plantenga, Low-energy electron-impact excitation spectra of acetylene, *Chem. Phys. Lett.* **38,** 493 (1976).

[V9] E. H. van Veen, Triplet $\pi \rightarrow \pi^*$ transitions in thiophene, furan and pyrrole by low-energy electron-impact spectroscopy, *Chem. Phys. Lett.* **41,** 535 (1976).

[V10] E. H. van Veen, W. L. Van Dijk and H. H. Brongersma, Low-energy electron-impact excitation spectra of formaldehyde, acetaldehyde and acetone, *Chem. Phys.* **16,** 337 (1976).

[V11] E. H. van Veen, Low-energy electron impact spectroscopy on ethylene, *Chem. Phys. Lett.* **41,** 540 (1976).

[V12] K. Vasudevan, S. D. Peyerimhoff, R. J. Buenker and W. E. Kammer, Theoretical study of the electronic spectrum of diimide by ab initio methods, *Chem. Phys.* **7,** 187 (1975).

[V13] K. Vasudevan and W. E. Kammer, An investigation on the vertical electronic spectra of *trans*-difluorodiimide (N_2F_2), diazirine (N_2CH_2) and difluorodiazirine (N_2CF_2), *Chem. Phys.* **15,** 103 (1976).

[V14] K. Vasudevan and F. Grein, Theoretical study on the vertical electronic spectrum of carbonyl fluoride, F_2CO, *Int. J. Quantum Chem.* **14,** 717 (1978).

[V15] K. Vasudevan and F. Grein, Configuration interaction studies on low-lying excited states of acetyl fluoride, *Chem. Phys. Lett.* **83,** 526 (1981).

[V16] R. V. Vedrinskii, A. P. Kovtun, V. V. Kolesnikov, Yu. F. Migal, E. V. Polozhentsev and V. P. Sachenko, L-absorption spectra of sulfur in the SF_6 molecule in the cluster approximation, *Bull. Acad. Sci. SSSR, Phys. Ser.* **38/3,** 8 (1974).

[V17] H. Venghaus, Determination of the dielectric tensor of pyrene ($C_{16}H_{10}$) by means of electron energy loss measurements, *Phys. Status Solidi* **54b,** 671 (1972).

[V18] H. Venghaus and H.-J. Hinz, Electron energy loss measurements on fluorene ($C_{13}H_{10}$) vapor and single crystals, *Phys. Status Solidi* **65b,** 239 (1974).

[V19] H. Venghaus and H.-J. Hinz, Electron energy loss measurements on *p*-terphenyl ($C_{18}H_{14}$) single crystals and vapor, *J. Chem. Phys.* **62,** 4937 (1975).

[V20] H. Venghaus and H.-J. Hinz, Electron energy loss measurements on pyrene ($C_{16}H_{10}$) vapor, *J. Chem. Phys.* **64,** 30 (1976).

[V21] G. J. Verhaart, W. J. Van der Hart and H. H. Brongersma, Low energy electron impact on chlorofluoromethanes and CF_4: resonances, dissociative attachment and excitation, *Chem. Phys.* **34,** 161 (1978).

[V22] G. J. Verhaart, P. Brasem and H. H. Brongersma, Low-energy electron-impact spectroscopy on bridged [10] annulenes and other 10 π-electron systems, *Chem. Phys. Lett.* **62,** 519 (1979).

[V23] G. J. Verhaart and H. H. Brongersma, Triplet $n \rightarrow \pi^*$ and $\pi \rightarrow \pi^*$ transitions in glyoxal and biacetyl by low-energy electron-impact spectroscopy, *Chem. Phys. Lett.* **72,** 176 (1980).

[V24] G. J. Verhaart and H. H. Brongersma, Electronic excitation (5–9 eV) in ethylene and some haloethylenes by threshold electron-impact spectroscopy with an improved energy resolution, *Chem. Phys.* **52,** 431 (1980).

[V25] S. K. Vidyarthi, C. Willis and R. A. Back, The ultraviolet absorption spectrum of methyldiimide vapor, *Can. J. Chem.* **55,** 1396 (1977).

[V26] A. S. Vinogradov, T. M. Zimkina, V. N. Akimov and B. Shlarbaum, Structure of the absorption spectra of the molecules N_2, O_2 and NF_3 near the ionization thresholds of the inner atomic shells, *Bull. Acad. Sci. USSR, Ser. Phys.* **38,** 69 (1974).

[V27] A. S. Vinogradov, A. Yu. Dukhnyakov, T. M. Zimkina, V. M. Ipatov, I. V. Karunina, D. E. Onopko, A. A. Pavlychev, S. A. Titov and E. O. Filatova, Quasimolecular character of x-ray absorption in the compounds KPF_6 and NH_4PF_6, *Sov. Phys. Solid State* **22,** 1517 (1980).

[V28] R. Vitenberg, B. Katz and B. Scharf, The assignment of the benzene first ionization Rydberg spectrum via J-T splittings involving linearly inactive modes, *Chem. Phys. Lett.* **71,** 187 (1980).

[V29] D. Vocelle, A. Dargelos, R. Pottier and C. Sandorfy, Photoelectron and far-ultraviolet absorption spectra of nonaromatic azomethine compounds, *J. Chem. Phys.* **66,** 2860 (1977).

[V30] J. Vogt, M. Jungen and J. L. Beauchamp, Triplet states of ketene by trapped-electron spectroscopy, *Chem. Phys. Lett.* **40,** 500 (1976).

[V31] L. Vuskovic and S. Trajmar, Electron impact excitation of methane, *J. Chem. Phys.* **78,** 4947 (1983).

[W1] W. R. Wadt, W. A. Goddard, III and T. H. Dunning, Jr., The electronic structure of pyrazine. Configuration interaction calculations using an extended basis, *J. Chem. Phys.* **65,** 438 (1976).

[W2] W. R. Wadt and W. A. Goddard, III, The low-lying excited states of water, methanol, and dimethyl ether, *Chem. Phys.* **18,** 1 (1976).

[W3] W. R. Wadt, personal communication, 1983.

[W4] I. C. Walker, A. Stamatovic and S. F. Wong, Vibrational excitation of ethylene by electron impact: 1−11 eV, *J. Chem. Phys.* **69,** 5532 (1978).

[W5] C. M. Walmsley and W. P. Watson, Very high Rydberg states ($n \approx 600$) of carbon in the interstellar gas, *Astrophys. J.* **255,** L123 (1982).

[W6] H.-t. Wang, W. S. Felps and S. P. McGlynn, Molecular Rydberg states. VII. Water, *J. Chem. Phys.* **67,** 2614 (1977).

[W7] N. Washida, M. Suto, S. Nagase, U. Nagashima and K. Morokuma, Emission spectra of CF_3 radicals. IV. Excitation spectra, quantum yields and potential energy surfaces of the CF_3 fluorescences, *J. Chem. Phys.* **78,** 1025 (1983).

[W8] M. Watanabe, H. Kitamura and Y. Nakai, Vacuum ultraviolet absorption spectra of ice, *in* "Vacuum Ultraviolet Radiation Physics," E.-E Koch, R. Haensel and C. Kunz, *eds.*, Pergamon, 70 (1974).

[W9] F. H. Watson and M. W. Nycum, The Rydberg doublets of ethylene: intensity stealing in mono-olefins, *Spectrosc. Lett.* **8,** 223 (1975).

[W10] J. K. G. Watson, Assignment of the Schüler band of the ammonium radical to the 3p $^2F_2$ - 3s $^2A_1$ electronic transition, *J. Mol. Spectrosc.* **107,** 124 (1984).

[W11] A. W. Weiss and M. Krauss, Bound state calculation of scattering resonance energies, *J. Chem. Phys.* **52,** 4363 (1970).

[W12] H. R. Wendt and H. E. Hunziker, The UV spectra of primary, secondary and tertiary alkyl radicals, *J. Chem. Phys.* **81,** 717 (1984).

[W13] R. L. Whetten, K.-J. Fu and E. R. Grant, Ultraviolet two-photon spectroscopy of benzene: a new gerade Rydberg series and evidence for the 1^1E_{2g} valence state, *J. Chem. Phys.* **79,** 2626 (1983).

[W14] R. L. Whetten and E. R. Grant, Direct observation of nonlinear Jahn-Teller effects in the $1^1A_{1g} \rightarrow 3s\ ^1E_g$ two-photon spectrum of cyclohexane, *J. Chem. Phys.* **80,** 1711 (1984).

[W15] R. L. Whetten and E. R. Grant, Vibronic structure of nonadiabatic and fluxional states: two-photon absorption spectroscopy of jet isolated 3s $^1E'$ *sym*-triazine, *J. Chem. Phys.* **81,** 691 (1984).

[W16] E. A. Whittaker, B. J. Sullivan, G. C. Bjorklund, H. R. Wendt and H. E. Hunziker, ND_4 Schüler band absorption observed by laser FM spectroscopy in a photochemical reaction, *J. Chem. Phys.* **80,** 961 (1984).

[W17] K. B. Wiberg, K. S. Peters, G. B. Ellison and J. L. Dehmer, Rydberg states of butadiene, *J. Chem. Phys.* **66,** 2224 (1977).

[W18] J. M. Wiesenfeld and B. I. Greene, Femtosecond relaxation dynamics of molecular Rydberg states using time-resolved multiphoton ionization, *Phys. Rev. Lett.* **51,** 1745 (1983).

[W19] G. R. Wight and C. E. Brion, K-shell excitation of CH_4, NH_3, H_2O, CH_3OH, CH_3OCH_3 and CH_3NH_2 by 2.5 KeV electron impact, *J. Electron Spectrosc. Related Phenomena* **4,** 25 (1974).

[W20] G. R. Wight and C. E. Brion, K shell and valence shell excitations in CF_4 by 2.5 eV electron impact, *J. Electron Spectrosc. Related Phenomena* **4,** 327 (1974).

[W21] G. R. Wight and C. E. Brion, Carbon K-shell excitation in acetone by 2.5 KeV electron impact, *J. Electron Spectrosc. Related Phenomena* **4,** 347 (1974).

[W22] G. R. Wight and C. E. Brion, Estimation of the excitation and ionization energies of NH_4, H_3O and H_2F radicals using core analogies applies to K-shell electron energy loss spectra, *Chem. Phys. Lett.* **26,** 607 (1974).

[W23] D. G. Wilden, P. J. Hicks and J. Comer, An electron impact energy-loss study of triplet states of acetylene, *J. Phys. B: At. Mol. Phys.* **10,** L403 (1977).

[W24] D. G. Wilden and J. Comer, High-resolution electron energy-loss spectroscopy of ethylene: analysis of the vibrational structure of the triplet ($\pi \rightarrow \pi^*$) state, *J. Phys. B: At. Mol. Phys.* **12,** L371 (1979).

[W25] D. G. Wilden and J. Comer, High resolution electron impact studies of electric dipole-forbidden states of benzene, *J. Phys. B: At. Mol. Phys.* **13,** 627 (1980).

[W26] D. G. Wilden and J. Comer, Rydberg states of C_2H_4 and C_2D_4: assignments using the technique of low-energy electron energy-loss spectroscopy, *J. Phys. B: At. Mol. Phys.* **13,** 1009 (1980).

[W27] D. G. Wilden, J. Comer and S. Taylor, Energy-loss spectroscopy of the higher excited states of acetylene, *J. Phys. B: At. Mol. Phys.* **13,** 2849 (1980).

[W28] G. R. J. Williams and P. W. Langhoff, Photoabsorption in H_2O: Stieltjes-Tchebycheff calculations in the time-dependent Hartree-Fock approximation, *Chem. Phys. Lett.* **60,** 201 (1979).

[W29] M. W. Williams, R. N. Hamm, E. T. Arakawa, L. R. Painter and R. D. Birkhoff, Collective electron effects in molecular liquids, *Int. J. Radiat. Phys. Chem.* **7,** 95 (1975).

[W30] R. T. Williams, D. J. Nagel, P. H. Klein and M. J. Weber, Vacuum ultraviolet properties of beryllium fluoride glass, *J. Appl. Phys.* **52,** 6279 (1981).

[W31] H. F. Winters and M. Inokuti, Total dissociation cross section of CF_4 and other fluoroalkanes for electron impact, *Phys. Rev.* **25A,** 1420 (1982).

[W32] K. Wittel, W. S. Felps, L. Klasinc and S. P. McGlynn, Molecular Rydberg transitions. VI. *Trans*-dibromoethylene. The relation between vacuum ultraviolet and photoelectron spectroscopy, *J. Chem. Phys.* **65,** 3698 (1976).

[W33] K. Wittel, J. L. Meeks and S. P. McGlynn, Electronic structure of the cyanato and thiocyanato groups, ground state and excited states, *in* "Chemistry of the Cyanates and their Thio Derivatives", S. Patai, *ed.*, Wiley, Chichester, Eng., 1 (1977).

[W34] S. F. Wong and G. J. Schulz, Vibrational excitation in benzene by electron impact via resonances: selection rules, *Phys. Rev. Lett.* **35**, 1429 (1975).

[W35] K. V. Wood and J. W. Taylor, A photoionization mass spectrometric study of autoionization in ethylene and *trans* 2-butene, *Int. J. Mass Spectrom. Ion Phys.* **30**, 307 (1979).

[W36] C. W. Worrell, The photoelectron and absorption spectra of allyl halides, *J. Electron Spectrosc. Related Phenomena* **3**, 359 (1974).

[W37] C. Y. R. Wu, L. C. Lee and D. L. Judge, Photoabsorption cross sections of CH_3F, CHF_3, CH_3Cl, and CF_2Cl_2 from 175 to 760 nm, *J. Chem. Phys.* **71**, 5221 (1979).

[W38] C. Y. R. Wu and D. L. Judge, Lyman-α fluorescence from hydrogen photofragments of CH_4 and H_2O, *J. Chem. Phys.* **75**, 172 (1981).

[Y1] D. L. Yeager and V. McKoy, Equations of motion method: excitation energies and intensities in formaldehyde, *J. Chem. Phys.* **60**, 2714 (1974).

[Y2] K. Yoshikawa, M. Hashimoto and I. Morishima, Photoelectron spectroscopic study of cyclic amines. The relation between ionization potentials, basicities and s character of the nitrogen lone pair electrons, *J. Amer. Chem. Soc.* **96**, 288 (1974).

[Y3] V. Y. Young and K. L. Cheng, The photoelectron spectra of halogen substituted acetones, *J. Chem. Phys.* **65**, 3187 (1976).

[Y4] H. T. Yu, A. Sevin, E. Kassab and E. M. Evleth, A comparative theoretical analysis of the photochemistry of the methyl radical and related systems, *J. Chem. Phys.* **80**, 2049 (1984).

[Z1] G. D. Zeiss, W. J. Meath, J. C. F. MacDonald and D. J. Dawson, Dipole spectrum of water vapor and its relation to the energy loss of fast-charged particles, *Radiat. Res.* **63**, 64 (1975).

[Z2] G. Ya. Zelikina, V. V. Bertsev and T. G. Meister, Temperature dependence of the position of the long-wavelength Rydberg bands of methyl iodide, trifluoroiodomethane and carbon disulfide dissolved in liquified gases, *Opt. Spectrosc.* **42**, 575 (1977).

[Z3] G. Ya. Zelikina and T. G. Meister, Interpretation of the long-wavelength Rydberg bands of methyl iodide, trifluoroiodomethane and carbon disulfide dissolved in liquified gases on the basis of the exciton model, *Opt. Spectrosc.* **42**, 621 (1977).

[Z4] G. Ya. Zelikina and T. G. Meister, Influence of cryogenic solvents on electronic bands of different character ($\pi\pi^*$ and Rydberg bands), *Opt. Spectrosc.* **43**, 46 (1977).

[Z5] L. D. Ziegler and B. S. Hudson, Resonance Raman scattering of ethylene: evidence for a twisted geometry in the V state, *J. Chem. Phys.* **79**, 1197 (1983).

[Z6] L. D. Ziegler and B. Hudson, Resonance rovibronic Raman scattering of ammonia, *J. Phys. Chem.* **88**, 1110 (1984).

[Z7] T. M. Zimkina and A. S. Vinogradov, Photoionization absorption in the ultrasoft x-ray region by atoms in polyatomic molecules, *Bull. Acad. Sci., Phys. Ser.* **36**, 229 (1972).

Index

D

E

F

I

J

K

L

M

N

Q

R

S

T

U

V

W

X